T0333939

Atomic Physics: Precise Measurements and Ultracold Matter

Atomic Physics: Precise Measurements and Ultracold Matter

Massimo Inguscio
Department of Physics and Astronomy
European Laboratory for Nonlinear Spectroscopy (LENS)
University of Florence, Italy

and

National Institute of Metrological Research (INRIM), Italy

Leonardo Fallani
Department of Physics and Astronomy
European Laboratory for Nonlinear Spectroscopy (LENS)
University of Florence, Italy

OXFORD
UNIVERSITY PRESS

OXFORD
UNIVERSITY PRESS

Great Clarendon Street, Oxford, OX2 6DP,
United Kingdom

Oxford University Press is a department of the University of Oxford.
It furthers the University's objective of excellence in research, scholarship,
and education by publishing worldwide. Oxford is a registered trade mark of
Oxford University Press in the UK and in certain other countries

First Edition published in 2013
Impression: 2
Reprinted with corrections 2015

Published in the United States of America by Oxford University Press
198 Madison Avenue, New York, NY 10016, United States of America

British Library Cataloguing in Publication Data
Data available

Library of Congress Control Number: 2013940814

ISBN 978–0–19–852584–4

Printed and bound by
Clays Ltd, St Ives plc

a Giovanna, Alessandro, Bianca Maria Serena e Agostino Pietro Maria

a Manuela

Foreword

In 1958, Arthur L. Schawlow and Charles H. Townes published a seminal paper that triggered the race to build the first laser. The advent of the laser ushered in a period of extraordinary advances in science and technology. Atomic physics moved again to the centre stage. No less than 21 Nobel Prizes have since been awarded for laser-related research, many of them recognizing achievements in precision spectroscopy (Bloembergen, Schawlow, Hall, Hänsch) and in laser cooling (Chu, Cohen-Tannoudji, Phillips, Cornell, Wieman, and Ketterle). The most recent Nobel Prize in Physics has been awarded to Haroche and Wineland "for ground-breaking experimental methods that enable measuring and manipulation of individual quantum systems". This book by Massimo Inguscio and Leonardo Fallani illuminates the extraordinary evolution of atomic physics during the past decades, and it leads the reader to the fast-moving frontier of current research. The text conveys the fascination and excitement of the field through the eyes of pioneering researchers, so that it can provide inspiration to students and seasoned colleagues alike.

Reading this book made me realize how fortunate I have been to witness the development of this field as an active participant from the beginning. Here and there, I could even influence the evolution with contributions of my own. Essential parts of my work have been motivated by the goal of precision spectroscopy of the simple hydrogen atom, which permits unique confrontations between experiment and fundamental theory. This continuing quest has inspired inventions such as the first highly monochromatic, widely tunable dye laser, techniques for Doppler-free laser spectroscopy, and the frequency comb technique for measuring the frequency of laser light with extreme precision. Recent laser measurements of the Lamb shift in muonic hydrogen have unveiled a "proton size puzzle" that may hint at a possible dent in the armour of quantum electrodynamics. Future precision spectroscopy of antihydrogen might detect small differences between matter and antimatter.

Laser cooling was originally invented as a tool for precision spectroscopy. It has enabled the construction of microwave caesium fountain clocks, which form the basis for the definition of the second and for international time keeping. Today, laser-cooled trapped atoms or ions serve as pendulums in optical atomic clocks, which have reached a reproducibility of better than 10^{-17}. Such clocks enable laboratory experiments that can search for slow changes of fundamental constants and they permit stringent new tests of special and general relativity.

Laser cooling followed by evaporative cooling has been the key to Bose–Einstein condensation of ultracold atoms. Today, experimental studies of degenerate quantum gases have developed into a rich field of research at the interface between atomic physics and condensed matter physics. Ultracold atoms in optical lattice potentials offer a particularly intriguing playground. Starting with the observation of a quantum phase transition from a wave-like superfluid state to a strongly correlated crystal-like Mott insulator state, it has become possible to simulate many-body quantum phenomena from solid-state physics to particle physics and astrophysics.

The evolution of atomic physics along the path outlined in this book has led to a veritable Renaissance in science, engaging researchers from many different fields, including atomic physics, condensed matter physics, particle physics, astrophysics, chemistry, mathematics, and computer science. What can we expect from atomic physics in the future? We can safely predict more Nobel Prizes for surprising discoveries. However, we cannot predict the evolution of atomic physics. As in the past, the most important research results will be those that make all planning obsolete.

Munich, January 2013

Prof. T. W. Hänsch

Preface

A long story of atoms and light

The observation that atoms and molecules only absorb or emit electromagnetic waves at discrete frequencies is probably the most striking experimental evidence that led to the birth of quantum mechanics. Measuring the "colour" of atomic lines, which is the basic aim of spectroscopy, gives us an exceedingly powerful instrument to understand Nature. The history of atomic physics has been shaped by a constant quest for more and more precision. Indeed, precision atomic spectroscopy of simple atoms (such as hydrogen or helium) can be used to measure fundamental physical constants, to perform extremely precise measurements of feeble effects, to validate existing physical models, to search for possible discrepancies that could hint at novel exotic theories. It could also reveal whether the quantities that we use to treat as fundamental constants are really constant or whether they are instead very slowly changing in time, which could implicate major revisions in our understanding of the Universe.

The history of atomic physics is marked by fundamental discoveries as well as by revolutionary technological advances. Among these, the invention of lasers 50 years ago has disclosed unprecedented horizons of precision in atomic spectroscopy, providing sources of very monochromatic light for probing atomic or molecular structures. A second revolution started in the 1970s, when it was realized that lasers are not only the primary tools that allow us to probe atomic spectra but can also be used to control the motion of the atoms, slowing them down to amazingly low temperatures, less than a millionth of a degree above absolute zero, indeed much lower than what is possible with traditional cryogenic techniques. The realization of ultracold atomic gases opened new horizons in spectroscopy thanks to the possibility of eliminating all the limiting effects connected to the motion of the atoms. Laser cooling was indeed the key to detect tinier and tinier effects and to realize extremely precise atomic clocks which are so precise as to lose or gain less than one second over the entire age of the Universe! These accurate measurements have been possible also thanks to recent important developments in laser physics: the production of extremely monochromatic lasers with ultra-narrow linewidth and the revolutionary invention of the *frequency comb*, which has fundamentally changed the field of spectroscopy by allowing direct accurate measurements of optical frequencies.

And when matter becomes ultracold, new fascinating effects emerge. Ultracold atoms move so slow that, when viewed at a sufficiently small length scale, it becomes impossible to distinguish them one from another, owing to the Heisenberg uncertainty principle. As a consequence, quantum statistics becomes important. At a sufficiently small temperature, a gas of bosonic atoms undergoes Bose–Einstein condensation in which the atoms suffer an "identity crisis" and they collectively occupy the same quantum state. This state of matter, predicted by A. Einstein more than 80 years ago and observed only in the 1990s with ultracold atomic dilute gases, share many similarities with laser light, since it manifests as the macroscopic occupation of the same quantum state, and it is accompanied by the same properties of macroscopic coherence as superfluids and superconductors. With similar experimental techniques, dilute gases of fermionic atoms can be cooled down to Fermi degeneracy, which is characterized by the unity occupation of all the single-particle energy levels up to the Fermi energy of the system.

Lasers can also confine atoms in very small traps, which are used to provide an ideal setting for high-precision spectroscopy. Atoms trapped in *optical lattices*, i.e. periodic arrays of laser traps, are isolated from the environment and can be trapped for long times in the absence of perturbing fields: furthermore, their motion can be controlled and even frozen to the ground state of the trap in such a way as not to cause any perturbation to their spectroscopy. Besides constituting valuable tools for the realization of accurate optical clocks, optical lattices have opened a new field of research which goes beyond spectroscopy and is founded on the close analogy between neutral atoms trapped in periodic structures made of light and electrons moving through the periodic crystalline structure of a solid. Ideal solid-state physics models can be experimentally realized and fundamental effects such as Bloch oscillations, Anderson localization, and conductor–insulator quantum phase transitions can be very finely investigated, thus making atoms powerful and precise *quantum simulators* of quantum dynamics and many-body systems.

Why this book?

There are many excellent textbooks of atomic physics; why another one? Fifty years of international conferences in atomic physics have shown that during this time atomic physics has been characterized by a continuous redefinition of its meaning: from pre-laser spectroscopy (described in classic Oxford books such as the ones by G. W. Series and A. Corney), to the success of laser and non-linear spectroscopy in the 1970s, to more recent milestones represented by the possibility of using the laser to manipulate atomic velocities (which means colder and colder samples for more and more precise measurements) and by the possibility of performing direct frequency measurements in the optical domain. Nowadays atomic physics has moved from the investigation of atomic structures to a field in which the aforementioned possibilities of control on both the internal and the external degrees of freedom allow us to use atoms to test the validity of quantum theories, to measure fundamental constants, to build precise atomic clocks, to realize quantum simulators of condensed-matter systems. We have decided to write this book to illustrate this change of paradigm in atomic physics, from conventional laser spectroscopy to laser manipulation of the atomic motion and its implications for precise measurements: in a sense, this book takes a snapshot of what atomic physics has now become and what the new challenges are.

This is not a classical atomic physics textbook, in which the focus is on the theory of atomic structures and light–matter interaction. This book focuses on the experimental investigations, both illustrating milestone experiments and key experimental techniques, and discussing the results and perspectives of current research activity. Emphasis is given on the investigation of precision Physics: from the determination of fundamental constants of Nature to tests of General Relativity and Quantum Electrodynamics, from the realization of ultra-stable atomic clocks to the precise simulation of condensed-matter theories with ultracold gases. The book discusses these topics while tracing the evolution of experimental atomic physics from traditional laser spectroscopy to the revolution introduced by laser cooling.

Book structure

The first part of the book is structured following the increase of complexity in atoms. As most of the books in atomic physics, Chapter 1 starts with hydrogen, the simplest of the atoms, which is one of the main doors that allowed physicists to discover the quantum world and a primary testing ground for fundamental physical theories. The history of hydrogen spectroscopy will be used to illustrate fundamental steps in laser spectroscopy, from Doppler-free spectroscopy to direct frequency measurements with optical frequency combs. Chapter 2 is devoted to alkali atoms, which have a very similar electronic structure to hydrogen: their simple and accessible transitions were used for the first demonstration of laser cooling, that will be introduced in this chapter together with its applications to microwave atomic clocks (that currently provide the definition of the SI second) and atom interferometry. The ultimate frontiers of cooling will be discussed in Chapter 3, which is devoted to the investigation of ultracold quantum degenerate gases, with a particular focus on the properties and applications of Bose–Einstein condensates. In Chapter 4 we will move to the physics of helium, which shows qualitatively different properties from hydrogen and alkali atoms, owing to the presence of two electrons. Important applications of helium will be discussed, from high-precision spectroscopy for the determination of fundamental constants to the applications of degenerate helium gases for experiments of quantum atom optics. Chapter 5 starts with alkaline-earth atoms, which feature the same two-electron structure as helium: these atoms are of utmost importance in frontier atomic physics, since their extremely narrow transitions enable the realization of ultra-accurate optical atomic clocks. In this chapter we will also discuss laser cooling and spectroscopy of ions, which represent the ultimate frontier of accuracy (below 10^{-17}!) in the realization of optical frequency standards, as well as the most precise measurements in physics.

The last two chapters will be devoted to the physics which arises when ultracold atoms are trapped into optical lattices. In Chapter 6 we will focus on the physics of quantum transport in optical lattices, which both provides a testing ground for ideal solid-state physics and allows very promising applications for the determination of fundamental constants and for the precise measurements of forces where ultracold atoms are used as sensors. Chapter 7 extends this possibility to the emerging field of *quantum simulation*, in which ultracold atoms are used to experimentally realize basic condensed-matter models to precisely investigate their properties and their quantum phase transitions in an ultimately clean setting where decoherence or unwanted interactions with the environment can be avoided.

Acknowledgements

This book is the result of exciting years of research and teaching carried out at LENS (European Laboratory for Nonlinear Spectroscopy) and at the University of Florence. Since its foundation 22 years ago, the LENS research in atomic physics has spanned from high-precision laser spectroscopy to the new frontiers of ultracold quantum gases, following the evolution of the field which is described in this book.

We would like to thank the many researchers, visitors, and students who have walked along this path with us. In particular, we acknowledge the friendship of important LENS colleagues who have been participating in this adventure since the early days: Marco Bellini, Pablo Cancio Pastor, Francesco S. Cataliotti, Jacopo Catani, Paolo De Natale, Marco Fattori, Francesca Ferlaino, Gabriele Ferrari, Chiara Fort (to whom we are particularly grateful for critical suggestions and careful reading of the manuscript), Giovanni Giusfredi, Francesco Marin, Francesco Minardi, Giovanni Modugno, Francesco S. Pavone, Marco Prevedelli, Leonardo Ricci, Giacomo Roati, and Guglielmo M. Tino. We are particularly indebted to two distinguished scientists, Theodor W. Hänsch (who wrote the foreword to this book) and Eric A. Cornell, for their inspiring advice and continuous support in pushing the LENS research in atomic physics to frontier topics in high-precision spectroscopy and ultracold matter. Profound discussions with theorists like Franco Dalfovo, Michele Modugno, Lev P. Pitaevskii, Augusto Smerzi, and Sandro Stringari are also acknowledged.

Finally, we would like to thank our Oxford University Press publisher Sonke Adlung, without whose constant encouragement this book would have never been published.

Firenze, January 2013

Massimo Inguscio
Leonardo Fallani

Contents

1
Hydrogen

Never measure anything but frequency!
A. Schawlow

High-precision spectroscopy of simple atoms is a wonderful tool for precision tests of fundamental physics. Among the simple atoms hydrogen has an undisputed prime role. Its minimal internal structure, only a proton and an electron bound together, provides the possibility of formulating extremely precise theoretical predictions, which can be compared with the results coming from high-precision spectroscopy. Nowadays, atomic transition frequencies can be measured with spectacular accuracy. As an example of this possibility, the most precise experimental determination of the hydrogen $1s-2s$ transition frequency reported up to now is

$$\delta\nu(1s-2s) = 2\,466\,061\,413\,187\,035\,(10)\ \mathrm{Hz}\ , \tag{1.1}$$

which was obtained in a recent measurement performed by the group of T. W. Hänsch (Nobel Prize for Physics in 2005) at MPQ (Garching) (Parthey *et al.*, 2011). This amazing precision of 4.2×10^{-15} is the result of decades of advances in atomic physics, which have been marked by outstanding scientific discoveries and technological progress, leading to a continuous increase in the measurement accuracy, as illustrated in Fig. 1.1. Among the milestones in this journey to the land of precision one cannot avoid mentioning the invention of lasers, the development of nonlinear spectroscopic techniques, and the recent possibility of performing direct frequency measurements of light with optical frequency combs. In this first chapter we will present a brief history of hydrogen spectroscopy, evidencing the most important steps and focusing on the state of the art and on future perspectives and implications of this research.

1.1 The hydrogen spectrum

Since the dawn of quantum mechanics, the hydrogen spectrum has represented a benchmark for testing the predictions of quantum theories (Series, 1957; Series, 1988). The problem of two quantum particles at a distance r interacting with a Coulomb $\sim 1/r$ potential is indeed one of the few relevant physical situations for which quantum mechanics can provide results in analytic form. Its solution, which the reader can find discussed in any general quantum mechanics textbook, dates back to a famous paper by E. Schrödinger in 1926 (Schrödinger, 1926). The Schrödinger equation for the two-body

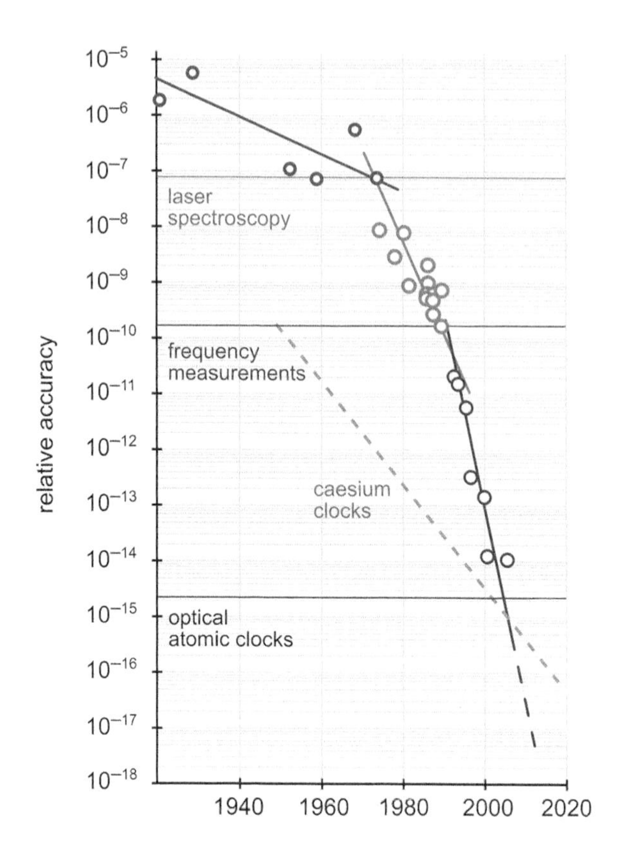

Fig. 1.1 Historical evolution of the accuracy in the measurement of optical hydrogen transitions (not including the last 2011 MPQ measurement). Reprinted with permission from Hänsch (2006). © American Physical Society.

problem of an electron bound to a proton through the Coulomb interaction gives the well-known Bohr energies

$$E_n = -hcR_\infty \frac{1}{n^2}\,, \tag{1.2}$$

where h is the Planck constant, c is the speed of light, n is the atomic principal quantum number, and R_∞ is the *Rydberg constant*, which can be expressed in terms of other fundamental constants as

$$R_\infty = \frac{m_e e^4}{8\epsilon_0^2 h^3 c} \simeq 1.097 \times 10^7 \text{ m}^{-1}\,, \tag{1.3}$$

where m_e is the electron mass, e is the elementary charge, and ϵ_0 is the permittivity of free space.

The hydrogen spectrum is represented in Fig. 1.2, together with a sketch of some possible transitions between different energy levels. These transitions are traditionally grouped in series, which correspond to sets of lines involving the same low-energy level. Historically, the first identified series was the Balmer series (from J. Balmer, who

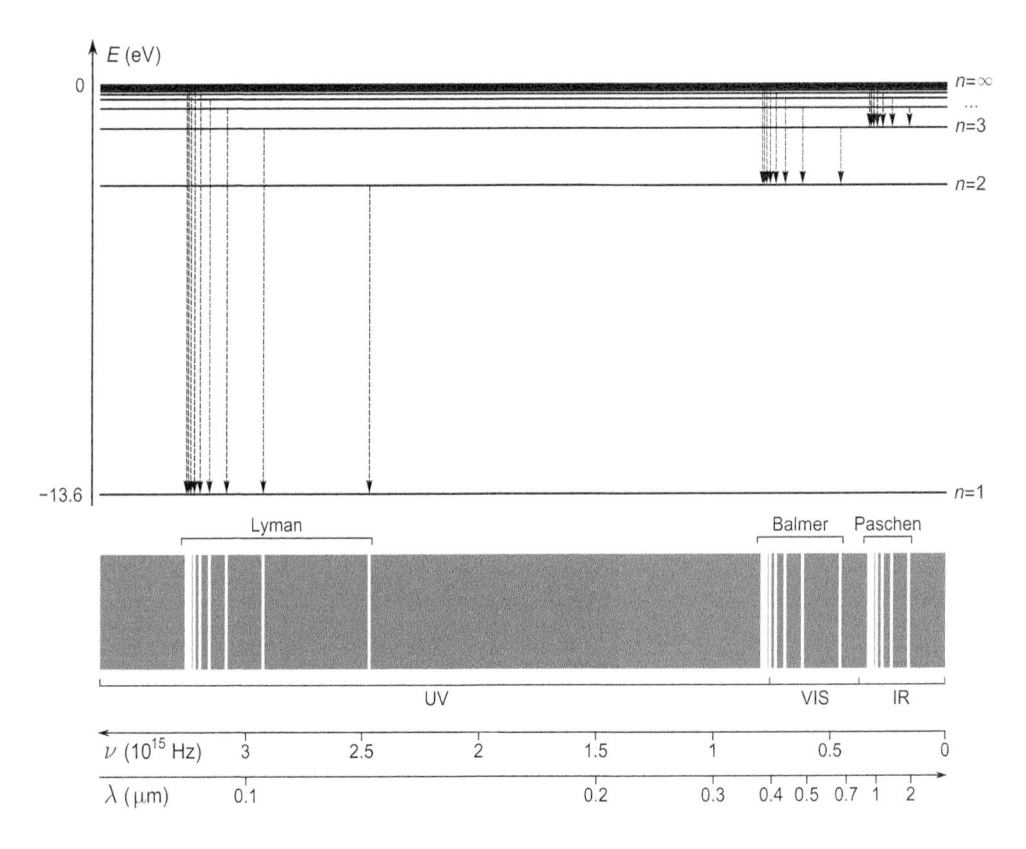

Fig. 1.2 Scheme of hydrogen levels and different groups of transitions: Lyman (in the ultraviolet), Balmer (largely in the visible spectrum) and Paschen (in the infrared).

discovered regularities in the wavelength of the lines as early as 1885), which includes all the hydrogen transitions having the $n = 2$ state as a lower state. Balmer lines were discovered before the others as they are the only hydrogen transitions lying in the visible region of the spectrum, as is shown in Fig. 1.2, with the longest-wavelength transition Balmer-α (or H_α) from $n = 3$ to $n = 2$ corresponding to a wavelength of 656 nm.[1] The Lyman series (from T. Lyman, who discovered it in 1906–1914) includes all the hydrogen lines in which the ground state $n = 1$ is involved: all the lines of this series are lying in the ultraviolet region of the spectrum, with the longest-wavelength transition Lyman-α (or Ly_α) from $n = 2$ to $n = 1$ corresponding to a wavelength of 122 nm.

[1]The Balmer-α line was already identified by J. von Fraunhofer in the early years of the nineteenth century during his studies on the dark lines in the solar spectrum (Fraunhofer C line). The connection between dark lines in the solar spectrum and absorption by the different chemical elements was established later after the work of Kirchoff and Bunsen in the mid nineteenth century.

1.2 Balmer-α: from Bohr to QED

After the invention of lasers, laser spectroscopy of hydrogen naturally started with the Balmer lines, in particular with the Balmer-α line at a wavelengthof 656 nm, which was easily accessible to tunable dye lasers. In order to excite this transition, molecular hydrogen H_2 has to be dissociated to atomic H and then excited to populate the $n = 2$ state. This process is typically realized in a discharge tube by impact with highly energetic electrons, which are produced and accelerated by a voltage difference across two electrodes. As we will further discuss in Sec. 1.3, atoms excited to the $2s$ state are metastable, with a lifetime of $\approx 1/7$ seconds.

1.2.1 Fine structure

In Fig. 1.3a (centre) we show a spectrum of the Balmer-α (H_α) line at 656 nm, measured in 1972 at Stanford by T. W. Hänsch, I. S. Shaning, and A. L. Schawlow with a narrow-band tunable dye laser (Hänsch *et al.*, 1972). The curve shows the absorption coefficient of hydrogen as the excitation laser is scanned over a ≈ 20 GHz frequency interval across the atomic resonance. Instead of just a single line, a structured lineshape with two resolved maxima appears. This structure, which is not accounted for by the nonrelativistic Schrödinger theory of hydrogen, is called *fine structure* and it naturally results from a relativistic treatment of the quantum two-body problem.

A successful theoretical description of relativistic spin-1/2 particles is given by the *Dirac equation*, which, differently from the Schrödinger equation, satisfies Lorentz invariance. The Dirac theory of hydrogen predicts a more complicated form of the spectrum:

$$E_{nj} = m_e c^2 \left[1 + \left(\frac{\alpha}{n - (j + \frac{1}{2}) + \sqrt{(j + \frac{1}{2})^2 - \alpha^2}} \right)^2 \right]^{-\frac{1}{2}} , \tag{1.4}$$

where α is the *fine structure constant*, which is defined as:

$$\alpha = \sqrt{\frac{2hcR_\infty}{m_e c^2}} \simeq \frac{1}{137} . \tag{1.5}$$

This dimensionless constant is connected to the ratio between the energy scale of the atomic binding energies hcR_∞ and the rest energy of the electron $m_e c^2$. It also corresponds to the ratio v/c between the ground-state root-mean-square electron velocity $v = \hbar / m_e a_0$ (where a_0 is the Bohr radius) and the speed of light c. Not only does it measure the "relativisticness" of the hydrogen electron: α is a fundamental constant of physics, since it ultimately describes the strength of the electromagnetic interaction, and its measurement will be discussed in detail in Sec. 4.5. Owing to the small value of the fine structure constant, the energies in eqn (1.4) can be conveniently expanded in a series of α. At order α^4 this expansion yields

$$E_{nj} = m_e c^2 - hcR_\infty \frac{1}{n^2} - hcR_\infty \alpha^2 \frac{1}{n^3} \left(\frac{1}{j + \frac{1}{2}} - \frac{3}{4n} \right) + \cdots , \tag{1.6}$$

where the first term is the rest energy of the electron, the second term is the hydrogen Bohr energies of eqn (1.2) and the third term represents the leading-order relativistic

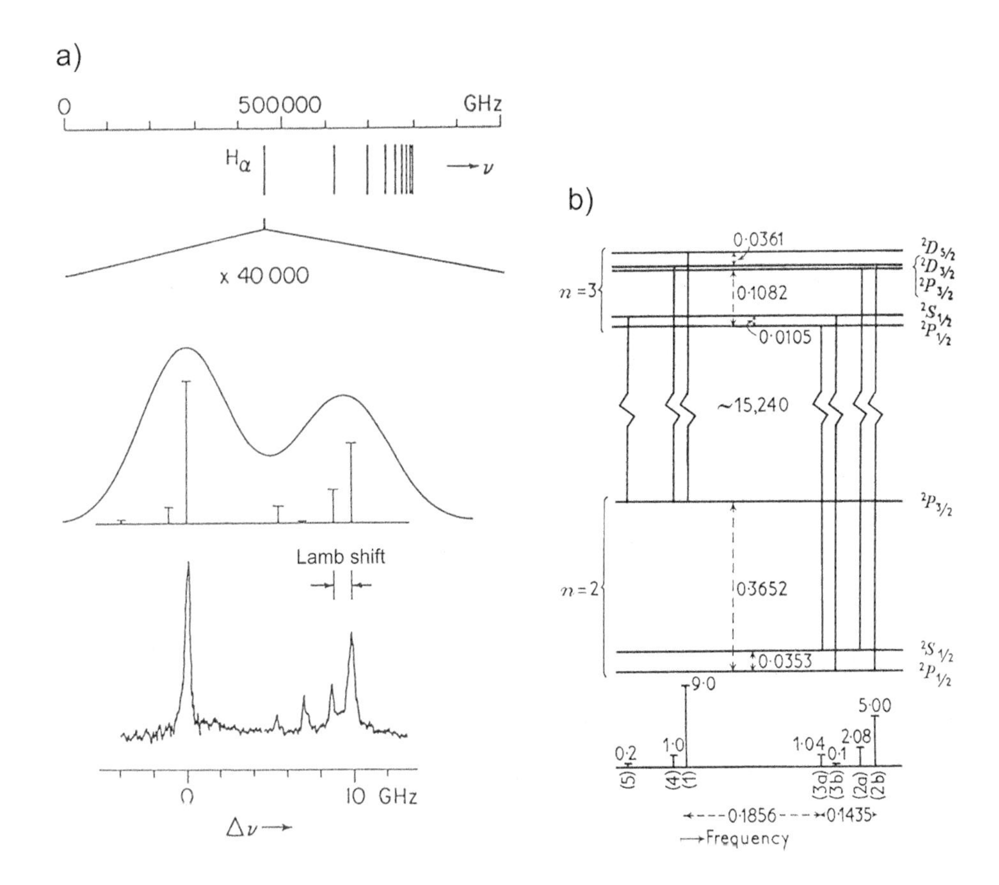

Fig. 1.3 Spectroscopy of the hydrogen Balmer-α line. a) Energies of the lines in the Balmer series (top), Doppler spectrum of the Balmer-α line at 656 nm (centre), and Doppler-free spectrum of the same transition obtained with saturation spectroscopy in Hänsch *et al.* (1972) (bottom). Reprinted from the Proceedings of "The Hydrogen Atom" symposium held in Pisa in 1988 (Hänsch, 1989). b) Fine structure of the Balmer-α line, including Lamb shift (energy intervals are expressed in spectroscopic units, with 1 cm$^{-1} \simeq h \times 30$ GHz). Reprinted from the classical G. W. Series book *Spectrum of Atomic Hydrogen* (Series, 1957).

fine-structure corrections. The latter term, of order $\alpha^2 \approx 10^{-4}$ to the Bohr energies, is due to the combined effects of three different corrections to the nonrelativistic hydrogen theory:

1. leading-order relativistic corrections $3p^4/8m_e^3c^2$ to the classical kinetic energy $p^2/2m_e$ (where $\mathbf{p}$ is the electron momentum);
2. a *spin–orbit* interaction term, proportional to $\mathbf{l} \cdot \mathbf{s}$, arising from the interaction between the electron spin $\mathbf{s}$ and the effective magnetic field generated by the proton charge rotating about the electron (in the electron frame) because of the orbital angular momentum $\mathbf{l}$;

3. a Darwin interaction term, with no classical analogue, which can be related to the *zitterbewegung* rapid motion of the electron arising from the Dirac theory.[2]

As evident from eqn (1.4), fine-structure corrections depend on the additional quantum number j, which arises from the quantization of the total angular momentum $\mathbf{j} = \mathbf{l} + \mathbf{s}$. As a result of this dependence on j, fine-structure corrections to the hydrogen spectrum partially lift the degeneracy between $n = 2$ levels: the $2p_{3/2}$ state with $j = 3/2$ has a higher energy than the $2s_{1/2}$ and $2p_{1/2}$ states with $j = 1/2$, the shift being $\simeq 11$ GHz. The fine structure of the first three levels of hydrogen is schematically shown in the central column of Fig. 1.4.

The two maxima in the Balmer-α spectrum of Fig. 1.3a (centre) originate from the fine-structure separation between the $n = 2$ levels predicted by the relativistic Dirac theory of hydrogen. Fine-structure separations had already been observed decades before the invention of lasers,[3] as spectrometers of high-resolving power (based on diffraction gratings or Fabry–Perot interferometers) were already in use at the beginning of the last century. The major limitation for spectroscopy before the advent of lasers was not in the wavelength measurement, but in the spectral width of the light emitted or absorbed by the atoms.

1.2.2 Doppler effect and saturation spectroscopy

The main contribution to the spectral width observed in Fig. 1.3a comes from the *Doppler effect*. According to special relativity, if light with frequency ν and wavevector $\mathbf{k}$ illuminates an atom moving with velocity $\mathbf{v}$, the effective frequency of the light experienced by the atom in its rest frame is:

$$\nu' = \nu \left(1 - \frac{\hat{k} \cdot \mathbf{v}}{c} + \frac{v^2}{2c^2} + \cdots \right), \tag{1.7}$$

where the second term in parentheses represents the first-order Doppler shift ($\hat{k}$ is the photon direction), while the following term in the expansion is the second-order Doppler shift, arising from relativistic time dilation.

We start considering the effects of the first-order Doppler shift for a gas of atoms in thermal equilibrium at temperature T. Owing to the Maxwell–Boltzmann thermal distribution of velocities, the shift is different for each atom in the gas, which results in an inhomogeneous *Doppler broadening* of the atomic line. For hydrogen at room temperature the Doppler-broadened linewidth, defined as the full width at half maximum (FWHM) of the lineshape, amounts to

$$\Delta\nu = \nu \sqrt{\frac{8 \ln(2)\, k_B T}{mc^2}} \approx 6 \text{ GHz}. \tag{1.8}$$

[2]This effect, which can be explained as an interference effect between solutions to the Dirac equation with positive and negative energy, has never been observed for free particles, but has been recently simulated using trapped ions (see Sec. 7.3.4 devoted to quantum simulation experiments of relativistic quantum mechanics).

[3]For heavier atoms relativistic corrections become more important. In hydrogen-like alkaline atoms these corrections are of order $(Z\alpha)^2$, where Z is the atomic number. A famous example of fine-structure splitting is given by the sodium D1-D2 doublet centred at 589.3 nm. The structure of alkaline atoms will be discussed in more detail in Sec. 2.1.

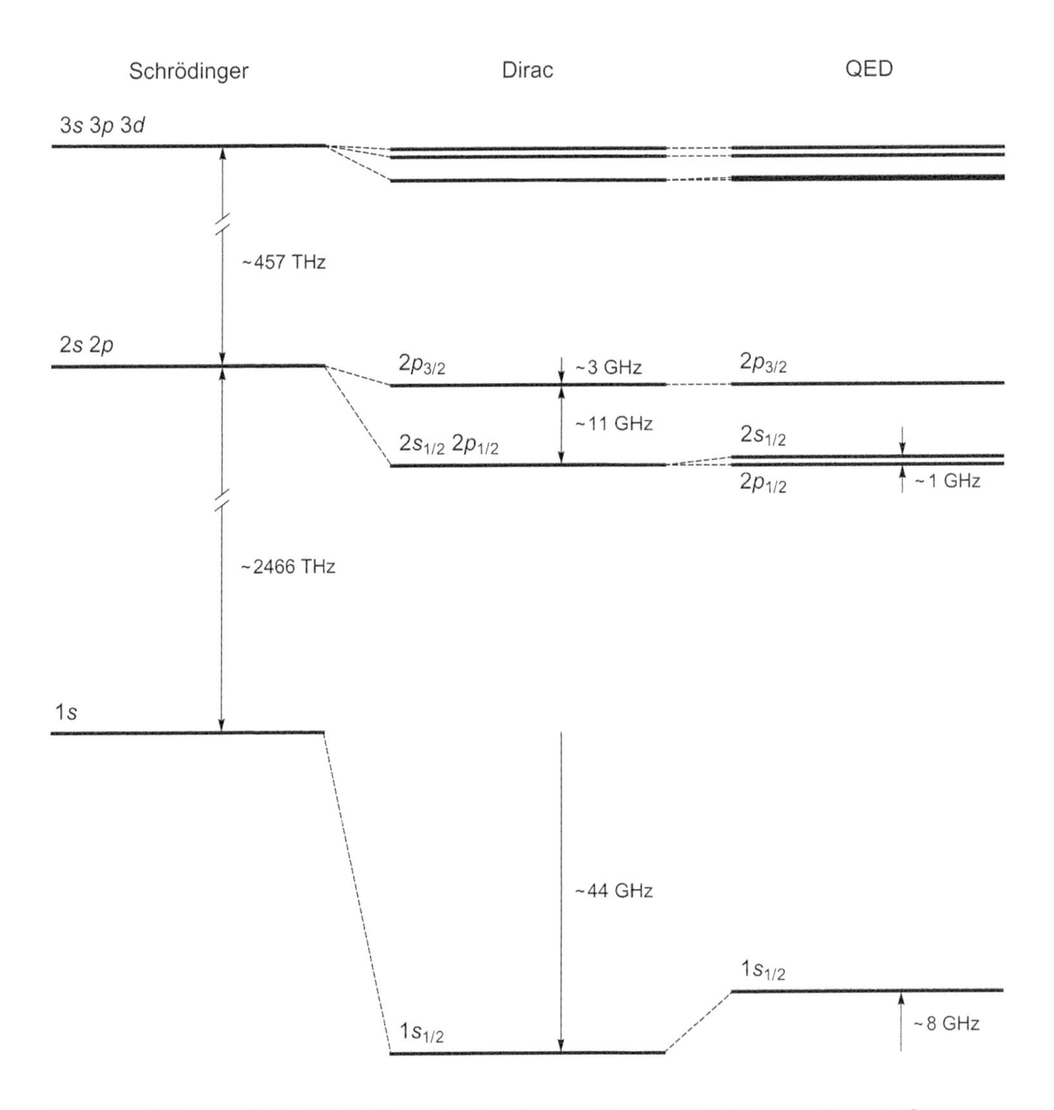

Fig. 1.4 Scheme of relativistic (fine-structure) corrections and QED corrections to the energy of hydrogen $n = 1$ and $n = 2$ levels.

This width is small enough to allow the observation of the fine-structure splitting in the $n = 2$ state, approximately 11 GHz (see Fig. 1.3a), but is too large to observe the fine structure of the upper $n = 3$ state, on the order of 3 GHz (see Fig. 1.3b).

The advent of lasers set the premises for the development of new revolutionary spectroscopic techniques. One of the distinctive features of laser light is the high degree of monochromaticity, which offers the possibility of achieving a very high spectral density (defined as intensity per unit of frequency). As a consequence, with lasers it became possible to achieve new regimes of light–matter interaction, where nonlinear effects become important. One of the most important manifestations of nonlinear

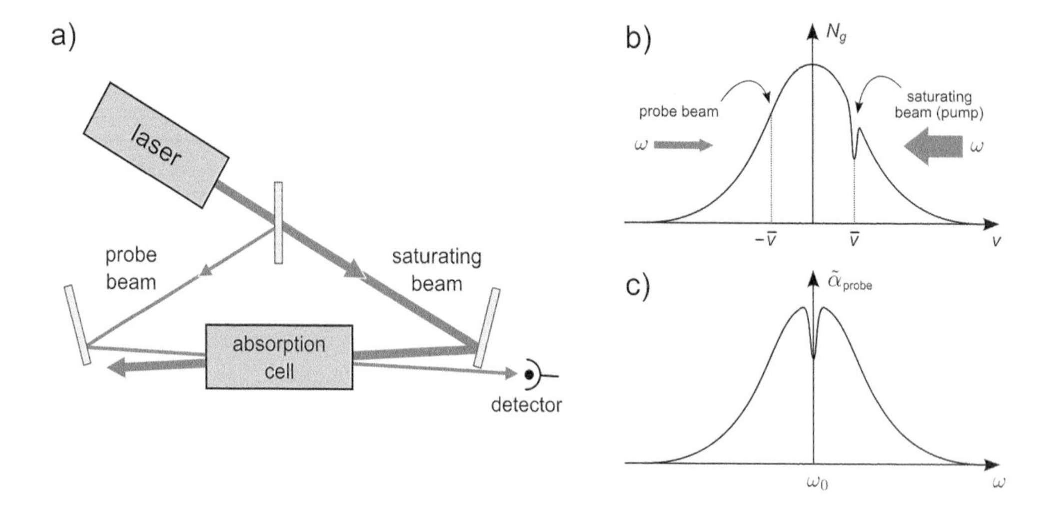

Fig. 1.5 Absorption saturation spectroscopy. a) Scheme of the experimental setup. b) Number of atoms in the ground state N_g as a function of atom velocity v for a thermal gas of atoms probed by two counterpropagating (probe and saturating) laser beams at frequency ω: the saturating beam digs a hole in the curve in correspondence with the class of atomic velocities $\bar{v}$ which are resonant. c) Probe absorption coefficient as a function of ω, featuring a saturation Lamb dip at the line centre, where the atoms are resonant with both the beams.

light–matter interaction is *saturation* (which is extensively discussed in Appendix A devoted to the interaction between coherent radiation and a two-level atom). At low spectral densities the intensity of the light absorbed by the atoms (and, consequently, the intensity of the light emitted by fluorescence as well) is proportional to the intensity of the incident light. At high spectral densities, when saturation sets in, the intensity of light absorbed or re-emitted by the atoms does not grow linearly with the incident intensity and, at very high power, it saturates to a constant value. This happens because, at steady state, a large fraction of the atoms is pumped from the ground state to the excited state, which makes the atomic sample less absorptive.

This effect is at the basis of Doppler-free *saturation spectroscopy*, a clever spectroscopic technique developed in the late 1960s after the first studies on saturation of the laser gain medium (Schawlow, 1982), and perfected by T. W. Hänsch (Smith and Hänsch, 1971; Hänsch *et al.*, 1971) and C. Bordé (Bordé, 1970) in 1970–1971, who applied this concept to spectroscopy of gases placed externally to the laser cavities. In the classical scheme of saturation spectroscopy, shown in Fig. 1.5a, the output beam of a tunable laser is split into an intense saturating pump beam and a weaker probe beam at the same frequency[4] ω, crossing the sample one opposite to the other. The

[4]Following a widespread practice, in this book we will use the term *frequency* to denote both real frequencies, indicated with ν and defined as the inverse of the oscillation period (measured in Hz), and angular frequencies, indicated with ω (measured in s^{-1} or rad/s) and related to the real frequencies ν by the well-known relation $\omega = 2\pi\nu$. Frequency values expressed in hertz will always refer to real frequencies ν.

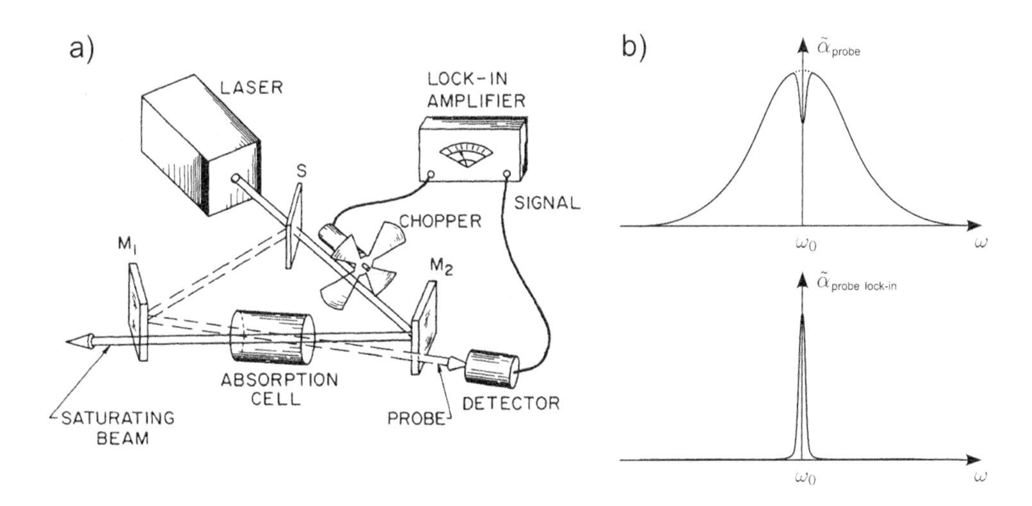

Fig. 1.6 Saturation spectroscopy with amplitude modulation of the pump beam. a) Scheme of the experimental setup. Reprinted with permission from Schawlow (1982). © American Physical Society. b) Lock-in detection of the probe absorption signal (upper graph) at the pump modulation frequency yields a zero-background signal in which the contribution of atoms with nonzero velocity component is removed (lower graph).

pump beam saturates the atomic transition causing a depletion of the number N_g of atoms in the ground state. Since the transition is broadened by the Doppler mechanism, saturation only occurs for a narrow class of axial atomic velocities centred around $\overline{v} = (\omega_0 - \omega)/k$ (where ω_0 is the transition frequency and k is the laser wavenumber), as shown in Fig. 1.5b. For $\omega \neq \omega_0$ the absorption of the counterpropagating probe is not affected by the pump-induced saturation, since the probe is resonant with atoms moving with opposite velocity $-\overline{v}$. When $\omega \simeq \omega_0$, however, both the pump and the probe interact with the same atoms having axial velocity $\overline{v} \simeq 0$, and the probe experiences a reduction in the absorption caused by the pump-induced saturation. This effect can be observed in Fig. 1.5c, in which the probe absorption coefficient is plotted as a function of the laser frequency: a saturation *Lamb dip* appears at the centre of the absorption Doppler-broadened profile.

The width of the Lamb dip is typically much smaller than the Doppler linewidth, since only a restricted class of atomic velocities contributes to it. An improved version of saturation spectroscopy was introduced in Hänsch *et al.* (1971) by T. W. Hänsch, M. D. Levenson, and A. L. Schawlow in order to dramatically enhance the intensity of the Doppler-free signal over a nearly zero background. In this scheme, represented in Fig. 1.6a, the pump beam is amplitude-modulated by a chopper, which makes the sample periodically saturated or non-saturated depending on whether the pump light passes through the chopper blades or not. In this case the probe absorption is modulated as well according to the periodic saturation of the sample induced by the pump beam. By detecting the probe absorption with a photodetector and demodulating the signal with a lock-in amplifier driven at the pump modulation frequency, only the signal coming from the interaction with the Doppler-unshifted zero-velocity class is left (Fig. 1.6b). Applying this scheme to hydrogen spectroscopy, T. W. Hänsch *et al.* could obtain a saturation spectrum of the Balmer-α line in which the narrow fine structure of the hydrogen $n = 3$ state could be eventually detected, as shown in the bottom of Fig. 1.3a.

Looking at the saturated-absorption spectra in more detail, a broader Gaussian pedestal could still be identified under the Doppler-free peaks. This pedestal can be attributed to the effect of velocity-changing collisions, i.e. elastic collisions that redistribute the atomic excitation induced by the pump beam among a larger class of atomic velocities (Smith and Hänsch, 1971). Variations on the saturated-absorption technique illustrated above can be used to eliminate this effect, e.g. with *polarization spectroscopy*. In this technique the pump beam is circularly polarized and induces an orientation of the atomic spin: instead of measuring the absorption of the probe beam (which is sensitive to the population difference), the birefringence of the sample is detected by measuring the rotation of the probe beam polarization by the oriented sample, according to the Faraday effect. Velocity-changing collisions destroy the sample orientation and do not contribute to the pedestal. More details on this and other implementations of saturation spectroscopy can be found e.g. in Demtröder (2003).

1.2.3 Lamb shift

The spectacular increase in spectral resolution provided by nonlinear saturation spectroscopy marked the beginning of a new era in laser spectroscopy, allowing more precise measurements and the observation of new effects. An example of this possibility is provided by the same Balmer-α spectrum shown in Fig. 1.3a. In addition to the fine structure, saturation spectroscopy allowed the detection of one more line which is not expected from the relativistic Dirac theory of hydrogen.

According to the result of the Dirac theory in eqn (1.6), the $2s_{1/2}$ and $2p_{1/2}$ states of hydrogen should have the same energy. Instead, it is experimental evidence that the $2s_{1/2}$ state has a higher energy than the $2p_{1/2}$ state. This shift, which amounts to $\simeq 1$ GHz, is explained by *quantum electrodynamics* (QED), which is the relativistic quantum field theory which describes the electromagnetic interaction. This correction to the hydrogen fine structure, denoted by the name of *Lamb shift*, mostly comes from the interaction of the electron with the vacuum fluctuations of the electromagnetic field.[5] According to a simple semiclassical model, the interaction of the electron with the vacuum field causes a fluctuation in the electron position, leading to a shift of the binding energy to the proton.[6] Only the s states are significantly affected by the

[5] According to the quantum theory of light, even in the absence of photons the electromagnetic field is characterized by zero-point fluctuations, as it happens for the ground state of a harmonic oscillator, in which position and momentum fluctuations lead to a finite zero-point energy $\hbar\omega/2$ (where ω is the frequency of the oscillator, in this case the frequency of one particular electromagnetic field mode). The interaction of the atom with this "vacuum" electromagnetic field is important since it is responsible for the spontaneous emission process as well.

[6] In the QED description of the interaction between light and matter, the Lamb shift emerges from the evaluation of different interaction terms involving the exchange of virtual photons and/or the interaction with virtual particles. The hydrogen Lamb shift is almost entirely due to the electron *self-energy*, which arises from the emission and re-absorption of a virtual photon, leading to a renormalization of the electron mass. The second most important contribution in hydrogen ($\sim 3\%$)

Lamb shift, since for s states the atom has a nonzero probability to be at the nucleus position: this increased sensitivity to the electron position makes the effect of vacuum fluctuations on the proton–electron binding energy more important. QED corrections for the first three levels of hydrogen are shown in the right column of Fig. 1.4 as corrections to the Dirac energies.

While the one reported in Hänsch *et al.* (1972) and shown in Fig. 1.3a is the first observation of the Lamb shift directly in the optical domain, it is worth remembering that the Lamb shift had been first observed in a milestone experiment in the history of spectroscopy by W. E. Lamb and R. C. Retherford already in 1947 (Lamb Jr and Retherford, 1947). The two physicists used radiofrequency (RF) spectroscopy to induce transitions between the $2s_{1/2}$ state and the $2p_{1/2}$, $2p_{3/2}$ states. By taking advantage of the long lifetime of the $2s_{1/2}$ state ($\sim 1/7$ s), they were able to measure the atomic absorption by detecting the number of metastable $2s_{1/2}$ atoms remaining after the excitation. Radiofrequency transitions have the advantage of a much reduced Doppler broadening with respect to optical transitions, since the Doppler shift $\delta\nu$ is directly proportional to the photon frequency ν, as shown in eqn (1.7). This allowed Lamb and Retherford to precisely measure the frequencies of the $2s_{1/2} \rightarrow 2p_{1/2}$,$2p_{3/2}$ transitions and detect a splitting of ≈ 1 GHz between the $2s_{1/2}$ and $2p_{1/2}$ states not accounted for by the Dirac theory. This shift, caused by the QED corrections to the $2s_{1/2}$ state, represented a significant motivation for the development of the quantum field theory of the electromagnetic interaction.

More recent microwave measurements of the $n = 2$ hydrogen Lamb shift, based on evolutions of the Lamb–Retherford technique, yielded an average value

$$\nu(2s_{1/2}) - \nu(2p_{1/2}) = 1\,057\,843.9(7.2)\ \text{kHz} \tag{1.9}$$

limited primarily by the finite lifetime of the $2p_{1/2}$ state (Biraben, 2009) and in excellent agreement with the predictions of QED. More recently, optical measurements on narrow two-photon transitions (which will be the topic of the next section) have provided improved precisions on the measurement of the hydrogen Lamb shift: the reader can refer to Karshenboim (2005) for an excellent review on experiments and theory of the hydrogen Lamb shift and other QED tests with simple atoms.

1.3 $1s - 2s$: a quest for precision

Although nonlinear spectroscopy allows the circumvention of Doppler broadening, a more fundamental limit on the width of spectral lines emerges. The ultimate limit is given by the natural linewidth $\Delta\nu$ of the transitions, which is determined by the lifetime τ of the atomic states through the relation $\Delta\nu = 1/\tau$. Therefore, for high-resolution spectroscopy, e.g. for testing fundamental theories or measuring fundamental constants, lines with intrinsically small width have to be chosen in order to achieve a better resolution in the frequency measurement.

Among the possible transitions between hydrogen levels, the transition between the $1s$ and the $2s$ levels has a particular significance. This transition offers the possibility of

comes from the *vacuum polarization* effect, in which the interaction between the electron and the proton is modified by the creation of virtual electron/positron pairs creating an effective polarization of the vacuum. A detailed discussion of the hydrogen Lamb shift can be found in Series (1988).

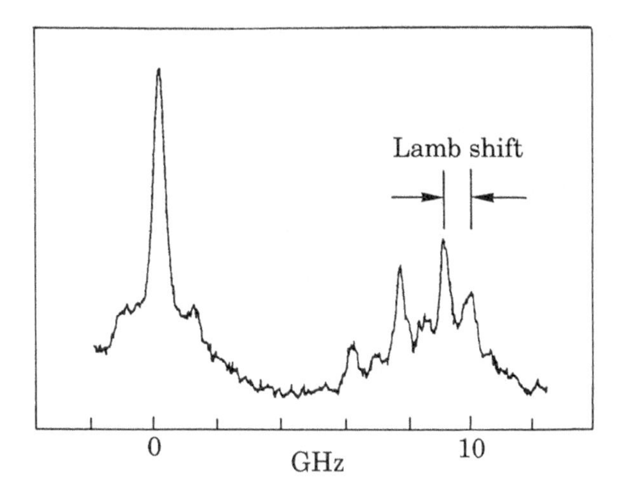

Fig. 1.7 Since the invention of the laser, the progress in high-precision spectroscopy has been promoted by important advances in laser technology. Semiconductor diode lasers have now become the most popular laser sources for spectroscopy in the visible or near-infrared region, owing to their simplicity of use, low mantainance, transportability, tunability, and inexpensiveness. As an example, the figure shows the same saturation Balmer-α spectrum of Fig. 1.3a, here recorded with a semiconductor diode laser at MPQ. Reprinted from the Proceedings of the CXX International Enrico Fermi school on "Frontiers in Laser Spectroscopy" (Inguscio, 1994).

very accurate measurements, since it is characterized by a large frequency $\nu_0 \simeq 2.5\times10^{15}$ Hz and a very small natural linewidth $\Delta\nu \simeq 1.3$ Hz. The reason for this small linewidth lies in the fact that the $2s$ level is metastable, since the only decay channel available is towards the ground state $1s$, which is highly forbidden by exchange of only one photon.[7] Using the terminology usually employed for oscillators and resonators, we can associate a very large *quality factor* $Q = \nu_0/\Delta\nu \simeq 2 \times 10^{15}$ to this transition. This large quality factor, combined with the very accurate theoretical predictions which are possible for such a simple atomic system as hydrogen (see e.g. Biraben (2009)), makes the hydrogen $1s - 2s$ transition a powerful tool for the measurement of fundamental constants as the Rydberg constant.

The investigation of the $1s - 2s$ transition in atomic hydrogen has been pursued by T. W. Hänsch for more than 30 years, with the first experimental demonstration of direct $1s - 2s$ excitation at Stanford in 1975 (Hänsch *et al.*, 1975). Below we discuss relevant aspects of the $1s - 2s$ spectroscopy in connection with the development of spectroscopic techniques. Later in the chapter we will discuss its implications for the measurement of the Rydberg constant and for tests of fundamental theories.

[7]Decay by electric dipole transition is forbidden by the selection rule $\Delta l = \pm1$ (see Appendix A.4), while decay by magnetic dipole transition is quite suppressed by the orthogonality of the $1s$ and $2s$ wavefunctions (in nonrelativistic approximation). The most important decay channel is two-photon decay, in which two photons are simultaneously emitted by an atom in $2s$, limiting the lifetime of this state to $\approx 1/7$ seconds, from which the small linewidth follows.

1.3.1 Two-photon spectroscopy

The same selection rule that suppresses the decay from the $2s$ level has a consequence on the excitation of the $1s - 2s$ transition, which cannot be induced by absorption of one photon only. This transition can only be excited by a higher-order process through the simultaneous absorption of two photons having an energy $h\nu_0/2$ corresponding to half of the transition energy $h\nu_0$. In order to have a large absorption probability for two-photon processes, large radiation intensities are required; therefore, two-photon transitions can be excited only by working with lasers, in which the power spectral density is much larger than in incoherent light sources, e.g. lamps.

Two-photon spectroscopy was proposed by V. P. Chebotayev and coworkers in Vasilenko *et al.* (1970), where it was shown that two-photon transitions induced by two counterpropagating laser beams result in spectra with very narrow and intense peaks, which do not suffer from Doppler broadening. This happens because of Doppler shift cancellation when the transition is driven by absorption of two photons coming from opposite directions: an atom with velocity v along the direction of the laser wavevector absorbs one photon with an effective frequency $\nu\,(1 + v/c)$ and a second counterpropagating photon with frequency $\nu\,(1 - v/c)$; therefore, their sum does not depend on the atom velocity and matches the atomic resonance frequency ν_0 for $2\nu = \nu_0$. Doppler-free two-photon spectroscopy was first demonstrated in Biraben *et al.* (1974), Levenson and Bloembergen (1974), and Hänsch *et al.* (1974) for sodium atoms, in which the two-photon excitation was favoured by the existence of a state at almost half the energy separation. Although this is not the case for hydrogen, the $1s - 2s$ two-photon transition can be excited as well, as first demonstrated in Hänsch *et al.* (1975). In this experiment a tunable dye laser with wavelength $\lambda = 486$ nm and frequency ν was frequency-doubled to a wavelength $\lambda = 243$ nm and frequency 2ν with nonlinear optics techniques of second-harmonic generation (see Appendix B.3 for an introduction to nonlinear optics). The lower panel of Fig. 1.8a shows the Doppler-free two-photon spectrum of the hydrogen $1s - 2s$ transition as the laser was scanned across the resonance (Hänsch *et al.*, 1975).

The measurement of the $1s$ Lamb shift represents an important QED test. As previously discussed in Sec. 1.2.3, in the case of the $2s$ state the shift could be measured from the removal of degeneracy with the nearby $2p$ state, which could be determined both by direct microwave excitation and by the fine structure of the optical transitions. Instead, for $n = 1$ no level splittings are availabile and the $1s$ shift can only be deduced from the corrections to the frequency predicted by the Dirac theory for the ultraviolet $1s - 2s$ transition. For this task a precise calibration of the laser frequency (or wavelength) is needed, which is a major problem in high-precision spectroscopy. This problem has been eventually fully solved with the invention of the optical frequency comb, as we shall discuss in more detail in Sec. 1.4. In the pioneering work of Hänsch *et al.* (1975) the problem was solved with a “self-referenced” measurement in which hydrogen itself was used as a reference for the laser calibration. This was possible thanks to the $\sim 1/n^2$ dependence of the Bohr energies, which determines the almost coincidence (neglecting fine structure and QED corrections) between the $1s - 2s$ transition energy $E_{1s-2s} = \frac{3}{4}hcR_\infty$ (121.5 nm wavelength) and four times the energy of the Balmer-β transition (from $n = 2$ to $n = 4$) $E_{\mathrm{H}_\beta} = \frac{3}{16}hcR_\infty = E_{1s-2s}/4$ (486 nm

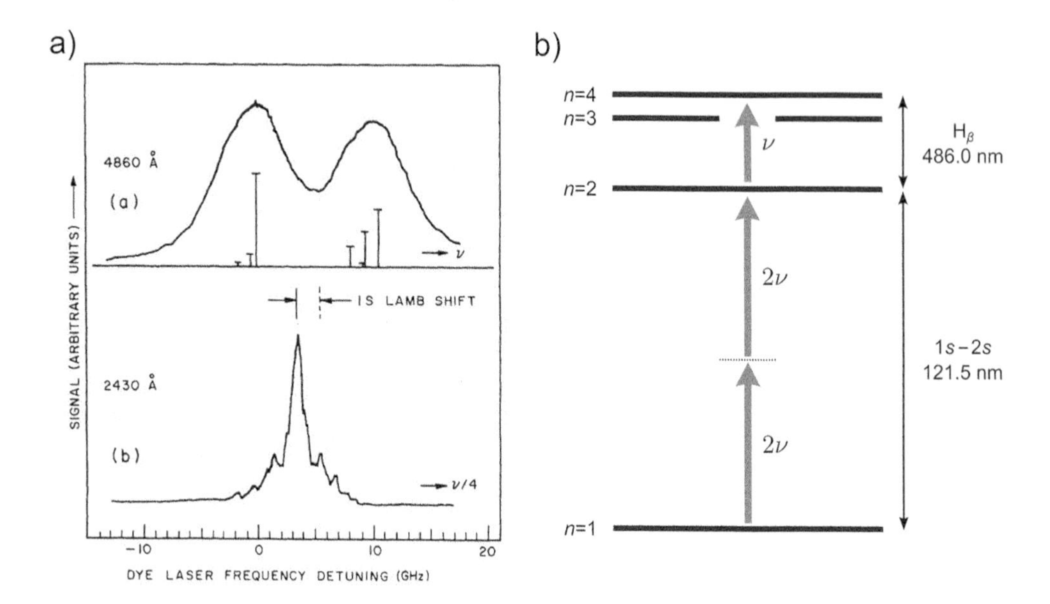

Fig. 1.8 a) The upper graph is the spectrum of the hydrogen Balmer-β line from $n = 2$ to $n = 4$, showing the superposition of different Doppler-broadened fine-structure components as the laser frequency ν is varied. The Balmer-β transition is also used as a frequency reference for the $1s - 2s$ transition shown in the lower graph. The two-photon transition is driven by radiation at twice the frequency 2ν. The shift indicated in the figure is the Lamb shift of the $1s$ state (note that the frequency scale of this graph has to be multiplied by a factor 4). Reprinted with permission from Hänsch *et al.* (1975). © American Physical Society. b) Scheme of the transitions.

wavelength), as represented in the level scheme in Fig. 1.8b. In Hänsch *et al.* (1975) these two transitions were simultaneously excited by using radiation from the same laser source at frequency ν and wavelength $\lambda = 486$ nm. The laser beam at frequency ν was used to directly excite the Balmer-β line, as shown in the upper panel of Fig. 1.8a. Second-harmonic generation of light from the same laser source was used to provide the radiation at $\lambda = 243$ nm for the excitation of the two-photon $1s - 2s$ transition, as shown in the lower panel of Fig. 1.8a. The simultaneous recording of the two spectra allowed the measurement of the energy shifts from the Bohr prediction and, knowing the fine-structure corrections to the levels, resulted in a first spectroscopic determination of the Lamb shift of the ground state $1s$ (indicated in the figure).

We note that the width of the two-photon spectrum of the $1s - 2s$ transition in Fig. 1.8a is much smaller than the width of the Doppler-broadened one-photon spectrum of the Balmer-β line. Similarly to other Doppler-free spectroscopic techniques (such as saturation spectroscopy), two-photon spectroscopy can probe the atoms *as if* they are not moving. But this is true only at first order. Two-photon spectroscopy is very effective in circumventing the first-order Doppler broadening, but it cannot cancel the second-order Doppler effect (see eqn (1.7)), which is always positive and does not

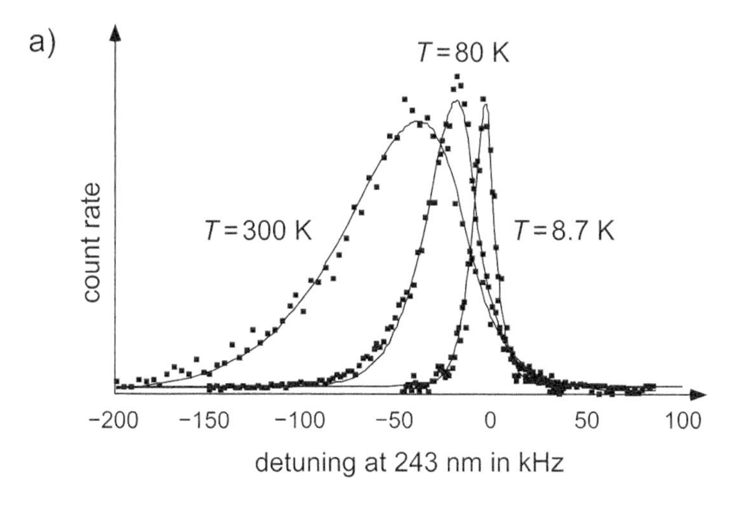

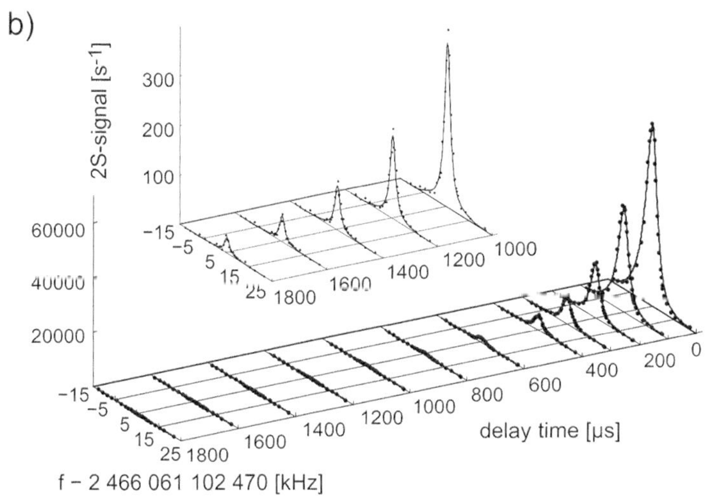

Fig. 1.9 Spectra of hydrogen $1s - 2s$ transition measured at MPQ. a) Cooling the atomic beam the line shifts and gets narrower owing to the second-order Doppler shift. b) Changing the delay between excitation and detection allows for extrapolating the line-centre for vanishing atomic velocity. Reprinted from the Proceedings of "The Hydrogen Atom: Precision Physics of Simple Atomic Systems" symposium held in Castiglione della Pescaia in 2000 (Biraben *et al.*, 2001).

depend on the relative orientation of the atomic velocity and photon wavevector. This property is particularly important since it results in *both* a broadening of the line *and* a shift of its centre, according to the atomic velocity distribution. For hydrogen at room temperature the second-order Doppler shift of the $1s - 2s$ transition for the rms atom velocity $v_{rms} = \sqrt{3k_BT/m}$ amounts to

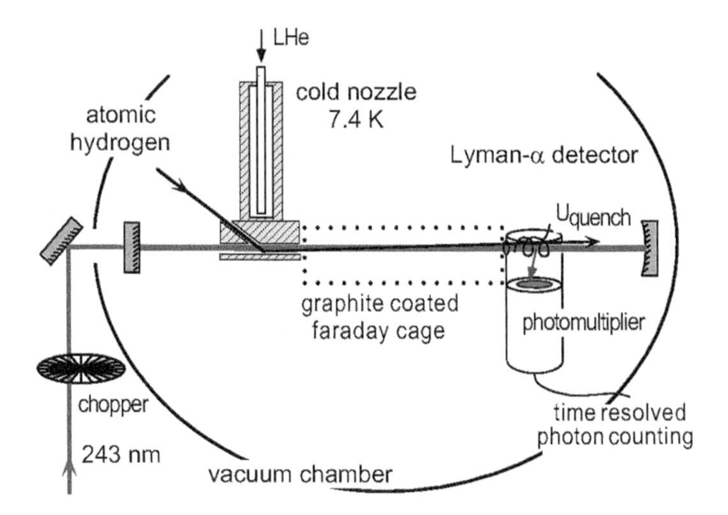

Fig. 1.10 Setup of the hydrogen spectrometer used at MPQ by the group of T. W. Hänsch in the late 1990s. Sub-Doppler two-photon spectroscopy of the $1s - 2s$ transition is performed inside an optical resonator in which 243 nm laser light forms a standing wave, exciting the hydrogen atoms from two counterpropagating directions. A chopper and a time-resolved detector allow the selection of the signal coming from atoms with different velocities. Reprinted with permission from Huber *et al.* (1999). © American Physical Society.

$$\Delta\nu = \nu\frac{v_{rms}^2}{2c^2} = \frac{k_B T}{2mc\lambda} \simeq 30 \text{ kHz} , \tag{1.10}$$

which is about 10^5 times smaller than the first-order Doppler broadening.

Since the second-order Doppler shift is not direction-dependent, it cannot be cancelled with two-photon spectroscopy. This effect can be observed in the spectra reported in Fig 1.9a, which were recorded at MPQ in the late 1990s using the setup shown in Fig. 1.10. Here spectra of the hydrogen $1s - 2s$ transition are reported for different temperatures of the nozzle injecting hydrogen atoms into the interaction zone. When the atoms travel through the nozzle, their temperature changes owing to collisions with the nozzle walls (cooled with a liquid helium cryostat). It is striking to observe how the line becomes narrower and moves towards higher frequencies as the atoms are cooled down. The latter is a direct consequence of the reduction of the second-order Doppler shift as the atom velocities are made smaller.[8]

Despite the enormous utility of nonlinear Doppler-free spectroscopy for resolving atomic structures, when more and more accurate measurements are involved, eventually one must face the limits imposed by the atomic motion. In order to improve the accuracy

[8]The second-order Doppler effect manifests itself as a redshift in the spectra of Fig. 1.9a, since in the atom rest frame the effective frequency of the laser is higher than the value measured in the laboratory frame (see eqn (1.7)); therefore, a *smaller* laser frequency is required to excite the atoms. We also note that the lineshape is asymmetric, since the probability distribution function for the atomic squared velocity v^2 (entering the second-order Doppler shift) is not symmetric around the most probable velocity (in a gas at thermal equilibrium and in an atomic beam as well).

of the measurements the atoms must be *really* slowed down, as it is clearly shown by the spectra in Fig 1.9a. Laser cooling, allowing temperatures much lower than those achievable with standard cryogenic techniques, was primarily invented to answer this need for more and more precision. In Chapter 2 we will show how the Doppler effect, which causes the main obstacle for high-precision spectroscopy, is itself the solution that allows laser cooling to work.

Selection of slow atoms by time-of-flight detection. Laser cooling of hydrogen, however, is challenging from a technical point of view, since it requires ultraviolet laser light on the Lyman-α transition at 121.5 nm (see Sec. 1.5.2). In the MPQ experiment a trick was used to "observe" only the coldest atoms with velocity close to zero. The idea stems from the fact that in the atomic beam setup used in the experiment the excitation region is spatially separated from the detection region by a distance L. This allows the possibility of implementing a *time-delayed detection*, in which the excitation is pulsed (by a chopper placed on the path of the excitation beam, see Fig. 1.10) and the atoms are detected only after a time of flight Δt following the excitation pulse. Owing to the atomic beam configuration, this means that only atoms which have travelled with a longitudinal velocity $v = L/\Delta t$ can reach the detection region and thus contribute to the signal. In this way, by selecting long delays Δt, it becomes possible to probe only the small-velocity tail of the Doppler velocity distribution, for which the second-order Doppler shift is less and less important (Huber *et al.*, 1999).

In Fig. 1.9b two-photon spectra of the $1s - 2s$ transition of atomic hydrogen are reported for different times of flight Δt. As the delay time increases, atoms with lower and lower velocities are probed, which results in a less shifted and much smaller signal (since fewer and fewer atoms are present in the beam when the detection velocity approaches zero). Extrapolation of the observed line centre to infinite delay time allows the estimation of the Doppler-unshifted transition frequency.

1.4 Optical frequency measurements

But how to measure the transition frequency? The very small linewidth of the $1s - 2s$ hydrogen transition and the possibility of controlling first- and second-order Doppler effect (as well as other systematic sources of line broadenings and shifts, which become more and more important as the resolution is increased) would be useless in the absence of adequate techniques for measuring the transition wavelength or frequency. In traditional spectroscopy, line centres were measured with optical *wavelength* measurements, which can be performed with spectrometers (based on prisms or diffraction gratings) and interferometers (Michelson, Fabry–Perot, ...). These techniques, however, are affected by systematic sources of uncertainties (mostly geometric wavefront deformations and the dependence of the measured wavelength on the index of refraction) which ultimately limit the precision of the measurement around 10^{-10}. A direct measurement of frequency is much more preferable, since it is less affected by systematic effects. Furthermore, a frequency can be measured more precisely than any other physical quantity: as a matter of fact, the direct measurement of a frequency is an intrinsically "digital" process, in which the cycles of the oscillation are counted one by one. However, light oscillates much faster than the response time of any electronic instrument!

1.4.1 Frequency chains

Very complex dedicated *frequency chains* had been built starting from the 1970s in order to perform absolute frequency measurements. These devices allowed a direct link between an optical frequency (hundreds of THz) and an electronically synthesizable microwave (referenced to the primary Cs time standard). The link was provided by intermediate frequency standards which were mutually phase-locked thanks to several mixing processes and frequency multiplications achieved in cascade, by generation of harmonics in nonlinear electronic components at microwave frequencies, and by nonlinear optical processes in the infrared-visible domain (see Appendix B.3). These chains required an extremely large effort to be constructed and maintained (owing to the large number of phase-locked electronic and laser sources), and their operation was limited to a very narrow final frequency range, owing to the limited tuning range of some of the intermediate standards.

As an example to illustrate the complexity of these devices, we show in Fig. 1.11 the scheme of an early frequency chain built by K. Evenson *et al.* at NBS (now NIST), which allowed the absolute frequency measurements of two CO_2 laser lines at 9.3 μm and 10.2 μm and of the CH_4-stabilized He–Ne laser line at 3.39 μm, which were directly linked to the primary caesium frequency standard (Evenson *et al.*, 1973). This frequency chain was used in a milestone experiment in which the speed of light was measured at the 1 m/s level (fractional accuracy 3×10^{-9}) by direct multiplication of the absolute frequency ν of the methane-stabilized He–Ne laser times its wavelength λ measured with interferometric techniques (Evenson *et al.*, 1972),[9] following the fundamental relation $c = \nu\lambda$. A key element in the development of the frequency chain used in Evenson *et al.* (1972) was represented by point-contact diodes as mixing devices. In these diodes a tiny metal tip is brought into contact with a semiconductor or metallic bulk through a thin layer of insulating material. These devices, based on electron tunnelling across the junction, are characterized by a highly nonlinear response (making the generation of high-order harmonics possible) and, especially for metal–insulator–metal (MIM) diodes (Evenson *et al.*, 1985), by an extremely large bandwidth, up to the THz region, which justified their early use as detectors in the far-infrared region of the electromagnetic spectrum (Hocker *et al.*, 1968).

The implementation of more evolved ad-hoc frequency chains allowed the possibility of performing absolute frequency measurements in hydrogen. The F. Biraben group performed the absolute frequency measurement of the 778 nm hydrogen $2s - 8d$ two-photon transition at the few kHz level (6×10^{-12} accuracy) by measuring the frequency difference with respect to a diode laser stabilized on a two-photon rubidium transition, the absolute frequency of which was measured by a dedicated frequency chain referenced to the caesium standard (de Beauvoir *et al.*, 1997).

[9] At the time of this experiment the metre was still defined according to its 1960 SI definition as equal to 1 650 763.73 wavelengths of the orange emission line of krypton-86 in vacuum. Given the advances in the measurement of c (Evenson *et al.*, 1972), which was ultimately limited by the precision in the definition of the metre, in 1983 the speed of light was defined exactly as 299 792 458 m/s and the metre was redefined on the basis of the SI standard of time as "the length of the path travelled by light in vacuum during a time interval of 1/299 792 458 of a second".

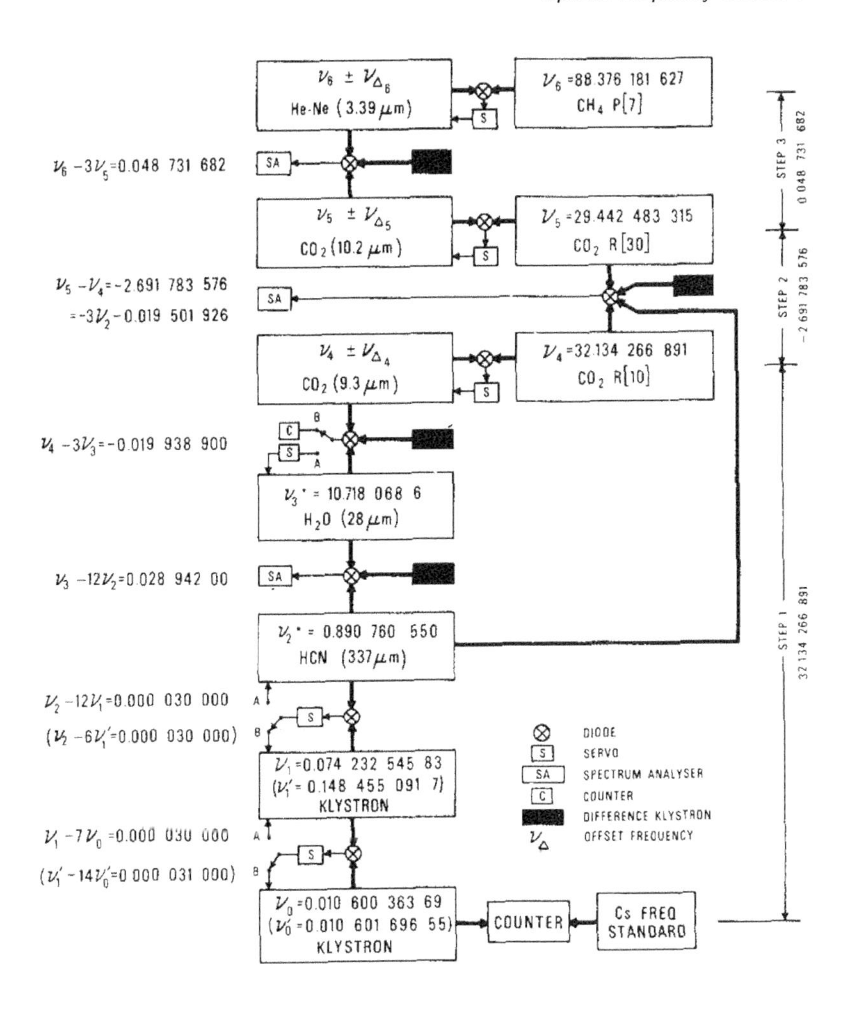

Fig. 1.11 Chain linking by frequency-multiplication the primary caesium standards to two stabilized CO_2 infrared lasers and a methane-stabilized He-Ne laser at 3.39 μm. Reprinted with permission from Evenson *et al.* (1973). © American Institute of Physics.

In the T. W. Hänsch labs an absolute frequency measurement of the hydrogen $1s-2s$ two-photon transition was carried out at the sub-kHz level (3.4×10^{-13} accuracy) by implementing a frequency chain that compared the hydrogen frequency with the 28th harmonic of the methane-stabilized 3.39 μm He-Ne laser, the frequency of which had been previously measured by another frequency chain which referenced the measurement to the primary caesium clock (Udem *et al.*, 1997). This frequency chain, shown in Fig. 1.12a, was based on a cascade of *frequency-divider* steps (represented in the figure as grey boxes): with this system the frequency difference between two lasers is reduced by a factor 2 at each step, bringing an initial frequency difference $\Delta \simeq 2.1$ THz in the far-infrared region to a signal $\Delta/32$ in the microwave region, where it can be detected and processed. The basic scheme of each frequency-divider step, involving an

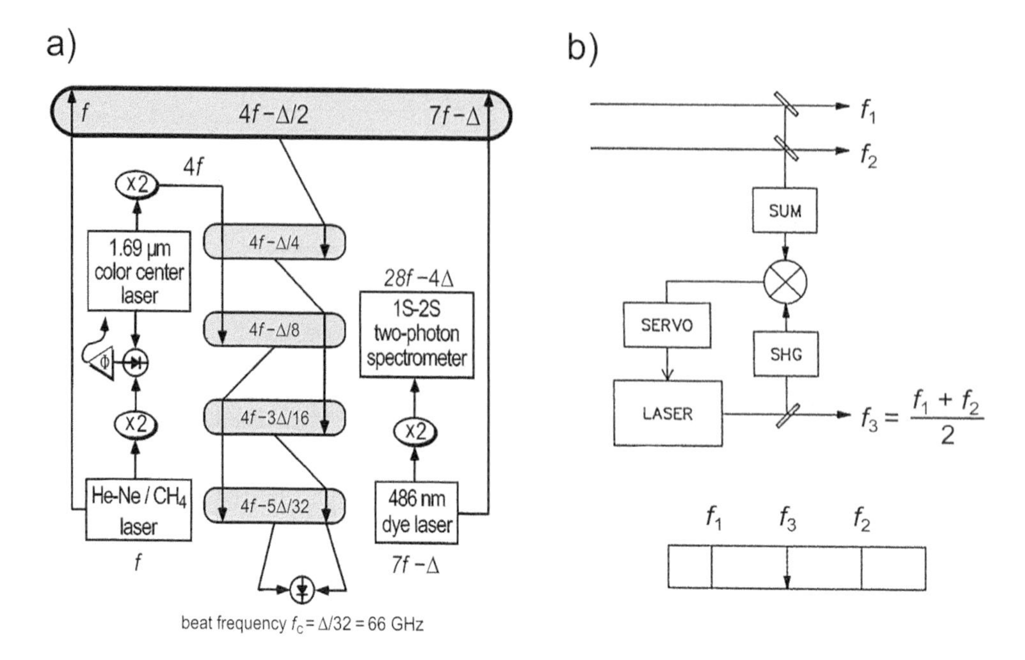

Fig. 1.12 a) Frequency-divider chain used in Udem *et al.* (1997) for the 1997 MPQ absolute frequency measurement of the hydrogen $1s-2s$ two-photon transition. Each of the grey rounded boxes represents a frequency-divider step which generates coherent radiation at a frequency (indicated in the centre of the box) which is the average of the two inputs' frequencies (indicated at the sides of the boxes). Reprinted with permission from Udem *et al.* (1997). © American Physical Society. b) Scheme of the elementary frequency-divider step: a laser at frequency f_3 is frequency-doubled and locked to the sum-frequency radiation generated by the two input beams at frequencies f_1 and f_2, in such a way that $f_3 = (f_1 + f_2)/2$. Reprinted from the Proceedings of "The Hydrogen Atom" symposium held in Pisa in 1988 (Hänsch, 1989).

auxiliary laser, which is frequency-doubled and mixed with the sum-frequency radiation generated by the two input beams, is sketched in Fig. 1.12b.

1.4.2 Frequency combs

A revolution in the field of optical frequency measurements was started in the late 1990s with the invention of the *optical frequency comb*, which was one of the motivations for the 2005 Nobel Prize in Physics awarded to J. L. Hall and T. W. Hänsch (Hall, 2006; Hänsch, 2006). This technique relies on the spectral properties of *pulsed lasers*, in which light is emitted as a train of short wavepackets. Because of the short duration of the pulses (from nanoseconds down to few femtoseconds), pulsed lasers are often used for studying nonlinear optical processes requiring large peak intensities or for performing time-resolved spectroscopy of very fast chemical reactions, relaxation processes, or molecular dynamics (Demtröder, 2003). However, they have found important applications in the field of high-precision spectroscopy as well. As an example of this

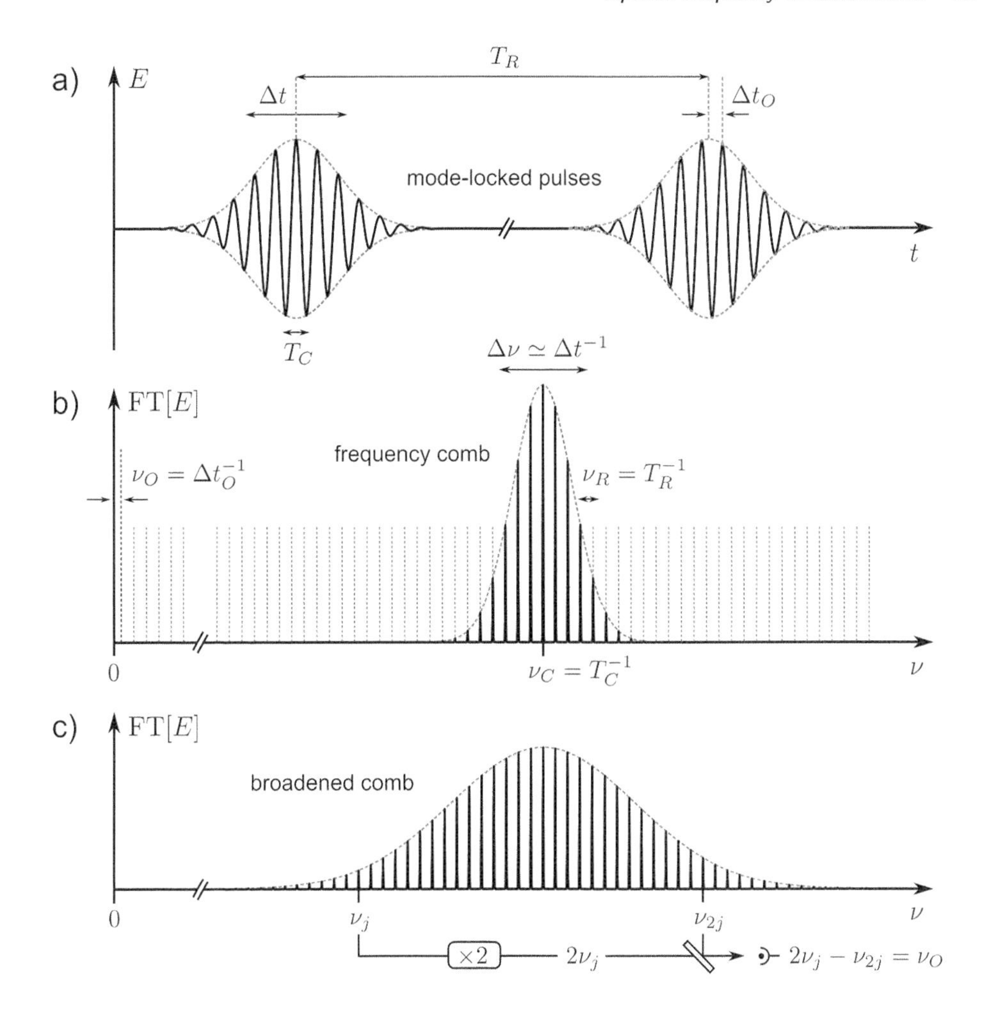

Fig. 1.13 Frequency comb. The plots schematically show: a) the time-domain representation of the laser pulses emitted by a mode-locked laser, b) its frequency spectrum, c) the comb after spectral broadening in a photonic crystal fiber, which serves for the comb self-referencing stabilization technique described in the text.

possibility, the broad spectrum of pulsed lasers was taken advantage of as early as in the 1970s for the development of laser *quantum beats* spectroscopy, in which short light pulses excite a coherent superposition of atomic levels and the time-resolved detection of fluorescence shows a beating signal at the frequency difference between the levels, allowing the measurement of hyperfine or fine structures (an early review of this spectroscopic technique can be found in Haroche (1976)).

The (pulsing) heart of a frequency comb is represented by a femtosecond mode-locked laser, typically a titanium-sapphire (Ti:Sa) ring laser, emitting laser pulses of duration Δt and characterized by a spectral width $\Delta\nu \simeq \Delta t^{-1}$, which is remarkably larger than the spectral structures which have been discussed in the previous sections

($\Delta\nu \simeq 30$ THz for a $\Delta t = 30$ fs-long pulse). As sketched in Fig. 1.13a, the laser pulses are emitted at regular time intervals, corresponding to the round-trip time T_R of the photons in the laser cavity, which equals the inverse of the free spectral range ν_R ($\nu_R = T_R^{-1} \simeq 300$ MHz for a typical round-trip cavity length 1 m, see Appendix B.2). In a mode-locked laser these pulses are emitted coherently, which means that there is a well-defined phase relation between the optical oscillation of the carrier wave and the pulse envelope. The constant carrier-envelope time shift Δt_O (Fig. 1.13a) from one pulse to the following one is related to the dispersion within the laser cavity, which makes the phase velocity v different from the group velocity v_g.[10] The spectrum of this train of phase-locked laser pulses emitted at regular intervals is a comb of narrow lines starting at the offset frequency $\nu_O = \Delta t_O^{-1}$ and separated by the frequency interval ν_R, with an overall width $\Delta\nu$ (see Fig. 1.13b).

Spectrum of a mode-locked laser. The spectrum of a mode-locked pulsed laser can be demonstrated starting from its description in time domain and evaluating the Fourier transform of its electric field $E(t)$. We first consider the spectrum of a single laser pulse $E(t) = A(t)e^{i2\pi\nu_C t}$, characterized by a carrier frequency ν_C and an envelope $A(t)$. By using well-known theorems for the Fourier transform, the spectrum of the pulse is given by the Fourier transform of the envelope $\tilde{A}(\nu)$ centred at the carrier frequency ν_C:

$$
\begin{aligned}
\mathrm{FT}\left[E(t)\right](\nu) &= \mathrm{FT}\left[A(t)e^{i2\pi\nu_C t}\right](\nu) \\
&= \mathrm{FT}\left[A(t)\right](\nu - \nu_C) \\
&= \tilde{A}\left(\nu - \nu_C\right) \,.
\end{aligned} \tag{1.11}
$$

When an infinite train of phase-coherent laser pulses is considered, assuming a constant pulse-to-pulse phase slip $\phi_O = 2\pi\nu_C\Delta t_O$ (see Fig. 1.13a), the spectrum becomes:

$$
\begin{aligned}
\mathrm{FT}\left[E(t)\right](\nu) &= \mathrm{FT}\left[\sum_{n=-\infty}^{\infty} A(t-nT_R)e^{i[2\pi\nu_C(t-nT_R)+n\phi_O]}\right](\nu) \\
&= \sum_{n=-\infty}^{\infty} e^{in\phi_O}\,\mathrm{FT}\left[A(t-nT_R)e^{i[2\pi\nu_C(t-nT_R)]}\right](\nu) \\
&= \sum_{n=-\infty}^{\infty} e^{in\phi_O}e^{-i2n\nu\pi T_R}\,\mathrm{FT}\left[A(t)e^{i2\pi\nu_C t}\right](\nu) \\
&= \tilde{A}\left(\nu-\nu_C\right)\sum_{n=-\infty}^{\infty} e^{in(-2\pi T_R\nu+\phi_O)} \\
&= \tilde{A}\left(\nu-\nu_C\right)\sum_{j=-\infty}^{\infty} \delta(\nu - j\nu_R - \nu_O)\,,
\end{aligned} \tag{1.12}
$$

where the δ function in the last line comes from the fact that, in the limit of an infinite series of coherent pulses, the sum of exponential factors in the line above does

[10]The velocity at which the envelope of a laser pulse propagates is given by the group velocity $v_g = \partial\omega/\partial k$. In a dispersive medium in which $\omega = \omega(k)$ the group velocity is generally different from the phase velocity $v = \omega/k$.

not vanish only if the arguments are multiples of 2π. Equation (1.12) shows that the spectrum of a mode-locked pulsed laser is made up of a comb of equispaced lines (Fig. 1.13b) centred at frequencies

$$\nu_j = j\nu_R + \nu_O \ , \qquad (1.13)$$

where ν_R is called the *repetition frequency* and $\nu_O = \phi_O \nu_R / 2\pi$ is the *offset frequency*, which are both in the radiofrequency range.[11]

This comb provides a frequency ruler, which can be used to perform measurements of optical frequency differences. As a matter of fact, the spacing between the comb lines can be determined quite precisely by measuring the beating between them with a fast photodiode. First experiments succeeded in measuring the absolute frequency of atomic transitions using the frequency comb to bridge large frequency intervals of tens of THz between the transition frequencies and existing frequency standards calibrated with frequency chains (Udem *et al.*, 1999).

Absolute frequency measurements. In order to increase the range of the measurement, the spectral width of the comb has to be increased. In principle, it could be possible to enlarge the comb spectrum by producing shorter laser pulses, which however is not obvious since it requires an accurate compensation of dispersion within the laser cavity. It is much simpler to take advantage of nonlinear effects (four-wave mixing, harmonic generation, ...) which occur after the interaction of intense ultrashort laser pulses with matter (see Appendix B.3). A fundamental advance in this direction happened when the possibility of broadening the comb spectrum without destroying the coherence between the modes was reported. The first experimental demonstration was made by M. Bellini and T. W. Hänsch (Bellini and Hänsch, 2000), who measured the coherence properties of supercontinuum white light generated by mode-locked laser pulses focused onto a CaF_2 glass substrate. Owing to several complex nonlinear processes (including self-phase modulation caused by Kerr nonlinearities in the medium) the extremely high pulse peak intensities produced a continuum of wavelengths spanning from the near UV to near IR. In Bellini and Hänsch (2000) it was demonstrated that independent white light pulses, generated by the same mode-locked laser focused on different points of the plate, after recombination showed clear stable interference fringes. In other words, the complex nonlinear processes happening within the substrate did not destroy the phase coherence between the modes of the comb all across the spectrum.

An important technical advance in the supercontinuum generation was provided by the production of specially designed air-silica *photonic crystal fibers*, which allowed the spectrum of a passively mode-locked Ti:Sa laser to be broadened over a full optical octave already at moderatively low pulse intensities (Ranka *et al.*, 2000). This possibility was particularly important because, once the comb spectrum covers an octave (i.e. the spectrum in eqn (1.13) includes observable lines with both j and $2j$), it becomes possible to control the offset frequency ν_O by *self-referencing* the comb. In this technique, the red-side of the (broadened) comb spectrum is frequency-doubled (with second-harmonic generation techniques) and superimposed with the blue-side of the same spectrum on a fast photodiode, as shown in Fig. 1.13c. From eqn (1.13), the beat note between the

[11] Of course, in a real laser there is a finite time over which the coherence between pulses is maintained, which causes the comb lines to acquire a finite linewidth.

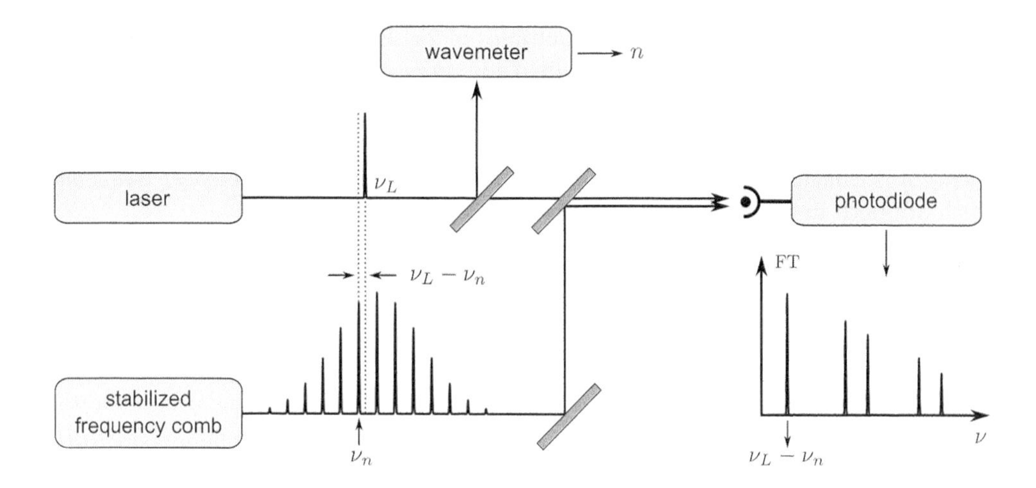

Fig. 1.14 Simplified scheme for the absolute measurement of an unknown laser performed with an optical frequency comb (stabilized onto a frequency standard, e.g. a caesium clock). A wavelength meter is used to perform a coarse frequency measurement in order to identify the index n of the comb frequency ν_n which is the closest to the laser frequency ν_L to be determined. The lowest-frequency signal in the beat note between the laser and the frequency comb (detected by a fast photodiode) yields the frequency difference between ν_L and the comb frequency ν_n (in the radio range).

frequency-doubled mode $2\nu_j$ and the mode ν_{2j} of the same comb occurs exactly at the offset frequency:

$$2\nu_j - \nu_{2j} = 2\,(j\nu_R + \nu_O) - (2j\nu_R + \nu_O) = \nu_O\ , \tag{1.14}$$

which can be measured with a spectrum analyser or a frequency counter.

With a complete control on the repetition frequency ν_R and offset frequency ν_O, which can be both referenced to the primary time standard, it is possible to precisely know the frequency of each comb line. By beating the unknown laser with the frequency comb one can measure the frequency difference (in the radio range) between the unknown frequency and the closest comb tooth and therefore perform an *absolute frequency measurement*, as sketched in Fig. 1.14. For more information on the operation of the frequency comb and its applications see Ye and Cundiff (2005) and the review articles: Cundiff and Ye (2003), and Maddaloni *et al.* (2009).

Frequency comb extension. After the first implementations of optical frequency-comb synthesizers based on Ti:Sa lasers, which cover the visible (VIS) and near-infrared (NIR) regions, several schemes were devised to extend them to operate in other regions of the electromagnetic spectrum.

The extension of the frequency comb to the medium infrared is particularly important for applications in molecular spectroscopy (see Schliesser *et al.* (2012) for an updated review). The use of highly nonlinear fibers allowed extension of traditional frequency combs up to wavelengths of about 2.3 μm (Thomann *et al.*, 2003). However, being the strongest molecular rovibrational transitions well beyond 2 μm wavelength, in the so-called "fingerprint region", different approaches based on nonlinear optics had to be developed. Frequency-comb synthesis schemes based on difference-frequency-generation (DFG) proved to be the most successful: absolute frequency measurements on transitions of CO_2 around 4.3 μm could thus be performed by phase-locking two laser sources on a Ti:Sa-based comb (Mazzotti *et al.*, 2005) and a specially designed Ti:Sa femtosecond laser provided a comb at 3.4 μm with about 270-nm spectral coverage (Foreman *et al.*, 2005). More direct down-conversion of a fibre-based comb, centred at telecom frequencies, was obtained in a nonlinear crystal by DFG with a Yb-fiber-amplified laser (Maddaloni *et al.*, 2006), which also proved useful for demonstrating high-precision spectroscopic applications (Malara *et al.*, 2008). More recently, a full spectral coverage from 2.6 to 6.1 μm has been achieved (Leindecker *et al.*, 2012).

Generation of a free-space-propagating THz comb has been demonstrated very recently in the far-infrared range, by down-conversion of radiation from a mode-locked laser around 800 nm. Such a comb proved also to be sufficiently intense to allow phase-lock of a 2.5 THz quantum-cascade laser on one of the comb teeth (Consolino *et al.*, 2012).

On the opposite side of the electromagnetic spectrum, frequency combs have also been recently used to extend the realm of high-precision frequency measurements towards the vacuum and extreme ultraviolet (VUV and XUV) regions. Here, the most accessible sources of coherent radiation are based on the process of high-order harmonic generation (Corkum, 1993). However, the ultrashort duration of the harmonic pulses has always limited their application to high spectral resolution studies, until it was demonstrated that Ramsey-type techniques, based on the use of delayed coherent XUV exciting pulses (Cavalieri *et al.*, 2002; Liontos *et al.*, 2010; Eramo *et al.*, 2011), could achieve this goal. Phase-stabilized comb oscillators were first used in this context only as a precise way to generate accurate delays between the exciting XUV pulses (Witte *et al.*, 2005; Kandula *et al.*, 2010). A significant step forward was made when the entire train of phase-locked pulses emitted from a stabilized mode-locked laser was directly converted to the XUV by means of cavity-enhanced harmonic generation (Gohle *et al.*, 2005; Jones *et al.*, 2000). In this case the whole comb structure characteristic of the visible or near-IR laser spectrum could potentially be translated to the XUV, as was recently confirmed by the single-photon spectroscopy of both an argon transition at 82 nm and a neon transition at 63 nm with unprecedented (in this spectral region) ten-megahertz linewidths (Cingöz *et al.*, 2012).

1.4.3 The Rydberg constant

The possibility of using the frequency comb for absolute frequency measurements allowed striking advances in the precise determination of the hydrogen transition frequencies. These experimental values are the primary input data for the determination of the Rydberg constant, which is the most accurately measured fundamental constant of Physics. The current value, determined by the Committee on Data for Science and Technology (CODATA) in the last adjustment of the fundamental constants, is

$$R_\infty = 10\,973\,731.568\,539(55)\ \mathrm{m}^{-1}\,, \tag{1.15}$$

with a fractional uncertainty of 5.0×10^{-12} (Mohr *et al.*, 2012). The most important contributions to this determination come from hydrogen spectroscopy, in particular from the $1s-2s$ transition frequency measured at MPQ with a fractional uncertainty of 1.4×10^{-14} (Fischer *et al.*, 2004). Other important contributions come from hydrogen and deuterium measurements performed at LKB-SYRTE on the $2s-8s, 8d$ and $2s-12d$ two-photon transitions, with relative uncertainties down to 7.7×10^{-12} (de Beauvoir *et al.*, 1997; Schwob *et al.*, 1999). Contributions from microwave measurements of the Lamb shift and other optical measurements on hydrogen and deuterium are included as well (Mohr *et al.*, 2008).

The determination of R_∞ is, of course, not only a matter of performing more and more precise experiments. An important effort from theory is needed, since a number of QED corrections have to be considered to provide a theoretical expression for the transition frequencies (with R_∞ as free parameter) at the same or better level of precision than the experimental value. The current uncertainty of experiments ($\approx 10^{-14}$) is at the moment well below the uncertainty in the hydrogen theory. One of the main limitations comes from the knowledge of an important quantity for the evaluation of hydrogen QED corrections coming from the finite size of the nucleus: the proton charge radius (Biraben, 2009). Very recent work on the experimental determination of this quantity by spectroscopy of exotic hydrogen will be discussed in Sec. 1.5.1.

1s–3s hydrogen spectroscopy. High-resolution frequency measurements of different hydrogen transitions starting from the $1s$ ground state are important resources for a deeper insight into the hydrogen level energies. In particular, comparing different $1s - ns$ transition frequencies would allow a better knowledge of the hydrogen Lamb shift and a more precise determination of the Rydberg constant. Experiments on the $1s - 3s$ transition with two-photon spectroscopy at the UV wavelength $\lambda = 205$ nm are currently being performed at LKB and MPQ.

Recently, the first absolute frequency measurement of the hydrogen $1s-3s$ transition has been reported by the LKB group in Arnoult *et al.* (2010). The experiment was performed on a thermal beam of atomic hydrogen, the velocity distribution of which (responsible for the second-order Doppler shift) was characterized by measuring the quadratic Stark effect induced by a static magnetic field, appearing as an electric field in the rest frame of the atoms. The frequency-comb-assisted measurement of the $1s - 3s$ transition frequency resulted in a value with 4.5×10^{-12} fractional accuracy, mostly determined by the statistical error (which could be reduced by increasing the UV laser power). This result represents the best measurement of an optical transition in hydrogen after the MPQ measurement on the $1s - 2s$ transition. Although so far it does not improve significantly the value of the Rybderg constant, a future reduction of its uncertainty will provide an important input for a better measurement of the hydrogen Lamb shift and Rydberg constant.

1.5 New frontiers of hydrogen

1.5.1 Spectroscopy of exotic hydrogen

Exotic hydrogen-like atoms made of particles different from the proton and/or the electron share the same attractive features as hydrogen and, owing to the simple

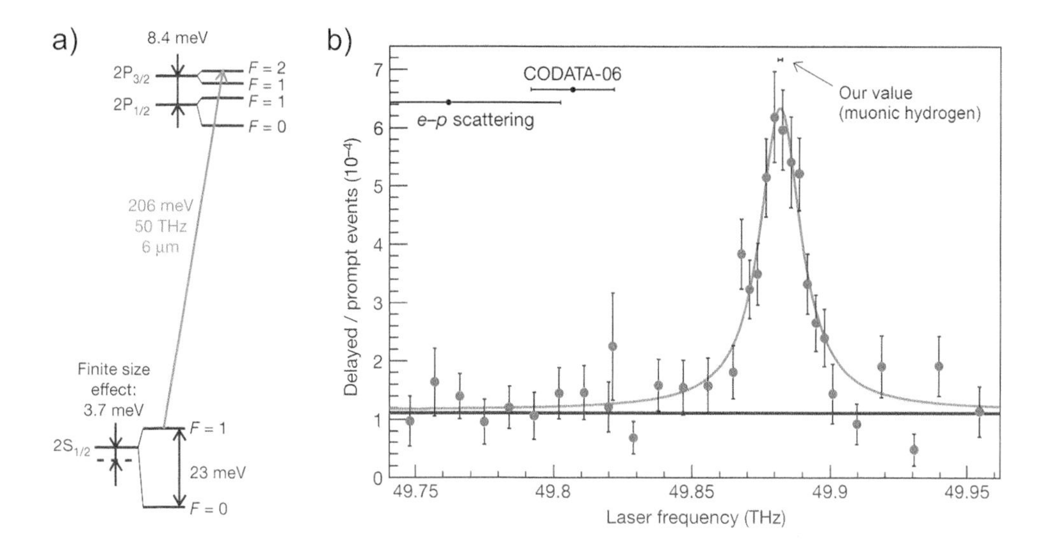

Fig. 1.15 Spectroscopy of muonic hydrogen μp. a) Scheme of the $n = 2$ fine structure and Lamb shift corrections. b) Spectrum of the $n = 2$ Lamb shift. The top intervals show the resonance centre and the expected centres based on the CODATA value for the rms proton charge radius and on the value from electron–proton scattering. Adapted with permission from Pohl *et al.* (2010). © Macmillan Publishers Ltd.

theoretical description of the two-body bound state, offer new possibilities of performing QED tests and to measure important physical quantities.

One of the most interesting systems for the spectroscopy of exotic hydrogen is muonic hydrogen μp, which is the bound state between a proton and a negatively-charged muon μ^-. The most important difference with respect to hydrogen is the muon mass m_μ, which is approximately 200 times larger than the electron mass m_e. Here we recall that the level energies E_n and the Bohr radius a_0 of a hydrogen-like atom scale with the electron mass m_e as[12]

$$E_n = -\frac{\alpha^2 m_e c^2}{2n^2} \propto \frac{m_e}{n^2} \tag{1.16}$$

$$a_0 = \frac{\hbar}{m_e c\alpha} \propto \frac{1}{m_e} \ . \tag{1.17}$$

Owing to the different mass, in muonic hydrogen the transition energies are larger by a factor $m_\mu/m_e \simeq 200$ with respect to the twin transitions in atomic hydrogen: this means that "electronic" transitions in muonic hydrogen can be excited by soft X-rays (instead of visible or near-UV light for atomic hydrogen), while smaller Lamb shift intervals lie in the infrared region (instead of microwaves), as shown in Fig. 1.15a. The

[12]Actually, the reduced mass $\mu_e = m_e m_p/(m_e + m_p)$ of the electron–proton system (or the muon–proton reduced mass $\mu_\mu = m_\mu m_p/(m_\mu + m_p)$ in the case of muonic hydrogen) should appear in these equations in the place of m_e.

increased mass is also responsible for the smaller size of the orbitals, again by a factor m_μ/m_e with respect to the electronic Bohr radius, which makes the energy levels of muonic hydrogen much more sensitive than hydrogen to the internal structure of the nucleus. In particular, the Lamb shift of the $n = 2$ level has a relatively large ($\sim 2\%$) contribution coming from the finite "size" of the proton, which can be characterized by an effective spherical charge distribution with rms radius r_p.

A precise measurement of the muonic hydrogen $n = 2$ Lamb shift has been carried out recently at the proton accelerator of the Paul Scherrer Institute in Switzerland (Pohl *et al.*, 2010), where muonic hydrogen atoms were created by impact of a low-energy beam of negative muons onto a molecular hydrogen target. The Lamb shift was measured by direct excitation of the muonic hydrogen $2s_{1/2} - 2p_{3/2}$ transition with short infrared laser pulses (see Fig. 1.15a).[13] From the measured Lamb shift 49.88188(76) THz the authors deduced an rms proton charge radius

$$r_p = 0.84184(67) \text{ fm} , \tag{1.18}$$

which was ten times more precise than the previous CODATA 2006 determination of this quantity $r_p = 0.8768(69)$ fm (Mohr *et al.*, 2008), which was obtained mostly from the comparison between hydrogen spectroscopy and bound-state QED calculations, but is smaller than the latter by 4%, corresponding to five standard deviations of the previous value. The muonic hydrogen value is also not consistent with non-spectroscopic measurements relying on electron-proton scattering experiments, which yield a charge radius $r_p = 0.897(18)$ fm (Sick, 2003; Blunden and Sick, 2005). Figure 1.15b shows a spectrum of the infrared μp resonance detected in Pohl *et al.* (2010), together with the expected line centres based on the CODATA r_p value and on the value obtained from electron-proton scattering.

Very recently, new measurements performed on muonic hydrogen reinforced this "proton radius puzzle", confirming the results of the 2010 measurement (Antognini *et al.*, 2013). Laser spectroscopy of different hyperfine components of the $2s_{1/2} - 2p_{3/2}$ transition in muonic hydrogen, combined with a reanalysis of the previous measurements, allowed a determination of both charge and magnetic radius of the proton.[14] While the former is in agreement with previous determinations, the latter resulted in a more precise value for the proton charge radius

$$r_p = 0.84087(39) \text{ fm} , \tag{1.19}$$

in agreement with the previous determination of Pohl *et al.* (2010) and 1.7 times more precise, but still in strong disagreement with the new CODATA value $r_p = 0.8775(51)$ fm (not taking into account the muonic hydrogen result) by 7 standard deviations (Mohr *et al.*, 2012).

The origin of this large discrepancy is not yet known. If the spectroscopic measurements are not affected by hidden systematics, the disagreement with previous data

[13] A critical experimental sequence is needed because of the short lifetime of muonic hydrogen, which is limited by the ≈ 2.2 μs lifetime of the muon.

[14] The magnetic radius of the proton accounts for the finite-sized distribution of magnetic moment inside the proton.

from hydrogen spectroscopy and electron–proton scattering could be attributed to yet-unidentified problems in bound-state QED, possibly including errors or underestimates of terms in the QED energies of hydrogen-like atoms. Or it might point at some difference between the quantity measured in atomic spectroscopy and the one probed in high-energy electron–proton scattering experiments (in which an extrapolation of the form factor to zero momentum transfer is required, being the momentum transfer several orders of magnitude larger than the one implied in spectroscopic measurements). A different value of the proton charge radius would have implications on the determination of the Rydberg constant R_∞ as well (see Sec. 1.4.3) and, if it is confirmed by independent investigations, could lead to a change in the value of R_∞. Although the puzzle is still unsolved (see Antognini *et al.* (2013) for a discussion of different scenarios and for the relevant references), it clearly shows the importance of high-precision spectroscopy as a primary tool for the test of quantum theories and for the determination of fundamental physical quantities.

1.5.2 Spectroscopy of antimatter

Alongside the Equivalence Principle (see Sec. 5.5.2), a cornerstone of our knowledge of Nature is based on the concept of CPT invariance. According to this fundamental symmetry, all physical phenomena should be invariant under simultaneous spatial inversion, time reversal, and charge inversion. One of the consequences of CPT invariance is that "*anti-atoms*" made of antimatter must have the same identical energy levels as the corresponding atoms. For instance, antihydrogen, made of a positron and an antiproton bound together, should have exactly the same transition frequencies as hydrogen. If the frequencies were found to be even slightly different, this would imply a violation of CPT invariance. Considering the extreme levels of precision achieved with hydrogen spectroscopy, a similar spectroscopic investigation of antihydrogen could constitute a strong model-independent test of CPT.

In recent years striking progress has been made in the controlled production and trapping of antihydrogen. Production of antihydrogen in a cryogenic environment was reported in 2002 by both the ATHENA (Amoretti *et al.*, 2002) and the ATRAP (Gabrielse *et al.*, 2002) collaborations operating at the CERN antiproton decelerator. The ALPHA collaboration at CERN demonstrated magnetic trapping of cold antihydrogen in 2010 (Andresen *et al.*, 2010) and then showed that the trapped antihydrogen is in its ground state and can be kept magnetically confined for extremely long times on the order of 1000 seconds (Andresen *et al.*, 2011). In 2012 the first spectroscopy experiments on cold magnetically trapped antihydrogen were performed by the ALPHA collaboration, who demonstrated the possibility of inducing antihydrogen spin-flip transitions with resonant microwave radiation (Amole *et al.*, 2012).

Towards (anti)hydrogen cooling. Future developments of antihydrogen experiments will involve attempts to perform laser cooling of antihydrogen, which is an important requirement in order to confine the antiatoms in smaller volumes, which would be very beneficial for reducing systematic effects on the $1s-2s$ spectroscopy. Important technological advances will be required for this task, since optical cooling of (anti)hydrogen requires coherent radiation at 121.5 nm on the allowed $1s-2p$ transition. The generation of a coherent source at this vacuum-ultraviolet wavelength is extremely challenging from a technological point of view, since light in this spectral region is strongly absorbed by air and by most dielectric materials. This is different from the case of alkali atoms, in which laser cooling is performed on visible or near-infrared transitions (see Chapter 2) for which plenty of laser power is available.

An early success in generating 121.5 nm light allowed the first observation of laser-cooled magnetically trapped hydrogen already in the 1990s (Setija *et al.*, 1993) by short pulses of Lyman-α radiation obtained with cascade triple- and second-harmonic generation from a dye laser. Although an efficient laser cooling of hydrogen is still lacking, promising results for the generation of continuous-wave 121.5 nm light (which could result in a higher cooling efficiency) have been recently obtained in Kolbe *et al.* (2012) by an improved four-wave mixing scheme in atomic mercury vapours.

In the 1980s, when hydrogen was believed to be the best candidate for Bose–Einstein condensation, these obstacles to laser cooling represented a major motivation for the development of alternative cooling and trapping techniques, which in turn resulted in being fundamental for the achievement of Bose–Einstein condensation in dilute gases, as it will be discussed in Chapter 3. This is the case of evaporative cooling (see Sec. 3.1.2), which was first demonstrated for hydrogen (see Sec. 3.4.1) and since then it has represented the key cooling technique for all the existing Bose–Einstein condensation experiments. This is an example of the guiding role that hydrogen has always had in atomic physics. Not only is it the simplest atom but it is also the most relevant one for high-precision spectroscopy and tests of quantum theories: its investigation has constantly set the pace of atomic physics, stimulating fundamental developments and opening new directions.

2 Alkali atoms and laser cooling

In the previous chapter we have evidenced the effects of atomic motion, and consequently of temperature, on high-precision spectroscopic measurements. Whereas first-order Doppler broadening can be effectively cancelled by taking advantage of nonlinear effects in the atom–light interaction (e.g. in saturation spectroscopy or two-photon spectroscopy), the direction-insensitive second-order Doppler shift cannot be reduced unless the atoms are really slowed down. Cryogenic techniques can be used to make a gas of atoms, or an atomic beam, colder, which has been successfully employed in the case of hydrogen spectroscopy (see Sec. 1.3). However, there are much more efficient ways to cool atomic ensembles, and these methods rely on the manipulation of atoms with laser light.

This chapter describes this new paradigm in the use of lasers for atomic physics experiments: from sources of coherent and ultra-stable light for precision spectroscopy, to powerful tools to control the atomic motion. In this chapter we illustrate the basic concepts of laser cooling, focusing on its implementation with alkali atoms, which were the first elements to be cooled in experiments, thanks to their simple electronic structure and to the availability of laser sources. After early proposals in the 1970s, laser cooling was demonstrated in the 1980s and enabled researchers to reach extremely low temperatures (indeed much smaller than the expected ones!) on the order of a few millionths of kelvin above absolute zero.

After an illustration of the most used laser-cooling techniques, we will consider important applications of laser-cooled alkali atoms to the development of precise *atomic clocks* for the definition of time and of *atom interferometers* for the accurate measurement of forces.

2.1 Alkali atoms

Alkali atoms are multi-electron atoms with a remarkably simple level structure, which arises from the complete filling of all the inner electronic shells but the most external one, which is occupied only by one electron. The ground state of an alkali atom with atomic number Z has the electronic configuration $\{Z-1\}\,ns$, where $\{Z-1\}$ indicates the electronic configuration of the noble gas which precedes it in the periodic table and the valence electron occupies the first available level with principal quantum number n and angular momentum $l=0$. The spectrum of an alkali atom in the visible or near-visible range is determined entirely by the excited levels occupied by the valence electron, which are well described by a hydrogen-like formula

Table 2.1 Some values for the stable isotopes of the alkali atoms (the negative $\Delta_{\rm hfs}$ for ^{40}K indicates that hyperfine level $F+1/2$ has a lower energy than $F-1/2$).

	n	Z	λ_2 (nm)	$\gamma/2\pi$ (MHz)	$\Delta_{\rm fs}$ (THz)	I	$\Delta_{\rm hfs}$ (GHz)
^{6}Li	2	3	671.0	6	0.01	1	0.23
^{7}Li	"	"	"	"	"	3/2	0.80
^{23}Na	3	11	589.2	10	0.52	3/2	1.77
^{39}K	4	19	766.7	6	1.73	3/2	0.46
^{40}K	"	"	"	"	"	4	-1.29
^{41}K	"	"	"	"	"	3/2	0.25
^{85}Rb	5	37	780.2	6	7.12	5/2	3.04
^{87}Rb	"	"	"	"	"	3/2	6.83
^{133}Cs	6	55	852.3	5	16.61	7/2	9.19

$$E_{nl} = -hcR_\infty \frac{1}{(n-\delta_l)^2}\,, \tag{2.1}$$

differing from the hydrogen Bohr energies in eqn (1.2) by the *quantum defect* δ_l, an empirical quantity describing the shielding effect of the nucleus charge by the electrons of the inner shells (Foot, 2005). Since the latter determine a deviation of the effective electron–nucleus attraction from the $\sim 1/r$ dependence of the Coulomb potential, the energies depend on the angular momentum quantum number l as well.[1] The lowest energy levels of an alkali atom are sketched in Fig. 2.1. The first excited state has electronic configuration $\{Z-1\}\,np$ and it is split into two fine-structure levels $^2P_{1/2}$ and $^2P_{3/2}$. The fine-structure separation $\Delta_{\rm fs}$ of this doublet increases with increasing atomic number, from only a few GHz in lithium to 16 THz in caesium.[2] The transitions from the ground state to these two levels are traditionally referred to as the D1 line for the $^2P_{1/2}$ and the D2 line for the $^2P_{3/2}$. These transitions, lying in the visible-near infrared spectrum, are allowed by dipole selection rules and their typical linewidth is in the 5–10 MHz range. The values of these quantities for all the stable alkali isotopes are reported in Table 2.1.

All the energy levels of alkali atoms, including the ground state $^2S_{1/2}$, are characterized by an additional level splitting, which is determined by the *hyperfine interaction* between the electronic angular momentum $\mathbf{J}$ and the nuclear spin $\mathbf{I}$ (Arimondo *et al.*, 1977). The effect of this interaction (relevant for the energy levels of hydrogen as well, although it was not explicitly mentioned in the previous chapter) can be described by an interaction term proportional to $\mathbf{J}\cdot\mathbf{I}$, which determines a splitting of the levels into states with defined values of the total angular momentum $\mathbf{F}=\mathbf{J}+\mathbf{I}$. In the ground state with $J=1/2$, the possible values for the total angular momentum quantum number F are $I-1/2$ and $I+1/2$, as shown in Fig. 2.1. The energy separation between

[1]The quantum defect δ_l is a positive quantity which effectively describes the deviation of the energies from the completely shielded case in which the electron interacts with a nucleus of charge $+e$. It depends on the atomic number (its value for s levels increases monotonically from 0.41 in lithium to 4.13 in caesium) and on the angular momentum quantum number (it decreases with increasing l, describing the lower penetration of the valence electron wavefunction inside the core of the internal electrons).

[2]Fine-structure separations increase with the atomic number Z approximately as $\sim Z^2$.

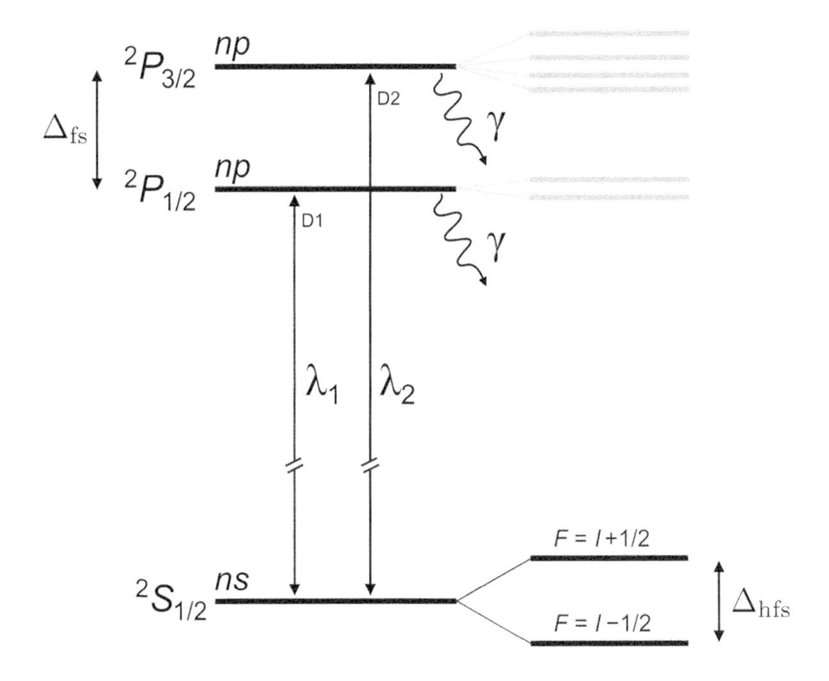

Fig. 2.1 Simplified level structure of an alkali atom.

these two states varies from several hundreds of MHz to several GHz depending on the different isotopes, as reported in Table 2.1. These two states can be treated as two different ground states: as a matter of fact, the upper hyperfine state $F = I + 1/2$ can decay to the absolute ground state $F = I - 1/2$ by a magnetic dipole transition (supporting $\Delta l = 0$, see Appendix A.4.2), but the small energy separation causes the decay time to be very much longer than the timescale of any spectroscopic experiment.[3]

2.2 Atomic clocks

The hyperfine structure of the alkali atoms has a very important metrological significance. As a matter of fact, the current definition of the SI unit of time, the second, is based on the frequency of the (magnetic dipole) transition between the $F = 3$ and $F = 4$ hyperfine ground states of ^{133}Cs, which amounts (now by definition) exactly to

$$\nu_{\mathrm{Cs}} = 9\,192\,631\,770 \text{ Hz} . \tag{2.2}$$

[3] As discussed in Appendix A.2, the natural linewidth γ depends on the transition frequency ω_0 and on the transition matrix element μ_{eg} as $\gamma \propto \omega_0^3 \mu_{eg}^2$. For this microwave transition the frequency is $\sim 10^5$ smaller than the frequency of an optical transition, while the transition matrix element of a magnetic dipole transition is typically $\sim 1/\alpha \sim 10^2$ times smaller than the one for an electric dipole transition (where α is the fine structure constant). As a result, the natural linewidth of this transition is $\sim (10^5)^3 (10^2)^2 = 10^{19}$ smaller than the one of a typical optical transition.

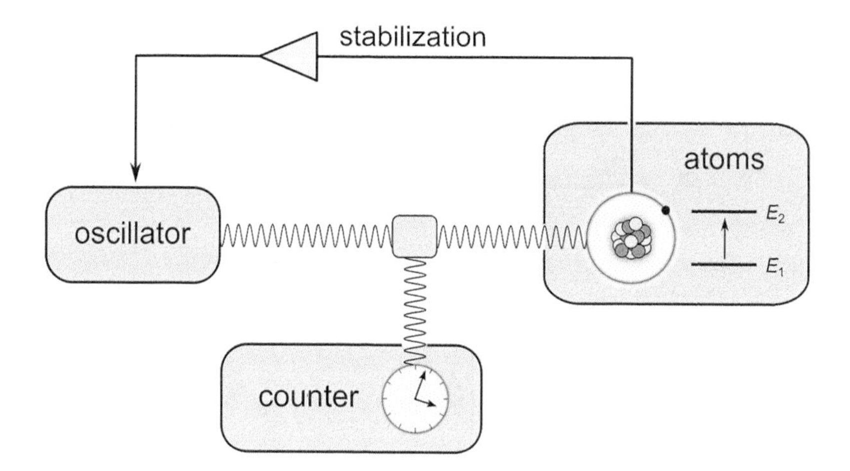

Fig. 2.2 Basic scheme of an atomic clock. An oscillator is stabilized onto a frequency reference given by the energy difference between two atomic levels $\nu_0 = (E_2 - E_1)/h$ and the number of oscillations is measured by a counter.

This definition dates back to 1967 and followed the first demonstration of a microwave Cs atomic clock[4] in 1955 by L. Essen and J. V. L. Parry (Essen and Parry, 1955).

The basic scheme of a generic atomic clock is shown in Fig. 2.2. An oscillator generates electromagnetic radiation (either in the microwave, IR, visible or UV domains), the frequency of which is locked to an atomic reference provided by the energy difference between two atomic levels probed in a spectroscopic experiment. The quality of the frequency lock is determined by the $\nu_0/\delta\nu$ Q-factor of the transition lineshape, where ν_0 is the transition frequency and $\delta\nu$ is the linewidth (see Sec. 5.3 for a more extended discussion on the stability of atomic clocks). The stable output of the oscillator can be analysed by a counter (an electronic device for a microwave clock, or a frequency comb for an optical clock) which "counts" the number of oscillations per unit of time.

In this section we will focus on microwave atomic clocks, while optical clocks will be discussed in Sec. 5.3.

2.2.1 Microwave atomic clocks

In a typical microwave clock configuration, as the one shown in Fig. 2.3a, an atomic beam is created by heating the alkali metal in an oven and filtering the hot vapours by a set of apertures, in such a way as to produce a collimated stream of atoms moving in the same direction with small angular divergence. The beam is then prepared in a well-defined hyperfine state $|1\rangle$ by spatially filtering the atomic trajectories with magnetic Stern–Gerlach deflection (or, in more recent implementations, by using optical-pumping

[4]Caesium has the largest hyperfine splitting among the alkali atoms and hydrogen (1.4 GHz) as well, and offers some technical advantages, including its very low melting point, which results in high room-temperature vapour pressures.

techniques).[5] The beam then passes through a microwave cavity, where the excitation occurs, and the efficiency of the excitation is measured by the number of atoms detected in the hyperfine state $|2\rangle$.

In a microwave transition the Doppler effect, which so profoundly affects high-resolution optical spectroscopy, has less dramatic consequences, since the Doppler broadening is proportional to the frequency of the transition, as is evident from eqn (1.8). In addition, a careful setup of the atomic-beam experiment allows a small beam divergence, which results in small "effective" temperatures in the transverse direction, along which the microwave photons travel.[6] Since the natural linewidth can be neglected as well (as discussed in the previous section), other forms of line broadening become important in the microwave domain, most prominently the *interaction-time broadening* (or *time-of-flight* broadening) $\Delta\omega \sim 1/\Delta t$, which is determined by the finite interrogation time Δt (see Fig. 2.3a). In an atomic-beam setup, for atoms travelling at a velocity v, this time is limited by the size of the region of space Δl in which the microwave field is present by $\Delta t = \Delta l/v$.[7] In order to increase the precision of the measurement, the interaction time can be enlarged either by reducing the atomic velocity or by building larger microwave cavities.

2.2.2 Ramsey spectroscopy

A key contribution in the development of microwave atomic clocks was given by N F Ramsey, who developed the "separated oscillating fields" technique (Ramsey, 1950), then re-named as *Ramsey spectroscopy*, which was recognized with the award of the Nobel Prize in Physics in 1989 (Ramsey, 1990). The Ramsey technique allows the interaction-time broadening to be reduced by exciting the atoms in two different microwave cavities separated (in principle) by an arbitrary large distance. This spectroscopic technique is also one of the simplest examples of *atom interferometer* (see Sec. 2.5), where the interference between *internal states* of the atom is considered.

We denote the hyperfine ground states of an alkali (e.g. caesium) atom as $|1\rangle$ and $|2\rangle$ and their energy separation as $\hbar\omega_0$. In the Ramsey technique an atomic beam is produced and a first excitation pulse is applied in a first interaction region. For the sake of illustration, we consider a so-called "$\pi/2$ pulse", in which the pulse duration Δt and the microwave intensity are chosen in such a way as to excite the atoms from the state $|1\rangle$ to a superposition state $(|1\rangle - i|2\rangle)/\sqrt{2}$ (this happens when the pulse duration is a quarter of the Rabi period, see Appendix A.1.3). As the atoms leave the interaction region, their internal state evolves accordingly to a Hamiltonian unitary evolution of the state $\left(|1\rangle - ie^{-i\omega_0 t}|2\rangle\right)/\sqrt{2}$, in which the relative phase of the superposition changes in time according to the frequency ω_0 corresponding to the energy difference between the two levels. After a time T the atoms reach the second interaction region,

[5] An initial state-selection operation is required, since the energy separation between the hyperfine ground states corresponds to $\sim 10^{-3}k_B T$ (where k_B is the Boltzmann constant and T is the temperature), which causes the two hyperfine states to be almost-equally thermally populated.

[6] For a typical Cs atomic beam, the longitudinal atomic velocity is $v \simeq 200$ m/s. A beam divergence of 10 mrad implies a transverse velocity spread $v_\perp = 2$ m/s, which results in a Doppler broadening $\Delta\nu = \nu_{Cs}(v/c) \simeq 60$ Hz.

[7] For the same Cs atoms with velocity $v \simeq 200$ m/s travelling through a microwave cavity with $\Delta l = 5$ cm length, the interaction-time broadening is $\Delta\nu \simeq 1/\Delta t = v/\Delta l = 4$ kHz.

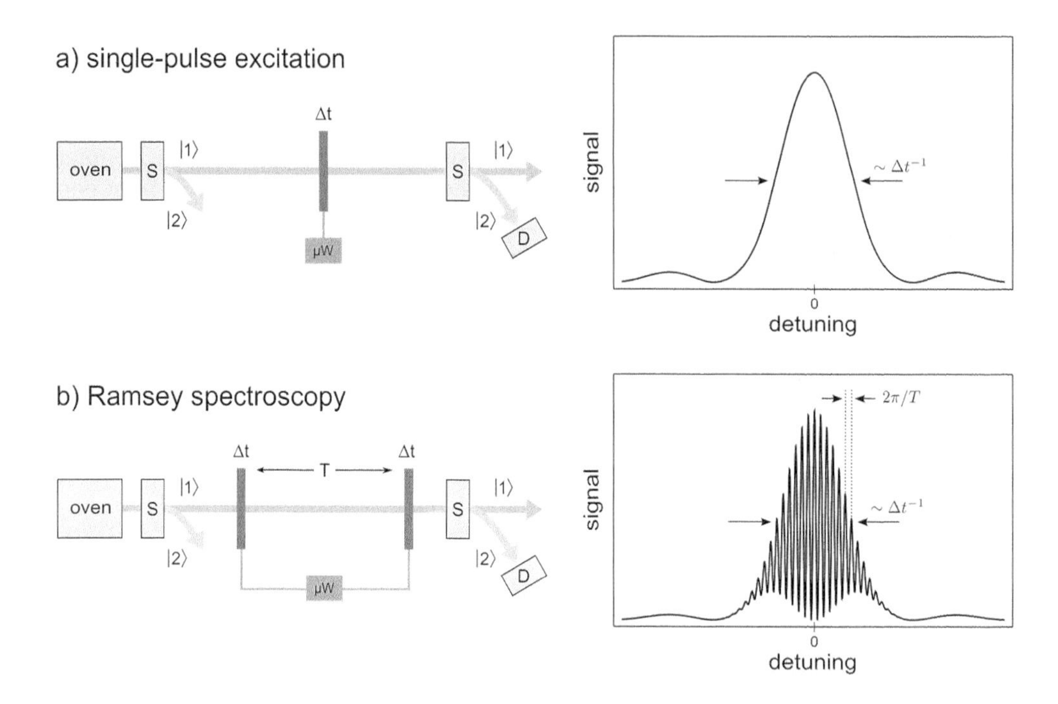

Fig. 2.3 a) Time-of-flight broadening in a microwave clock transition: the width of the observed resonance is proportional to the inverse of the interrogation time Δt. b) Ramsey spectroscopy setup with a double-pulse excitation: the spacing between the fringes is given by $2\pi/T$, where T is the time separation between the pulses.

where they are probed again by a second $\pi/2$-pulse induced by a microwave field in phase with the first pulse (the same microwave source has to be used).

Notably, the effect of the second pulse depends on the relative phase accumulated between the microwave field and the internal atomic superposition. The probability of measuring the atom in the state $|2\rangle$ is plotted in Fig. 2.3b as a function of the detuning $\delta = \omega - \omega_0$, where ω is the frequency of the microwave field. The coherent evolution of the atomic state during the travel time T between the two excitations is reflected in the appearance of interference fringes (*Ramsey fringes*) spaced by a (angular) frequency difference $2\pi/T$. The envelope of the fringes follows the same interaction-time broadened line profile of the single-pulse excitation.[8] At the highest maximum for $\delta = 0$ the microwave field has remained in phase with the evolution of the atomic state between the two pulses, therefore the effect of the second pulse adds up constructively as if the atom had interacted with a twice-long pulse. Instead, for $\delta \neq 0$ the microwave field runs out of phase with respect to the atomic superposition, which

[8]There is a close analogy in optics with the Young two-slit experiment: there the spacing between the fringes is determined by the spacing between the two slits, while the envelope of the fringes is given by the diffraction figure of each aperture.

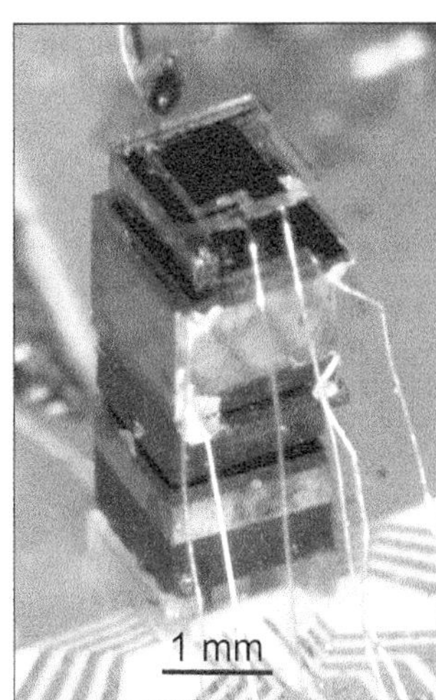

Fig. 2.4 Left) Picture of L. Essen and J. V. L. Parry with the first caesium atomic clock built by them in 1955 (Essen and Parry, 1955). Right) Recent technological progresses resulted in the miniaturization of atomic clocks, which are now available on the market for their integration on portable electronic devices: the picture shows the prototype of a millimetre-sized on-chip atomic clock developed at NIST. Reprinted with permission from Kitching *et al.* (2005). © IOP Publishing Ltd.

results in a fast decrease of the excitation probability and periodic revivals at multiples of $2\pi/T$ when, again, the field and the atomic superposition rephase together.

Ramsey spectroscopy is a powerful spectroscopic technique, that finds a number of applications in atomic physics experiments. Not only it is a precious tool for atomic clocks: it is commonly used in atom interferometry (see Sec. 2.5) as well as in quantum optics experiments (e.g. as a non-destructive detection technique for the state of the electromagnetic field in the cavity-QED experiments (Haroche and Raimond, 2006) which led S. Haroche to the Nobel Prize in Physics 2012). In microwave clocks the maximum of the central Ramsey fringe is used as a definition of the transition frequency. The precision of the measurement, with respect to a single excitation, is improved by a factor $T/\Delta t$, as if the atoms had interacted with the microwave field for the whole distance travelled between the two interaction regions. The precision of the measurement can thus be improved by using a longer T, which means either larger distances between the two pulses or slower atoms.

Laser cooling, that will be described later in this chapter, allows the realization of ultraslow atomic beams, with atoms moving with velocities as slow as a few cm/s. At these extremely slow velocities, however, the effect of gravity becomes important and the atoms stop to travel along a straight line, which complicates the experimental setting and represents the limiting factor for the Ramsey separation T. In order to better control this effect and increase the clock performances, different strategies can be implemented, which include the realization of *atomic fountains* and the planned operation of laser-cooled atomic clocks in space (see Sec. 2.4).

2.2.3 Masers

Before introducing laser cooling and illustrating more recent evolutions in the context of atomic clocks, we discuss here the operation of *masers* as frequency standards. A maser (acronym for *Microwave (or Molecular) Amplification by Stimulated Emission of Radiation*) is an oscillator based on atomic or molecular transitions, capable of generating highly coherent microwave radiation. The invention of the maser in 1954, recognized by the 1964 Nobel Prize in Physics to C. H. Townes, N. G. Basov, and A. M. Prokhorov, was fundamental for the development of the *laser*, proposed in a famous 1958 article by C. H. Townes and A. L. Schawlow (Schawlow and Townes, 1958) and demonstrated experimentally two years later by T. H. Maiman (Maiman, 1960).

As a matter of fact, the maser already contains the two main ideas that allowed the operation of the laser (for a historical review of the maser/laser invention, the reader can refer e.g. to Lamb *et al.* (1999)). First, it relies on the *population inversion* $N_2 > N_1$ between two internal atomic or molecular levels with energies $E_2 > E_1$, which allows the stimulated emission rate $B\rho N_2$ to be larger than the absorption rate $B\rho N_1$ (where B is the Einstein coefficient of absorption/stimulated emission and ρ is the density of energy at the transition frequency $\nu_0 = (E_2 - E_1)/h$). If spontaneous emission can be neglected,[9] this condition causes the medium to *amplify* the microwave radiation instead of absorbing it.[10] In addition, the use of a *resonant cavity* for the radiation provides the feedback mechanism necessary to establish oscillation and turns the microwave amplifier into a source of coherent microwave radiation. The first maser by C. H. Townes was based on the inversion transition of ammonia NH_3 at 24 GHz (Gordon *et al.*, 1954; Gordon *et al.*, 1955; Townes, 1972). This transition arises from a quantum-mechanical effect related to the pyramidal geometry of the ammonia molecule, with the N atom lying out of the plane formed by the three H atoms: the existence of two stable energy configurations for the N atom, above or below the H plane, and the possibility for it to tunnel through the plane, are responsible for a splitting of the lowest vibrational level into an inversion doublet (Townes and Schawlow, 1975). Emission of narrow-band radiation at 24 GHz was observed in Gordon *et al.* (1954) as the molecules, selected in their upper energy state, entered the resonance microwave cavity, where oscillation could be established.

The first atomic maser, built in 1960 by H. Goldenberg, D. Kleppner, and N. Ramsey (Goldenberg *et al.*, 1960; Ramsey, 1965), was based on hyperfine transitions between the two hydrogen ground states $F = 0$ and $F = 1$, separated by an energy interval $\delta E \simeq h \times 1.42$ GHz.[11] The scheme of the hydrogen maser is reported in Fig. 2.5. A

[9] The spontaneous emission coefficient A scales as $A \propto \nu^3$ with the radiation frequency ν. While spontaneous emission can be generally neglected for microwave transitions, it is important for electronic transitions in the visible spectrum.

[10] Of course, this condition can be mantained only with an active pumping mechanism, since at equilibrium the populations are given by the Boltzmann law $N_2/N_1 = \exp[-(E_2 - E_1)/k_B T]$ and a population inversion $N_2/N_1 > 1$ would correspond to a negative temperature T.

[11] This transition (often referred to as the "21 cm" line of hydrogen, because of its wavelength) has a remarkable importance in modern astronomy. The analysis of microwave signals from space, including

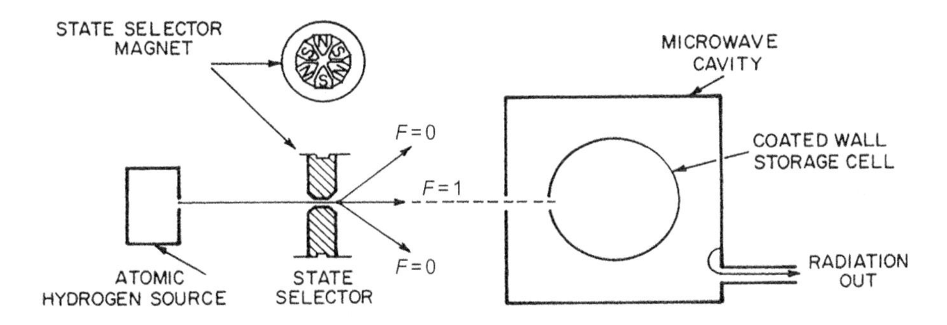

Fig. 2.5 Scheme of the hydrogen maser. Reprinted with permission from Ramsey (1990). © American Physical Society.

beam of atomic hydrogen is produced by dissociation of molecular hydrogen H_2 in a discharge and by the subsequent collimation of the atomic gas. A population inversion is created by selecting the atoms in the upper $F = 1$ state with a strong inhomogeneous magnetic field which deflects the $F = 0$ atoms out of the beam axis and focuses the $F = 1$ atoms inward. The atomic beam then enters a glass cell, which is placed within a microwave cavity tuned to be resonant with the hydrogen hyperfine splitting.[12]

The linewidth of the maser emission is small, thanks to the long permanence time of the atoms inside the cell, typically on the order of the second (much larger than the transit time of an atom in an atomic beam setup). The main systematical effect causing a shift of the line is represented by the interaction of the atoms with the cell walls, which can shift the frequency up to a few parts in 10^{11}, but this effect can be accurately characterized and well controlled. Thanks to these properties, soon after its realization the hydrogen maser was used to perform accurate measurements of the properties of hydrogen itself, primarily its hyperfine splitting

$$\Delta\nu = 1\,420\,405\,751.7667(9)\ \text{Hz} \tag{2.3}$$

with a fractional precision of 6.3×10^{-13} (Ramsey, 1993). This measurement of the ground state splitting, combined with recent accurate measurements of the hyperfine structure of the $1s - 2s$ transition (see Sec. 1.3), allows precise QED tests where the effect of the poor knowledge of the proton structure is mostly cancelled out (Kolachevsky *et al.*, 2009*a*).

Because of the well-defined emission frequency, hydrogen masers are still used as secondary standards of frequency. In particular, the performances of hydrogen masers

the 1.42 GHz hydrogen radiation, provides us with more and complementary information with respect to shorter-wavelength radiation emitted by extraterrestrial bodies, since microwave radiation can penetrate large clouds of interstellar cosmic dust more easily than visible light. The "21 cm" line is also the only accessible transition that allows us to detect the presence of atomic hydrogen in the cold interstellar medium, where the low temperature forbids the population of electronically excited states.

[12]The population inversion is maintained by the continuous arrival of new atoms in the upper $F = 1$ state, while other atoms already in the cell escape through the same entrance aperture. At steady state, the energy radiated out of the maser is compensated by the energy increase due to the excited atoms entering the cavity.

are quite good in terms of short-term stability ($< 10^{-12}$ after 1 second, $\approx 10^{-15}$ after 1 day), which is typically better than commercial caesium standards, and become worse in the long term mostly because of changes in the cavity resonance frequency over time.

2.3 Laser cooling

It has been well known from the beginning of spectroscopy that light can be used to study the *internal* degrees of freedom of atoms and molecules. The whole idea of spectroscopy is based on the application of *energy conservation* to the atom–light interaction: when a photon is absorbed (or emitted), its energy is given to (or removed from) the atom, which changes its internal state accordingly in order to maintain the total atom + photon energy constant. By analyzing the energy of the photons absorbed or emitted by the atoms (measuring either their wavelength or their frequency) it is possible to extract information on the atom's internal structure.

What about the photon momentum? In atom–light interaction *momentum conservation* has to be considered as well.[13] The momentum given or taken by the photon modifies the *external* (translational) degree of freedom of the atom, which recoils after an absorption or emission event, in such a way that the total atom + photon momentum is constant. Since there is a momentum exchange, this means that the light can exert *forces* onto the atoms. These forces allow light to be used as a powerful tool for the manipulation of the motional degrees of freedom of atoms. Transfer of momentum in the interaction between atoms and light is at the heart of the laser cooling and trapping techniques developed in the last thirty years and acknowledged by the 1997 Nobel Prize in Physics to S. Chu, C. Cohen-Tannoudji, and W. D. Phillips (Chu, 1998; Cohen-Tannoudji, 1998; Phillips, 1998).

There are essentially two kinds of radiative forces (Cohen-Tannoudji, 1992), associated with the absorptive and the dispersive properties of the interaction:

1. There is a dissipative force, also called *radiation pressure force*, associated with the transfer of momentum from light to atoms in a resonant scattering process. The atoms, *absorbing photons* from an incident light beam, are pumped into an excited state, then they spontaneously decay back to the ground state, emitting fluorescent photons in random directions. This dissipative force, characterized by an average momentum transfer $\hbar\mathbf{k}$ per atom for each absorption/emission cycle, with $\mathbf{k}$ wavevector of the laser light, is at the basis of the most common laser-cooling techniques.
2. There is a conservative force, also called *optical dipole force*, arising from the non-absorptive interaction between light and atoms. On a macroscopic scale, this force is connected with the refraction of light (i.e. a change in the photon momentum) by a polarizable medium. For single atoms, it originates from the interaction between the electric field and the induced atomic electric dipole. In this process there

[13] As originally pointed out by J. C. Maxwell, the electromagnetic field transports both energy and momentum and both have to be conserved in an absorption process. In classical electrodynamics, an electromagnetic wave with electric field $\mathbf{E}$ and magnetic field $\mathbf{H}$ is characterized by an energy $\langle S \rangle$ and by a momentum $\langle S \rangle /c$ per unit of time and area, where $\mathbf{S} = \mathbf{E} \times \mathbf{H}$ is the Poynting vector and c is the speed of light.

is *no absorption* of photons and no excitation of atoms to higher-energy states, and mechanical effects are caused by a redistribution of the photons among the elementary plane waves in which a non-uniform radiation field can be expanded.

In the next sections of this chapter, we will discuss the origin and the most relevant applications of the radiation pressure force, as an introduction to the laser-cooling techniques used in atomic physics. We do not intend to cover here a full discussion of laser cooling: the interested reader could refer to other published works, e.g. Cohen-Tannoudji (1992), Cohen-Tannoudji and Guéry-Odelin (2011), Metcalf and van Der Straten (1999), and Foot (2005) for a more complete treatment of this matter. The dipole force will be introduced later in Sec 5.2 in the context of the realization of optical traps for spinless two-electron atoms and will be extensively discussed in Chapter 6 devoted to the physics of optical lattices.

2.3.1 Radiation pressure

In its simplest form laser cooling relies on the effect of radiation pressure, the force which originates from the atomic recoil that follows photon absorption. If N_{ph} photons are absorbed in a time Δt, the radiation pressure force on the absorbing body can be written as

$$\mathbf{F} = \hbar\mathbf{k}\frac{N_{ph}}{\Delta t}, \tag{2.4}$$

corresponding to the momentum $\hbar\mathbf{k}$ carried by one photon times the number of photons absorbed per unit of time. In ordinary life this force is ridicolously small and its effects on macroscopic objects can be neglected. However, by using laser light it is possible to achieve high spectral densities, that, directed onto narrow-band absorbers such as the atoms of a dilute gas, result in large absorption cross sections and considerable forces on very light objects. From the theory of the interaction between coherent radiation and a two-level atomic system (developed in Appendix A), it is possible to evaluate $N_{ph}/\Delta t$ and demonstrate that an atom with resonance frequency ω_0, illuminated by a narrow-line laser beam with frequency ω, wavevector $\mathbf{k}$, and intensity I, experiences a radiation pressure force

$$\mathbf{F} = \hbar\mathbf{k}\frac{\gamma}{2}\frac{\frac{I}{I_{sat}}}{1+\frac{I}{I_{sat}}+\frac{4\delta^2}{\gamma^2}}, \tag{2.5}$$

where $\delta = \omega - \omega_0$ is the detuning, γ is the natural linewidth, and I_{sat} is the atomic saturation intensity. This expression is the result of a classical (non-quantum) treatment of both the atomic motion and the radiation field, which are both described with classical variables. At this level of analysis quantum mechanics enters only when considering the atom internal structure and the description of light–matter interaction (which includes saturation of the transition). This approach is well justified for large radiation fluxes (a large number of scattered photons allows us to consider the average value of the quantum-mechanical force operator) and large atomic velocities (in order to treat the atomic velocity as a continuous variable, its value has to be much larger than the recoil velocity following the absorption of one photon, which is the unit change of atomic velocity that the radiation field can induce).

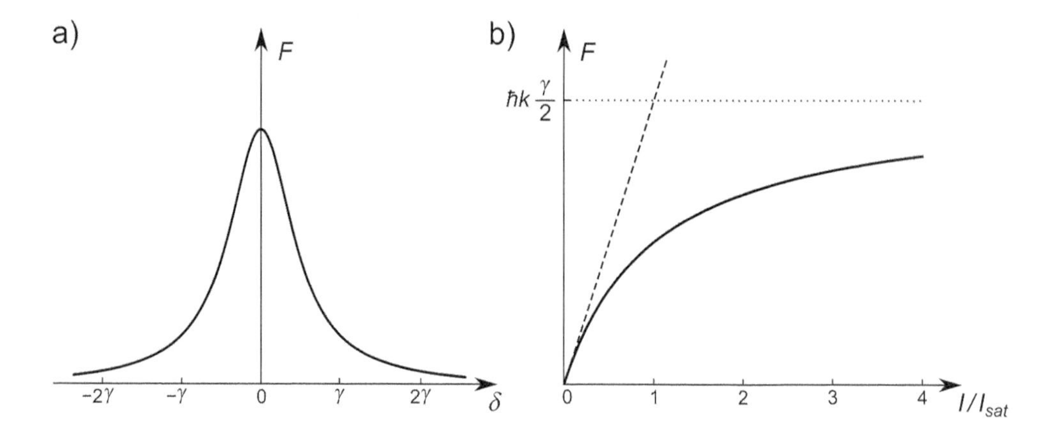

Fig. 2.6 Dependence of the radiation pressure force in eqn (2.5) on the detuning δ from the atomic resonance (a) and on the laser intensity I (b).

We note some basic properties of eqn (2.5): 1) the force has the same direction as the momentum of the absorbed photons; 2) it is maximum at resonance, i.e. when $\delta = 0$; 3) for weak fields it is proportional to the light intensity I; 4) for strong fields, above saturation intensity, it saturates to a limiting value $F = \hbar k\gamma/2$. The latter limit is particularly clear: above saturation, the atom is continuously absorbing photons and gains a recoil momentum $\hbar k$ at the maximum rate $\gamma/2$, which is limited only by the lifetime $1/\gamma$ of the excited state.[14] The dependency of the radiation pressure force on the laser detuning and intensity is plotted in Fig. 2.6.

For typical atomic velocities, the momentum carried by a photon is very small compared to the momentum of the atom (which justifies the classical treatment of the motion variables). As an example, we consider ^{87}Rb, which is a rather popular alkali atom for laser-cooling experiments. Rubidium atoms at room temperature $T = 300$ K move with a rms velocity $v_T = \sqrt{3k_BT/m} \approx 300$ m/s, while the velocity change induced by a single photon recoil at the resonant wavelength $\lambda = 2\pi/k = 780$ nm is only $v_R = \hbar k/m \approx 6$ mm/s, i.e. 5×10^4 times smaller. Nevertheless, the force exerted onto the atom by a resonant laser beam can be quite intense, provided that the number of absorbed photons per unit of time is made large. For ^{87}Rb the spontaneous decay rate is $\gamma \simeq 2\pi \times 6$ MHz (which means $\gamma/2 \simeq 2 \times 10^7$ photons scattered per second at saturation), which results in a saturation value for the force $F = \hbar k\gamma/2 = 1.6 \times 10^{-20}$ N, corresponding to an acceleration $a = F/m = 1.1 \times 10^5$ m/s^2, i.e. 10 000 times larger than the gravity acceleration on Earth!

A momentum transfer is also involved when the atom comes back to its lower energy state following a spontaneous emission process. Due to the stochastic nature of this process, a photon is emitted in a random direction, therefore the average momentum transfer is equal to zero. For this reason, spontaneous emission does not enter eqn (2.5).

[14] The factor 2 comes from the fact that for large intensities the population of the excited state approaches an upper bound of 1/2, owing to saturation of the transition (see Appendix A).

However, as we shall discuss in Sec. 2.3.3, this process results in nonzero fluctuations of the force, which produce a "random walk" in momentum space which is ultimately responsible for the finite temperature which can be obtained with Doppler laser cooling.

2.3.2 Atomic beam deceleration

One of the first applications of laser cooling concerned the deceleration of atomic beams. A resonant laser beam can be shone opposite to the atomic beam direction in order to reduce the longitudinal atomic velocity with the radiation pressure force of eqn (2.5). First, yet non-conclusive, indications of a beam deceleration were reported in Balykin *et al.* (1980) and Andreev *et al.* (1982) with the observation of a modified Doppler lineshape in the fluorescence spectrum of sodium atoms excited along the atomic beam axis.

In order to make the deceleration process efficient, a problem that had to be solved in the early 1980s was about finding a way to circumvent the Doppler effect (Phillips, 1998): as the atoms are slowed down by the counterpropagating beam, the change in the Doppler shift brings them out of resonance from the laser beam excitation, thus decreasing the deceleration force. For the ^{87}Rb case considered in the above section, an initial velocity $v = 300$ m/s corresponds to a Doppler shift $\Delta\nu = v/\lambda \simeq 390$ MHz, which is $\simeq 65$ linewidths away from the resonance. If one uses a detuned laser to start the deceleration process, after a modest change in velocity the atoms get out of resonance and no further deceleration occurs.

The solution that was found at NBS (now NIST, USA) by W. D. Phillips and H. Metcalf in Phillips and Metcalf (1982) was based on the following idea. A non-uniform magnetic field $B(x)$, spatially varying along the atomic beam axis $\hat{x}$, was used to shift the atomic levels in such a way as to keep the laser always in resonance with the atomic transition as the atoms are slowed down. For resonant light driving a transition between two states with differential magnetic moment μ, this happens when the spatially dependent Zeeman shift $\mu B(x)$ exactly compensates the decreasing Doppler shift $kv(t)$, i.e. when

$$kv(t) = \mu B(x) \ . \tag{2.6}$$

If the solenoid producing the magnetic field $B(x)$ is designed according to the above condition (see Fig. 2.7a–b), the atoms constantly absorb resonant light throughout their phase-space trajectory $\{x(t), v(t)\}$ and move with constant deceleration down to almost vanishing final velocity (provided that the solenoid is long enough[15]). The integration of the classical equation of motion for different initial velocities gives the phase-space trajectories in Fig. 2.7c: below a critical velocity v_0, all the atoms are "captured" in the same trajectory and reach the end of the solenoid with small final velocity. Figure 2.8 shows the longitudinal velocity distribution of a sodium atomic beam measured in early experiments at NIST (obtained from the Doppler spectrum of a longitudinal spectroscopy beam) in the absence and in the presence of the decelerating laser beam:

[15] There is a minimal length of the solenoid in order to "stop" the atomic beam. This limit stems from the fact that the radiation pressure force saturates at a value $F = \hbar k\gamma/2$. Therefore, if atoms with inital velocity v have to be slowed down to zero velocity, simple kinematic equations for the motion with constant deceleration $-F/m$ give a minimal stopping length $L = mv^2/2F = mv^2/\hbar k\gamma \simeq 1.6$ m for ^{87}Rb at an initial velocity $v = 300$ m/s.

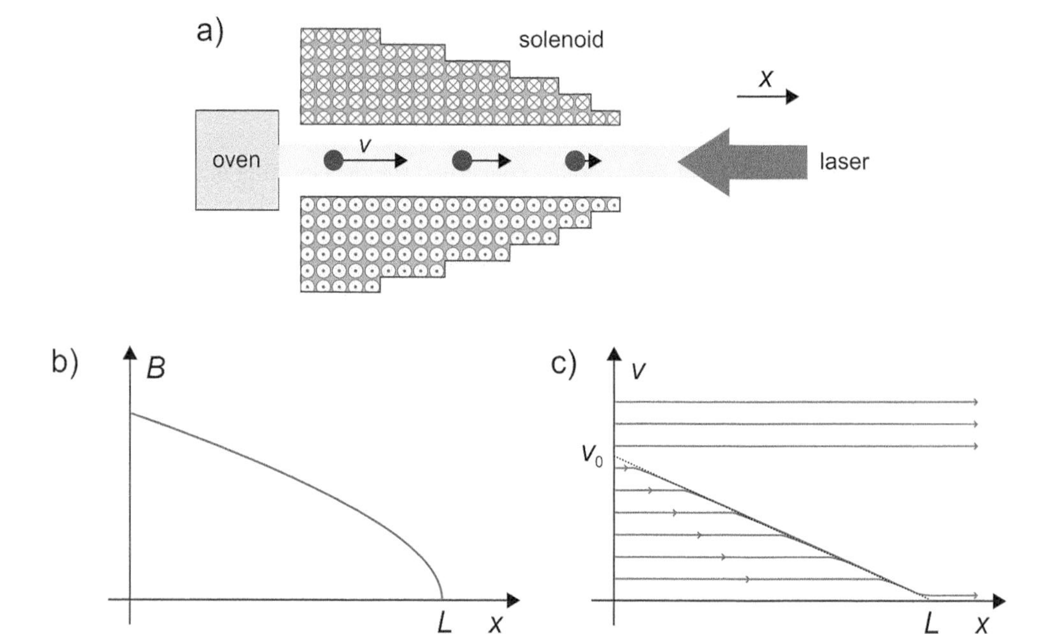

Fig. 2.7 Zeeman Slower for laser-deceleration of an atomic beam. a) A solenoid with varying number of windings produces a non-uniform axial magnetic field, necessary to keep the atoms in resonance with the slowing laser beam. b) Magnetic field intensity as a function of the position along the solenoid axis. c) Phase-space diagram of the atom trajectories for different initial velocities: below a capture velocity v_0 atoms are decelerated down to vanishing velocities.

it is evident how the atoms of the initial Maxwellian distribution below a velocity ≈ 1300 m/s are almost entirely decelerated into a narrow peak of slow atoms around 700 m/s (Phillips, 1998). Improvements of the deceleration and detection protocols then allowed the NIST group to achieve deceleration of atoms down to zero average velocity (Prodan *et al.*, 1985).

Though the Zeeman slower scheme is by far the most used configuration, other techniques of beam deceleration have been considered and implemented. In *chirp slowing*, originally proposed by V. Letokhov et al. in Letokhov *et al.* (1976), no magnetic fields are used and the changing Doppler effect is compensated by sweeping the frequency of the slowing laser in such a way that the atoms are always in resonance along their trajectory. Differently from the Zeeman slowing technique, in which all the atoms along the beam trajectory are simultaneously resonant with the laser beam, the chirp slowing technique only works in pulsed mode: the resonance condition is met only for localized bunches of atoms which are "followed" by a laser chirp synchronized with their trajectory. This technique was used at JILA in Ertmer *et al.* (1985) to stop bunches of atoms to zero average velocity and at Bell Labs as a preliminary step for loading optical molasses in Chu *et al.* (1985) (as discussed in the next section).

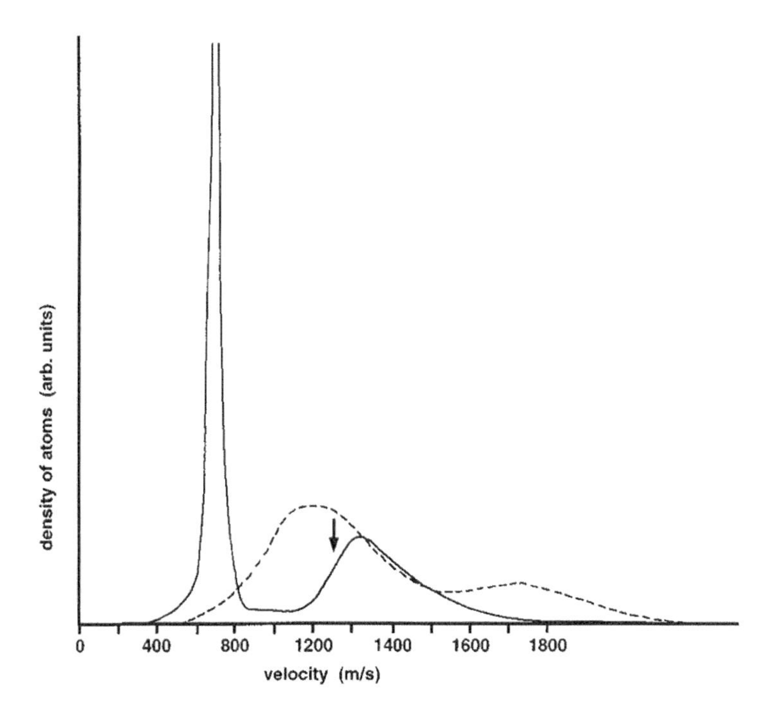

Fig. 2.8 Longitudinal velocity distribution of a sodium atomic beam before (dashed line) and after (solid line) Zeeman slowing (the second maximum in the dashed spectrum has to be attributed to atoms in a different hyperfine state, that are optically pumped out of that state during the slowing process). Reprinted with permission from Phillips (1998). © American Physical Society.

2.3.3 Doppler cooling

Doppler cooling, proposed in 1975 by T. W. Hänsch and A. Schawlow (Hänsch and Schawlow, 1975) for free atoms, and in a similar fashion by D. Wineland and H. Dehmelt for trapped ions (Wineland and Dehmelt, 1975), relies on the following argument. Consider a free atom with velocity $\mathbf{v}$ which is illuminated by two counterpropagating laser beams with the same (angular) frequency ω, aligned with their wavevectors along the same direction as $\mathbf{v}$ (see Fig. 2.9a). Owing to the Doppler effect, in the atom reference frame the effective frequency of the photon moving opposite to the atomic motion is upshifted to $\omega' = \omega(1 + v/c)$, while the effective frequency of the photon moving in the same direction as the atom is downshifted to $\omega'' = \omega(1 - v/c)$. If the laser frequency ω is smaller than the atomic resonance frequency ω_0, the effective frequency ω' is closer to resonance than ω'' and the atom has a higher probability of absorbing a photon from the counterpropagating beam. The recoil which follows the absorption then results in a decelerating force.

More quantitatively, if we consider the small intensity limit $I \ll I_{sat}$ (in order to neglect saturation effects and simplify the theoretical treatment) we can write the total force acting on the atom (grey solid line in Fig. 2.9b) simply as the sum of the

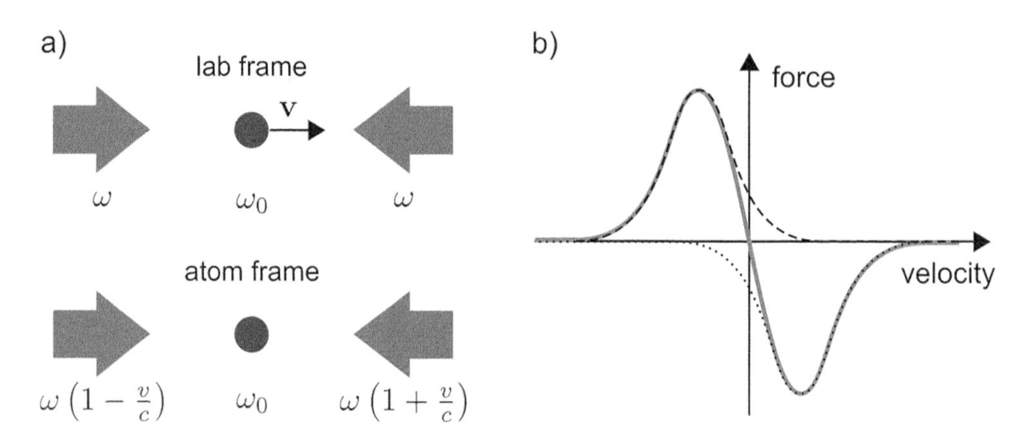

Fig. 2.9 Doppler cooling. a) Scheme of a 1D optical molasses configuration, illustrated both in the laboratory and in the atom reference frame. b) Radiation-pressure force as a function of velocity for $\omega < \omega_0$: the dashed line is the contribution of the left beam, the dotted line is the contribution of the right beam, the grey solid line is the total force.

forces exerted by the two beams separately. Starting from eqn (2.5) and including the Doppler shift, one finds

$$\mathbf{F} \simeq \hbar \mathbf{k} \frac{\gamma}{2} \left[\frac{\frac{I}{I_{sat}}}{1 + \frac{4(\omega-\omega_0-kv)^2}{\gamma^2}} - \frac{\frac{I}{I_{sat}}}{1 + \frac{4(\omega-\omega_0+kv)^2}{\gamma^2}} \right] , \tag{2.7}$$

where the first term in square parentheses is the force generated by the redshifted co-propagating beam (dashed line in Fig. 2.9b), while the second term is the force generated by the blueshifted counter-propagating beam (dotted line in Fig. 2.9b). A small velocity expansion of the equation above yields the simple result

$$F = -\beta v + O(v^3) , \tag{2.8}$$

which describes a friction force with damping coefficient

$$\beta = -8\hbar k^2 \frac{I}{I_{sat}} \frac{\frac{\delta}{\gamma}}{\left(1 + \frac{4\delta^2}{\gamma^2}\right)^2} , \tag{2.9}$$

which is positive for a detuning $\delta = \omega - \omega_0 < 0$. Owing to the friction character of the force in eqn (2.8), this configuration is traditionally called "optical molasses" (Chu, 1998). This force comes from two fundamental conservation principles of physics: momentum conservation on one side, which results in the existence of the radiation pressure force, and energy conservation, which is responsible for the non-symmetric absorption of the molasses beams. By generalizing the above derivation to the case of three orthogonal pairs of counterpropagating beams, it is possible to demonstrate that eqn (2.8) holds vectorially and the force has a friction character along any arbitrary direction. Therefore, an atom which enters the region of intersection of the molasses beams will slow down exponentially to vanishing velocity.

Of course eqn (2.8) cannot describe correctly the atom dynamics at small velocities.[16] The discrete nature of the absorption and emission processes, which becomes important at low velocities, determines a finite lower bound for the temperature that can be achieved in optical molasses. We note that, even if the atomic system is not in thermodynamic equilibrium (since it is continuously exchanging energy with the laser field), it is possible to define an effective temperature thanks to the equipartition theorem:

$$\frac{1}{2}k_B T = \frac{1}{2}m\left\langle v^2\right\rangle , \tag{2.10}$$

where $\left\langle v^2\right\rangle$ represents the average squared velocity of an ensemble of atoms moving in the optical molasses.[17] The lowest temperature attainable is determined by the equilibrium between two effects: the "cooling" described by the average force in eqn (2.8) and the "heating" which is caused by the force fluctuations induced both by spontaneous emission (which results in recoil kicks with random directions) and by fluctuations in the intensity of the laser beams (which have a Poissonian statistics in the number of photons). Both these heating contributions determine a diffusion in momentum space: similarly to what happens in Brownian motion, the atom makes a random walk in momentum space which, in the absence of the cooling mechanism, can be described by an average squared velocity growing linearly with time, according to

$$\left\langle v^2\right\rangle = v_R^2 R t , \tag{2.11}$$

where R is the scattering rate of photons and $v_R = \hbar k/m$ is the recoil velocity. By equating the cooling rate to the heating rate (Foot, 2005) it is possible to find a steady state which, in the low intensity limit, is characterized by a *Doppler temperature*

$$T_D = \frac{\hbar\gamma}{4k_B}\frac{1+\left(\frac{2\delta}{\gamma}\right)^2}{\left(\frac{2|\delta|}{\gamma}\right)} \tag{2.12}$$

which takes its minimum value for $|\delta| = \gamma/2$

$$T_D^{min} = \frac{\hbar\gamma}{2k_B} \tag{2.13}$$

dependent only on the linewidth γ of the cooling transition. For ^{87}Rb atoms, $\gamma \simeq 2\pi \times 6$ MHz and $T_D^{min} \simeq 150$ μK.

The first demonstration of optical molasses was realized in 1985 by S. Chu *et al.* by the group of A. Ashkin at Bell Labs (Chu *et al.*, 1985). In this pioneering experiment a beam of ^{23}Na atoms was decelerated with the "chirp slowing" method and then

[16] Otherwise it could be possible to cool atoms down to an arbitrarily small temperature after a sufficiently long time!

[17] It is possible to experimentally observe that the atomic velocities are indeed distributed according to a Maxwell–Boltzmann distribution, therefore the definition of temperature seems appropriate.

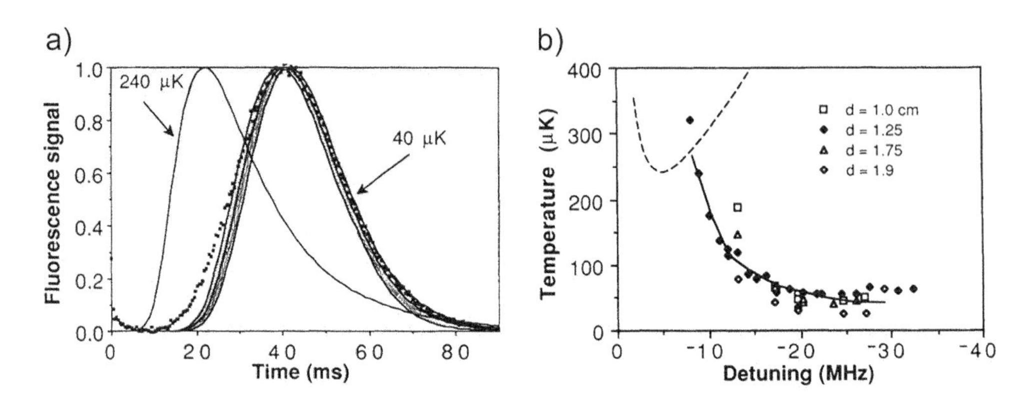

Fig. 2.10 Temperature of sodium atoms in optical molasses measured at NIST in 1988. A probe beam aligned under the molasses records the passage of the atoms as a function of the time from the molasses switch-off (see text). a) Spectrum of the atom arrival times, which is in agreement with a temperature $T \simeq 40$ μK much lower than the minimum Doppler temperature $T_D^{min} \simeq 240$ μK. b) Dependence of T on the molasses detuning: the dashed line is the prediction of the Doppler theory given by eqn (2.12). Reprinted with permission from Lett *et al.* (1988). © American Physical Society.

captured in the molasses region. In order to estimate the temperature of the atoms in the molasses, a time-of-flight measurement was used: the cooling beams were blocked for a variable amount of time δt, allowing the cloud to expand ballistically, and the atomic velocity was then inferred by measuring the fraction of atoms that could be recaptured once the cooling beams were switched on again after δt. The result of this measurement gave substantial agreement with the prediction of eqn (2.13) for sodium $T_D^{min} \simeq 240$ μK, with a density $n \approx 10^6$ atoms/cm^3 and a molasses lifetime of ≈ 0.1 s. Actually, S. Chu discussed later that in this first experiment they observed slightly lower temperatures than the Doppler limit, but they "made the cardinal mistake of experimental physics: instead of listening to Nature, we were overly influenced by theoretical expectations" (Chu, 1998). Indeed, as explained in the next section, when the internal structure of the ground state and conservation of angular momentum are considered, new effects come into play and new cooling (and trapping) possibilities emerge.

2.3.4 Sub-Doppler cooling

In 1988 the NIST group led by W. D. Phillips reported for the first time that the temperature of sodium atoms cooled in optical molasses was by far lower than the Doppler limit in eqn (2.13) (Lett *et al.*, 1988). Figure 2.10 shows the results of one method for the temperature measurement (out of four independent techniques used in Lett *et al.* (1988) as a confirmation of the results). A resonant probe beam was aligned below the molasses region and the fluorescence of the atoms falling through the probe after the molasses beams had been switched off was measured as a function of time: comparing the shape of the fluorescence signal with the result of a simple simulation of the atomic motion allowed P. Lett et al. to determine the temperature of the atoms.[18] The data in Fig. 2.10a are clearly inconsistent with the predictions based on the minimum Doppler temperature $T_D^{min} = 240$ μK expected for sodium atoms (thin line), instead they showed agreement with a much lower temperature $T \simeq 40$ μK. Figure 2.10b also shows that the dependence of the measured temperature on the molasses detuning was completely different from the predictions of the Doppler theory (dashed line), the temperature being lower and lower for increasing detuning.

These results, confirmed by other groups at Stanford, ENS, and JILA with both caesium and sodium atoms (Chu, 1998), showed that the Doppler theory introduced in Sec. 2.3.3 was not capturing the whole physics of laser cooling. It was soon realized that other cooling mechanisms were playing a role and these mechanisms were based on the fact that the atoms could not be treated as pure two-level systems. As realized by C. Cohen-Tannoudji and J. Dalibard, and independently by S. Chu (Chu, 1998; Cohen-Tannoudji, 1998), the presence of a Zeeman structure in the ground state, combined with the polarization gradients that are formed in optical molasses, was responsible for new *sub-Doppler cooling* mechanisms.

There are several schemes of sub-Doppler cooling, depending on the particular choice of the molasses beams polarization, as first described in Dalibard and Cohen-Tannoudji (1989) and Ungar *et al.* (1989). Here we introduce the mechanism of *Sisyphus cooling* as proposed by J. Dalibard and C. Cohen-Tannoudji in Dalibard and Cohen-Tannoudji (1989). This effect can be observed in optical molasses in the presence of a polarization gradient, such as the one that is obtained when two counterpropagating molasses beams have opposite linear polarization, as sketched in Fig. 2.11a: in this configuration the resulting polarization is spatially modulated with a period $\lambda/2$ and changes ellipticity going from right-circular to linear, then to left-circular and linear again (Fig. 2.11b). For Sisyphus cooling to work, the atomic ground state must be characterized by a nonzero angular momentum J, which is always the case for an alkali atom (see Sec. 2.1) but it is not always true for atoms with different electronic structure (e.g. the alkaline-earth atoms discussed in Sec. 5.1). The ground state is split in Zeeman sublevels with different angular momentum component m_J, the energy of which is shifted owing to the ac-Stark effect,[19] which depends on m_J and on the polarization state of the light: since the polarization is spatially dependent, also the energy of the Zeeman sublevels changes as a function of position.

This behaviour is exemplified in Fig. 2.11c for the particular case of an atom with angular momentum $J = 1/2$. The energies of the two Zeeman states $m_J = 1/2$ and $m_J = -1/2$ change sinusoidally in space with period $\lambda/2$ and are out-of-phase by π with respect to the other. In the regime of low light intensity the atoms spend most of their time in the ground state. An atom in the initial state $m_J = 1/2$ moving with positive velocity has to climb up a potential hill, which is caused by the spatially-dependent ac-Stark shift: the increase in potential energy is compensated by a decrease in the atom

[18] The atoms were so cold that the Doppler broadening was smaller than the natural linewidth of the transition, preventing a simple spectroscopic measurement from giving information on their temperature.

[19] This effect will be discussed with more detail in Sec. 5.2.1 concerning the possibility of using it for the realization of optical traps.

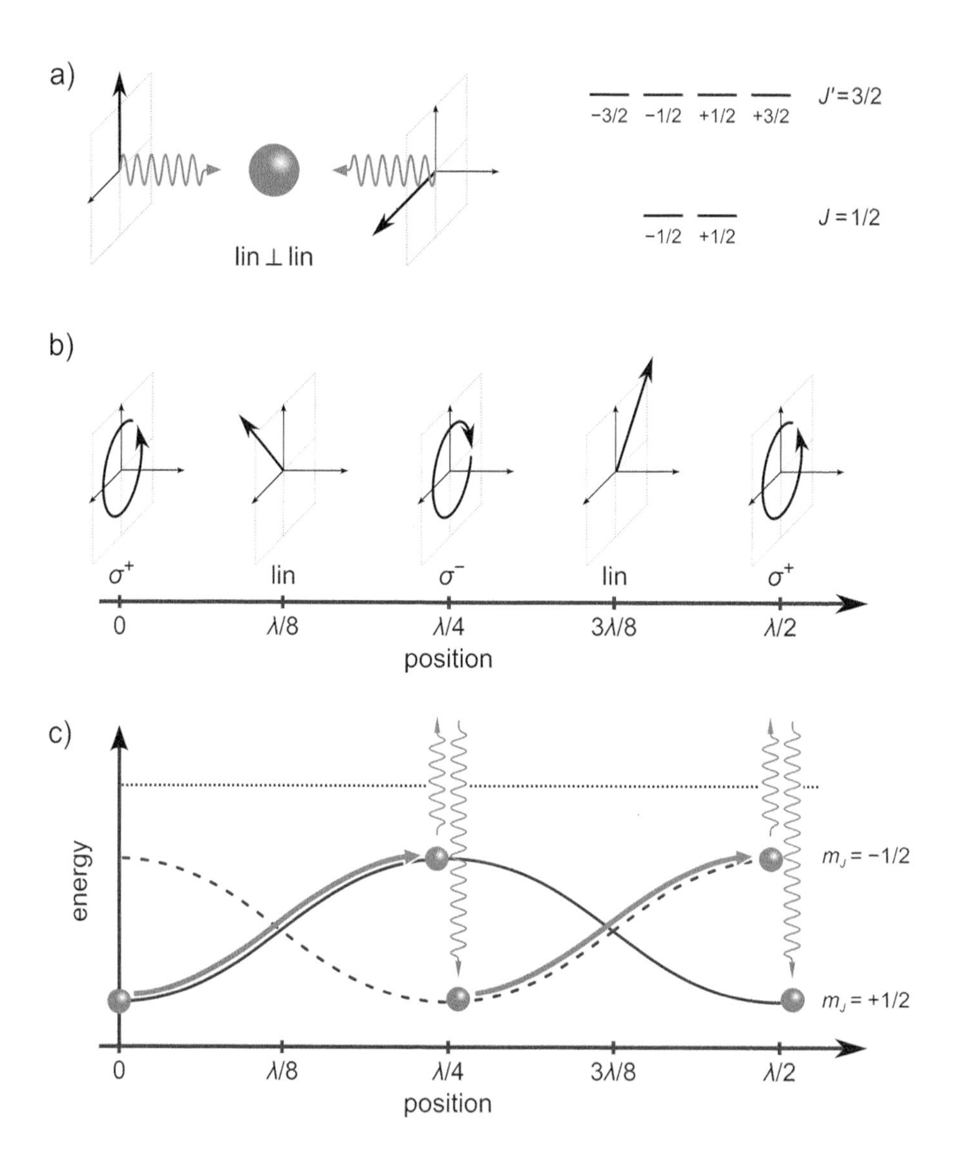

Fig. 2.11 Sketch of the sub-Doppler Sisyphus cooling mechanism proposed in Dalibard and Cohen-Tannoudji (1989). a) We consider a lin⊥lin molasses configuration for an atom with ground state $J = 1/2$ and excited state $J' = 3/2$. b) A polarization gradient with periodicity $\lambda/2$ is formed by the interference of the molasses beams. c) As a consequence of the spatially varying coupling induced by the polarization gradient, the two ground-state Zeeman sublevels $m_F = 1/2$ and $m_F = -1/2$ are characterized by a spatially modulated ac-Stark shift (the dotted line is the bare ground-state energy). Optical pumping in this polarization gradient forces the atom to continuously climb up a potential hill, decreasing their kinetic energy.

kinetic energy. When the atom arrives, with smaller velocity, on top of the potential hill at $x = \lambda/4$, if absorption takes place, the σ^- polarization of light favours optical pumping into state $m_J = -1/2$, which has a lower energy. The atom, once pumped into this state, has to climb up a new potential hill, which causes it to slow down again. On top of the new potential hill at $x = \lambda/2$ the atom may be optically pumped by the σ^+ polarization into state $m_J = 1/2$ and the story starts over again ... This mechanism, which enables researchers to slow down the atoms to much lower temperatures than the Doppler limit, as shown in detail in Dalibard and Cohen-Tannoudji (1989), was named by J. Dalibard and C. Cohen-Tannoudji as *Sisyphus cooling*, by analogy to the Greek mythological character who was forced by the punishment of the Gods to continuously roll an immense boulder up a steep hill.

2.3.5 Magneto-optical traps

Optical molasses provide a very efficient method to cool atomic gases. However, they do not provide any spatial confinement and, if new atoms are not continuously fed into the molasses, the atomic cloud expands diffusively and its density decreases (see Hodapp *et al.* (1995) for a detailed study of the diffusion dynamics).

A significant upgrade of the optical molasses scheme is represented by the *magneto-optical trap* (MOT), in which an inhomogeneous magnetic field is used to break the spatial symmetry and produce a resulting trapping force towards a well-defined trap centre. This force arises from a third fundamental conservation law of physics, in addition to the conservation of momentum and energy which are at the basis of Doppler cooling: it is the conservation of angular momentum, which determines the selection rules for the absorption of polarized light. The invention of the magneto-optical trap, originally proposed by J. Dalibard in 1986 (Chu, 1998), represented a milestone in the history of cold atoms, since it proved to be a very robust experimental technique allowing atoms to be both trapped and continuously cooled for arbitrarily long times. Trapping times (without continuous feeding of new atoms) can be as long as several minutes, limited by losses of atoms from the trap due to collisions with the room-temperature background gas: for this reason, a vacuum chamber evacuated down to the ultra-high vacuum regime ($< 10^{-9}$ mbar) is required in order to keep the cold atoms well isolated from the environment and ensure long trapping times.

In its first and more popular version, a MOT is composed by a spherical quadrupole magnetic field, typically created by a pair of coils placed in anti-Helmoltz configuration, and three pairs of counterpropagating red-detuned laser beams intersecting orthogonally in the region of zero magnetic field, in such a way that beams aligned along the same axis have opposite circular polarization.

A simplified 1D version of this scheme, which is useful to understand the principle of operation of a MOT, is represented in Fig. 2.12a. The two coils produce a magnetic field $\mathbf{B}(x) = (bx)\hat{x}$ with uniform field gradient b in the proximity of the trap centre at $x = 0$ and the two red-detuned laser beams counterpropagating along $\hat{x}$ have opposite circular polarization σ^+ and σ^- (the σ^+ beam comes from negative x). As a case study, we consider an atom having a ground state with angular momentum number $J = 0$

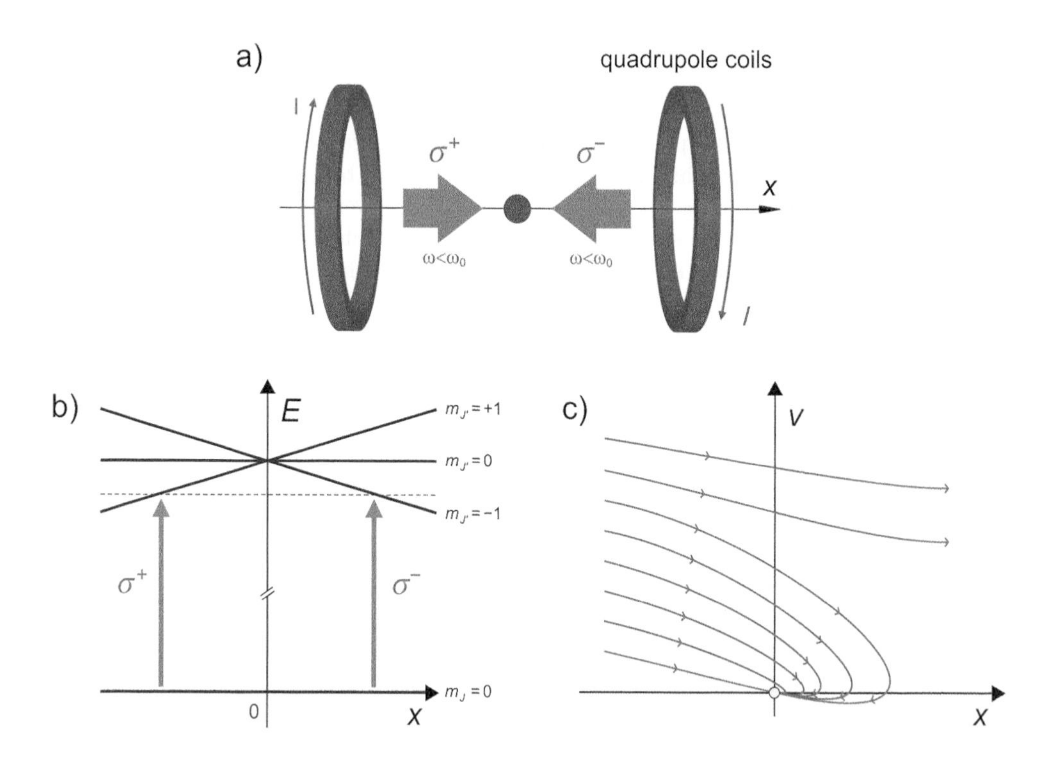

Fig. 2.12 Principle of operation of a magneto-optical trap (MOT) in a simplified 1D configuration. a) An atom interacts with two red-detuned counterpropagating laser beams having opposite circular polarization in the presence of a magnetic field gradient generated by two quadrupole coils. b) Energy levels of an atom having $J = 0$ ground state and $J' = 1$ excited state in the presence of the magnetic field. c) Phase-space trajectories of the atoms for different initial velocities: below a threshold capture velocity the atoms are slowed and trapped at the MOT centre position.

and an excited state $J' = 1$ with a Zeeman threefold $m_{J'} = -1, 0, 1$.[20] In Fig. 2.12b we show the energies of the atomic levels taking into account the position-dependent Zeeman shift of the excited state

$$\Delta E(x) = g_{J'} \mu_B m_{J'} b x \, , \tag{2.14}$$

where $g_{J'}$ is the Landé factor of the excited state (see Sec. 3.1.1 devoted to magnetic trapping), $\mu_B = e\hbar/2m_e$ is the Bohr magneton, and the field gradient b is chosen in such a way that $g_{J'} b > 0$. We note that the angular momentum component $m_{J'}$

[20]As will be further discussed in Sec. 2.3.6, this assumption is not realistic for alkali atoms, which have a ground state with nonzero angular momentum, but it exactly describes what happens in alkaline-earth atoms with zero nuclear spin (see Sec. 5.1).

and the circular polarization state of light are both defined according to the same $\hat{x}$ quantization axis.[21]

We start considering an atom located at $x > 0$. Because of the position-dependent Zeeman shift induced by the quadrupole field, the atom preferentially absorbs a photon from the σ^- field, because the red-detuned light is closer to resonance with the $m_J = 0 \to m_{J'} = -1$ transition. As a consequence of the momentum transfer in the atom–photon interaction, the atom experiences a radiation pressure force directed towards the trap centre. In the case of an atom located at $x < 0$ the situation is reversed and the most probable process is absorption from the σ^+ field, that still produces a net force directed towards the trap centre. In addition, since the two beams are red-detuned, this restoring force is accompanied by a viscous force, as the one produced in optical molasses. More quantitatively, in the low intensity limit, we can write down the total radiation pressure force including the Zeeman shift and adding the contributions of the two beams similarly to what has been done in eqn (2.7):

$$\mathbf{F} \simeq \hbar\mathbf{k}\frac{\gamma}{2}\left[\frac{\frac{I}{I_{sat}}}{1+\frac{4(\omega-\omega_0-kv-g_{J'}\mu_B bx/\hbar)^2}{\gamma^2}} - \frac{\frac{I}{I_{sat}}}{1+\frac{4(\omega-\omega_0+kv+g_{J'}\mu_B bx/\hbar)^2}{\gamma^2}}\right] . \tag{2.15}$$

Expanding the above expression for small displacements x and small velocities v, one finds that the total force exerted on the atom can be written as

$$F \simeq -m\omega^2 x - \beta v \,, \tag{2.16}$$

which describes the force acting on a damped harmonic oscillator. This result can be generalized for atoms with different electronic configurations and to the 3D case, when cooling beams are applied in the other two orthogonal directions as well: the qualitative results are the same and eqn (2.16) still holds vectorially, providing the possibility of three-dimensional cooling and trapping. Figure 2.12c shows the phase-space trajectories of atoms entering the MOT region: if the initial velocity is smaller than a typical MOT capture velocity,[22] the atoms describe damped trajectories towards the trap centre at $x = 0$.

The first operating MOT was realized in 1987 starting from a decelerated beam of ^{23}Na atoms (Raab *et al.*, 1987) and, a few years later, the possibility of trapping the atoms in a MOT directly from a room-temperature ^{133}Cs vapor was demonstrated (Monroe *et al.*, 1990). Nowadays, a magneto-optical trap represents the most popular way to produce cold trapped atomic gases: a standard MOT of an alkali atom, e.g. ^{87}Rb, typically contains $\approx 10^9$ atoms at a temperature of few tens of μK.[23]

[21] And not according to the magnetic field direction (which reverses crossing $x = 0$) or to the light propagation direction (which is opposite for the two beams), respectively.

[22] The finiteness of the MOT capture velocity is related to the saturation behaviour of the radiation pressure force and to the maximum MOT size, defined by the volume of intersection of the orthogonal pairs of laser beams.

[23] At low densities the temperature of the atoms in a MOT is similar to the ones reached in optical molasses. At larger densities collective mechanisms emerge, such as light-assisted two-body collisions or reabsorption of already scattered photons, which pose an upper limit to the achievable MOT densities and reduce the MOT cooling efficiency.

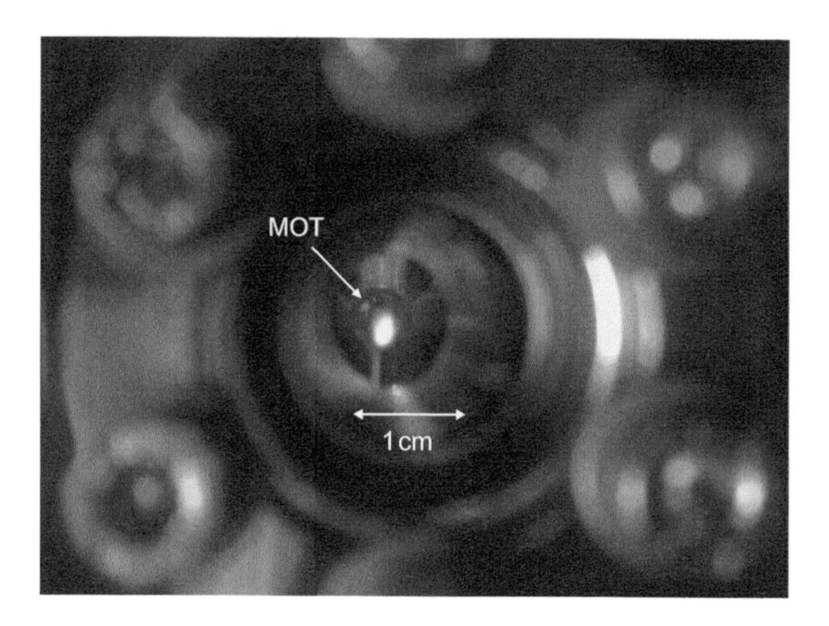

Fig. 2.13 Photo of a magneto-optical trap operating at LENS with ^{174}Yb atoms. The bright spot in the centre is the fluorescence of a cloud of $N \approx 10^9$ atoms at a temperature of $T \approx 40$ μK. The $J = 0 \rightarrow J' = 1$ transition used to cool ^{174}Yb corresponds to the ideal MOT model in Fig. 2.12 and allows a test of the Doppler cooling theory in the absence of sub-Doppler mechanisms.

2.3.6 Laser cooling in multi-level atoms

We conclude this section devoted to laser cooling by noting that the alkali atoms that were used in all the first laser-cooling experiments are not two-level atoms. In particular, as discussed in Sec. 2.1, alkali atoms with nuclear spin I have two different hyperfine ground states with total angular momentum $F = I - 1/2$ and $F = I + 1/2$. This hyperfine structure is present also in the excited state and is typically narrower (with a frequency separation between the manifold components of tens or hundreds of MHz) compared to the one in the ground state (with separation of several GHz). In order to make laser cooling possible in the presence of a multi-level electronic configuration, it is necessary to use a closed transition, i.e. a transition from a stable state $|g\rangle$ to an excited state $|e\rangle$ that has no other decay channel than the one to the state $|g\rangle$: in this way an atom can undergo as many absorption/emission cycles as needed to be slowed down to the limits of laser cooling. Owing to the hyperfine structure of alkalis, no completely closed transitions can be found, since there is always a finite probability of exciting a nearby open transition, which can optically pump the atoms in a dark state. As a consequence, additional *repumping lasers* are needed to remove atoms from the dark states and to re-populate the states involved in the cooling transition.

As an example, we show in Fig. 2.14 the hyperfine structure of the $5\,^2S_{1/2}$ and $5\,^2P_{3/2}$ states of ^{87}Rb, which are connected by the 780 nm D2 transition. Among the possible choices within the hyperfine manifold, the most favourable transition for laser

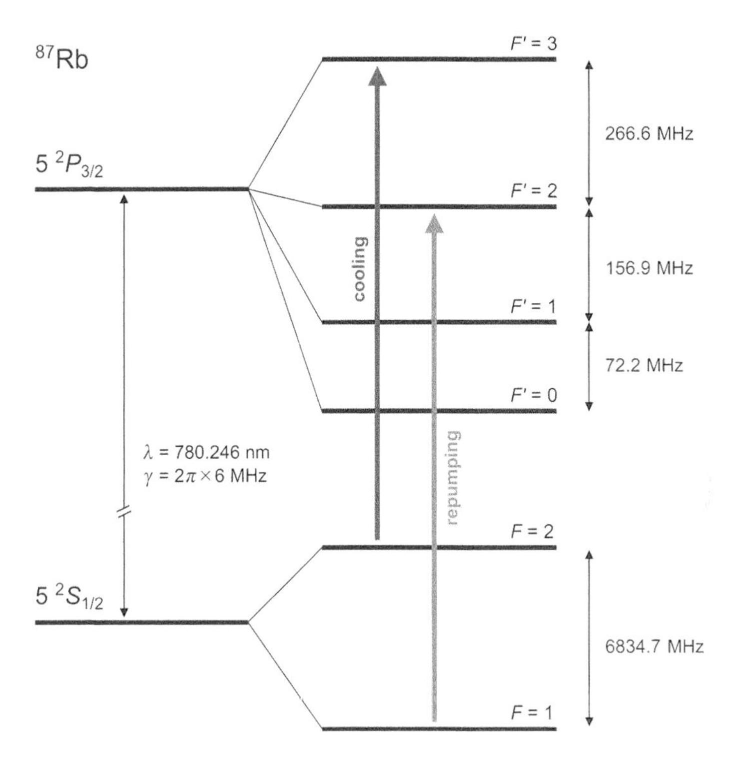

Fig. 2.14 Hyperfine structure of the $5\,^2S_{1/2}$ and $5\,^2P_{3/2}$ states of ^{87}Rb, including the transitions used for laser cooling and repumping.

cooling is the $F = 2 \to F' = 3$. This is a closed transition since atoms excited to $F' = 3$ can only decay to $F = 2$ (owing to the $\Delta F = \pm 1$ selection rule for dipole-allowed transitions) and no atoms are lost into the dark $F = 1$ state. However, because of the small hyperfine separation in the excited state, there is a finite probability of non-resonant excitation to the $F' = 2$ state, which has the dark ground state $F = 1$ as allowed decay channel.[24] A repumping laser operating on the $F = 1 \to F' = 2$ transition has to be used in order to "re-cycle" the atoms and make them available for new absorption from the cooling laser beams.

The hyperfine splitting in the electronic structure of alkali atoms may also prevent sub-Doppler cooling. This happens for those atoms in which the hyperfine splitting of the excited $^2P_{3/2}$ state is not much larger than the transition linewidth, as in the case of lithium (^{6}Li and ^{7}Li) and bosonic potassium (^{39}K and ^{41}K). In the case of potassium the total hyperfine separation of the $^2P_{3/2}$ state is ~ 5 times the linewidth: this small splitting causes the cooling laser to significantly excite transitions to different states in the excited manifold, inducing strong optical pumping and causing the cooling force to

[24] The probability of excitation scales as $1/(1+4\delta^2/\gamma^2) \simeq (\gamma/2\delta)^2 \approx 10^{-4}$ for a detuning $\delta = 2\pi \times 267$ MHz equal to the frequency difference between the $F' = 2$ and the $F' = 3$ states. Considering the scattering rate of photons at saturation $\gamma/2 \approx 10^7$ s^{-1}, this means that after a time on the order of a millisecond the atoms are lost into the non-absorbing $F = 1$ state.

noticeably depart from the one evaluated for an isolated transition (Fort *et al.*, 1998). A clear sub-Doppler cooling of bosonic potassium has been observed only very recently (Landini *et al.*, 2011) thanks to an optimized cooling strategy based on the control of optical pumping by a fine tuning of the laser parameters.

2.4 Laser-cooled atomic clocks

We have already anticipated the importance of the revolution that laser cooling has produced on high-precision spectroscopy. In the context of microwave atomic clocks, where the finite transit time is the most significant source of line broadening, laser cooling allows many orders of magnitude longer interrogation times. Using the techniques described in the previous section it is possible to slow the motion of atoms down to a few metres per second (even few cm/s in the case of optical molasses), resulting in a travel time along the apparatus on the order of one second, which means a transit-time broadened linewidth of only ~ 1 Hz (see Sec. 2.2.1). In this regime of small velocity the atomic trajectories are strongly deviated by the effect of gravity and, in order to further increase the interaction time in a more compact setup, *fountain atomic clocks* where the atomic beam is oriented vertically can be realized.

The concept of fountain clocks dates back to the 1950s with the early attempt of J. Zacharias (Ramsey, 1956), who proposed and tried to experimentally realize an atomic clock with a thermal beam of caesium atoms oriented upwards: the atoms were made to cross a microwave cavity, then the slowest of them were expected to follow a ballistic trajectory, invert their motion and cross again the same cavity in a downward direction, realizing a Ramsey experiment with a long time interval between the two excitations. The idea was based on the possibility of selecting the extreme low-velocity tail of the Maxwell–Boltzmann distribution in a "hot" atomic beam, but the signal turned out to be too small to be detected. The idea of fountain atomic clocks was reconsidered in the late 1980s and experimentally realized with laser-cooled atoms in Kasevich *et al.* (1989). The success of this new attempt, made possible by the much larger fluxes of slow atoms and their reduced temperature, turned out to be fundamental for the development of very precise atomic clocks and for the atom interferometry experiments discussed in Secs. 2.5 and 4.5.2.

The scheme of a typical atomic fountain clock setup is shown in Fig. 2.15, taken from the review article Wynands and Weyers (2005). The atoms are first accumulated and cooled in optical molasses or in a magneto-optical trap, which is formed in the lower part of the experimental setup. Then, they are launched upwards with a well-controlled velocity v by a *moving optical molasses* phase, where a pair of vertical molasses beams with slightly detuned frequencies ω and $\omega + \delta\omega$ cool the atoms in a reference frame moving at velocity $v = \delta\omega/2k$ (where k is the laser wavenumber) with respect to the laboratory frame: these moving molasses allow a very accurate control on the atom velocity v (typically few m/s) and contribute to reducing the temperature (typically $\simeq 1$ μK) of the launched cloud of atoms. After the launch, the atoms move ballistically according to their classical trajectory, reaching the turning point at a time v/g, after which they start to fall down again. In most of the experimental configurations the microwave cavity is placed right above the cooling and detection region, and the turning point is at a much higher level: in this scheme, the atoms pass twice through the same

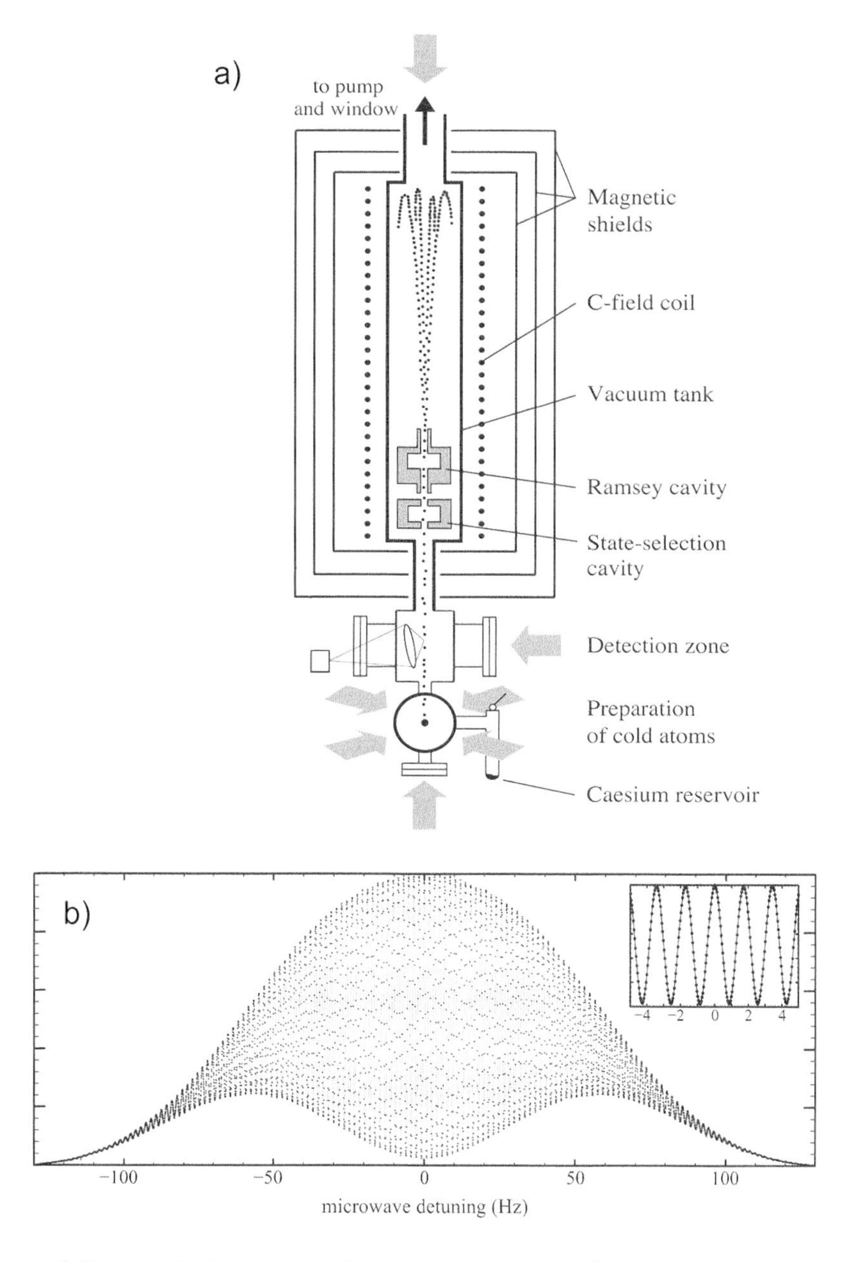

Fig. 2.15 a) Scheme of a laser-cooled fountain atomic clock. b) Ramsey fringes detected by a probe laser after the second passage through the microwave cavity (data from PTB-CSF1 fountain clock). Reprinted with permission from Wynands and Weyers (2005). © IOP Publishing Ltd.

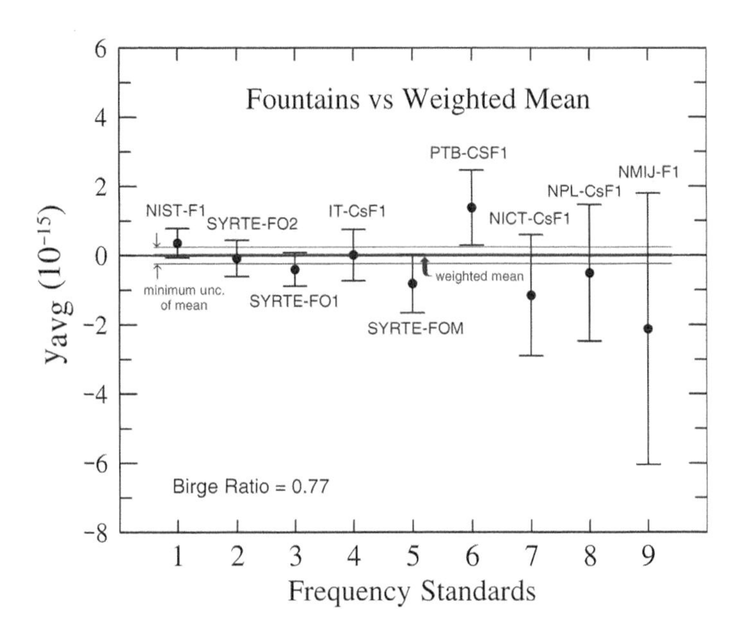

Fig. 2.16 Long-term comparison of atomic fountain clocks. The graph shows the average fractional frequency shift of the caesium hyperfine splitting measured by nine different atomic fountain clocks. Reprinted with permission from Parker (2010). © IOP Publishing Ltd.

microwave cavity (as in Zacharias' early proposal), which results in Ramsey fringes detected by laser excitation after the second passage. Figure 2.15 shows a typical experimental Ramsey signal featuring a fringe separation of ≈ 2 Hz (see inset).

The first laser-cooled caesium fountain demonstrating a better stability than traditional thermal-beam caesium clocks was reported in Clairon *et al.* (1995). Nowadays, caesium fountain clocks provide the most precise and accurate realization of the SI second. Figure 2.16 shows the results of a long-term comparison of nine atomic fountain clocks that are regularly reporting their measurements to the Bureau International des Poids et Measures (BIMP) for the determination of the accuracy of the International Atomic Time (TAI) (Parker, 2010). This comparison, based on the clocks' operation over several years, resulted in a very good agreement of the estimated Cs frequency within their quoted uncertainties. At the time of the comparison (2010) the fractional uncertainty of the weighted mean of the different fountain clocks was $\approx 4 \times 10^{-16}$.

Comparisons between clocks can also be made, in a more direct way, by physically transporting atomic clocks in the same place. This is also important for the absolute frequency measurement of optical transitions discussed in Sec. 1.4.2: for a stable and accurate frequency measurement, an optical frequency comb has to be locked to a high-quality radiofrequency signal, which can be provided by a microwave atomic clock. As an example, the absolute frequency measurements of the hydrogen $1s - 2s$ transition discussed in Sec. 1.3 have been performed thanks to a transportable fountain caesium clock developed at LKB-SYRTE and transported at MPQ in repeated measurement campaigns from 1999 to 2010.

2.4.1 Improving atomic fountain clocks

The accuracy of state-of-the-art atomic fountain clocks is currently limited by the evaluation of systematic uncertainties in the clock operation. At the $\sim 10^{-16}$ level of precision, many effects have to be taken into account, among these: Zeeman shifts, collisional shifts induced by atom–atom interactions, residual Doppler effects, ... While many of these effects can be controlled and corrected with high accuracy, some of them introduce important uncertainties in the overall clock uncertainty budget.

A major technical source of uncertainty is connected to the characterization of the microwave field, which can exhibit spatial inhomogeneities in its intensity and phase. In particular, a phase variation of the microwave inside the resonant cavity is responsible for a *distributed cavity phase shift* coming from the first-order Doppler shift of the atoms moving in the distorted wavefronts. This effect, indeed one of the limiting factors of the accuracy of fountain clocks, is heavily dependent on the cavity geometry and fountain alignment: although its accurate modelling requires particular care, recently its control at $< 10^{-16}$ level has been demonstrated (Guéna *et al.*, 2011).

A more fundamental source of uncertainty comes from *blackbody radiation shifts*, which are caused by the ac-Stark shift of the two hyperfine levels induced by the room-temperature blackbody radiation field (Itano *et al.*, 1982). For thermal radiation at room temperature $T = 300$ K the fractional blackbody shift is $\approx 2 \times 10^{-14}$ and exhibits the characteristic $\propto T^4$ dependence of the Stefan–Boltzmann law for the intensity of the blackbody spectrum. Corrections to the measured caesium frequency have to be applied to take this effect into account. However, in a real setup the environment only emits "pseudo-blackbody" radiation which can deviate from the ideal behaviour, therefore conservative estimates have to be made on the uncertainty of the blackbody shift correction. In order to suppress this effect, new fountain clocks operating at cryogenic temperatures have been recently built at NIST (USA) and INRIM (Italy): in these experiments the main body of the setup enclosing the microwave cavity and the atom trajectory between the Ramsey pulses is cooled by liquid nitrogen down to a temperature $T \simeq 80$ K, in such a way as to decrease the blackbody contribution by a factor $(300/80)^4 \simeq 200$. The clocks are now operating and the evaluation of their performances is currently under way (Levi *et al.*, 2010).

Quantum metrology. What about the statistical uncertainties associated with the clock operation? In state-of-the-art fountain clocks the main source of statistical uncertainty does not come from technical detection noise, but from a more fundamental noise associated with the measurement process itself. This is the so-called *quantum projection noise*, which comes from the projective nature of the quantum-mechanical measurement, in which the final atomic state $|\psi\rangle = c_1|1\rangle + c_2|2\rangle$ (with c_1 and c_2 depending on the relative phase between the atomic superposition and the microwave field at the moment of the second Ramsey pulse) is projected onto one of the two possible measurement outcomes $|1\rangle$ and $|2\rangle$. Considering e.g. state $|1\rangle$ as a possible outcome, the projection operator $\hat{P} = |1\rangle\langle 1|$ describing the measurement process is characterized by an expectation value $p_1 = |c_1|^2$ and by a variance $\sigma_1^2 = |c_1|^2(1 - |c_1|^2)$ (Itano *et al.*, 1993). If N uncorrelated atoms are detected, the ensemble-averaged measurement is characterized by fluctuations with standard error $\sigma_1^{(N)} = \sigma_1 N^{-1/2}$, scaling as the

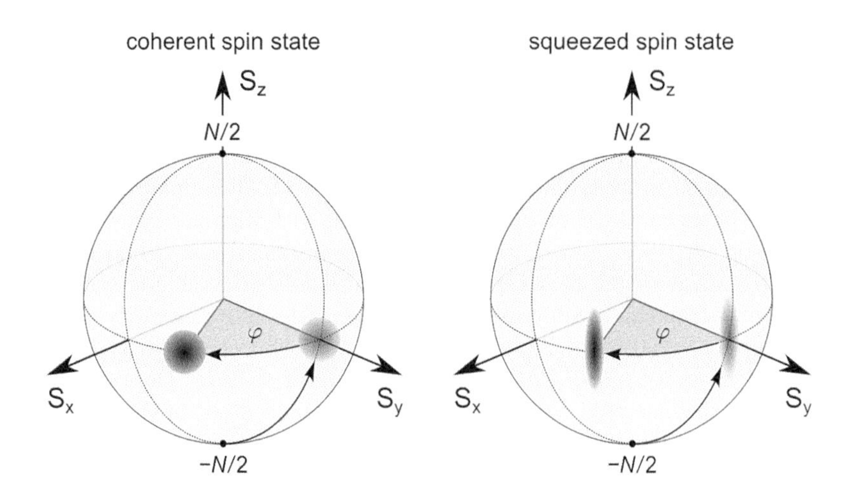

Fig. 2.17 The state of an ensemble of N two-level atoms maps onto a collective spin $\mathbf{S} = \sum_i \mathbf{s}_i$, where $\mathbf{s}_i$ is a 1/2-spin representing the internal state of the i-th atom ($s_{iz} = -1/2$ for state $|1\rangle$, $s_{iz} = +1/2$ for state $|2\rangle$). The figure shows the Bloch-sphere representation of the collective spin evolving with a phase φ increasing in time after a $\pi/2$ pulse. The left figure shows the Heisenberg uncertainty for a coherent spin state in which N uncorrelated atoms occupy the same state. The right figure shows a spin-squeezed state in which the uncertainty in one direction is suppressed due to quantum correlations between the atoms.

inverse square root of the atom number, as observed in Santarelli *et al.* (1999) with the first demonstration of an atomic fountain clock operating in the quantum-limited regime. Reducing this noise below the standard quantum limit $\sim N^{-1/2}$ is not fundamentally prohibited by quantum mechanics: it requires the presence of quantum correlations between the atoms, which can be induced by the generation of entangled atomic states.

The collective state of a system of N two-level atoms maps onto a collective $N/2$ spin, as shown in Fig. 2.17: N independent atoms are described by a coherent collective spin state (left), while certain types of entanglement result in spin-squeezed states, in which the uncertainty of one collective spin component is reduced at the expense of an increased uncertainty along the orthogonal direction (right) (Wineland *et al.*, 1994). These states are particularly promising for *quantum metrology* since they can lead to improvements in the atomic-clock precision (Giovannetti *et al.*, 2004). Recently, performances of atomic clocks below the standard quantum limit have been obtained in proof-of-principle experiments in Leroux *et al.* (2010), in which spin squeezing was achieved taking advantage of the collective interaction of the atoms with a common mode of an optical cavity, and in Louchet-Chauvet *et al.* (2010), where spin squeezing was obtained by applying a sequence of quantum non-demolition measurements creating entanglement between the atoms.

In a recent work, spin squeezing has been demonstrated also for Bose-condensed atoms trapped on atom chips (Riedel *et al.*, 2010), which is particularly promising for the development of a new generation of compact tranportable atomic clocks, which is under active development (Deutsch *et al.*, 2010). Further discussion on the advantages of multi-particle entangled states for an improvement of the measurement precision will be presented in Sec. 3.2.6 in the related context of quantum atom interferometry with Bose–Einstein condensates.

Atomic clocks in space. Finally, a further increase of precision in atomic clocks could be achieved by improving the quality of the frequency discriminant, i.e. reducing the linewidth of the Ramsey fringes by increasing the interrogation time. In a fountain clock setup this time is limited to ≈ 1 s by the value of the gravity acceleration on Earth and by the finite size of the experimental apparatus. This limit could be beaten by running experiments in reduced-gravity configurations. This is the goal of an ongoing project, called PHARAO ("Projet d'Horloge Atomique par Refroidissement d'Atomes en Orbit"), developed by a French research team operating in Paris at SYRTE and École Normale Supérieure. The project goal is the operation of a laser-cooled caesium clock in space, where the microgravity environment may allow atoms which are gently pushed out of optical molasses to travel along a vacuum tube at very small velocities for very long times without being deflected. The PHARAO project has passed the initial tests and aims at reaching a fractional accuracy of 10^{-16}, allowing accurate long-distance comparisons with (and between) ground-based clocks and studies of fundamental physics including Special and General Relativity tests (Laurent *et al.*, 2006). The clock will be operated on the International Space Station, together with a hydrogen maser, as a part of the ACES (Atomic Clock Ensembles in Space) mission of ESA (European Space Agency) which is scheduled to be launched in 2013 (Cacciapuoti and Salomon, 2009).

2.5 Atom interferometry

Ramsey spectroscopy, already discussed in Sec. 2.2.2, is an example of an atom interferometer operating on the space of the *internal* degrees of freedom. With an optical analogy (illustrated in Fig. 2.18), we can interpret the two internal states of the atom as the two paths in a Mach–Zehnder interferometer and the microwave pulses as the two beam splitters which first divide (coherently) and then recombine the light beam. During the time between the two microwave pulses the atom is in a quantum superposition of two different internal states $|1\rangle$ and $|2\rangle$, as the light in the Mach–Zehnder interferometer travels along both the paths: the atomic superposition initialized by the first microwave pulse is then analysed by the second pulse, the outcome of which depends on the relative phase of the superposition, as the light intensity at the two output ports of the interferometer depends on the relative phase acquired by the beams during the propagation in the two paths.

This idea of atom interferometry can be extended to consider the external degrees of freedom. This becomes a necessary extension of the above Ramsey experiment when optical photons are used instead of microwaves, since the transfer of momentum in the atom–photon interaction causes the atom in the two internal states to have significantly different momenta and to follow different trajectories in space, as originally pointed out by C. J. Bordé in a seminal work on optical Ramsey experiments (Bordé, 1989).

This spatial separation is at the basis of the possibility of building *atom interferometers* based on the interference in the *external* degrees of freedom. In *atom optics*

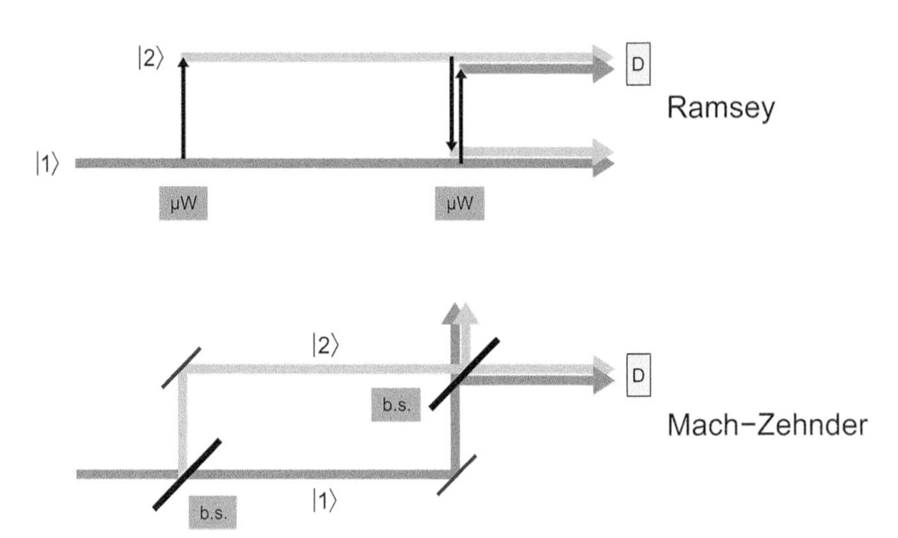

Fig. 2.18 Analogy between a Ramsey spectroscopy experiment and an optical Mach–Zehnder interferometer. In both cases a coherent splitting process (a microwave pulse in the first case, a beam splitter in the second) and the different phase acquired along two possible paths result in the observation of interference fringes in the output port (D) of the interferometer.

the atom is considered as a de Broglie matter wave $\psi(\mathbf{r})$ (the atom centre-of-mass wavefunction), with an associated quantum-mechanical phase $\phi = \mathbf{k}_{dB} \cdot \mathbf{r}$, where

$$\lambda_{dB} = \frac{2\pi}{k_{dB}} = \frac{h}{p} \tag{2.17}$$

is the de Broglie wavelength, dependent on the atomic momentum p. In an atom interferometer the atom is spatially delocalized along two different paths 1 and 2 and the two components of the atom wavefunction ψ_1 and ψ_2 may experience a differential phase shift induced by any effect coupling to the motional degree of freedom. When the two trajectories are made to overlap again, ψ_1 and ψ_2 interfere and the probability of detecting the atom, i.e. the squared modulus of the total wavefunction $|\psi_1 + \psi_2|^2$, depends on the relative phase they have acquired.

In the last 30 years of atomic physics a whole toolkit of atom-optical components has been developed to manipulate matter waves with almost the same degrees of freedom (and in some cases more) which are available in optics experiments with photons. The matter-wave counterpart of most optical components, such as mirrors, beam splitters, lenses, diffraction gratings, etc... can be realized with a variety of different techniques. Among these, the interaction of atoms with laser light has an undisputed prime role, thanks to the possibility of tuning the interaction strength and duration in order to finely control its effects (e.g. the quantity of momentum transferred to the atoms and the efficiency of the splitting process).

The first atom interferometers were built in 1991 and were based on different techniques: in Carnal and Mlynek (1991) the Young's double slit experiment was demonstrated for a beam of metastable helium impinging on microfabricated mechanical slits, in Keith *et al.* (1991) three microfabricated gratings were used to diffract and recombine a collimated beam of sodium, while in Riehle *et al.* (1991) and Kasevich and Chu (1991) the momentum transfer of optical pulses was used to split and recombine an atomic beam, with narrow one-photon resonances and two-photon Raman transitions respectively.

Atom interferometers can be used to precisely measure forces, external fields, accelerations, velocities. For instance, they can be used to measure gravity when the trajectories of the atom after the splitting are separated in height or to measure rotations when the interferometer is operated in a non-inertial rotating frame. The performances of atom interferometers are spectacular in term of sensitivities and in many applications they outperform their optical analogue. Many types of atom interferometers exist, based on different splitting and recombination techniques and different atomic sources (thermal atomic beams, laser-cooled beams, Bose–Einstein condensates).

The first developments of atom interferometry are covered in Berman (1997), while for an excellent review on the history, basic principles, and applications of atom interferometry the reader can refer to Cronin *et al.* (2009) and references therein.

2.5.1 Gravity measurements

A spectacular example of the opportunities offered by atom interferometry is related to the possibility of accurately measuring gravity. The gravity acceleration on Earth is not a constant, since it varies both with space and time. However, its precise measurement is very interesting for a number of reasons, which span from fundamental physics (tests of relativity and gravity theories) to more applied research (geophysics, metrology, and even search for oil, water, or other resources).

In 1999 the group of S. Chu at Stanford reported on the measurement of the local gravity acceleration g with a precision of 3×10^{-9} (Peters *et al.*, 1999; Peters *et al.*, 2001). Figure 2.19a shows a sketch of the experimental setup, which reminds us of the design of the fountain-based atomic clocks discussed in Sec. 2.4.[25] Similarly to the latter, also in this case a laser-cooled cloud of atoms, with very narrow momentum spread, is launched upwards along a vertical vacuum chamber and is detected again when it falls down. An important difference is in the presence of two counterpropagating laser beams oriented vertically (labelled as "Raman beams" in the figure), which induce a two-photon Raman transition between the two hyperfine states $F = 3$ and $F = 4$ of caesium atoms. Differently from the microwave transitions discussed previously, the momentum transfer in a Raman transition driven by counterpropagating optical photons with wavevector $\mathbf{k}$ and $-\mathbf{k}$ is given by twice the photon momentum $2\hbar\mathbf{k}$ and is $\approx 10^5$ times more than the momentum transfer by a microwave photon (Raman transitions will be discussed again in more detail in Sec. 4.5.2 devoted to the interferometric measurement of h/m for the determination of the fine structure constant).

[25]Interferometers based on atomic fountains are not the only class of atom interferometers for the measurement of gravity. In Sec. 6.4.3 we will discuss a different experimental approach based on Bloch oscillations of ultracold atoms trapped in optical lattices, which provide both high sensitivity and high spatial resolution.

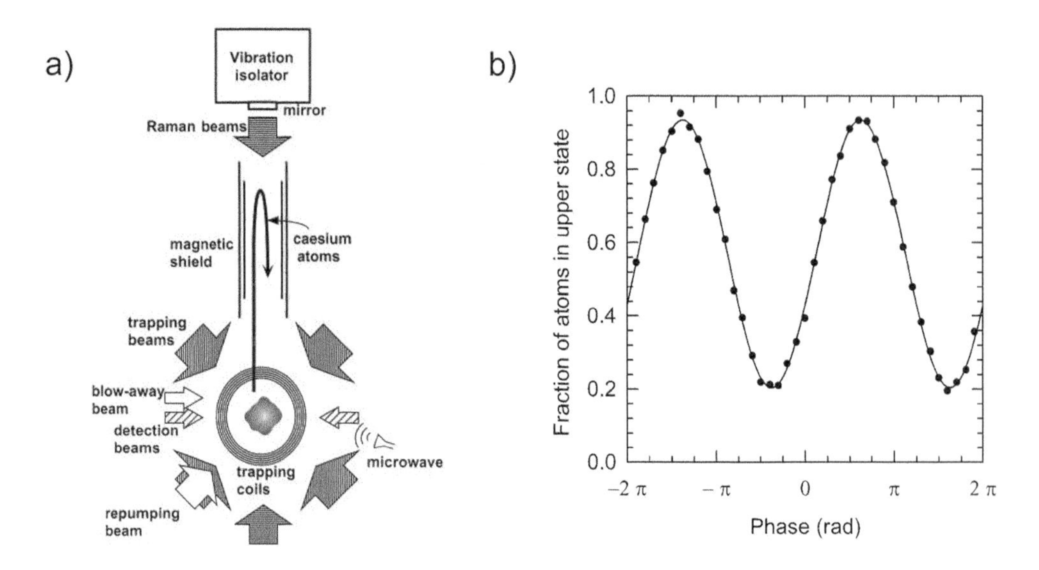

Fig. 2.19 a) Scheme of the atomic-fountain setup used in Peters *et al.* (1999) for the interferometric measurement of the gravity acceleration g. b) Plot of the measured interferometric signal: the absolute phase of the interferogram yields a value for g with a precision of 3×10^{-9}. Reprinted with permission from Peters *et al.* (1999). © Macmillan Publishers Ltd.

In the interferometric measurements of g these Raman beams are used to produce a sequence of three $\pi/2$, π, $\pi/2$ pulses, which represent the matter-wave analogue of a Mach–Zehnder interferometer for light (beam splitter / mirror / beam splitter). The interferometric sequence is shown in Fig. 2.20, where the two atomic trajectories (path A and path B) are represented in a plot showing the vertical position as a function of time: the straight lines represent the zero-gravity trajectories, while the curved lines represent the real trajectories in the presence of the gravitational field. The three Raman laser pulses are separated by the same time interval T. The fraction of atoms in one of the two hyperfine states (labelled as $|a\rangle$ and $|b\rangle$ in figure) is then detected.

The probability of detecting the atom in one of the two states depends on the differential phase shift acquired by the atom in the two possible paths.[26] A first contribution to the differential phase shift comes from the free evolution of the atom wavefunction between the laser pulses. According to Feynman's path integral formulation of quantum mechanics, this phase can be calculated as

$$\Delta\phi_{path} = (S_B - S_A)/\hbar \qquad (2.18)$$

where $S_{A,B}$ are the actions calculated along the classical trajectories as integrals of the Lagrangian for the evolution in a gravitational field $L[\dot{z}, z] = m\dot{z}^2/2 - mgz$:[27]

[26]We note that in this type of atom interferometer the internal state of the atom is entangled with its motional state and the former can be used to probe the external phase with common spectroscopic techniques.

[27]This approach is valid in the limit in which the classical action is much larger than $\hbar$.

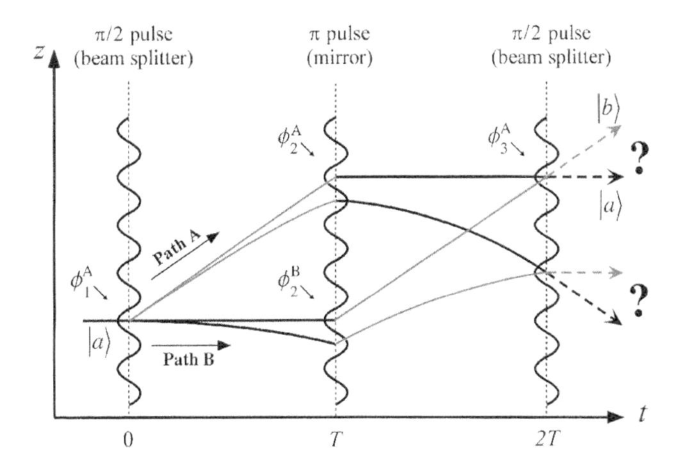

Fig. 2.20 Scheme of the interferometric sequence used in Peters *et al.* (1999). Three Raman pulses, applied at times $t = 0$, T and $2T$, are used to coherently manipulate the atom wavefunction by changing the atomic momentum along the vertical direction $\hat{z}$. After the first splitting, the atom follows two spatially separated trajectories (path A and path B), which recombine at the last $\pi/2$ pulse. Reprinted with permission from Peters *et al.* (2001).

$$S_{A,B} = \int_0^{2T} L\left[\dot{z}_{A,B}(t), z_{A,B}(t)\right] dt \; . \tag{2.19}$$

It is possible to show that if g is uniform the total shift in eqn (2.18) is zero, since the phase shifts $S_{A,B}/\hbar$ exactly cancel for the two trajectories.

There is another contribution to the phase shift which comes from the phase imprinted by the laser pulses when the atom changes its internal state at times $t = 0$, T, $2T$. When this phase shift is considered, the differential phase shift turns out to be

$$\Delta\phi_{light} = 2kgT^2 \; . \tag{2.20}$$

Basically the interferometer probes the differential phase of the light field at the two heights corresponding to the two trajectories at the time of the π pulses (Peters *et al.*, 2001; Cohen-Tannoudji and Guéry-Odelin, 2011). From eqn (2.20) it is evident that the sensitivity of the interferometer dramatically increases with the time separation between the pulses as T^2. The time separation $T = 160$ ms used in Peters *et al.* (1999), combined with the wavelength $\lambda = 2\pi/k = 852$ nm of the Raman lasers, yields a phase shift of 3.7×10^6 rad.

Figure 2.19b shows the measured fraction of atoms in one of the two internal states for a typical interferometric sequence in which the phase of one of the Raman pulses is arbitrarily changed across an interval of 4π, producing a clear fringe pattern. The absolute phase of the fringe pattern at the centre of the plot (no phase shift between the laser pulses) then provides the value of the phase shift induced by the gravitational field, which can be evaluated by comparing the measurement with eqn (2.20). From

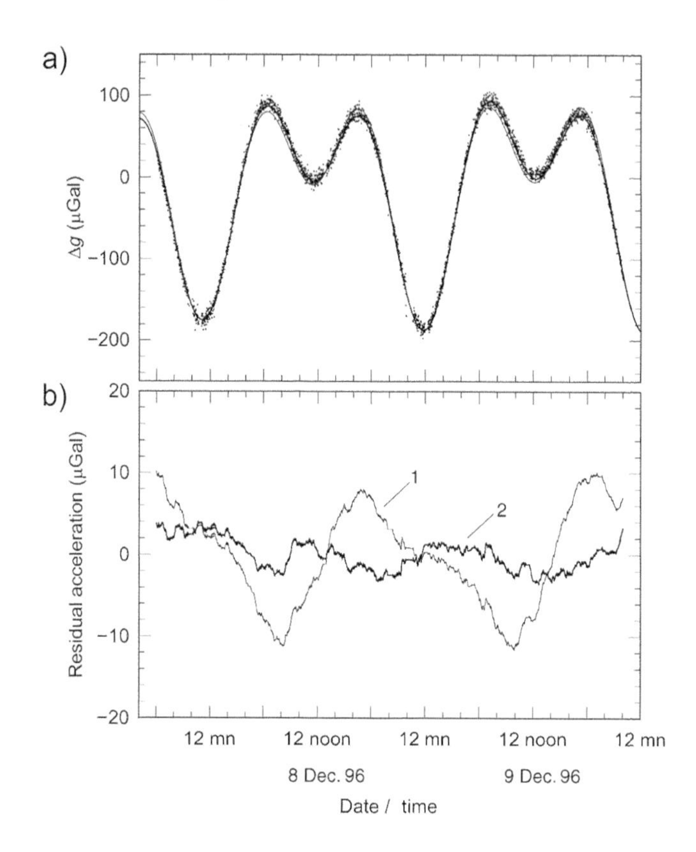

Fig. 2.21 a) Atom-interferometric measurement of g over a time interval of three days (1 μGal $= 10^{-8}$ m/s^2). b) Residuals of the fit of the experimental data to two tidal models, one considering the Earth as a solid elastic object (1), the other taking into account the ocean loading effect. Reprinted with permission from Peters *et al.* (1999). © Macmillan Publishers Ltd.

this analysis the authors of Peters *et al.* (1999) demonstrated a fractional uncertainty $\Delta g/g = 3 \times 10^{-9}$ after an integration time of 1 minute.

This interferometric technique for the measurement of g has also demonstrated a very good long-term stability. Figure 2.21a shows the results of a continuous measurement of g over a period of three days. The data clearly show a daily oscillation of the measured value of g, which has been fitted by the predictions of two different models of the Earth gravity. The residuals of the two fits are shown in Fig. 2.21b and clearly show that model 2, taking into account the effects of ocean loading of the Earth[28] fits better than model 1, where the Earth is modelled as a solid elastic object.

The authors of Peters *et al.* (1999) also compared the value of g obtained from the atom interferometer with the g measurement provided by a commercial gravimeter

[28]This is the deformation of the Earth's surface arising from tidal variations of the ocean pressure pushing on the ocean floor.

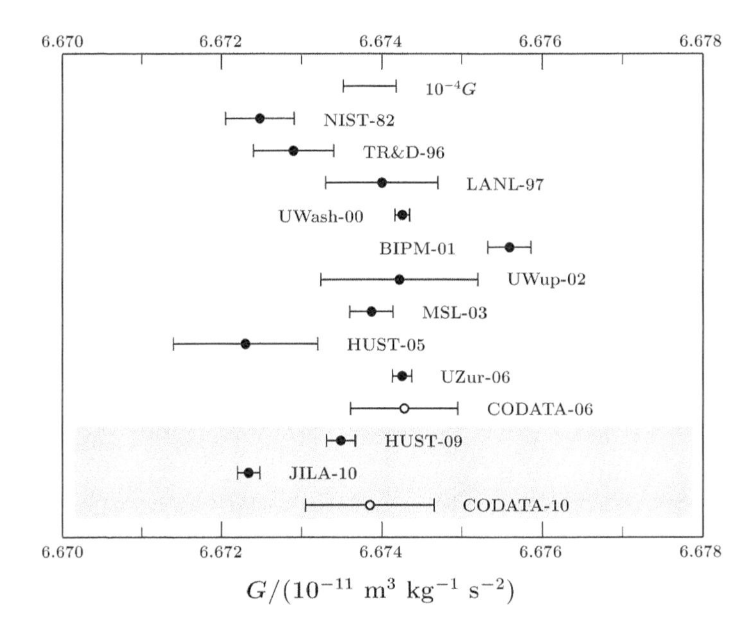

Fig. 2.22 Values of the Newtonian gravitational constant G considered by the CODATA in the latest adjustment of the fundamental constants. Reprinted with permission from Mohr *et al.* (2012). © American Physical Society.

based on the free fall of a corner cube in a Michelson-type optical interferometer. The comparison between the two values resulted in a relative difference of (7 ± 7) ppb (parts per billion), which is consistent with a zero difference between the acceleration of a macroscopic object and the acceleration of a quantum object like an atom. These measurements thus provide a precise test of the Galilean "universality of free fall" principle (also known as "weak equivalence principle"), which states that all the particles at the same time and position in a given gravitational field undergo the same acceleration, independent of their mass, composition, and properties.

Gradiometers and Newtonian constant. The amazing sensitivity of atom interferometers for the measurement of g allowed the possibility of measuring tiny variations in this quantity, arising either from the non-uniformity of the Earth gravitational field or from the presence of source masses modifying it.

A measurement of the Earth gravity gradient was performed by the group of M. A. Kasevich (Snadden *et al.*, 1998) by using a fountain interferometer in which two different cold atomic clouds were produced in two different positions, separated by ~ 1 m along the vertical direction. The two clouds were launched upwards together and were exposed to the same interferometric sequence driven simultaneously by the same Raman lasers, which made the system largely insensitive to systematic effects (e.g. vibrations or laser noise) thanks to common-mode rejection. The difference in absolute phase between the fringes observed for the upper and lower cloud allowed the authors of Snadden *et al.* (1998) to measure a local gravity gradient $\Delta g/\Delta z = (-3.37 \pm 0.18) \times 10^{-6}$ s^{-2}.

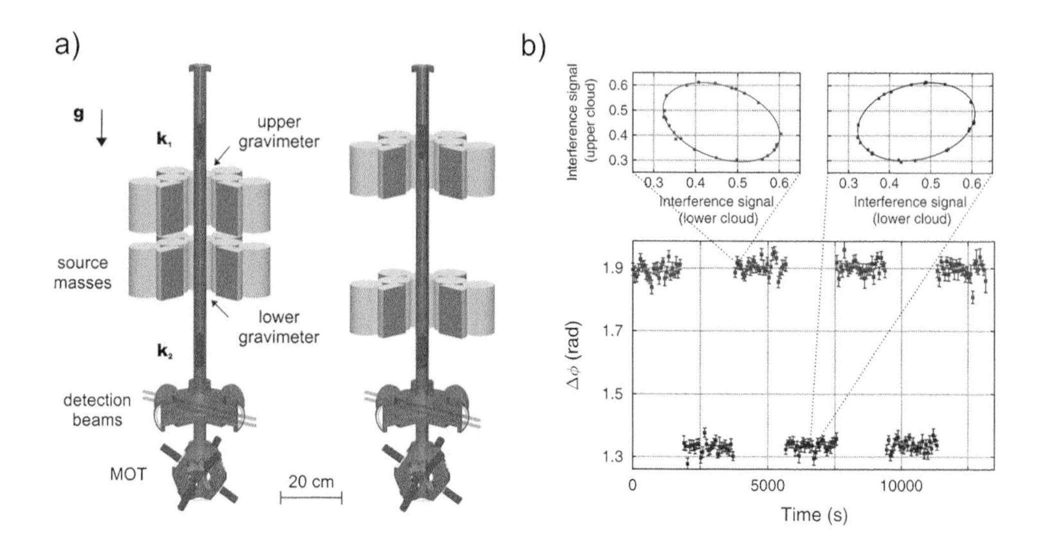

Fig. 2.23 a) Experimental setup for the measurement of the Netwonian graviational constant G: two source masses can be moved along the double gravimeter to determine differences in the local gravity acceleration g. b) Shift of the interferogram phase $\Delta\phi$ (lower graph) as the source masses are moved between two target positions along the interferometer axis. Reprinted with permission from Lamporesi *et al.* (2008). © American Physical Society.

Atomic gradiometers can also be used to measure the Newtonian constant of gravity G. Among the fundamental constants of Nature, G is the one known with least accuracy, owing to the extreme weakness of the gravitational force compared to the other fundamental interactions. The current CODATA value for the G constant is $G = 6.67384(\pm 0.00080)\ \mathrm{m}^3\ \mathrm{kg}^{-1}\ \mathrm{s}^{-2}$, obtained as the weighted mean of eleven different measurements performed with torsion balances and other mechanical devices (Mohr *et al.*, 2012). As evident from Fig. 2.22, the fractional uncertainty of only 1.2×10^{-4} is limited by the large discrepancy between existing measurements, which in several cases is much larger than the standard error of the single measurements, presumably owing to some hidden systematic effect. It is also interesting to observe that the new experimental values reported after the CODATA 2006 adjustment (shaded in grey in Fig. 2.22) are in disagreement with each other and with previous determinations of G, which led CODATA to increase the uncertainty of the 2010 value with respect to the 2006 one. For this reason different experimental techniques for the measurement of G are highly desirable.

First proof-of-principle experiments for the atom-interferometric determination of G were carried out by the group of G. M. Tino in Florence (Bertoldi *et al.*, 2006) and in the one of M. A. Kasevich at Stanford (Fixler *et al.*, 2007) by measuring the difference in gravity gradient induced by large Pb source masses moved at different heights along the interferometer chamber. Comparing the phase shifts measured at different positions of the masses (assuming the validity of the Newtonian gravitational law), the two groups were able to provide a value for the gravitational constant G with

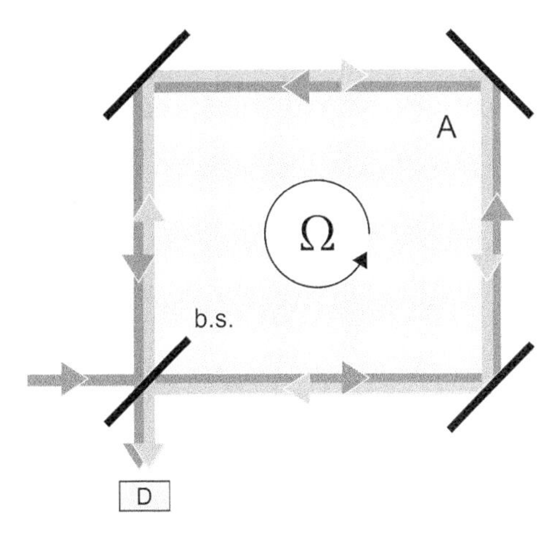

Fig. 2.24 Scheme of an optical Sagnac interferometer. If the interferometer plane is rotating at an angular velocity Ω, the two light beams travelling in opposite directions in the interferometer have different propagation times and arrive at the detector with different optical phases, which can be detected as a fringe shift.

a fractional uncertainty on the order of 1%, limited by both the statistical error and by the evaluation of systematic uncertainties affecting the measurement (mainly due to the calculation of the modified atom trajectories). A first precision measurement of G was reported by the Florence group in Lamporesi *et al.* (2008) using an improved experimental setup with different tungsten source masses (shown in Fig. 2.23). The geometry, composition, and position of the source masses were accurately characterized in order to precisely calculate the modification of the gravitational field experienced by the atoms. The comparison between this modelling and the measured phase shifts resulted in a value $G = 6.667(\pm 0.011)(\pm 0.003) \times 10^{-11}$ m^3 kg^{-1} s^{-2}, which was primarily limited by statistics. Even if the precision is currently about one order of magnitude worse than the CODATA value, these interferometric techniques are very promising and, in the near future, may provide new and competitive values of G with a target fractional accuracy below 10^{-4}.

2.5.2 Interferometers for inertial forces

An atom interferometer can be used to accurately measure real as well as inertial forces, which result from the non-inertiality of the reference frame in which the interferometer is operated. The accurate knowledge of inertial forces is very important for the operation of inertial navigation systems, in which the position of a moving object (a ship, an aeroplane, a spacecraft) is inferred from the continuous measurement of dynamical quantities, by means of accelerometers (sensing linear accelerations) and gyroscopes (measuring angular velocities).

High-performance gyroscopes often mounted on aeroplanes or spacecrafts are based on the optical *Sagnac effect* (Sagnac, 1913). This effect is based on the effective change of length that light experiences in an interferometer as it rotates around an axis which is orthogonal to the interferometer plane. An illustration of the Sagnac interferometer is given in Fig. 2.24. A light beam is divided by a beam splitter in two beams which run in a closed loop in opposite directions before they recombine again at the beam splitter, where a photodetector measures the total intensity. When the interferometer is put into rotation, it takes different times for the two beams to travel along the same path, depending whether they are travelling in the same or in the opposite direction with respect to the rotation. These different propagation times result in a phase shift, which is detected as a shift of the interference fringes forming at the detector. It is possible to demonstrate that the phase shift at the output port of the interferometer is given by

$$\Delta\phi = \frac{8\pi}{c\lambda}\mathbf{\Omega}\cdot\mathbf{A}\,, \tag{2.21}$$

where λ is the wavelength, c is the speed of light, $\mathbf{\Omega}$ is the angular velocity vector and $\mathbf{A}$ is the area enclosed by the interferometer (the direction of the vector being orthogonal to the interferometer plane).

A similar effect holds for matter waves as well, provided that one substitutes the speed of light with the atom velocity v and the laser wavelength with de Broglie wavelength $h/(mv)$:

$$\Delta\phi = \frac{8\pi m}{h}\mathbf{\Omega}\cdot\mathbf{A}\,. \tag{2.22}$$

This result illustrates the dramatic enhancement of sensitivity that matter-wave interferometers may provide on light interferometers for the measurement of inertial forces, which is connected with the massive nature of atoms and the typically much smaller associated wavelength. As a matter of fact, the phase shift in eqn (2.22) is larger by a factor $mc\lambda/h = mc^2/h\nu \approx 10^{11}$ (values for ^{133}Cs) than the phase shift for a laser interferometer enclosing the same area and rotating at the same speed. Despite the enclosed area of atom interferometers (typically a few tens of mm^2) being smaller than the area of fibre-optic gyroscopes (portable devices may have an effective area of several tens of m^2), atom inteferometers can still provide an increase in sensitivity for rotation sensing by several orders of magnitude.

The Sagnac effect for matter waves was first evidenced at PTB in 1991 (Riehle *et al.*, 1991) using thermal beams of Calcium excited on a narrow intercombination transition (see Chapter 5.1 for extended discussions on alkaline-earth atoms). The Sagnac effect was revealed by observing that the optical Ramsey fringes detected in the upper state population as a function of the laser frequency were offset when the system was put into rotation, as predicted in Bordé (1989). This effect was then applied to the demonstration of atomic gyroscopes in 1997 by the group of D. Pritchard (Lenef *et al.*, 1997) and by the group of M. A. Kasevich (Gustavson *et al.*, 1997).

The schematics of the experiment reported in Gustavson *et al.* (1997) is shown in Fig. 2.25a. A thermal beam of caesium atoms enters a Mach–Zehnder interferometer based on optical Raman splitting, such as the interferometer described in Sec. 2.5.1. However, differently from the latter, this interferometer lies in the horizontal plane and the optical pulses transfer momentum along a direction orthogonal to the atomic beam in order to give the inteferometer a finite area. In order to extract the Sagnac phase, the axis of the Raman pulses was put into rotation in order to simulate the effect of a

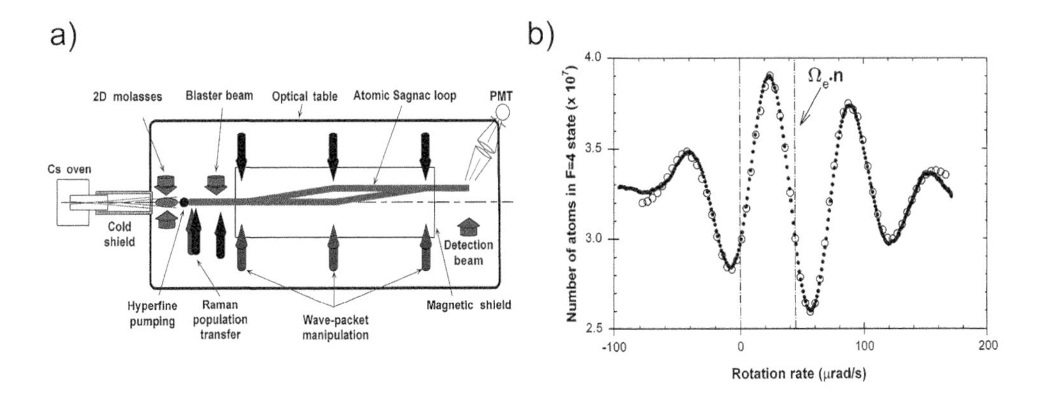

Fig. 2.25 Matter-wave gyroscope. a) Scheme of the experimental setup used in Gustavson *et al.* (1997): a thermal beam of caesium atoms enters a Mach–Zehnder interferometer based on optical Raman splitting and oriented along the horizontal plane. b) Interference fringes measured as a function of the angular velocity of the setup put in rotation along a vertical axis: the centre of the fringes envelope (indicated by the arrow) provides a measurement of the projection of the Earth rotation velocity along the normal to the interferometer plane. Reprinted with permission from Gustavson *et al.* (1997). © American Physical Society.

change in angular velocity and detect the absolute phase of the interferogram. Figure 2.25b shows the number of caesium atoms detected in the $F = 4$ hyperfine state as a function of the rotation rate of the laser setup. The interferogram is not centred at a zero rotation rate owing to the rotation of the Earth, which makes the lab frame non-inertial. By fitting the observed fringes with the result of a theoretical model the authors of Gustavson *et al.* (1997) were able to measure a rotation rate of 45 ± 3 μrad/s, which was consistent with the projection of the Earth angular velocity $\mathbf{\Omega}_e \cdot \mathbf{n} = 44.2$ μrad/s at the latitude of the experiment. This setup was then perfected in Durfee *et al.* (2006), where measurements on two counterpropagating atomic beams allowed several systematic effects to be cancelled thanks to common-mode rejection, resulting in a stability of 67 μdeg/h = 0.25 nrad/s with an integration time of 1.7×10^4 seconds, which was more than one order of magnitude better than commercial navigation-grade ring laser gyroscopes.

3

Bose–Einstein condensation

Besides the enormous opportunities disclosed to spectroscopy and atom interferometry, laser cooling has opened up the possibility of experimentally realizing and manipulating quantum degenerate gases. An ideal gas of identical quantum particles with mass m in equilibrium at a temperature T is characterized by a *thermal de Broglie wavelength*

$$\lambda_{dB} = \frac{h}{\sqrt{2\pi m k_B T}} , \tag{3.1}$$

which, in a simplified picture, quantifies the average spatial extent of the atomic wavepackets. Decreasing the temperature of the gas, the particles slow down and the thermal de Broglie wavelength increases. When λ_{dB} becomes comparable with the mean interparticle distance $n^{-1/3}$ (where n is the atomic density), the overlap between the wavepackets cannot be neglected and the quantum indistinguishability of the particles becomes important. In this limit of large phase-space density $n\lambda_{dB}^3 \sim 1$ quantum statistical mechanics predicts a quite different behaviour depending on the quantum nature of the particles, whether they are bosons or fermions,[1] as schematically shown in Fig. 3.1 for the case of an ideal gas trapped in a harmonic potential well. In the case of bosons, a critical temperature T_C exists below which the particles start to macroscopically occupy the lowest-energy state and a *Bose–Einstein condensate* (BEC) forms. In the case of polarized fermions (i.e. identical fermions in the same spin state), multiple occupancy of the motional states is ruled out by the Pauli exclusion principle and the particles form a *degenerate Fermi gas* in which they occupy all the lowest-energy states available until they reach the Fermi energy E_F.

In Sec. 3.1 we will review the experimental techniques that, starting from laser-cooled atomic clouds, allow the production and detection of degenerate quantum gases. Section 3.2 will be devoted to an illustration of the properties of Bose–Einstein condensates of alkali atoms, including some of their applications to precise measurements. Section 3.3 will contain an introduction to the physics of atomic Fermi gases. In Sec. 3.4 we will consider quantum degenerate gases belonging to different chemical families, while Sec. 3.5 will be devoted to an introduction to the ongoing research with ultracold molecules.

3.1 Experimental techniques

Laser cooling alone is not sufficient to achieve Bose–Einstein condensation. As a matter of fact, there is a lower bound to the temperatures achievable with laser cooling: the

[1] An atom behaves as a composite boson if it is formed by an even number of spin-1/2 particles (electrons + protons + neutrons) while it behaves as a fermion if the number of its constituent particles is odd.

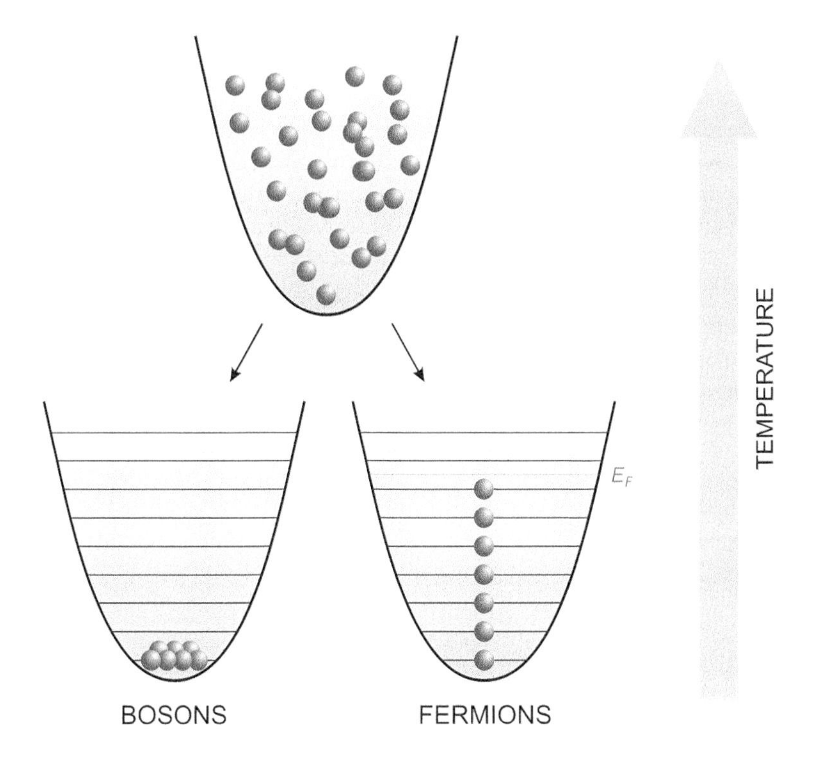

Fig. 3.1 Schematics of the level occupancy for a harmonically trapped quantum gas at ultralow temperatures. The ground state of the many-body system is completely determined by the quantum statistics. At zero temperature the bosons (even spin particles) all occupy the same single particle ground state forming a Bose–Einstein condensate. Spin-polarized fermions (odd spin particles), obeying the Pauli exclusion principle, pile up until they reach the Fermi energy E_F.

ultimate limit for cooling schemes based on radiation pressure from a single-photon excitation is on the order of the recoil temperature $T_R = \hbar^2 k^2 / m k_B$ (~ 350 nK for ^{87}Rb), corresponding to the minimal unit of kinetic energy acquired in a photon absorption process.[2] This limit, together with the low densities achievable with laser cooling ($\approx 10^{11}$ atoms/cm^3), makes a further cooling technique necessary for the achievement of quantum degeneracy.

This additional cooling stage is *evaporative cooling*, a technique initially developed for spin-polarized hydrogen (Masuhara *et al.*, 1988), which relies on the selective removal of the most energetic atoms of the sample and on the subsequent rethermalization of the remaining atoms at lower temperatures. This technique (described in Sec. 3.1.2) preliminarily requires the atoms to be trapped, with long storage times, in a confining

[2] More advanced laser-cooling techniques, such as Raman cooling or velocity-selective coherent population trapping (VSCPT) allow for even lower temperatures, but not as low as it would be needed to get to the quantum degenerate regime. For a discussion of these techniques, based on multi-photon transitions and coherent effects in multi-level systems, see e.g. Metcalf and van Der Straten (1999).

potential. The solution adopted in the first BEC experiments relied on the use of *magnetic traps*, which are based on the interaction between the atomic magnetic moment and an inhomogeneous magnetic field created by suitable arrangements of current-carrying coils, as will be described in the next section. A different approach to atom trapping will be presented in Sec. 5.2, devoted to the realization of *optical dipole traps*, earlier developed for magnetically untrappable atomic states, which offer a wider landscape of trapping possibilities also for atoms which can be magnetically confined.

3.1.1 Magnetic traps

Magnetic trapping is based on the interaction of the atomic magnetic dipole moment $\boldsymbol{\mu}$ with an inhomogeneous magnetic field $\mathbf{B}(\mathbf{r})$. From classical electrodynamics, this interaction is described by a potential energy

$$U(\mathbf{r}) = -\boldsymbol{\mu} \cdot \mathbf{B}(\mathbf{r}) \; . \tag{3.2}$$

All the alkali atoms can be magnetically trapped, since they all possess a permanent nonzero magnetic moment in the ground state, owing to their single-electron hydrogen-like structure. The magnetic moment can be written as

$$\boldsymbol{\mu} = -g_F \mu_B \frac{\mathbf{F}}{\hbar} \; , \tag{3.3}$$

where $\mathbf{F}$ is the total angular momentum, $\mu_B = e\hbar/2m_e$ is the Bohr magneton, and g_F is the hyperfine Landé factor, which in ground state $^2S_{1/2}$ alkali atoms can be written as a function of the nuclear spin I and of the total angular momentum F quantum numbers as[3]

$$g_F \simeq \frac{F(F+1) - I(I+1) + \frac{3}{4}}{F(F+1)} = \pm \frac{1}{I + \frac{1}{2}} \; , \tag{3.4}$$

where the $+$ sign holds for the $F = I + 1/2$ hyperfine state and the $-$ sign holds for the $F = I - 1/2$ state. Combining eqns (3.2) and (3.3) we obtain

$$U(\mathbf{r}) = g_F m_F \mu_B B(\mathbf{r}) \; , \tag{3.5}$$

where m_F is the quantum number denoting the projection of the total angular momentum along the quantization axis fixed by the magnetic field direction. If the field direction is spatially changing, the above equation still holds provided that the atom is moving slowly enough in such a way that m_F can be considered as a constant. Classically, this happens when the variation of the magnetic field orientation experienced by the atom in its moving frame takes place on a timescale much slower than the Larmor precession frequency $\omega_L = g_F e B/2m_e$, in such a way that the dipole precession axis can adiabatically follow the change in magnetic field orientation.

Not all the m_F states can be magnetically trapped. In classical electrodynamics it is possible to demonstrate that a region of space free from currents cannot support a local

[3]Here we neglect the contribution of the nuclear spin, which is coupled much more weakly to the magnetic field, owing to the larger mass of the nucleus, which is on the order of the proton mass m_p (the mass of the proton enters the definition of the nuclear magneton $\mu_N = e\hbar/2m_p$).

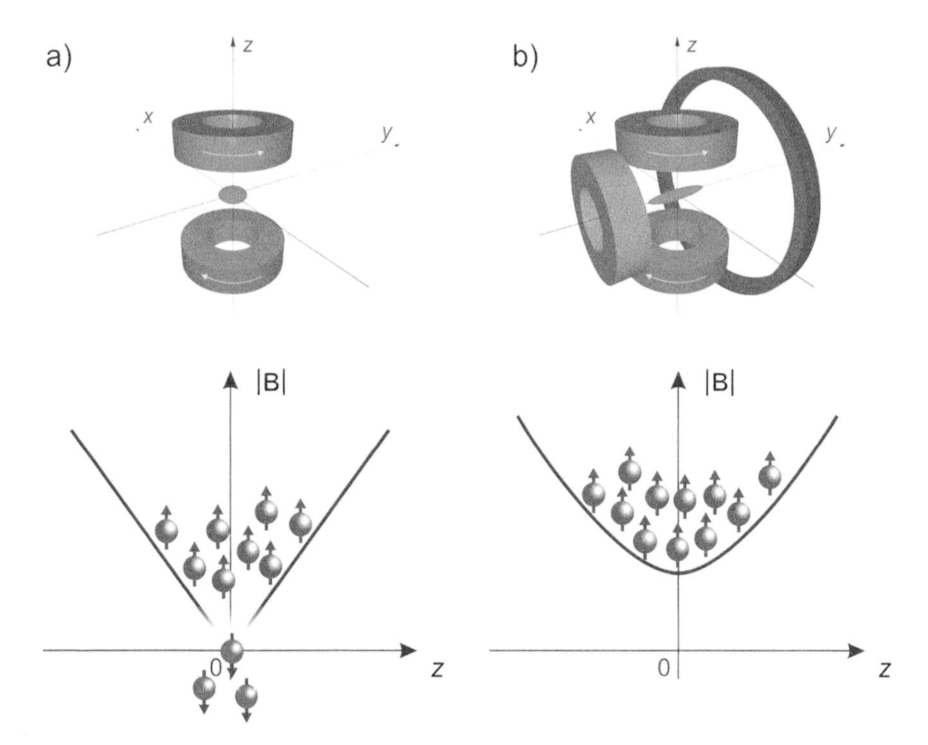

Fig. 3.2 Different schemes of magnetic traps. a) Quadrupole trap formed by two coils fed with oppositely circulating currents: the trap centre coincides with a vanishing magnetic field, representing a "Majorana hole" where losses of atoms due to non-adiabatic spin-flip transitions towards untrapped states can take place (see text). b) Four-coils magnetic trap used in one of the experiments at LENS (Florence): the modulus of the magnetic field varies harmonically around a nonzero field minimum (Ioffe–Pritchard configuration), which eliminates Majorana losses.

magnetic field maximum, but only a local field minimum (Wing, 1994). Hence, in order to be trapped in a minimum of the potential in eqn (3.5), the atoms should be polarized in a *low-field-seeking state* m_F chosen in such a way to make the product $g_F m_F$ positive and different from zero (corresponding to a magnetic dipole $\boldsymbol{\mu}$ *anti*-aligned with respect to the field $\mathbf{B}$).[4]

The simplest example of magnetic trap is the *quadrupole trap* shown in Fig. 3.2a, formed by two coils with the same axis $\hat{z}$ and fed with oppositely circulating currents (the same coil configuration which is used to create the field gradient in a magneto-optical trap). At the midpoint between the two coils the magnetic field is zero and its modulus increases linearly around this position along any direction, according to

$$B(\mathbf{r}) \simeq b\sqrt{2z^2 + x^2 + y^2}\,, \tag{3.6}$$

[4]For ^{87}Rb the magnetically trappable states are $|F = 1, m_F = -1\rangle$ (for which $g_F = -1/2$) and $|F = 2, m_F = 1\rangle$, $|F = 2, m_F = 2\rangle$ (for which $g_F = +1/2$), see eqn (3.4).

where b is the field gradient along a direction orthogonal to the $\hat{z}$ axis. This trap configuration was used in the first experimental demonstration of magnetic trapping, which was reported in Migdall *et al.* (1985) for a beam of sodium atoms optically decelerated in a Zeeman slower. A pre-cooling stage is indeed a fundamental requirement for loading magnetic traps, since the trap depth (i.e. the maximum energy for a particle to be trapped) is typically on the order of tens of mK with convenient magnetic fields.

Despite its simple experimental implementation, the quadrupole trap presents a disadvantage, which becomes particularly serious when the trapped atoms are very cold and move in close proximity to the trap centre. When the atoms move in a region with vanishing magnetic field, the adiabatic condition of (anti-)alignment between spin and magnetic field becomes more difficult to be maintained, since the magnetic field direction is inverted after crossing the trap centre. As a matter of fact, as the precession of the spin becomes slower (the Larmor precession frequency ω_L vanishes for $B \to 0$), the spin cannot adiabatically follow the inversion of magnetic field direction and loses its alignment. Quantum-mechanically speaking, this means that the atoms can make spin-flip transitions towards untrapped states with different m_F and, eventually, be expelled from the trap. This effect had been first addressed in 1932 by the Italian physicist E. Majorana in a seminal paper (Majorana, 1932) (recently translated and published in Bassani (2006)) devoted to the theoretical description of Stern–Gerlach experiments with rapidly varying magnetic fields. After his early contribution to the formulation and solution of this problem, losses of atoms from a quadrupole trap owing to breakdown of adiabaticity are called *Majorana losses*.

Different strategies can be implemented in order to solve this problem, which was significantly affecting the first experiments aimed at the achievement of Bose–Einstein condensation. In the JILA experiment E. A. Cornell and C. E. Wieman (Cornell and Wieman, 2002) eliminated Majorana losses by inventing a trap configuration called *TOP trap* (time-averaged orbiting potential), in which the centre of the quadrupole trap was rapidly moved around a circle by a time-dependent bias field (Petrich *et al.*, 1995). If the frequency of this motion is much larger than the oscillation frequency of the atoms in the quadrupole trap, the atoms experience an average trapping potential centred at a nonzero average magnetic field, avoiding the problem of non-adiabatic spin-flip transitions (Fig. 3.3 shows notes handwritten by E. A. Cornell containing the first ideas for the TOP trap). W. Ketterle and his group at MIT followed a different strategy (Ketterle, 2002): the "Majorana hole" was filled by shining a blue-detuned *plug* laser beam in the centre of the quadrupole trap, in such a way as to keep the atoms far from the zero of the magnetic field by the optical dipole force (that will be discussed in Sec. 5.2).

A different possibility consists in the realization of magnetostatic traps formed by more complex arrangements of coils in order to have a trap minimum at a magnetic field different from zero. An improved version of the quadrupole trap, which solves the problem of Majorana spin-flip losses, is shown in Fig. 3.2b. This kind of magnetostatic trap, used in one of the experiments at LENS (Florence), is a modification of the QUIC traps originally implemented in Söding *et al.* (1998) and Esslinger *et al.* (1998). The magnetic field produced in this trap shares the same features as the one obtained with the Ioffe–Pritchard configuration proposed in Pritchard (1983): it is a static field with

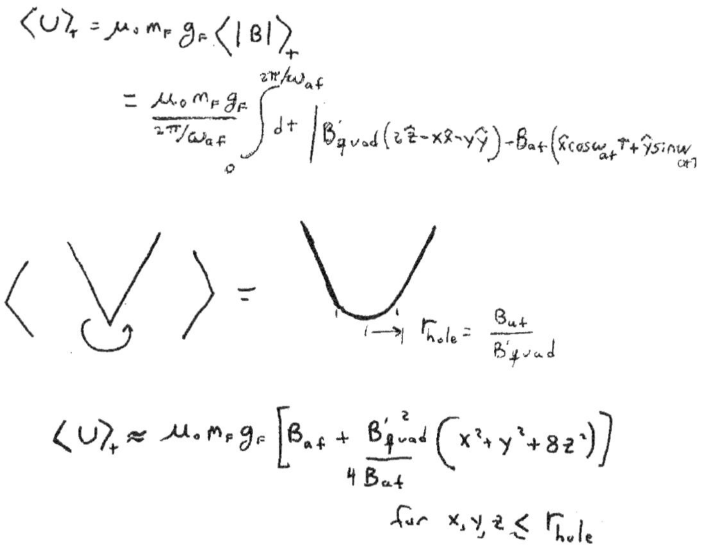

$$\langle U \rangle_t = \mu_0 m_F g_F \langle |B| \rangle_t$$

$$= \frac{\mu_0 m_F g_F}{2\pi/\omega_{af}} \int_0^{2\pi/\omega_{af}} dt \left| B'_{quad}(z\hat{z} - x\hat{x} - y\hat{y}) - B_{af}(\hat{x}\cos\omega_{af}t + \hat{y}\sin\omega_{af}t) \right|$$

$$r_{hole} = \frac{B_{af}}{B'_{quad}}$$

$$\langle U \rangle_t \approx \mu_0 m_F g_F \left[B_{af} + \frac{B'^2_{quad}}{4 B_{af}} \left(x^2 + y^2 + 8z^2 \right) \right]$$

for $x, y, z \lesssim r_{hole}$

Fig. 3.3 Handwritten notes by E. A. Cornell illustrating the principle of operation of the TOP trap. In this configuration a rapidly rotating quadrupole trap provides an average trapping potential centred around a nonzero magnetic field minimum, thus avoiding Majorana losses. Reprinted from Inguscio (2006) by courtesy of E. A. Cornell.

a well-defined direction and its modulus varies harmonically around a nonzero local minimum B_0, which avoids the complication of Majorana losses:

$$B(\mathbf{r}) \simeq B_0 + \frac{1}{2} m\omega_x^2 x^2 + \frac{1}{2} m\omega_y^2 y^2 + \frac{1}{2} m\omega_z^2 z^2 \,. \tag{3.7}$$

We note that magnetic trapping (as well as magneto-optical trapping, albeit in a less stringent way) must be performed in an ultra-high-vacuum environment (typically 10^{-10} mbar background pressure or below), in such a way as to ensure a perfect thermal isolation of the ultracold trapped samples by minimizing the collisions with the highly energetic room-temperature atoms of the background gas. This is an important requisite for the implementation of evaporative cooling.

3.1.2 Evaporative cooling

The final cooling stage in the route to Bose–Einstein condensation is *evaporative cooling*. This technique, initially successfully implemented on spin-polarized hydrogen (Masuhara *et al.*, 1988), relies on the forced removal of the atoms with the largest kinetic energy and on the subsequent rethermalization of the remaining atoms induced by atom–atom collisions. In a magnetic trap the forced evaporation can be obtained by using energy-selective radiofrequency (RF) or microwave (μW) transitions towards untrapped states. Since the atoms are trapped in an inhomogeneous magnetic field,

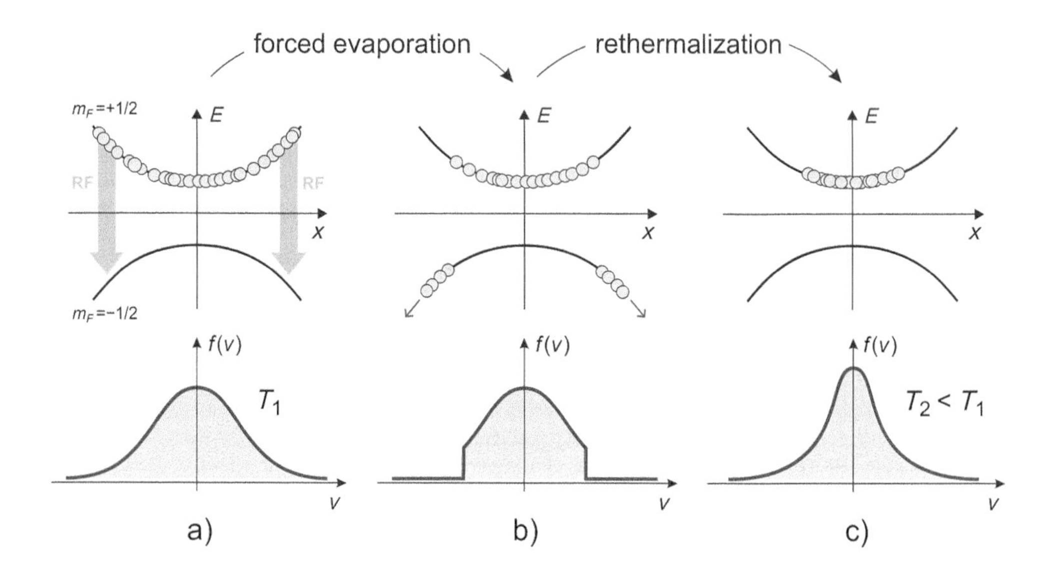

Fig. 3.4 Scheme of RF evaporative cooling in the simplified case of an atom with two Zeeman states $m_F = +1/2$ (low-field seeking) and $m_F = -1/2$ (high-field seeking). The graphs in the top row show the energy of the particles as a function of their position in the trap, while their velocity distributions are reported in the lower row. a) The warmest atoms are removed from the trap by selectively exciting them into the untrapped state. b) The resulting velocity distribution after this forced evaporation is a truncated Maxwellian. c) Collisions between atoms then allow a thermalization of the sample at a lower temperature.

their Zeeman shift is position-dependent (this is the principle of operation of any magnetic trap). Hence, a radiofrequency field with narrow linewidth can be used to excite the transition from a trapped Zeeman state to an untrapped state only in a given region of the trap, the one for which the energy of the RF photon matches the Zeeman shift between the levels. Initially, the velocity distribution of the trapped atoms is given by the classical Maxwellian distribution $f(v)$ for an ensemble in thermal equilibrium at $T = T_1$, as sketched in Fig. 3.4a. The most energetic atoms of the ensemble have higher mean velocities and can reach the outer regions of the trap potential. By applying a radiofrequency field resonant with the transition for atoms in the external part of the cloud, it is possible to remove only the most energetic ones. The resulting velocity distribution is an out-of-equilibrium Maxwellian distribution with truncated wings (Fig. 3.4b). The trapped atoms then re-thermalize via atom–atom elastic collisions to a new Maxwellian velocity distribution characterized by a lower temperature $T_2 < T_1$ (Fig. 3.4c). By decreasing the energy of the RF photons it is possible to iterate this procedure and lower the temperature of the system, at the expense of a loss of atoms. If the initial number of trapped atoms and phase-space density are high enough, the latter increases during the process (Ketterle and van Druten, 1996) and at the end of the evaporative ramp it can reach the conditions for the onset of quantum degeneracy.

3.1.3 Sympathetic cooling

The presence of collisional processes is fundamentally important for the rethermalization of the sample that has to be guaranteed during the forced evaporation. In the absence of atom–atom interactions evaporative cooling cannot work, because the atoms cannot redistribute their velocities and reach thermodynamic equilibrium. The same happens if the cross section for elastic collisions is not sufficiently large and inelastic processes leading to losses of atoms from the trap are dominant. In the case of spin-polarized fermions at low temperatures the situation is even more dramatic, since low-energy s-wave collisions are suppressed by the antisymmetrization of the wavefunction (see Sec. 3.1.4). In all these cases different cooling strategies have to be implemented in substitution of evaporative cooling.

The most successful technique to avoid such problems is *sympathetic cooling*, in which one uses atoms of a different kind as a refrigerator: cooling can be achieved provided that the "coolant" species can be evaporatively cooled and that the cross section for inter-species elastic collisions is sufficiently large to allow for efficient rethermalization between the two species. This technique was initially proposed for ion cooling in Wineland *et al.* (1978), then it was experimentally demonstrated with laser-cooled ions in Drullinger *et al.* (1980) and Larson *et al.* (1986) (one of its applications for ion spectroscopy will be discussed in Sec. 5.5).

Sympathetic cooling of neutral atoms was initially demonstrated in experiments at JILA with ^{87}Rb atoms placed in two different internal states (Myatt *et al.*, 1997); then it was applied to mixtures of different isotopes of the same element, e.g. for cooling fermionic ^{6}Li by thermal contact with evaporatively cooled bosonic ^{7}Li (Schreck *et al.*, 2001*a*; Truscott *et al.*, 2001) or for cooling a gas of bosonic ^{85}Rb (which naturally exhibits hostile collisional properties to evaporative cooling) "by sympathy" with ^{87}Rb (Bloch *et al.*, 2001).

Sympathetic cooling is not limited to mixtures of atoms of the same chemical species. Heteronuclear sympathetic cooling was successfully demonstrated at LENS with mixtures of potassium and rubidium atoms and has allowed the first Bose–Einstein condensation of ^{41}K (Modugno *et al.*, 2001). In the LENS experiment ^{87}Rb was the "coolant" species which was forcedly evaporated by microwave transitions in order to cool the potassium cloud. Figure 3.5a shows the temperature of the ^{87}Rb and ^{41}K trapped clouds at different stages during the evaporation of ^{87}Rb (from larger microwave frequencies to smaller ones): it is evident how collisions are very efficient in keeping the temperature of the two gases the same. In Fig. 3.5b it is possible to observe that, whereas the number of rubidium atoms decreases because of the forced evaporation, the number of potassium atoms remains largely the same: this possibility of cooling without losses of atoms is a major advantage offered by sympathetic cooling.

Not only has sympathetic cooling revealed itself as a powerful technique for atomic species with unfavourable collisional properties, it has also opened a novel field of research focused on the investigation of ultracold degenerate mixtures of different atomic species. It has provided the key mechanism for the simultaneous production of double-species ^{87}Rb–^{41}K condensates at LENS (Modugno *et al.*, 2002) and for the demonstration of double condensates with Feshbach tuning of the interspecies interaction (Thalhammer *et al.*, 2008) (see Sec. 3.1.4). It has also provided the key

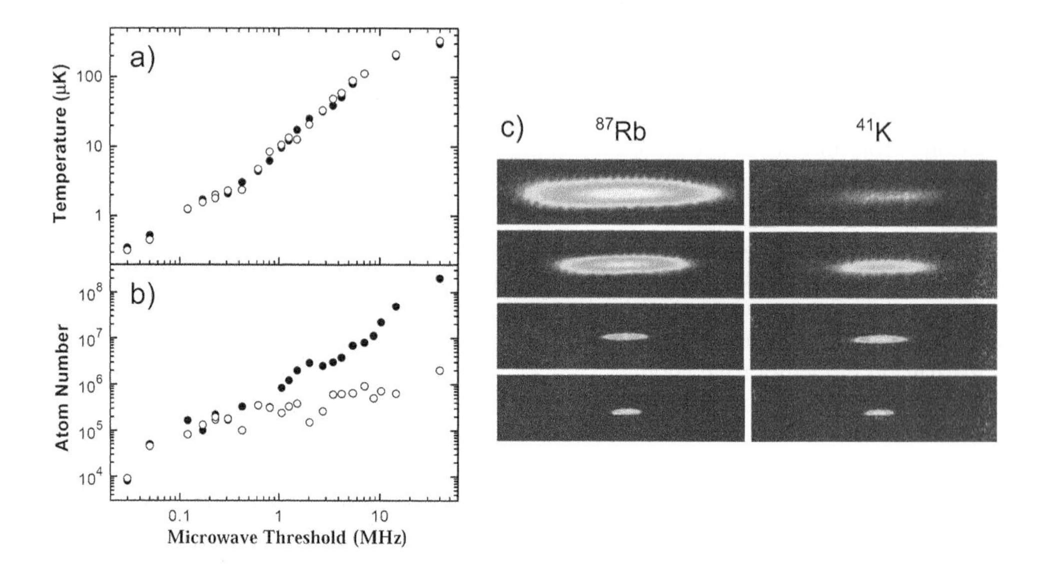

Fig. 3.5 Sympathetic cooling of ^{41}K with ^{87}Rb. a–b) Evolution of temperature and atom number during forced microwave evaporation of ^{87}Rb (solid circles) and sympathetic cooling of ^{41}K (empty circles). c) Absorption images of the trapped clouds at different stages of evaporative-sympathetic cooling (from top to bottom, 40 μK, 23 μK, 17 μK, and 0.9 μK). Adapted from Modugno *et al.* (2001).

cooling mechanism for the production of quantum degenerate Bose–Fermi mixtures with different atomic species, first realized with ^{23}Na–^{6}Li (Hadzibabic *et al.*, 2002) at MIT and with ^{87}Rb–^{40}K at LENS (Roati *et al.*, 2002).

3.1.4 Atom–atom interactions and Feshbach resonances

Evaporative and sympathetic cooling critically rely on the existence of interaction between the atoms, which permit the rethermalization of the gas as the atom removal proceeds. In neutral atomic gases no long-range Coulomb force is present and dipole–dipole interactions can be neglected in most of the experimentally relevant cases (with some exceptions that will be highlighted later in the text). The dominant interaction mechanism is represented by short-range forces that are described by a molecular potential such as the one depicted in Fig. 3.6a, featuring a hard-core repulsion at zero interatomic distance R and an attractive van der Waals $\sim 1/R^6$ tail at larger R. The typical range of the potential is on the order of several angstroms, which is much smaller than the average interparticle distance $d = n^{-1/3}$, on the order of 0.1 μm for a dense Bose–Einstein condensate with $n = 10^{14}$ atoms/cm^3. As a consequence, interactions can be modelled as collisions, occurring as "instantaneous" events when two atoms happen to be in the proximity of one another.

At the relatively low densities of cold trapped gases two-body interactions are the dominant interaction mechanism. Three-body interactions, becoming more and more important at larger densities, generally lead to inelastic collisions in which a molecule is

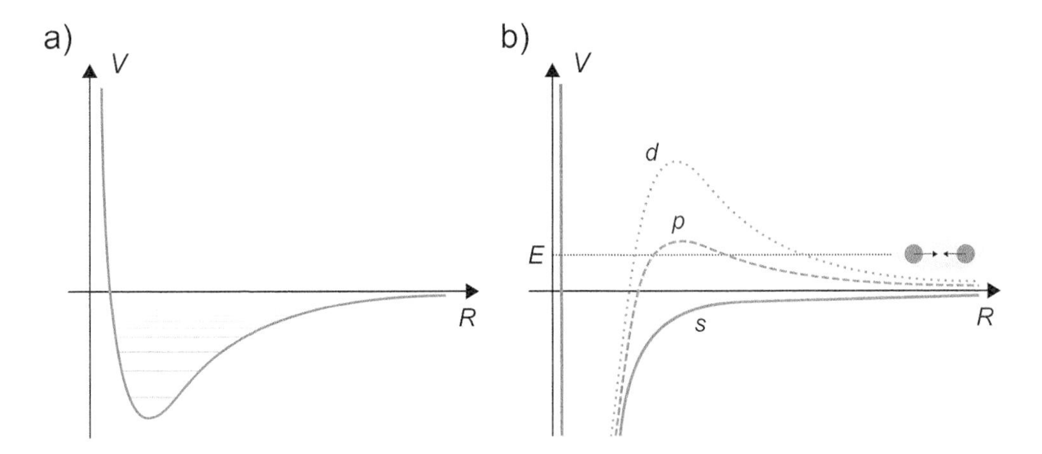

Fig. 3.6 a) Typical molecular potential $V(R)$ describing short-range interactions between two scattering atoms as a function of their distance R. b) Effect of the centrifugal barrier suppressing collisions at low temperature for partial waves with angular momentum $l > 0$.

formed between two collisional partners: the binding energy of the molecule is converted into kinetic energy of both the molecule and the third atom, which are all expelled from the trap (which is much shallower than the molecule binding energy). On the contrary, two-body interactions do not allow for molecule formation (since this violates conservation of energy and momentum), which explains why binary collisions between ground-state atoms are mostly elastic and generally do not result in atom losses.[5]

These elastic collisions can be described in terms of the quantum-mechanical theory of scattering (see e.g. Landau and Lifshitz (1977)). The wavefunction in the centre-of-mass coordinate system can be expanded into partial waves with different angular momenta l, usually denoted as s, p, d, ... Partial waves with nonzero angular momentum are responsible for an additional energy term in the Hamiltonian for the radial wavefunction, corresponding to the rotational kinetic energy of the atom pair $\hbar^2 l(l+1)/2\mu R^2$, where μ is the reduced mass of the two-body problem and R is the relative distance. This term represents a centrifugal barrier (shown in Fig. 3.6b for p and d waves) which limits the number of partial waves which are relevant for low-energy scattering. For ultracold atoms all the partial waves with $l \neq 0$ are energetically suppressed, so only s-wave scattering has to be considered. In the quantum-mechanical theory of scattering s-wave collisions are described by a scalar quantity a, the *scattering length*, which depends on the phase shift δ acquired by the wavefunction in the collision as $a = -\lim_{k\to 0}\tan(\delta(k))/k$ (where k is the relative momentum of the colliding particles). The scattering length quantifies the strength of the interactions, which, in the case of distinguishable particles, is described by a scattering cross section

$$\sigma = 4\pi a^2 \,. \tag{3.8}$$

[5]This is the case for two colliding atoms both in their absolute ground state. If one of them is in the upper hyperfine state, hyperfine-changing collisions can convert internal energy into kinetic energy, causing, again, escape of atoms from the trap.

The sign of the scattering length determines the character of the interactions: repulsive for $a > 0$, attractive for $a < 0$.

For distinguishable particles all partial waves are allowed to contribute to the partial wave expansion. For identical quantum particles, however, the wavefunction must obey the appropriate symmetry under exchange of the two particles, which coincides with the application of the parity operator to the wavefunction in the centre-of-mass coordinate system. Since partial waves have parity $(-1)^l$, it follows that identical bosons can scatter only in waves with even l (s, d, ...), while identical fermions can scatter only in waves with odd l (p, f, ...). Therefore, in the ultracold limit, where scattering in waves with $l \neq 0$ is energetically suppressed, identical bosons[6] are allowed to interact via s-wave collisions, while identical fermions are not:

$$\sigma(\text{bosons}) = 8\pi a^2 \tag{3.9}$$

$$\sigma(\text{fermions}) = 0 \,. \tag{3.10}$$

For this reason, Fermi gases can be cooled to the ultracold regime only by using mixtures of different spin states to make them distinguishable or by sympathetic cooling with a different species.

Feshbach resonances. Alkali atoms have a hyperfine structure, and the molecular potential describing their short-range interactions depends on their angular momentum quantum numbers $\{F, m_F\}$. When more collisional channels are considered, scattering resonances can arise from the resonant coupling of the incoming wave to quasibound molecular states in different collisional channels. These *Feshbach resonances*, originally predicted in the context of nuclear reactions (Feshbach, 1958) and by U. Fano in spectroscopy (Fano, 1961*a*), were proposed for ultracold atomic systems in Tiesinga *et al.* (1993) and now represent a powerful tool for controlling the interactions of ultracold gases, as clearly illustrated in the review paper (Chin *et al.*, 2010).

In Fig. 3.7a we consider a simple two-channel mode, in which the lower potential curve (*entrance* or *open* channel) refers to the atom internal state before the collision and the upper potential curve (*closed* channel) refers to different angular momentum quantum numbers. A Feshbach resonance occurs when the energy of a bound molecular state in the closed channel approaches the energy of the incoming atom pair in the entrance channel. When this happens, even in the presence of a very weak coupling between the two channels, the scattering properties of the atom pair are deeply modified and the scattering length exhibits a resonant behaviour. In magnetic Feshbach resonances the tuning is performed by applying an external magnetic field, which produces a relative energy displacement of the two potential curves in Fig. 3.7a, owing to their different magnetic moments and different Zeeman shift.[7] Figure 3.7b shows

[6]Because of the symmetrization of the wavefunction, for identical bosons the scattering cross section is amplified by a factor 2 with respect to the case of distinguishable particles.

[7]In *optical* Feshbach resonances the closed channel is represented by the molecular potential between a ground-state atom and an electronically excited atom. The optical Feshbach resonance can be scanned by changing the frequency of a coupling laser tuned in the proximity of a photoassociation resonance connecting the two channels (Fedichev *et al.*, 1996). Differently from magnetic Feshbach resonances, however, optical Feshbach resonances are accompanied by much larger losses due to spontaneous decay via the molecular state (Theis *et al.*, 2004).

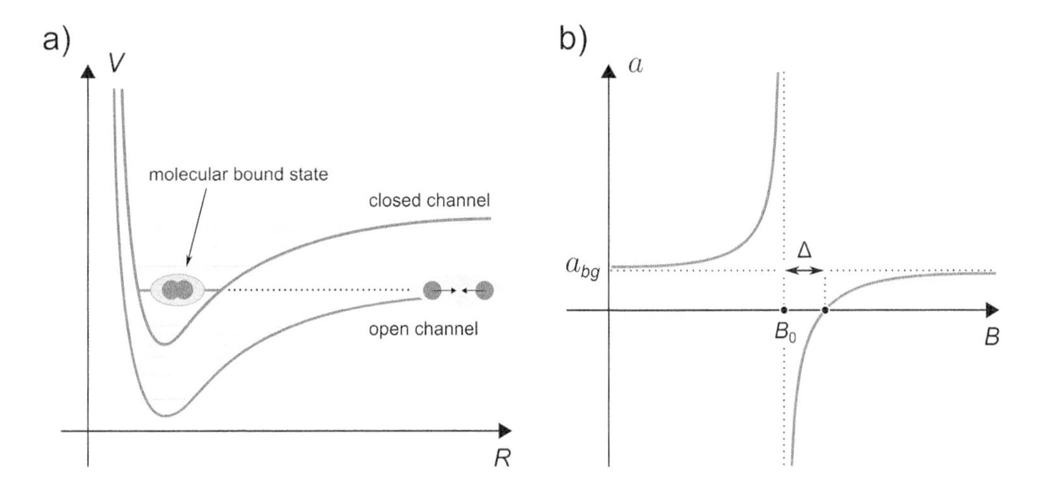

Fig. 3.7 a) A Feshbach resonance occurs when the energy of the scattering atom pair becomes quasi-resonant with the energy of a molecular bound state in a different scattering channel. b) As a consequence, the scattering length a shows a resonant behaviour as a function of the magnetic field B, which is the control parameter used for tuning the energy of the bound state in and out from resonance.

the behaviour of the scattering length around a Feshbach resonance as a function of the magnetic field B:

$$a(B) = a_{bg}\left(1 - \frac{\Delta}{B - B_0}\right) , \qquad (3.11)$$

where a_{bg} is the non-resonant background scattering length (associated to the open channel), B_0 is the magnetic field at the resonance, and Δ characterizes the resonance width (it corresponds to the difference between B_0 and the magnetic field value at which the scattering length is zero).

The first observation of Feshbach resonances in ultracold quantum gases was reported in Inouye *et al.* (1998) for a Bose–Einstein condensate of sodium atoms. The ultracold atoms were trapped in an optical dipole trap (see Sec. 5.2), which allows the possibility of adjusting freely the magnetic field used to access the Feshbach resonance. The scattering length value was determined by measuring the interaction energy of the cloud of atoms by performing time-of-flight detection (see Sec. 3.2.1). Figure 3.8b shows the measured value of the scattering length a as a function of the applied magnetic field, evidencing the expected resonant behaviour around a magnetic field value of 907 G. In addition to the change in scattering length, strong losses of atoms near the resonance were observed (Fig. 3.8a) as a result of inelastic collisions the rate of which is incremented as the two-body scattering length is increased.

3.1.5 Imaging ultracold atoms

The interaction between atoms and light is by far the most powerful tool providing information on the properties of atomic ensembles. The most immediate way to probe the state of ultracold gases is direct imaging of the atomic cloud, although spectroscopic

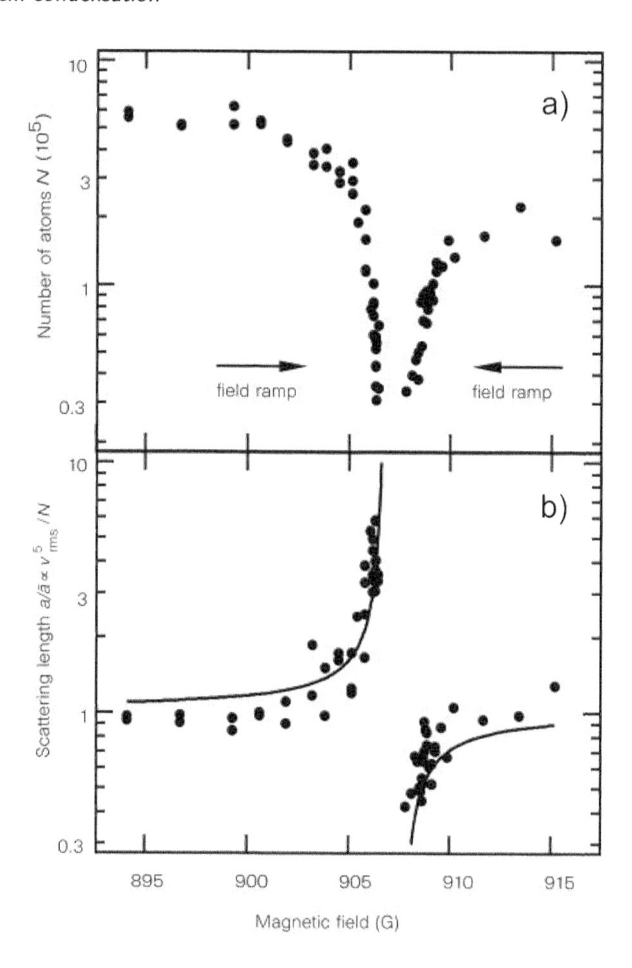

Fig. 3.8 Feshbach resonance in a sodium BEC. The graphs show the number of atoms remaining in the trap (a) and the measured value of the s-wave scattering length (b) for different values of the applied magnetic field. Reprinted with permission from Inouye *et al.* (1998). © Macmillan Publishers Ltd.

techniques can be implemented as well (see e.g. the paragraph devoted to hydrogen BEC in Sec. 3.4.1). Among the different imaging techniques, we can distinguish between *destructive* and *non-destructive* imaging. In the first class of techniques the atoms are illuminated by a beam of resonant light with intensity I_0, which is absorbed according to the usual Beer–Lambert law for the transmitted intensity I_t:

$$I_t = I_0 e^{-\sigma_0 \tilde{n}(z)} , \tag{3.12}$$

where $\sigma_0 = 3\lambda^2/2\pi$ is the resonant absorption cross section for a two-level atom (see Appendix A.3.1) and $\tilde{n}(z) = \int n(z)dz$ is the integral of the atomic density $n(z)$ along the beam propagation axis $\hat{z}$. In *absorption imaging* the transmitted beam with intensity I_t is then detected by a CCD camera, which takes a picture of the shadow cast by the

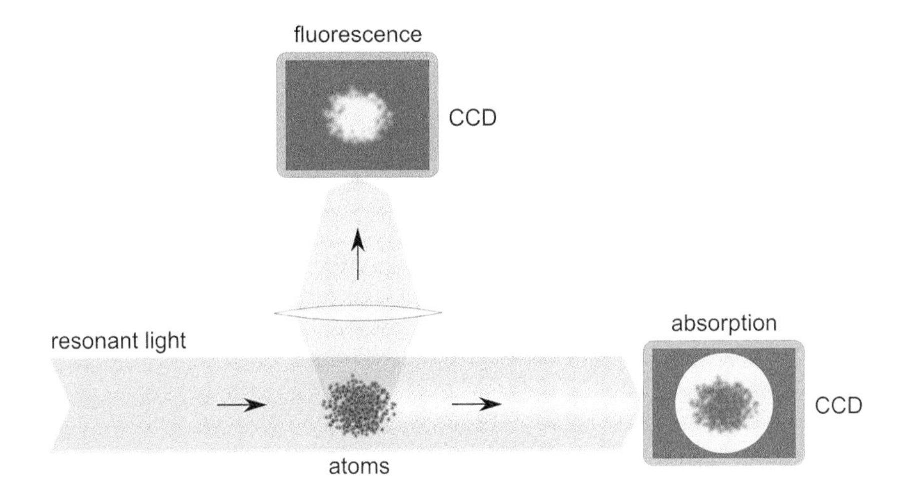

Fig. 3.9 Imaging of cold atoms with resonant light. The atoms can be imaged onto a CCD either by looking at the shadow left in the transmitted beam (absorption imaging) or collecting the fluorescent photons emitted in a different direction (fluorescence imaging).

atoms on the spatial beam profile (see Fig. 3.9), allowing a direct measurement of $\tilde{n}(z)$. Absorption imaging is by far the most used imaging technique for ultracold atoms, but *fluorescence imaging* can be performed as well: instead of detecting the absence of photons in the excitation direction, the fluorescence photons spontaneously emitted along a different direction can be detected. Although the signal is weaker (only a small fraction of the fluorescence light can be detected, since it is emitted on the whole solid angle) this technique has the advantage of having a zero background, which allows an easier detection of dilute samples (for which absorption can be very weak).

Due to their nature, both absorption and fluorescence imaging are destructive techniques, since the absorption of a few resonant photons is enough to cause a major heating of the atomic cloud. Nondestructive imaging can be performed by exploiting the non-absorptive interaction of atoms with off-resonant light: even in the absence of absorption, the index of refraction of the atomic cloud (see Appendix A.3.2) imprints a phase on the imaging light and the information contained in this phase is then reconverted into intensity by interference with a reference beam. These techniques generally yield a lower signal-to-noise ratio, but have the advantage of not being destructive (since no photons are absorbed), making a time-resolved imaging of the same atomic sample possible, which is particularly useful for the study of dynamics. For more information on the experimental implementation of these techniques, the reader can refer to the review article Ketterle *et al.* (1999).

3.2 Bose–Einstein condensates

Bose–Einstein condensation is a pure quantum phenomenon consisting in the macroscopic occupation of a single-particle state by an ensemble of identical bosons in thermal equilibrium at finite temperature. The occurrence of this phase transition in a gas of atoms was first predicted by A. Einstein in Einstein (1924) and Einstein (1925), following the ideas contained in a paper by S. N. Bose devoted to the statistical description of the quanta of light (Bose, 1924). Following the Maxwell–Boltzmann statistics, at $T = 0$ even the distinguishable particles of a classical gas would naturally occupy the lowest-energy state, since at $T = 0$ the thermal entropic contribution to the free energy vanishes. The intriguing feature of Bose–Einstein condensation is the macroscopic occupation of a single-particle state at $T > 0$, when the thermal energy $k_B T$ of the system is much larger than the level spacing and the classical Maxwell distribution would predict occupancy of a quasi-continuum of levels.

For a free ideal gas of identical bosons the BEC phase transition happens at a critical value of the phase-space density

$$\left(n\lambda_{dB}^3\right)_C = 2.612 \tag{3.13}$$

which results in a critical temperature

$$T_C = 3.313\frac{\hbar^2}{mk_B}n^{2/3}\,. \tag{3.14}$$

In the experimentally relevant case Bose–Einstein condensation is obtained in a gas of harmonically trapped bosonic atoms. In this case (extensively discussed in Appendix C), the critical temperature can be written as

$$T_C = 0.94\frac{\hbar\omega_{ho}}{k_B}N^{1/3}, \tag{3.15}$$

where N is the total number of particles and $\omega_{ho} = (\omega_x\omega_y\omega_z)^{1/3}$ is the geometric average of the trapping angular frequencies ω_i.

For a gas of ^{87}Rb atoms at normal conditions ($P = 10^5$ Pa, $T = 300$ K) $\lambda_{dB} \simeq 0.01$ nm and $n\lambda_{dB}^3 \approx 10^{-8}$. The eight orders of magnitude in the phase-space density separating a room-temperature gas from the quantum degenerate regime could be filled either by increasing the density or by decreasing the temperature. However, an important issue arises regarding the thermodynamic stability of the system and forces us to work with low-density gases. Indeed, at normal pressure and sufficiently low temperature, all the known interacting systems, with the exception of helium, undergo a phase transition to the solid phase well before reaching quantum degeneracy. In Fig. 3.10 (adapted from Pitaevskii and Stringari (2004)) we show a typical pressure–temperature diagram, indicating the boundaries between the solid, liquid, and gaseous phases. The dashed line corresponds to the BEC phase transition for an ideal gas. From this diagram it is clear that, in order to observe Bose–Einstein condensation, one has to get rid of all the processes that could bring the system into the thermodynamically stable solid phase. This can be achieved by working with very dilute samples, in which the probability of inelastic three-body recombination can be neglected and the system may reach a metastable Bose-condensed phase.

After early attempts with hydrogen (that will be discussed in Sec. 3.4.1), Bose–Einstein condensation in dilute gases of neutral atoms was experimentally achieved in 1995 by the group led by E. A. Cornell and C. E. Wieman at JILA working with ^{87}Rb

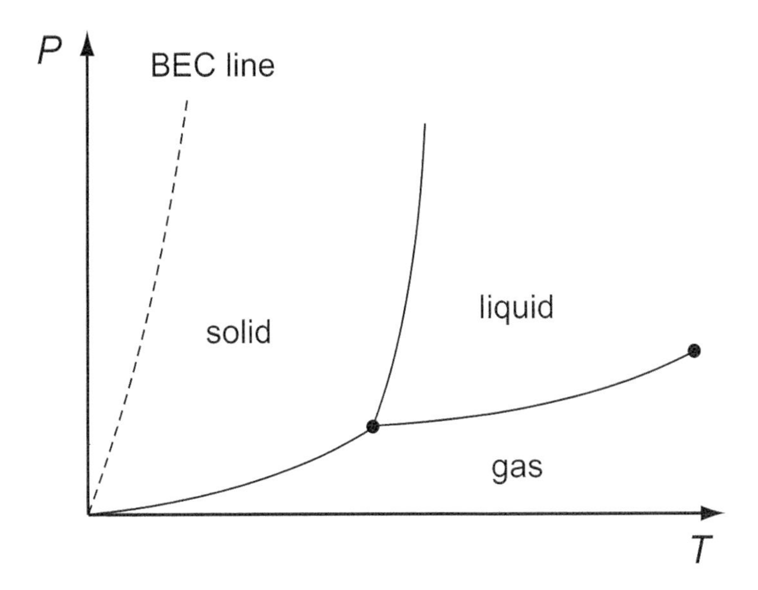

Fig. 3.10 A typical pressure–temperature phase diagram. The dashed line corresponds to the BEC transition for an ideal gas.

gases (Anderson *et al.*, 1995) and, after a few months, by the group of W. Ketterle at MIT with ^{23}Na (Davis *et al.*, 1995). In the same period first evidence of Bose–Einstein condensation in ^{7}Li gases was obtained also by the group of R. Hulet at Rice University (Houston, TX, USA) (Bradley *et al.*, 1995; Bradley *et al.*, 1997*b*; Bradley *et al.*, 1997*a*). This achievement, recognized by the Nobel Prize in Physics in 2001 to E. A. Cornell, C. E. Wieman, and W. Ketterle (Cornell and Wieman, 2002; Ketterle, 2002), is a spectacular result of the progress made in atomic physics in the last few decades with the development of advanced techniques for cooling and trapping neutral atoms. At the same time, the realization of this new state of matter marked the beginning of a new and interdisciplinary field of research, connected with the investigation of fundamental properties of quantum fluids, such as superfluidity and superconductivity, and to the development of a novel kind of atom optics based on coherent matter waves. For an excellent review of the theory of Bose–Einstein condensation of atomic gases the reader may refer to Pitaevskii and Stringari (2004), while comprehensive information on the early years of experimental investigations can be found in the contributions to the Proceedings of the CXL International Enrico Fermi School organized in 1998 in Varenna (Inguscio *et al.*, 1999).

In the following sections we will present the main properties of atomic Bose–Einstein condensates and fundamental experiments performed with trapped BECs of alkali gases. Then in Secs. 3.2.5 and 3.2.6 we will discuss the application of BECs to precise measurements and atom interferometry. Later in this chapter, Sec. 3.4 will be devoted to the discussion of Bose–Einstein condensation of atoms belonging to different chemical families.

3.2.1 BEC transition

The first and most striking experimental signature of Bose–Einstein condensation in a trapped gas of neutral atoms is the modification of the density distribution crossing the phase transition (Cornell and Wieman, 2002; Ketterle, 2002). Because of the small size (typically in the 10–100 μm range) and extremely large optical density $\sigma_0 \tilde{n}$ (even above 10^2–10^3), the diagnostic of ultracold trapped clouds (see Sec. 3.1.5) is typically performed with *time-of-flight absorption imaging*. In this technique the cloud is imaged after a free-expansion time t_{TOF} (typically a few tens of ms) following the removal of the trapping potential. This procedure, besides allowing a distortion-free imaging of a larger cloud with smaller optical density, maps the initial momentum distribution of the cloud $n_p(\mathbf{p}; 0)$ onto a density distribution in coordinate space $n(\mathbf{r}; t_{\mathrm{TOF}})$. In the far-field regime[8]

$$n(\mathbf{r}; t_{\mathrm{TOF}}) \simeq n_p(m\mathbf{r}/t_{\mathrm{TOF}}; 0) \,. \tag{3.16}$$

In the upper row of Fig. 3.11a we show three absorption images of expanded atomic clouds crossing the critical temperature for Bose–Einstein condensation. When lowering the temperature of the sample below the critical value T_C in eqn (3.15), the gas starts to condense in a low-momentum state and a pronounced peak in the expanded density distribution appears. Besides this increase in optical density, the distribution itself changes shape. The density of an expanded thermal cloud above T_C is well described by a Gaussian function, in accordance with the Maxwell–Boltzmann prediction

$$n_{th}(\mathbf{r}) \propto \exp\left(-\frac{r^2}{2\sigma^2}\right) \tag{3.17}$$

with a width $\sigma = t_{\mathrm{TOF}}\sqrt{k_B T/m}$ dependent on the cloud temperature T. A Bose–Einstein condensate below T_C is described by a different density distribution, owing to the effects of interactions between the atoms in the condensed gas. As detailed in Appendix C.2, in the weakly interacting limit (which is usually well satisfied by trapped Bose–Einstein condensates) the interactions are treated within the mean-field Gross–Pitaevskii theory, predicting a nonlinear wave equation for the condensate wavefunction $\psi(\mathbf{r})$

$$\left[-\frac{\hbar^2\nabla^2}{2m} + V_{trap}(\mathbf{r}) + g|\psi(\mathbf{r})|^2\right]\psi(\mathbf{r}) = \mu\psi(\mathbf{r}) \,, \tag{3.18}$$

where μ is the chemical potential, $V_{trap}(\mathbf{r})$ is the trapping potential, and $g = 4\pi\hbar^2 a/m$ is the strength of the mean-field interaction potential, where a is the scattering length. We consider the case of repulsive atom–atom interactions ($a > 0$), since this is the regime in which stable trapped BECs can be formed.[9] In the *Thomas–Fermi*

[8]This regime, equivalent to the definition of far-field in optics, holds when the size of the cloud after expansion is much larger than the initial size of the cloud in the trap. For a gas initially trapped in a harmonic trap with frequencies ω_i this happens when $t_{\mathrm{TOF}} \gg 1/\omega_i \; \forall i$.

[9]Attractive interactions cause a shrinking of the BEC and eventually its collapse, as first observed in Bradley *et al.* (1997*a*) and, by tuning the interactions with a Feshbach resonance, in Roberts *et al.*

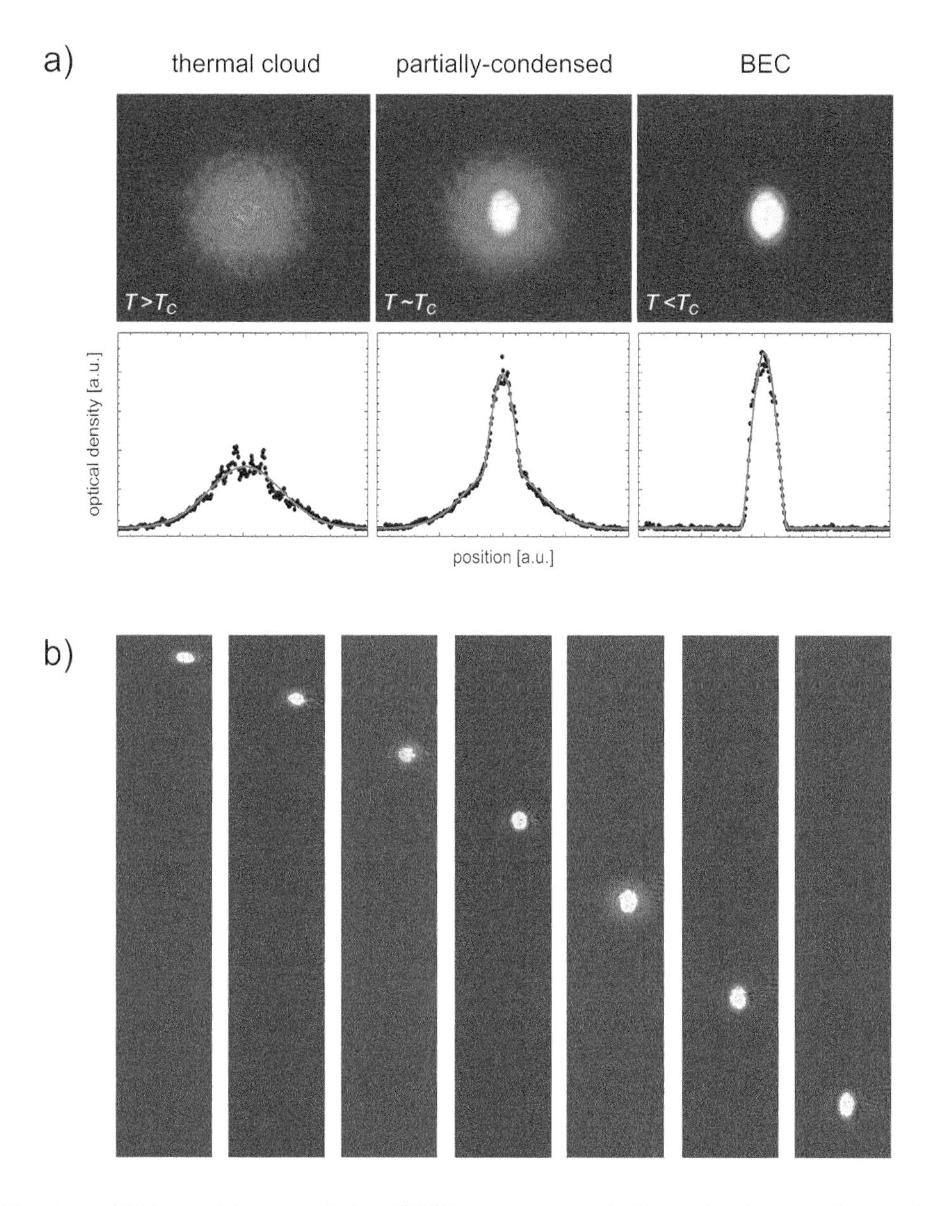

Fig. 3.11 BEC transition recorded in LENS experiments. a) Absorption images of expanded clouds of ^{87}Rb atoms for different temperatures T across the BEC transition at $T_C \simeq 150$ nK. From left to right: a thermal sample, a partially condensed cloud and a pure BEC with $N \simeq 4 \times 10^5$ atoms. The lower graphs show horizontal cross sections of the density distribution fitted with a Gaussian, a bimodal distribution, and an inverted parabola, respectively (see text). b) Absorption images for increasing time-of-flight $t_{\rm TOF}$ from 0 to 28 ms showing the aspect-ratio inversion of the BEC. The different vertical position is due to the free fall of the atoms following the removal of the confining potential.

limit of large number of atoms (see Appendix C.2), the kinetic energy term can be neglected and the density distribution takes the shape of the confining potential $n(\mathbf{r}) = |\psi(\mathbf{r})|^2 = (\mu - V_{trap}(\mathbf{r}))/g$. In the case of a harmonic trap, the condensate density is an inverted parabola

$$n_c(\mathbf{r}) \propto \max\left[1 - \left(\frac{x}{R_x}\right)^2 - \left(\frac{y}{R_y}\right)^2 - \left(\frac{z}{R_z}\right)^2, 0\right] \tag{3.19}$$

the shape of which is maintained during the expansion as well (although with increasing radii R_i (Castin and Dum, 1996)). The lower row of Fig. 3.11a shows a fit of the experimental pictures in the upper row, respectively with a Gaussian (for $T > T_C$), an inverted parabola (for $T < T_C$), and a bimodal Gaussian + parabola around the critical temperature, from which the condensate fraction and the temperature (from the wings of the Gaussian component) can be determined.

In the case of BECs confined in elongated traps (as the ones produced in many experiments), the expansion is strongly anisotropic, being faster along the direction where the confinement is initially stronger. For a noninteracting BEC, this is a consequence of the Heisenberg uncertainty principle $\Delta p_i \Delta x_i \sim \hbar$ holding for the BEC wavefunction: the tighter the confinement in the $\hat{x}_i$ direction, the bigger the momentum spread Δp_i and the faster the condensate expands along that direction. In the case of BECs with repulsive interactions this effect is amplified, since the expansion is mostly driven by atom–atom interactions, which push the atoms farther apart in the direction in which they are initially more strongly confined. This anisotropic expansion can be observed in Fig. 3.11b, where a BEC, after release from a horizontally elongated trap, inverts its aspect ratio and becomes elongated vertically. On the contrary, in the case of a thermal cloud, the far-field density distribution is spherically symmetric and no evidence of the trap anisotropy can be found, which further helps the identification of a condensed component on top of a thermal cloud (see Fig. 3.11a).

In addition to time-of-flight absorption imaging, the BEC formation can also be studied *in situ* with non-destructive imaging, which allows the same ultracold sample to be studied at different times, with negligible heating or atom losses. This technique was applied in Miesner *et al.* (1998) to investigate the dynamics of BEC formation when a bosonic cloud, slightly above T_C, was non-adiabatically cooled by a fast evaporation ramp. Fig. 3.12 shows a sequence of images at time intervals of 13 ms, revealing the onset of condensation and the dynamics of the BEC growth after this "shock cooling".

3.2.2 BEC excitations

Collective excitations. Atomic Bose–Einstein condensates provide a very clean system where it is possible to investigate the BEC transition without the complexities associated with strongly interacting condensed-matter systems. Atomic BECs are dilute gases, in which interactions can be accurately modelled in the Gross–Pitaevskii mean-field description, differently from superfluid helium, where interactions are much stronger and a satisfactory theoretical treatment cannot be easily given.

(2001). Only for small number of atoms (typically 10^3) can the kinetic energy associated with the trap confinement compensate the negative interaction energy and can a stable attractive BEC form (see also Appendix C.2.1).

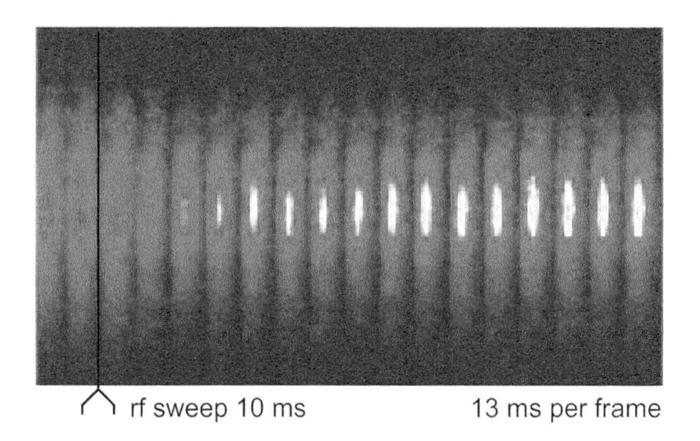

Fig. 3.12 *In-situ* nondestructive detection of the formation of a Bose–Einstein condensate after a fast evaporative cooling. Reprinted with permission from Miesner *et al.* (1998). © AAAS.

As always in Physics, frequency measurements are the most precise kind of measurements that it is possible to perform. Indeed, the measurement of collective mode frequencies was immediately recognized to be a fundamental and very precise tool to investigate the superfluid properties of atomic Bose–Einstein condensates. Low-energy collective modes manifest as long-wavelength shape oscillations of the trapped BEC. In a noninteracting ideal gas (both in the normal and in the Bose-condensed state), all the collective modes have frequencies which are integer multiples of the harmonic trapping frequencies. In an interacting gas, however, shape oscillations happen at characteristic frequencies which are non-integer multiples of the trapping frequencies and can be used as very sensitive probes of the state (normal or superfluid) of the gas.

Breathing excitations of trapped BECs can be induced quite easily by modulating the trap parameters and their frequency can be measured quite accurately by observing the evolution of the BEC size in time. The frequency of BEC collective excitations was first measured at JILA (Jin *et al.*, 1996), where frequencies deviating from the ideal gas behaviour were measured, evidencing the effect of interactions in the condensed cloud, and a longer damping time than for a normal gas was observed. Collective mode frequencies in the Thomas–Fermi regime were measured at MIT (Mewes *et al.*, 1996) and later at LENS (Fort *et al.*, 2000), confirming the validity of the seminal analytic treatment derived by S. Stringari in the hydrodynamic regime (Stringari, 1996). The shape oscillations are clearly visible in Fig. 3.13, together with a plot of the aspect ratio (i.e. the ratio between the width and the height of the cloud) as a function of time.

Sound propagation. Shape oscillations of a trapped BEC are connected to the excitation of long-wavelength phonons, the spectrum of which reduces to discrete energies owing to the finite BEC size. Higher-energy phonons with smaller excitation wavelength can be excited with different methods and studied in order to detect the complete BEC excitation spectrum. The energy-momentum dispersion relation of the excitations is

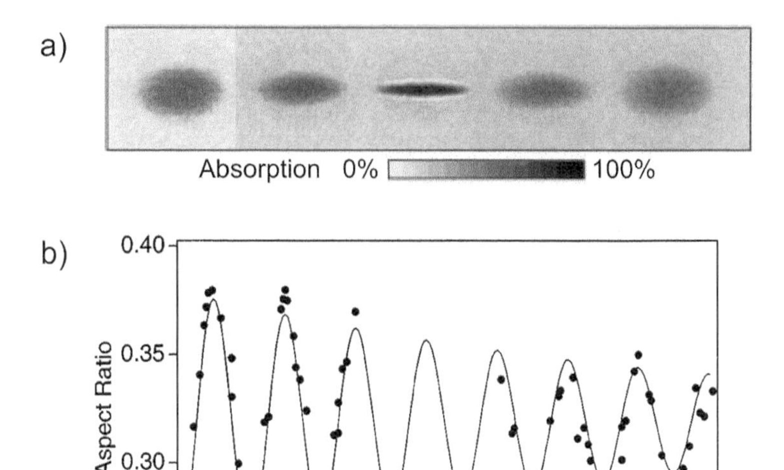

Fig. 3.13 Collective oscillations of a trapped Bose–Einstein condensate. a) Absorption images of a BEC performing breathing oscillations. b) Sinusoidal fit of the BEC aspect ratio as a function of the evolution time in the trap. Reprinted with permission from Mewes *et al.* (1996). © American Physical Society.

characterized by a linear branch at low energies

$$\omega \simeq ck \tag{3.20}$$

the slope of which is given by the sound velocity $c = \sqrt{gn/m}$, which depends on the condensate density n, the atomic mass m, and the strength of interactions $g = 4\pi\hbar^2 a/m$ which is proportional to the scattering length a.

The excitation of sound inside a condensate was performed at MIT (Andrews *et al.*, 1997*a*) by illuminating a BEC with a focused blue-detuned laser which created a hole inside the condensate, as shown in the lowest panel of Fig. 3.14a (see Sec. 5.2.1 for a detailed explanation of the optical dipole force which is responsible for this effect). The sudden creation of the hole produced a shock wave inside the BEC, characterized by twin density peaks moving towards the condensate edges. This is visible in the higher panels of Fig. 3.14a, which show the time evolution of the density profile measured *in situ* after the excitation: the dashed lines evidence the propagation of the density peaks, which move linearly at a velocity which turned out to be in agreement with the predicted speed of sound c inside the condensate.

The full excitation spectrum of the BEC (and of other ultracold quantum phases as well) can be measured very precisely using a powerful tecnique which is based on scattering of light from the condensate. This technique, called *Bragg spectroscopy*

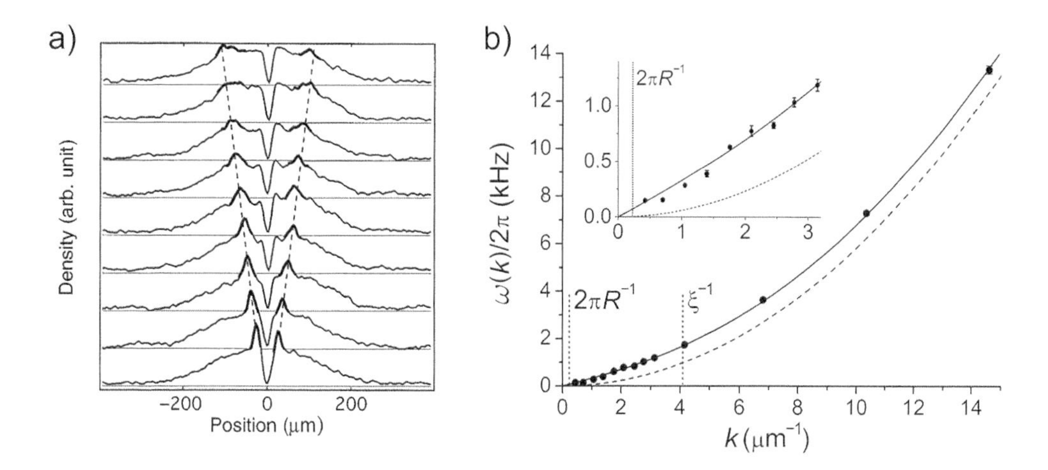

Fig. 3.14 a) In-situ density profiles showing the propagation of sound inside a Bose–Einstein condensate: time increases from bottom to top and the dashed lines mark the position of the density peaks created by the initial perturbation. b) Full excitation spectrum of a trapped BEC measured with Bragg spectroscopy. Reprinted with permission from Andrews *et al.* (1997*a*) (a) and Steinhauer *et al.* (2002) (b).

(Stenger *et al.*, 1999), is reminiscent of the scattering experiments usually performed in condensed-matter systems (e.g. with neutron beams), where the analysis of inelastic scattering is used to extract information on the excitations of the samples. In Bragg spectroscopy a BEC is illuminated with two non-resonant laser beams with frequency and wavevector $(\omega_1, \mathbf{k}_1)$ and $(\omega_2, \mathbf{k}_2)$, respectively. These beams can induce a Raman transition connecting two different atomic motional states and leaving the internal state unchanged (Kozuma *et al.*, 1999). The effective Raman excitation, which results from the redistribution of one photon from one laser beam to the other beam, is characterized by an energy transfer $\hbar\omega = \hbar(\omega_1 - \omega_2)$ and a momentum transfer $\hbar\mathbf{k} = \hbar\,(\mathbf{k}_1 - \mathbf{k}_2)$, which can be independently adjusted by controlling the frequency of the two laser beams and their propagation direction. At zero temperature and in the linear response regime (i.e. for weak perturbations) the number of induced excitation quanta is proportional to the *dynamic structure factor* $S(k, \omega)$ (Stenger *et al.*, 1999; Brunello *et al.*, 2001). For a weakly interacting BEC the dynamic structure factor is $S(k, \omega) \propto \delta\,[\omega - \omega_B(k)]$, where $\omega_B(k)$ describes the Bogoliubov spectrum of excitations

$$\hbar\omega_B = \sqrt{\frac{\hbar^2 k^2}{2m}\left(\frac{\hbar^2 k^2}{2m} + 2mc^2\right)} \tag{3.21}$$

which can be derived from the mean-field description of interactions used in the Gross–Pitaevskii equation (Dalfovo *et al.*, 1999; Pitaevskii and Stringari, 2004). This spectrum predicts a linear dependence on momentum $\omega_B \simeq ck$ at small k and a parabolic dependence $\omega_B \simeq \hbar k^2/2m + mc^2/\hbar$ at larger k. These two different dependencies reflect two different kinds of condensate excitations: collective phonons at low momentum (responsible for sound propagation) and single-particle-like excitations at larger momentum.[10]

The full excitation spectrum of a BEC was measured with Bragg spectroscopy by the group of N. Davidson at the Weizmann Institute (Steinhauer *et al.*, 2002). In the experiment the momentum of the excitations was varied by changing the relative angle between the two Bragg excitation beams. For each momentum the frequency difference of the beams was scanned to measure the position of the resonance. The points in Fig. 3.14b show the position of these resonances as a function of momentum. The measured spectrum is in excellent agreement with the Bogoliubov prediction in eqn (3.21), which is illustrated by the solid line. The data exhibit the expected linear dependence at small momenta (the slope of which yields a measurement of the sound velocity c) and clearly deviate from the parabolic energy/momentum relation of the single particle, which is illustrated by the dashed line.

Bragg spectroscopy is a very powerful technique and can be used to probe the properties of different ultracold systems, including atoms in optical lattices (see Secs. 6.4.1 and 7.1.3), strongly interacting fermions (Veeravalli *et al.*, 2008) and phase-fluctuating one-dimensional BECs (Richard *et al.*, 2003). Recently, it has been used at JILA to probe the excitation spectrum of ^{85}Rb Bose–Einstein condensates in the regime of strong interactions between the atoms, which was achieved by tuning the scattering length around a Feshbach resonance. In this regime interactions cannot be treated with a mean field theory and strong deviations from the Bogoliubov predictions were observed in the experimental data (Papp *et al.*, 2008). It is remarkable to note that a clear theoretical understanding of this strongly interacting regime is still missing. Further experimental and theoretical investigations on strongly interacting bosonic gases would be particularly important since they could help to elucidate the mechanisms which originate the roton minimum in the dispersion relation of superfluid helium (Noziéres and Pines, 1990). Rotonic excitations are not supported by the weakly interacting dilute Bose gas and are believed to appear only for superfluids with strong spatial correlations.[11]

3.2.3 Superfluidity

Critical velocity. The measurement of collective mode frequencies discussed in the previous section was the first evidence of the superfluidity of Bose-condensed gases. It was soon observed experimentally that superfluidity is not preserved for arbitrary condensate velocity, in accordance with the Landau criterion for superfuidity (Landau, 1941). Stirring a blue-detuned laser within a Bose–Einstein condensate at different speeds $\mathbf{v}$ (see Fig. 3.15a), the group of W. Ketterle at MIT demonstrated the existence of a critical velocity for the onset of dissipative flow in the condensate (Raman *et al.*, 1999). While at small velocities the condensate could be stirred without any heating

[10]The characteristic momentum value separating these two regimes corresponds to the inverse of the *healing length* $\xi = (8\pi na)^{-1/2}$ (where n is the density and a the scattering length), which is the typical length scale over which the condensate density can change (Pitaevskii and Stringari, 2004).

[11]Rotonic spectra have been predicted for Bose–Einstein condensates dressed by laser beams in order to induce long-range dipole–dipole interactions (O'Dell *et al.*, 2003) or for pancake-shaped BECs with static dipolar interactions (Santos *et al.*, 2003) (see also Sec. 3.4.3).

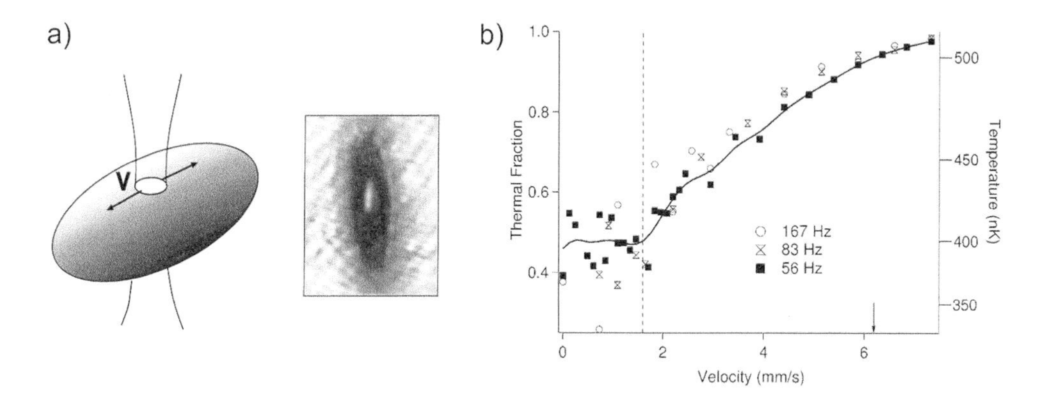

Fig. 3.15 a) A blue-detuned laser beam, stirred at speed $\mathbf{v}$, creates a moving hole in a Bose–Einstein condensate. b) Temperature and normal fraction of the trapped gas as a function of the stirring speed: above a critical speed strong dissipation is observed, in accordance with the Landau criterion for superfluidity. Reprinted with permission from Raman *et al.* (1999). © American Physical Society.

(corresponding to a stable superflow at velocity $-\mathbf{v}$ around an obstacle), at velocities larger than a critical value a strong dissipation with reduction in the condensate fraction was observed, as shown in Fig. 3.15b. The results of this experiment were in agreement with the Landau theory of superfluidity, which predicts a dissipationless flow provided that the superfluid velocity (relative to any obstacle) is below the speed of sound: above this point the excitation of phonons inside the BEC becomes energetically favoured and this growth of excitations in turn depletes the ground-state BEC.[12]

Vortices. Another important proof of superfluidity of atomic Bose-condensed gases is the formation of *quantized vortices*. The formation of vortices is a remarkable feature of quantum systems described by a macroscopic wavefunction $\psi(\mathbf{r}) = \sqrt{\rho(\mathbf{r})}\exp[i\theta(\mathbf{r})]$. The condensate velocity field

$$\mathbf{v}(\mathbf{r}) = \frac{\hbar}{m}\nabla\theta(\mathbf{r}) \tag{3.22}$$

is irrotational for all points in which the density ρ does not vanish and the phase θ is well defined. The superfluid, in order to rotate, has to accomodate vortices, i.e. its density should vanish at some points, around which the circulation of the velocity field is nonzero. The existence of a well-defined phase for the wavefunction then implies that the circulation around these singular points has to be quantized according to

$$\oint \mathbf{v}(\mathbf{r})\cdot d\mathbf{r} = \frac{\hbar}{m}\oint \nabla\theta(\mathbf{r})\cdot d\mathbf{r} = \frac{\hbar}{m}2\pi n = n\frac{h}{m}\,, \tag{3.23}$$

[12]When an optical lattice is considered, there are different kinds of instabilities which can destroy the superflow. Dynamical instability arising from the interplay between nonlinearity and periodicity will be discussed in Sec. 6.4.

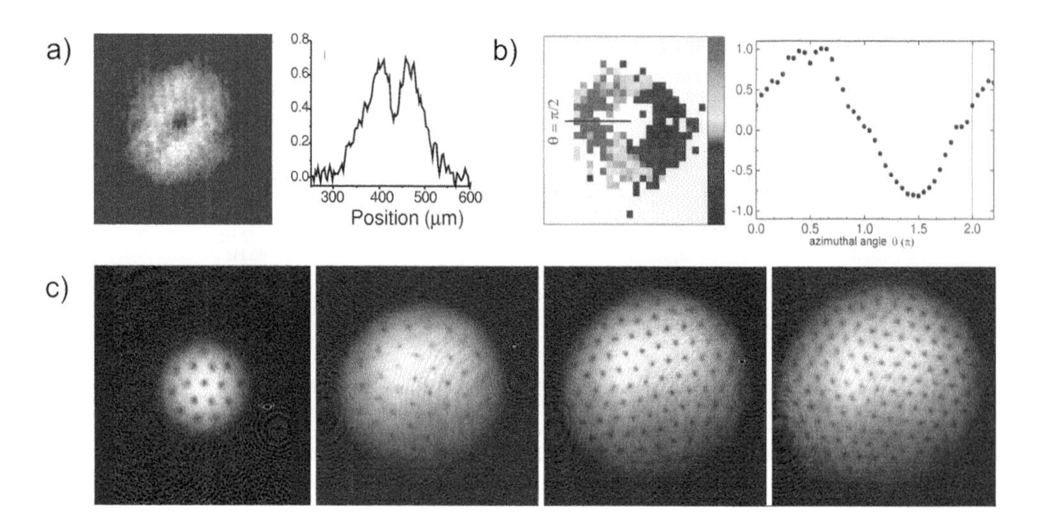

Fig. 3.16 a) Vortex created by stirring a blue-detuned laser beam inside a BEC. Reprinted with permission from Madison *et al.* (2000). © American Physical Society. b) Phase image of the condensate wavefunction, evidencing the 2π phase twist around a vortex. Reprinted with permission from Matthews *et al.* (1999). © American Physical Society. c) Vortex lattices created for large BECs in fast rotating traps. Reprinted with permission from Abo-Shaeer *et al.* (2001). © AAAS.

where n is an integer corresponding to the *vortex charge* and h/m is the quantum of circulation. This quantization condition corresponds to n windings of the condensate phase by 2π around the vortex core in which the density vanishes.

Vortices were created and observed for the first time by the group of E. A. Cornell at JILA by using a phase-engineering technique in which controlled microwave transitions were used to coherently couple the condensate to a different internal state (Matthews *et al.*, 1999). By suitably choosing the microwave parameters, the BEC atoms could be excited to a metastable state with a hole in the centre and a phase twist of 2π around the singularity, which could be probed with a phase-sensitive imaging technique (see Fig. 3.16b).

In following experiments, vortices were created with different techniques, primarily by creating time-dependent anisotropies in the trapping potential. This technique was first implemented by the group of J. Dalibard at ENS, in which a blue-detuned laser beam was used to "stir" the atomic cloud at different angular velocities ω along its symmetry axis (Madison *et al.*, 2000). The BEC was produced by evaporative cooling in the presence of the stirring beam, which is equivalent to studying the BEC transition in a rotating reference frame. Vortices were detected only above a critical stirring velocity (an example is shown in Fig. 3.16a) and their number increased with increasing rotation speed. This experiment is the analogue of the famous "rotating bucket" experiment with superfluid helium, in which no motion of helium can be observed for small velocities, while at higher velocities vortex filaments appear as singularity lines of vanishing density.

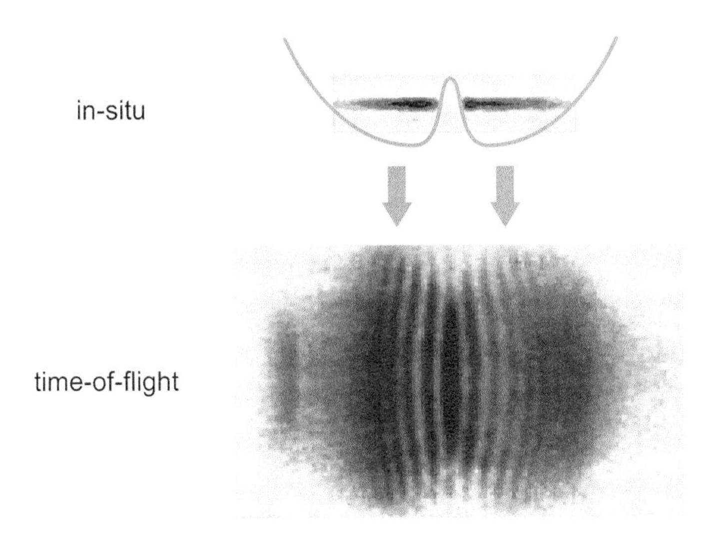

Fig. 3.17 Interference fringes produced by two separate overlapping Bose–Einstein condensates. Reprinted with permission from Andrews *et al.* (1997*b*). © AAAS.

The group of W. Ketterle at MIT observed vortex lattices by stirring large Bose–Einstein condensates containing up to 5×10^7 sodium atoms (Abo-Shaeer *et al.*, 2001). A very large number of vortices (up to 130) were observed and it was found that their arrangement followed a highly ordered triangular lattice structure, as shown in the images of Fig. 3.16c. This structure is reminiscent of the Abrikosov lattices which are formed by the magnetic flux lines which penetrate type-II superconductors. The formation of a distributed vorticity in the superfluid (which can be described as arising from an effective repulsion between vortices) is favoured energetically with respect to the creation of multiply-charged vortices: it can be shown that, after cross-graining over length scales larger than the vortex lattice spacing, the circulation of the velocity vector yields the same result expected for a classical fluid in which vorticity is uniformly distributed.

3.2.4 Phase coherence

Interference. Superfluidity is connected with the existence of a macroscopic order parameter, which is the condensate wavefunction. Macroscopic quantum coherence is indeed a key feature of Bose–Einstein condensation. Spatial coherence was directly demonstrated for the first time by W. Ketterle at MIT by observing interference fringes from two condensates initially produced in two separate traps (Andrews *et al.*, 1997*b*), as shown in Fig. 3.17. After removing the confining potential and allowing for a free expansion of the BEC wavefunctions, interference fringes were observed by time-of-flight absorption imaging in the overlap region between the two condensates, demonstrating the presence of spatial coherence.

It is noteworthy to stress that, differently from ordinary optical interference of two independent laser beams, which does not result in a detectable interference pattern since there is no phase coherence between them, for BECs interference fringes are observed even if the two condensates do not possess any initial phase relation. This apparently different behaviour can be explained by noting that in BEC interference experiments no *average* is made on the phases of the two condensates: the observation of interference fringes does not mean that coherence is present *between* the two condensates, it just implies that *each* of the two condensates can be described by an order parameter with a well-defined macroscopic phase.[13] When an average over multiple realizations of the experiment is performed, interference fringes fade out because the absolute phase of the fringe pattern changes randomly from one experiment to the following one. Experiments performed at ENS by the group of J. Dalibard on the interference of *several* Bose–Einstein condensates evidenced that single-shot interference fringes can be clearly observed even with ≈ 10 independent interfering BECs, although the visibility of the fringe pattern rapidly decreases as the number of condensates increases, in accordance with the single-shot interference signal expected for random initial phases (Hadzibabic *et al.*, 2004).

The investigation of BEC interference is a powerful tool that has allowed the investigation of many effects connected with the BEC transition and with the coherence properties of bosonic gases under different conditions, e.g. when trapped in *optical lattices* (see Chapter 6 and Sec. 7.3) or in low dimensions. The latter case is particularly important for the investigation of the BEC transition, which is profoundly affected by dimensionality.[14] When the motion of the gas is restricted in one or two directions (by tightly confining traps or optical lattices), the gas effectively behaves as 2D or 1D, respectively. Very interesting effects can be measured in these regimes by characterizing the phase fluctuations of low-dimensional bosonic gases, in connection with the Berezinskii–Kosterlitz–Thouless (BKT) transition in 2D (Stock *et al.*, 2005; Hadzibabic *et al.*, 2006) and with the onset of quasi-condensation in 1D (Dettmer *et al.*, 2001; Hellweg *et al.*, 2003; Hugbart *et al.*, 2005; Hofferberth *et al.*, 2007; Hofferberth *et al.*, 2008).

3.2.5 BEC for precision measurements

In the last few sections we have discussed the experimental observation of the BEC transition and the measurement of fundamental properties of atomic condensates, mostly connected to matter-wave coherence and superfluidity. In this and in the next section we show how Bose–Einstein condensates can be used as high-resolution probes for the detection of tiny quantum effects or for the precise measurement of external fields (see also Sec. 4.4.1 for an application of BEC in high-resolution spectroscopy).

Atom–surface interactions and the Casimir–Polder force. Collective oscillations of Bose–Einstein condensates can be used as a very sensitive tool for precise measurements: since the oscillations of a trapped BEC depend on the trap frequency, if the latter is modified by the presence of external forces, the frequency of the collective mode is shifted

[13] Even two independent laser beams, if observed for a time smaller than the reciprocal of their frequency difference, would show a clear interference signal, which, however, is washed out after averaging on typically much longer times.

[14] In free space Bose–Einstein condensation occurs only in 3D, while no true BECs form in 2D and 1D.

accordingly. As an example, accurate frequency measurements of BEC oscillations close to surfaces allow the investigation of *atom–surface interactions*, which is an interesting topic for both technological applications and fundamental studies (see e.g. Scheel and Hinds (2011) for a recent review). When a neutral atom is placed at a distance r from a metallic surface, it experiences an attractive van der Waals (Lennard-Jones) potential $U \sim r^{-3}$, which can be understood classically as the interaction energy of the fluctuating atomic electric dipole with its mirror image in the metallic surface. At distances larger than the wavelength of the atomic resonance, retardation effects have to be considered: a quantum electrodynamics formalism for this problem was developed by H. B. G. Casimir and D. Polder (Casimir and Polder, 1948), who showed that in this regime the attractive potential acquires a different $U \sim r^{-4}$ dependence.[15] It can be shown that the Casimir–Polder force is a purely quantum effect originating from quantum fluctuations of the electromagnetic field (Meschede *et al.*, 1990): the presence of the surface imposes boundary conditions for the vacuum of the electromagnetic field around the atom, resulting in a net attraction of the atom to the surface, which can be interpreted (in a simplified picture) as the result of a spatially varying Lamb shift originating from the modified vacuum state.

Experimental evidence that the proximity of surfaces strongly modifies the properties of atoms was already obtained in early cavity-QED experiments (see Haroche and Raimond (2006) for an excellent book on this topic), e. g. in the early observation that the natural transition linewidths are affected by the modified density of states of the electromagnetic field inside the cavity. Mechanical effects in the atom–surface interaction have been quantitatively studied starting from the 1990s by high-resolution spectroscopy experiments or by studying the deflection of atomic beams (as a matter of fact, these forces provide a valuable resource, e.g., for implementing atom-optical devices in atom interferometers (Cronin *et al.*, 2009), see Sec. 2.5). The effects of the van der Waals interaction were evidenced in Sandoghdar *et al.* (1992), where the Lennard-Jones $\sim r^{-3}$ attraction of an atom with its electric image in a micron-sized optical cavity made of plane conductors was measured spectroscopically from the energy shift of the atomic levels in a beam of sodium. The existence of the Casimir–Polder force was demonstrated with a similar setup in Sukenik *et al.* (1993) by measuring the transmission of the atomic beam for different cavity spacings: comparing the experimental results with a theoretical model of the beam deflection for different atom–surface potentials provided a quantitative evidence for the existence of the attractive $\sim r^{-4}$ retarded Casimir–Polder potential. The effects of this force were then observed in "atomic mirror" experiments: after first indications reported in Landragin *et al.* (1996) for ultracold atoms classically reflected by a laser evanescent wave at a dielectric wall (see also Sec. 5.2.2), the Casimir–Polder force was quantitatively investigated by studying the quantum reflection of a laser-cooled beam of metastable neon atoms from the attractive potential generated by different kinds of dielectric surfaces in Shimizu (2001).

[15]The Casimir–Polder force between an atom and a metallic surface is strictly related to the attractive Casimir force between two macroscopic metallic surfaces. A similar effect appears for the interaction of a neutral atom with a dieletric surface as well, as derived by E. M. Lifshitz in Lifshitz (1956).

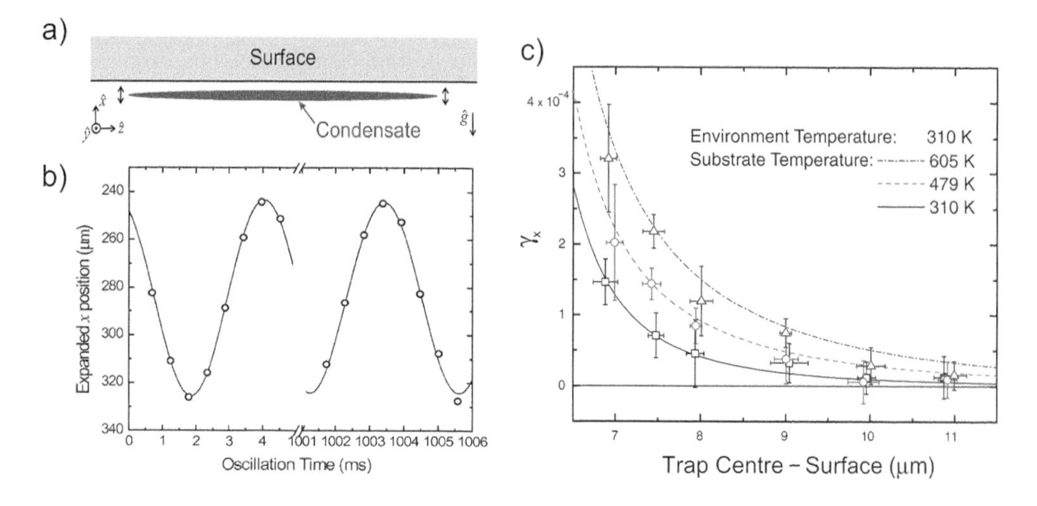

Fig. 3.18 Measurement of Casimir–Polder effects with ultracold atoms. a) Experimental scheme: a Bose–Einstein condensate is trapped at a few µm distance from a surface. b) Centre-of-mass oscillations of the BEC in the trap. c) Relative shift of the oscillation frequency as a function of surface temperature and surface–atoms distance. Reprinted with permission from Harber *et al.* (2005) (a,b) and Obrecht *et al.* (2007) (c). © American Physical Society.

Recent theoretical work (Antezza *et al.*, 2005; Antezza *et al.*, 2006) pointed out the role of thermal fluctuations (due to the nonzero temperature of the surface) in modifying the properties of the Casimir–Polder force, proposing its investigation with ultracold atoms trapped in proximity of material substrates. The first quantitative experimental study of the dependence of the Casimir–Polder force on the distance and the temperature of the surface was carried out by the group of E. A. Cornell at JILA (Harber *et al.*, 2005; Obrecht *et al.*, 2007). A Bose–Einstein condensate was confined in a harmonic magnetic trap centred at a few µm distance from a dielectric surface, as sketched in Fig. 3.18a, and center-of-mass oscillations of the BEC in the trap were studied by looking at the position of the atoms after different time intervals, as reported in Fig. 3.18b. The presence of the Casimir–Polder force slightly modifies the shape of the magnetic trap and, consequently, leads to a frequency shift of the oscillation. This frequency change was investigated as a function of the temperature of the surface and its distance from the atoms. The fractional change in the oscillation frequency is reported in Fig. 3.18c, together with the theoretical lines (with no adjustable parameters): it is evident that the frequency shift becomes larger as the cloud approaches the surface and as the temperature increases, in accordance with the nonequilibrium theory of the Casimir–Polder effect at finite temperature (Antezza *et al.*, 2005).

High-resolution magnetometry. Bose–Einstein condensates of alkali atoms can be used as magnetic field sensors in which high spatial resolution and high sensitivity can be simultaneoulsy achieved, outperforming the noise level of more traditional magnetometers based on superconducting quantum-interference devices (SQUIDs) in the µm range.

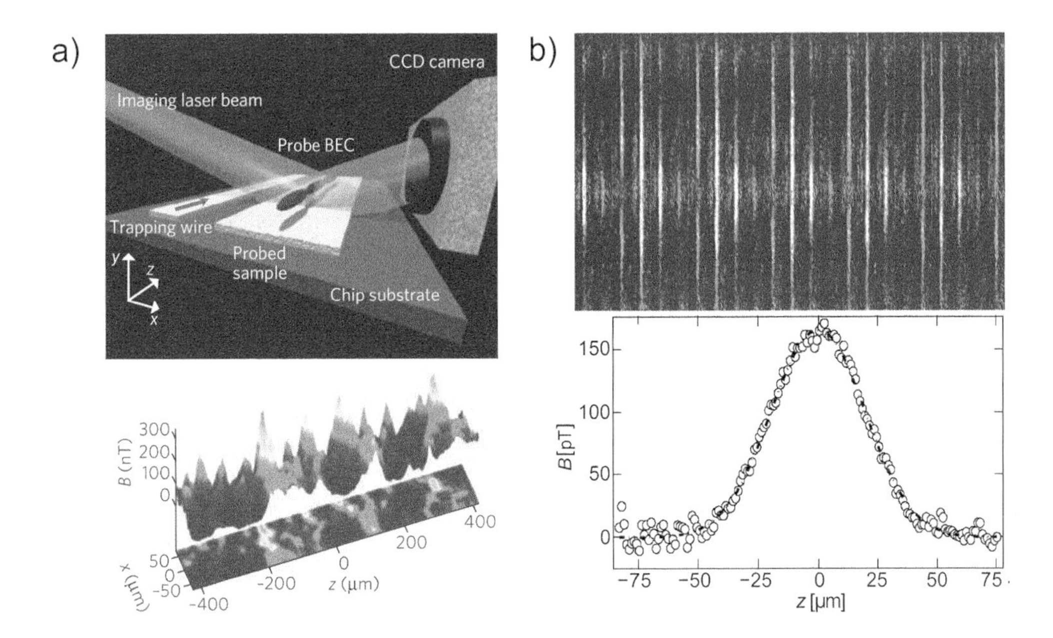

Fig. 3.19 Two different examples of precision magnetometry with BECs. a) A BEC is trapped on an atom chip in proximity of the probed sample (top). From the BEC density fluctuations a map of the magnetic field can be extracted (lower plot). Reprinted with permission from Wildermuth *et al.* (2005). © Macmillan Publishers Ltd. b) Time-resolved map of the BEC magnetization in the presence of an external inhomogeneous magnetic field (top): from the analysis of the local Larmor frequencies a precise measurement of the magnetic field can be derived (bottom). Reprinted with permission from Vengalattore *et al.* (2007). © American Physical Society.

Trapped BECs of alkali atoms are extremely sensitive to small magnetic field perturbations, even in the presence of large homogeneous bias fields. An inhomogeneous magnetic field $\mathbf{B}(\mathbf{r})$ produces a deformation of the magnetic trap potential $\delta U(\mathbf{r}) \sim \mu_B B(\mathbf{r})$ (see Sec. 3.1.1), which can be probed by imaging *in-situ* variations in the density of the trapped BEC. Energy variations on the scale of the BEC chemical potential, typically below 10^{-30} J, can be easily detected and provide local measurements of the magnetic field with nT (nanotesla) sensitivity, as demonstrated in Wildermuth *et al.* (2005). The top panel of Fig. 3.19a shows a scheme of the experiment, with a BEC trapped in proximity of the probed sample and an absorption imaging setup used to measure the BEC density fluctuations. The bottom panel of the same figure shows a two-dimensional map of the reconstructed magnetic field landscape, with a sensitivity of 4 nT and a resolution of 3 μm.

In the experiment described in Wildermuth *et al.* (2005) the BEC was produced on an *atom chip*, i.e. a microfabricated device in which small current-carrying wires, lithographically patterned on top of a substrate, provide the source of magnetic field for BEC trapping (Fortágh and Zimmermann, 2005). This technology, first demonstrated in 2001 independently by the groups of C. Zimmermann (Ott *et al.*, 2001) and T. W. Hänsch and J. Reichel (Hänsel *et al.*, 2001), allows a large simplification of the BEC experimental setup. Atom chips can be placed inside the vacuum chamber and the centre of the trap can be set very close to the chip surface, resulting in stronger magnetic field gradients and a much tigher confinement of the atoms than in a conventional magnetic trap. As a consequence, the atom density is larger and thermalization is faster, which speeds up evaporative cooling and relaxes the requirements on the atom lifetime in the trap: the vacuum requirements are less stringent, and the whole experimental cycle (from MOT to BEC) can be performed in the same chamber. For these reasons, atom chip experiments occupy less laboratory space than traditional setups based on standard magnetic traps, which makes them more portable, an important requirement for the applications of the proof-of-principle magnetometry demonstrated in Wildermuth *et al.* (2005). The ultimate example of portability of atom-chip setups is represented by a recent experiment in which a BEC apparatus was operated in free fall inside the 146-m-high drop tower of the Center of Applied Space Technology and Microgravity (ZARM) in Bremen (Germany) (van Zoest *et al.*, 2010). Under microgravity conditions, such as the ones experienced in free fall, the free evolution of the BEC can be observed for much longer times than in Earth-bound experiments, which is a fundamental requirement in view of the realization of atom-interferometry tests of relativistic effects with BECs.[16]

Coming back to the measurement of magnetic fields, a different magnetometric technique was employed in Vengalattore *et al.* (2007), where improved field sensitivities of a few pT (picotesla) were demonstrated. This technique is based on the Larmor precession of the atomic spin which is driven by the presence of a transverse magnetic field. The BEC was trapped in an optical dipole trap (see Sec. 5.2), which allows simultaneous trapping of different spin states, and was probed at different evolution times with a polarization-dependent phase-contrast imaging technique, which allows the local magnetization to be measured non-destructively with spatial resolution (Higbie *et al.*, 2005). The top panel of Fig. 3.19b shows maps of the BEC magnetization for different evolution times (increasing from left to right — the BEC is elongated vertically), in the presence of a small test magnetic field applied in the center of the atomic cloud. An accurate analysis of the local Larmor precession frequencies — performed pixel by pixel — yields the value of the local magnetic field. The bottom panel of Fig. 3.19b shows the results of this analysis and the excellent performance of this technique in terms of both spatial resolution and sensitivity.

3.2.6 Interferometry with BECs

Bose–Einstein condensates constitute the coldest kind of matter waves that can be produced in a laboratory. The macroscopic occupancy of the motional ground state makes them behave largely as a laser, i.e. a single-mode, coherent, bright source in which all the photons occupy the same state of the electromagnetic field. For this reason the achievement of BECs was welcomed as a revolution in the field of atom

[16] Larger BEC expansion times mean an increased sensitivity of the interferometer: as an example, the sensitivity to accelerations scales as the square of the time spent in the interferometer (see Sec. 2.5.1).

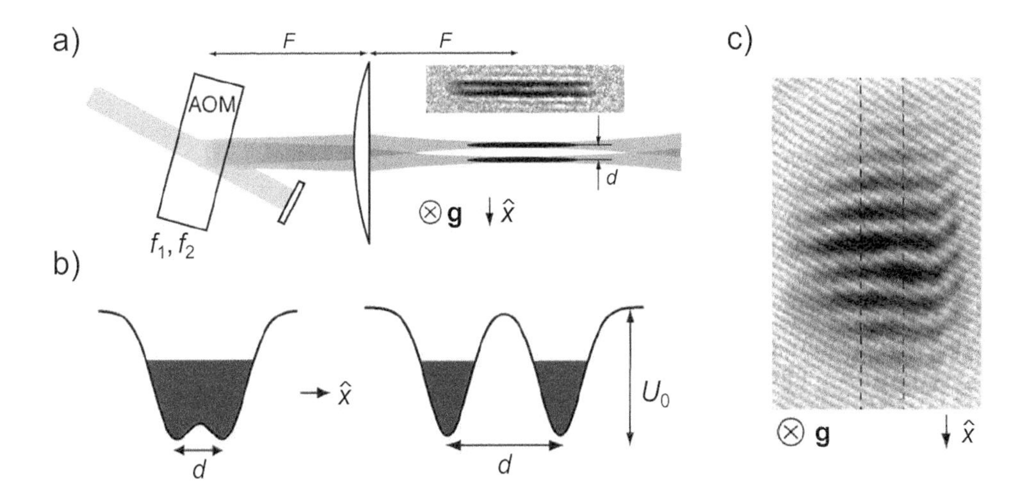

Fig. 3.20 Double-well trapped BEC interferometer. a) Optical setup for the double-well potential realized in Shin *et al.* (2004). An acousto-optic modulator (AOM) diffracts a laser beam into two beams producing two optical dipole traps: the distance d between the two potential wells is controlled by the radiofrequencies driving the AOM. The inset shows two separated condensates confined in the double-well potential. b) Scheme of the splitting procedure. c) Interference fringes formed by the two BECs in time-of-flight absorption imaging after release from the trap. Reprinted with permission from Shin *et al.* (2004). © American Physical Society.

interferometry, since it was thought to provide bright, almost monochromatic, sources of atoms for standard atom interferometers such as the ones described in Sec. 2.5, where a free-propagating cloud is split and recombined by appropriate beam-splitting devices.

In addition to the striking coherence properties, there is another precious advantage in using Bose–Einstein condensates: differently from photons, BECs can be easily trapped and held for relatively long times, which results in novel interferometric applications. In trapped BEC interferometers the atomic wavefunction is coherently split and recombined by raising a potential barrier. The location of the atom wavefunction can be known very precisely, which is an essential ingredient in experiments requiring high spatial resolution. These include the study of spatially dependent fields or interactions, as, for instance, the Casimir atom–surface force discussed in Sec. 3.2.5. The external force to be measured is encoded in the relative phase shift among two (or more) wells, the phase signal being proportional to the holding time which, in the split trap, can be very long.

Trapped BEC interferometers. An example of trapped interferometer with Bose–Einstein condensates was realized by the group of W. Ketterle and D. Pritchard at MIT as an evolution of the first pioneering experiments on BEC interference. The authors of Shin *et al.* (2004) reported on the realization of a trapped atom interferometer with Bose–Einstein condensates confined in an optical double-well potential. Differently from the experiment discussed in Sec. 3.2.4, this double-well trap was created dynamically by using two focused laser beams (see Sec. 5.2 for a discussion of optical trapping), the relative distance of which was controlled in real-time by deflecting a laser beam with an acousto-optic modulator driven by two radiofrequencies,[17] as sketched in Fig. 3.20a. This technique allowed the authors to study the phase evolution of a condensate initially produced in a single well and then coherently split into two parts by adiabatically deforming the single well into the double well potential (see Fig. 3.20b). The distance of the two wells after the splitting was large enough (13 μm) to allow the split condensates to be optically addressed individually (see an absorption image of the two BECs in the inset of Fig. 3.20a) and to have independent phase evolution. This requirement is guaranteed by the large distance between the two wells, which makes quantum tunnelling through the barrier negligible on the timescale of the experiments.[18] The authors of Shin *et al.* (2004) demonstrated that this splitting procedure is coherent: each realization of the experiment resulted in interference fringes (Fig. 3.20c) with the same spatial phase, which means that the splitting procedure produces two BECs with a well-defined relative phase. This situation is different from the observation of Andrews *et al.* (1997*b*) discussed in Sec. 3.2.4, in which two condensates were independently produced in separated traps and no deterministic phase relation was observed between them.

The maintenance of coherence during the splitting process has been further investigated in Schumm *et al.* (2005) and Hofferberth *et al.* (2006) in atom-interferometry experiments with one-dimensional BECs trapped on atom chips. Here the double-well potential was produced with a different technique which takes advantage of the combination of static magnetic fields and RF couplings (Zobay and Garraway, 2001). To illustrate this scheme, we consider two different Zeeman states with opposite magnetic moments, which are represented in Fig. 3.21a. In the presence of a static magnetic field, as the one produced in a Ioffe–Pritchard-like trap, one state is trapped in the field minimum and the other state is untrapped. If a strong RF coupling is induced between the two states, the atomic dressed states are shifted in energy one opposite to the other,[19] with a maximum shift around the minimum of the trap, where their initial separation is smaller. As a result, the atoms feel an adiabatic potential which features two minima at a distance that can be controlled in real-time by varying the RF parameters. Figure 3.21b shows the distribution of phases observed in time-of-flight interference for a series of identical experiments either adiabatically splitting an initially prepared BEC into two traps separated by ≈ 3 μm, or evaporatively cooling the atoms to BEC in the already split trap: while in the former case a well-defined relative phase is observed (thanks to the coherence of the splitting process), in the second case the phase is randomly distributed (Hofferberth *et al.*, 2006). This trapped-atom beam splitter has been then used in experiments by the same group to study the equilibration

[17]An acousto-optic modulator is a device which uses sound waves to diffract a laser beam (the deflection angle depends on the frequency of the sound wave).

[18]A different situation will be described in Chapter 6 devoted to the physics of optical lattices, in which the quantum tunnelling between different lattice sites is responsible for quantum transport in the lattice and for the emergence of a band structure in the energy spectrum.

[19]Dressed states in a two-level system will be further discussed in Sec. 5.2.1 devoted to the explanation of the optical dipole force in terms of the ac-Stark shift effect.

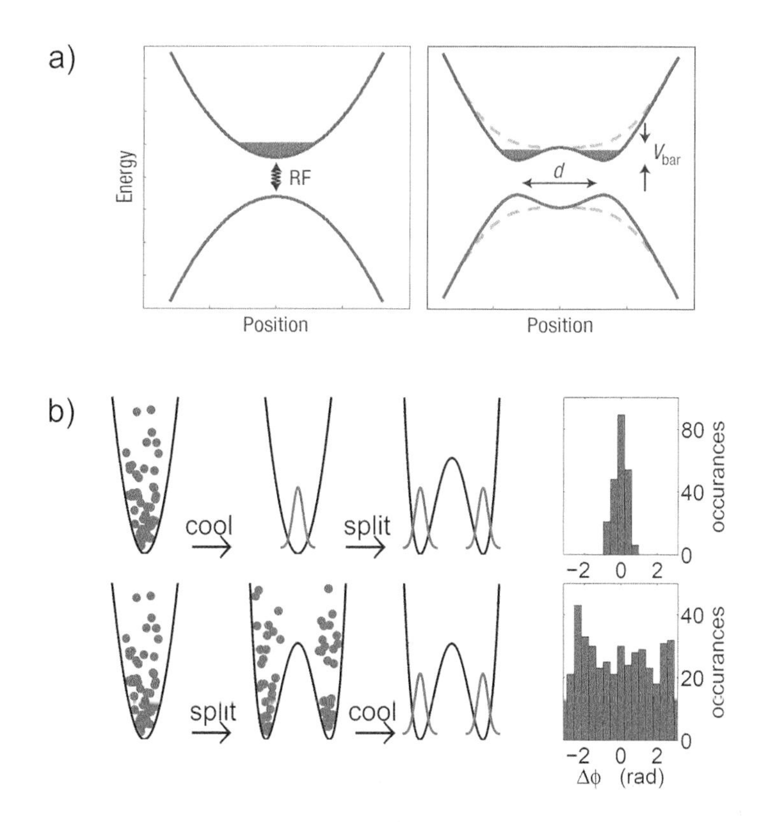

Fig. 3.21 Double-well trapped BEC interferometer based on adiabatic potentials. a) Scheme for the production of adiabatic potentials based on the energy of the dressed states in a magnetic trap in the presence of a strong RF coupling. b) Relative phase of two spatially separated Bose–Einstein condensates after coherent splitting of a single trap into a double-well potential (above) or after independent production in two separated traps (below). Reprinted with permission from Schumm *et al.* (2005) (a) and Hofferberth *et al.* (2006) (b). © Macmillan Publishers Ltd.

dynamics and the coherence properties of interacting 1D Bose gases (Hofferberth *et al.*, 2007; Hofferberth *et al.*, 2008).

Effects of interaction. Atom interferometers with trapped BECs provide high contrast and high spatial resolution, but there is a downside. Differently from photons, atoms interact among themselves. Trapped condensed clouds have a much larger density ($n \simeq 10^{14}$ atoms/cm^3) than the atomic beams or the clouds of laser-cooled atoms ($n \simeq 10^9$ atoms/cm^3) employed in traditional atom interferometers. As a consequence, whereas atom–atom interactions can be usually disregarded (or cause only a minor effect) in fountain-based atom interferometers, they have a significant role in experiments with trapped BECs, causing a deviation of the BEC interferometer operation from the ideal case of a collection of N independent particles. In high-density trapped condensed clouds, interactions induce phase diffusion (Castin and Dalibard, 1997) and can cause systematic frequency shifts due to uncontrolled atomic density gradient, thus seriously limiting the performances of a BEC atom interferometer.

There are two possible solutions to this problem. A first solution relies on changing Bose–Einstein condensates for spin-polarized fermionic gases, which are naturally noninteracting (see Sec. 3.1.4): interaction phase shifts are cancelled, but at the price of a strong reduction of the interference contrast (since fermionic particles occupy different quantum states) and reduced spatial resolution. A second solution relies on the possibility of cancelling interactions between the atoms in a BEC with a Feshbach resonance (Sec. 3.1.4). While interactions are fundamental for the efficiency of evaporative cooling, after the BEC is formed they can be turned to zero by a proper magnetic field and the trapped interferometer can be operated with ideal Bose–Einstein condensates in which N phase-coherent particles occupy the same single-particle state with no interactions among them: in this way the interferometer contrast is maximized with no loss in either coherence time or spatial resolution. Both solutions will be discussed in Sec. 6.4.2 in the context of atom interferometry with quantum gases in optical lattices.

Interactions, however, do not have only a deleterious effect on the interferometer contrast. Their control can be used to engineer nonclassical atomic states for the enhancement of the interferometer sensitivity below the standard quantum limit (in the spirit of what has been discussed in Sec. 2.4.1 in the context of metrological measurements), as described below.

Quantum interferometry with BECs. In the double-well BEC interferometers described in the previous pages the condensates were released from the trap right after the phase accumulation, giving rise to an interference density pattern typical of a double-slit interferometer (see Figs. 3.17 and 3.20). Alternative schemes can be employed for the readout of the interferometer phase. Larger sensitivities can be obtained with trapped Mach–Zehnder atom interferometers, where the phase accumulation step is preceded and followed by two 50%–50% beam splitter operations (as in the atom interferometers discussed in Sec. 2.5) and the phase readout is obtained by counting the number of particles in the two output modes. The phase sensitivity obtained with noninteracting atoms is limited by the atomic detection shot noise and is proportional to $\sqrt{N}$, where N is the total atom number. This limit, however, is not fundamental, as we have already discussed in Sec. 2.4.1, and can be overcome by preparing the atoms in a proper entangled state (Giovannetti *et al.*, 2004; Pezzé and Smerzi, 2009). Entangled states can be obtained by taking advantage of the atom–atom interactions that naturally characterize a trapped atomic BEC and thus represents the most interesting aspect of BEC interferometry. The ultimate sensitivity limit, the so-called *Heisenberg limit*, scales as N, thus promising an increase of phase sensitivity of several orders of magnitude.

There are several schemes to achieve entangled states useful for reaching a sub-shot-noise sensitivity in a BEC Mach–Zehnder interferometer, and some of them have already been demonstrated experimentally. For instance, entangled states are obtained as the ground state of a split double-well potential (Milburn *et al.*, 1997; Cirac *et al.*, 1998; Raghavan *et al.*, 1999) in the presence of interactions (Pezzé *et al.*, 2005). Without interactions, the ground state of N two-mode bosons is $(\hat{a}^\dagger + \hat{b}^\dagger)^N|0\rangle$, where $\hat{a}^\dagger$ and

$\hat{b}^\dagger$ are creation operators for the two lowest modes of the double well and $|0\rangle$ is the vacuum state corresponding to the two empty modes. This state corresponds to a Poissonian distribution of particles among the two modes and is a separable state.[20] As repulsive interactions increase, number fluctuations decrease. When interactions dominate, the ground state is $\left[(\hat{a}^\dagger)^{N/2} + (\hat{b}^\dagger)^{N/2}\right]|0\rangle$, corresponding to a twin-Fock state of $N/2$ particles in each of the two modes, without number fluctuations. The indirect demonstration that the adiabatic splitting of an initially coherent condensate leads to reduced number fluctuations (number squeezing) was achieved in lattice (Orzel *et al.*, 2001) and double-well (Jo *et al.*, 2007) systems, and later investigated in different experiments (Gerbier *et al.*, 2006; Maussang *et al.*, 2010; Sebby-Strabley *et al.*, 2007).

To demonstrate useful entanglement, however, it is necessary not only to have number squeezing, i.e. $[\Delta(N_a - N_b)]^2 \leq N$, but also a sufficient amount of coherence in order to fulfill the *spin squeezing*[21] condition $\xi^2 = N(\Delta N)^2/\langle \hat{a}^\dagger\hat{b} + \hat{b}^\dagger\hat{a}\rangle < 1$ (Wineland *et al.*, 1994). Spin squeezing in a BEC trapped in double-well and multiple-well potentials was first demonstrated by the group of M. K. Oberthaler at Heidelberg (Estève *et al.*, 2008). There, spin squeezing was measured directly by counting the number of particles in the two modes of the split BEC, while coherence was directly measured by the visibility of the averaged interference density pattern obtained in time-of-flight absorption imaging. A second way to generate spin squeezing is to exploit the collisional dynamics of BECs in two hyperfine levels, as first suggested in Kitagawa and Ueda (1993) and later adapted to BECs in Sørensen *et al.* (2001). The dynamical creation of entanglement poses several experimental challenges related to the coherent control of inter- and intra-species interactions. This was achieved experimentally in two experiments, which succesfully demonstrated spin squeezing: in Gross *et al.* (2010) intra-species interactions were tuned to zero by using a proper magnetic field, while in Riedel *et al.* (2010) they were controlled by employing a state-dependent potential varying the overlap of the two hyperfine state wavefunctions. In Gross *et al.* (2010) this resource was used to obtain the first experimental demonstration of sub-shot-noise sensitivity in a BEC Ramsey interferometer.

More recently, *spin-changing collisions* have been used to create entangled states in spinor BECs, as first suggested in Pu and Meystre (2000) and Duan *et al.* (2000). In a spinor $F = 1$ ^{87}Rb BEC initially prepared in the $m_F = 0$ state, the collision of two particles may generate a correlated pair of atoms in the $m_F = \pm 1$ states.[22] Within this scheme, number squeezing between the $m_F = \pm 1$ states was demonstrated in Bookjans *et al.* (2011), Lücke *et al.* (2011), and Gross *et al.* (2011). Number squeezing was also observed with other schemes involving the creation of correlated pairs of atoms

[20] The behaviour of an interacting bosonic system in a double-well potential will be considered in detail in Sec. 7.1 devoted to the investigation of interacting bosons in multi-well periodic potentials.

[21] Spin squeezing is a sufficient condition for entanglement (Ma *et al.*, 2011; Sørensen *et al.*, 2001).

[22] Formally, this process is equivalent to photon pair creation by optical four-wave mixing in non-degenerate optical parametric down-conversion (see Appendix B.3 for an introduction to nonlinear optics). In this analogy, the large BEC in $m_F = 0$ resembles a coherent pump, the nonlinear interactions in the BEC provide the equivalent of the nonlinear crystal, and the produced atoms in $m_F = \pm 1$ play the role of signal and idler. Pair production leads to an exponential amplification of the population in $m_F = \pm 1$, which resembles the gain of an optical parametric amplifier.

(Jaskula *et al.*, 2010; Bücker *et al.*, 2011). In particular, by employing the state created by spin-changing collisions, a sub-shot noise sensitivity in the estimation of a rotation angle in the Bloch sphere was experimentally demonstrated in Lücke *et al.* (2011).

3.3 Fermi gases

After the realization of atomic Bose–Einstein condensates, important experimental efforts were directed towards the achievement of quantum degeneracy in atomic Fermi gases. The interest for fermionic systems is justified by the consideration that most of the fundamental particles of Nature (quarks, leptons, baryons) are fermions and that quantum effects in Fermi gases are fundamental for the physics of many systems: electrons in metals and semiconductors, superconductors, superfluid ^{3}He, nuclear and quark matter, neutron stars.

The experimental techniques for cooling, trapping, and detecting Fermi gases are similar to the ones used for the production of Bose–Einstein condensates. The main difference concerns the implementation of evaporative cooling. As a matter of fact, as already discussed in Sec. 3.1.4, identical fermions at low temperature stop interacting because of the suppression of $s-$wave scattering caused by the antisymmetrization of the wavefunction. Therefore direct evaporative cooling of spin-polarized fermions works inefficiently owing to the lack of thermalization. Different strategies, such as cooling fermionic particles in multiple internal states (to make them distinguishable) or sympathetic cooling with other isotopes/atoms (see Sec. 3.1.3), are required to achieve Fermi degeneracy.

Differently from the case of Bose–Einstein condensation, in which the onset of degeneracy can be easily detected from the appearance of a bimodal density distribution, in the case of spin-polarized fermions the effect is less striking, since no phase transition occurs. For an ideal free Fermi gas the Fermi energy E_F (i.e. the energy of the highest level occupied at $T = 0$) is given by

$$E_F = \frac{\hbar^2}{2m}\left(3\pi^2 n\right)^{2/3} , \tag{3.24}$$

where n is the density of the gas, and it defines the scale of temperature $T_F = E_F/k_B$ below which the level occupation starts to significantly deviate from the Maxwell–Boltzmann distribution. Similarly to the BEC case, this happens when the phase-space density is on the order of unity, more precisely from the equation above it follows that

$$n\lambda_{dB}^3 \simeq 1.504 . \tag{3.25}$$

Since no phase transition occurs, the detection of degeneracy can be performed by measuring deviations in the momentum distribution from the classical Maxwell–Boltzmann statistics or by observing a decrease in the cooling efficiency which results from Pauli blocking in the completely filled core of the Fermi gas.

The first quantum degenerate gas of fermionic atoms was realized in 1999 by B. DeMarco and D. S. Jin at JILA (DeMarco and Jin, 1999) using a mixture of two different internal states of ^{40}K. While early experiments on magneto-optical trapping of ^{40}K used potassium samples with natural isotopic composition (Cataliotti *et al.*, 1998), its very low natural abundance (only 0.012%) motivated the JILA group to develop novel atomic sources with enriched samples (DeMarco *et al.*, 1999). The other fermionic alkali atom ^{6}Li was degenerated in 2001 by the group of R. Hulet at Rice University (Truscott *et al.*, 2001) and by the group of C. Salomon at ENS (Schreck *et al.*, 2001*b*) using in both cases sympathetic cooling with bosonic ^{7}Li. Fermi gases of lithium and potassium were then produced in 2002 by sympathetic cooling in heterospecies atomic mixtures ^{23}Na–^{6}Li (Hadzibabic *et al.*, 2002) at MIT and ^{87}Rb–^{40}K at LENS (Roati *et al.*, 2002). More recently, non-alkali degenerate Fermi gases have been produced with ^{3}He* (McNamara *et al.*, 2006), ^{173}Yb (Fukuhara *et al.*, 2007), ^{171}Yb (Taie *et al.*, 2010), ^{87}Sr (DeSalvo *et al.*, 2010; Tey *et al.*, 2010), and ^{161}Dy (Lu *et al.*, 2012).

Soon after the demonstration of Fermi degeneracy in DeMarco and Jin (1999), the investigation of ultracold fermions focused mostly on the characterization of the thermodynamic properties of ideal trapped Fermi gases, including their application to precise measurements (see Sec. 6.4.2) and the demonstration of absence of interaction-induced clock shifts (Gupta *et al.*, 2003). Then the interest in ultracold fermions rapidly moved towards the manipulation of interactions in two-component fermionic mixtures with Feshbach resonances. A particularly interesting field of research is connected to the investigation of strongly interacting fermionic gases, in order to experimentally access the physics of fermionic superfluidity, i.e. the so-called *BEC–BCS* crossover, which is described below. More information on the history of the first years of investigation of ultracold Fermi gases can be found in the exhaustive contributions to the Proceedings of the CLXIV International Enrico Fermi School organized in 2007 in Varenna (Inguscio *et al.*, 2008).

3.3.1 Fermionic superfluidity

While the superfluid behaviour of a bosonic gas emerges together with its condensation, in the case of fermions the situation is different. Fermionic superfluidity emerges from the pairing of fermions, which typically happens at a much lower temperature than the Fermi temperature T_F: at $\approx 10^{-5}T_F$ in the case of the electron gas in conventional superconductors, at $\approx 10^{-3}T_F$ in the case of ^{3}He, at $\approx 10^{-2}T_F$ in the case of high-Tc superconductors. The mechanism of electron pair condensation in conventional superconductors was explained in 1957 by the celebrated BCS theory of J. Bardeen, L. N. Cooper, and J. R. Schrieffer (Bardeen *et al.*, 1957): spin-up/spin-down electrons on opposite points of the Fermi surface form a weakly bound pair thanks to an effective attraction mediated by the lattice phonons. While a similar mechanism holds for superfluid ^{3}He as well, high-Tc superconductivity is not yet understood from a theoretical point of view and experiments with cold fermionic gases could offer new insight in the understanding of the relevant processes.

As a matter of fact, ultracold-atoms experiments can access a large interval of parameters and explore fermionic pairing in different regimes of interactions. As first conjectured by A. J. Leggett (Leggett, 1980) there is a smooth crossover which links the formation of weakly bound Cooper pairs in a fermionic many-body system (BCS limit) to the creation of tightly bound bosonic molecules which then undergo Bose–Einstein condensation (BEC limit): accordingly, the size of the fermionic pairs changes smoothly from being much larger than the interparticle distance in the former limit, to being

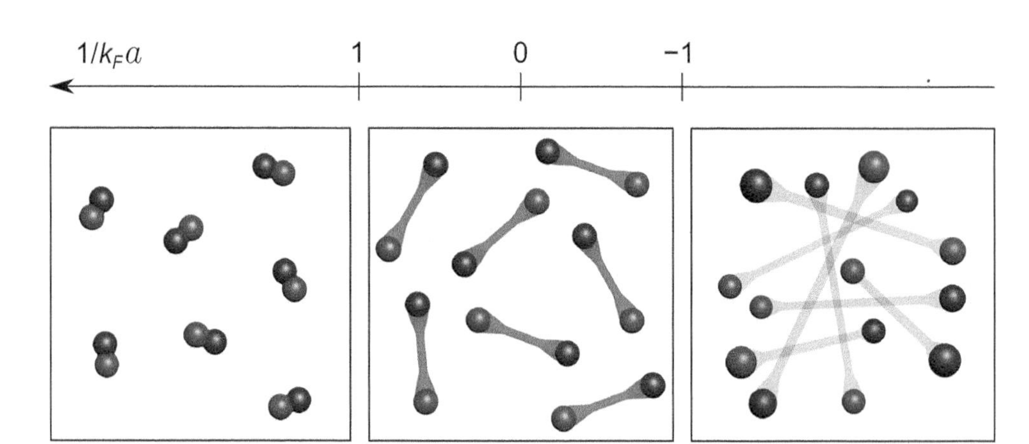

Fig. 3.22 Illustration of the BEC–BCS crossover. Changing the value of the $s-$wave scattering length a around a Feshbach resonance, a binary mixture of fermions in two spin states evolves from the condensation of tightly bound bosonic molecules (on the BEC side) to the formation of weakly bound Cooper pairs with a size much larger than the interparticle distance (on the BCS side). Reproduced from Ketterle and Zwierlein (2008) with kind permission of SIF © 2008.

much smaller in the latter, where a real molecular bound is formed, as illustrated in Fig. 3.22. The parameter which controls this crossover is $1/k_F a$, where a is the scattering length and $k_F = \sqrt{2mE_F}/\hbar$ is the Fermi momentum. The experimental investigation of the different regimes of the BEC–BCS crossover can be realized by means of interaction tuning around a Feshbach resonance (as described in Sec. 3.1.4): as the magnetic field is swept across the resonance the molecules on the BEC side for $1/k_F a > 1$ adiabatically transform into Cooper pairs on the BCS side for $1/k_F a < -1$.

Figure 3.23 shows a theoretical calculation of the critical temperature for fermionic superfluidity as a function of the interaction parameter $1/k_F a$ along the BEC–BCS crossover (Haussmann *et al.*, 2007): while condensation of pairs is easily accessible in the strongly interacting regime and on the BEC side, the condensation of Cooper pairs in the weakly attractive BCS limit $1/k_F a \ll -1$ is hardly achievable in ultracold experiments, because of the exponentially small critical temperatures $T_C \sim \exp(-\pi/2k_F|a|)$ for long-range pairing on the BCS side (dashed line). A regime of particular interest which can be experimentally accessed is the *unitarity* regime, which is realized when the scattering length becomes larger than the interparticle spacing, i.e. when $1/k_F|a| < 1$. In the limit of $|a| \to \infty$ the behaviour of the system does not depend on the strength (and sign!) of the interatomic potential and the only relevant scale is fixed by the Fermi momentum k_F: the behaviour of the Fermi gas at unitarity is therefore universal. For more information the reader can refer to Giorgini *et al.* (2008) for a review on the theory of ultracold fermionic gases and to Ketterle and Zwierlein (2008) for a comprehensive review of techniques and experimental results connected to the investigation of fermionic superfluidity in atomic systems.

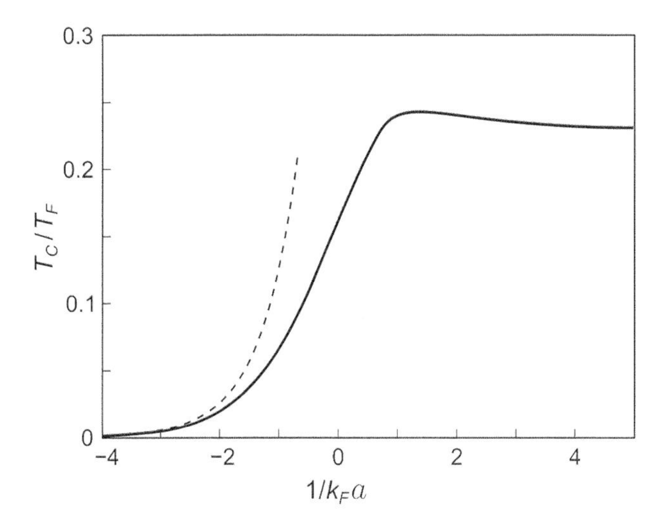

Fig. 3.23 Critical temperature for fermionic superfluidity (in units of the Fermi temperature) as a function of the interaction parameter $1/k_F a$ across the BEC–BCS crossover. The dashed line shows the results of the BCS theory. Reprinted with permission from Haussmann *et al.* (2007). © American Physical Society.

Evidence of strong interactions in two-component Fermi gases was obtained with the first observation of anisotropic expansion, similar to the one observed for Bose–Einstein condensates, in 2002 (O'Hara *et al.*, 2002). It was then realized that this effect is not by itself evidence of superfluidity, since it can be observed also for a strongly interacting normal gas in the collisional hydrodynamic regime. Bose–Einstein condensation of molecules on the BEC side was observed by several groups in 2003 (Greiner *et al.*, 2003; Jochim *et al.*, 2003; Zwierlein *et al.*, 2003) thanks to the onset of a bimodal distribution similar to the one observed for a bosonic gas. It is interesting to note that Feshbach molecules formed by two fermions have a much longer lifetime than molecules formed by two bosonic atoms: despite being formed in a highly excited vibrational state,[23] the Pauli exclusion principle strongly suppresses collisional relaxation (Petrov *et al.*, 2004). Later, Bose–Einstein condensation of fermionic pairs was observed on the BCS side of the crossover as well (Regal *et al.*, 2004; Zwierlein *et al.*, 2004), although far from the perturbative BCS limit. Other experiments with strongly interacting fermions investigated collective excitations (Kinast *et al.*, 2004; Bartenstein *et al.*, 2004), expansion (Bourdel *et al.*, 2004), radiofrequency excitation spectra (Chin *et al.*, 2004), and thermodynamics (Kinast *et al.*, 2005) across the BEC–BCS crossover, evidencing hints at the superfluid behaviour of the interacting Fermi gas.

As already discussed for Bose–Einstein condensates in Sec. 3.2.3, the most important evidences for superfluidity are connected with the existence of a critical velocity and to the quantization of vortices. A smoking-gun evidence for fermionic superfluidity

[23]See Sec. 3.5.1 for a discussion of recent experimental advances which have resulted in the production of ultracold molecules in their rovibrational ground state.

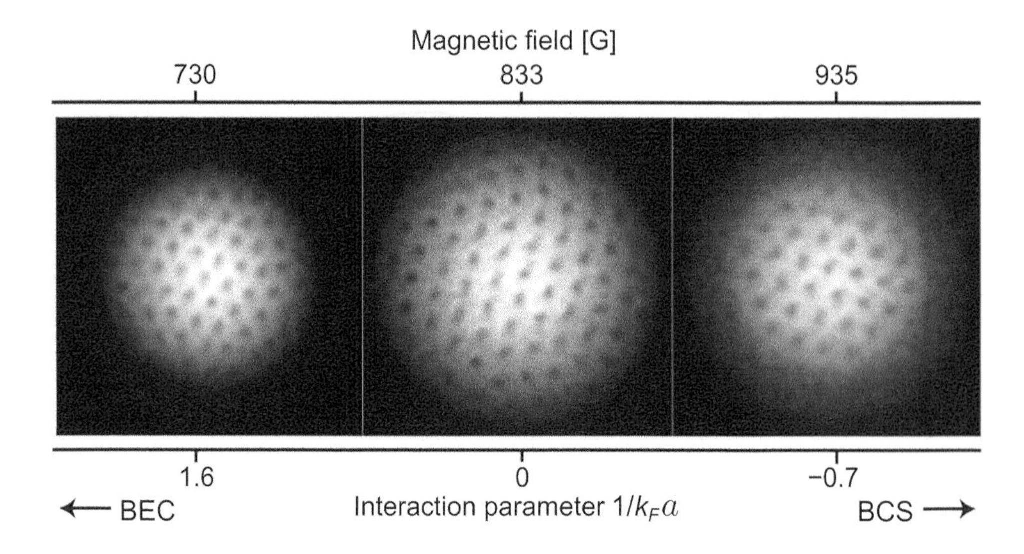

Fig. 3.24 Observation of vortex lattices in strongly interacting ultracold fermions at the BEC–BCS crossover (Zwierlein *et al.*, 2005). Reproduced from Ketterle and Zwierlein (2008) with kind permission of SIF © 2008.

across the whole BEC–BCS crossover was obtained at MIT in 2005 (Zwierlein *et al.*, 2005) with the observation of vortex lattices in two-component ^{6}Li gases. Figure 3.24 shows time-of-flight pictures of two-component Fermi gases put into rotation by a pair of stirring laser beams rotating around an axis intersecting the centre of the cloud. The formation of a vortex lattice is evident for a large range of interaction parameters, starting from the BEC side, through the strongly interacting unitary limit, all the way to the BCS side. The uniformity of the lattice spacings is a signature of the quantization of the vortex charge, which in turn is a strong evidence for fermionic superfluidity.

Imbalanced superfluidity. In the conventional theory of fermionic superfluidity, pairing is obtained between two Fermi gases with equal population, as in the case of the spin-up/spin-down components of the electron gas in a superconductor. Fermionic pairing in the presence of polarization, i.e. a mismatch between the chemical potentials of the two Fermi components, is important in relevant physical situations, e.g. in magnetized superconductors or, notably, in high-density quark matter at the core of compact stars. From the theoretical point of view, the phase diagram of imbalanced Fermi gases has been the subject of a long and yet not completely solved debate. Exotic forms of superfluidity have been predicted, including the Fulde–Ferrell–Larkin–Ovchinnikov (FFLO) phase, in which the mismatch of the Fermi surfaces is expected to favour the BCS-like pairing of two fermions in a Cooper pair with nonzero momentum, with a spontaneous breaking of spatial symmetry (see e.g. Casalbuoni and Nardulli (2004) for a review).

Fermionic superfluidity in imbalanced ultracold Fermi gases was investigated with unequal spin mixtures of ^{6}Li in Zwierlein *et al.* (2006), where the persistence of

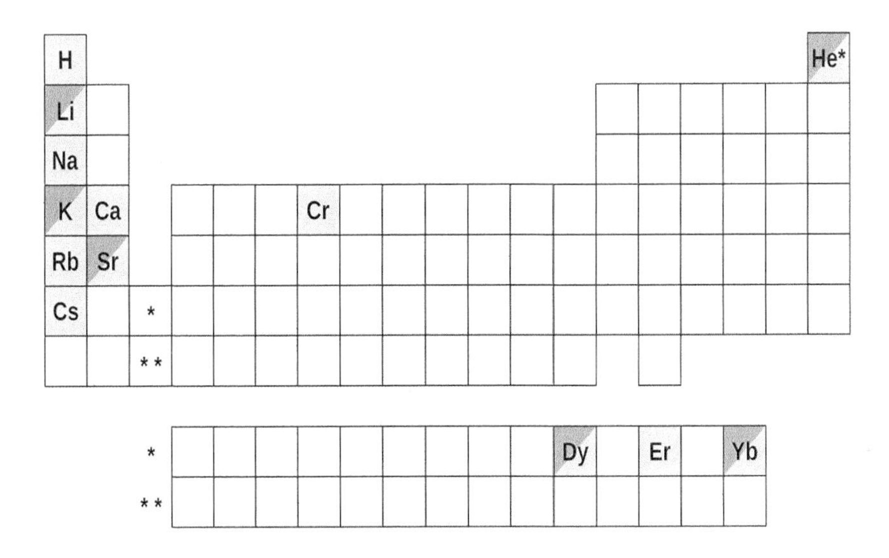

Fig. 3.25 Periodic table showing the elements for which quantum degeneracy has been demonstrated. Light grey refers to the elements which have been Bose-condensed, dark grey refers to the elements with fermionic isotopes for which a Fermi gas has been reported.

superfluidity was demonstrated by measuring the condensed fraction and probing the formation of vortex lattices for a large range of population imbalance. A similar system was studied in Partridge *et al.* (2006), where *in-situ* imaging was used to investigate the onset of phase separation between a mixed-component part and a majority-component part and in Shin *et al.* (2006), where phase-contrast *in-situ* imaging was used to show that the densities of the two components in the paired region were precisely the same. These observations have demonstrated the formation of a conventional superfluid state with balanced population in the centre of the trap, spatially separated from the excess atoms of the majority component lying externally to the superfluid core, all the way from the BEC to the BCS limit of the BEC–BCS crossover. No evidence has been reported so far of exotic forms of fermionic superfluidity such as the long-searched FFLO phase.

3.4 Non-alkali BECs

Although most of the running BEC and Fermi gas experiments are based on alkali atoms, several other elements belonging to different chemical families have been brought to quantum degeneracy, as highlighted in the periodic table in Fig. 3.25. We summarize here below the key features offered by these non-alkali quantum gases.

3.4.1 Hydrogen

Hydrogen has always had a guiding role in the search for Bose–Einstein condensation in atomic gases. The history of hydrogen trapping and cooling has been marked by fundamental developments, such as magnetic trapping and evaporative cooling, which have later allowed the achievement of BEC in alkali gases. This long and creative story is narrated in the proceedings of the lecture given by D. Kleppner at the CXL Enrico Fermi School on "Bose–Einstein condensation in atomic gases" in 1998 (Kleppner *et al.*, 1999), just a few weeks after the achievement of BEC in hydrogen at MIT.

A proposal of observing Bose–Einstein condensation in atomic hydrogen appeared as early as 1959 (Hecht, 1959) with the conjecture that spin-polarized hydrogen was expected to remain a gas down to absolute zero, then confirmed in 1976 with the publication of many-body calculations (Stwalley and Nosanow, 1976). Hydrogen has also the advantages of being the lightest atom, which means the highest critical temperature for quantum degeneracy (see eqn (3.14)), and to exhibit much weaker atom–atom interactions than in superfluid helium, making it a cleaner system than helium for the observation of Bose–Einstein condensation. Research on hydrogen started in the late 1970s and continued for the whole of the following two decades. Spin-polarized hydrogen was first stabilized by I. F. Silvera and J. T. M. Walraven in 1980 (Silvera and Walraven, 1980) and confined in a magnetic-bottle trap in the same year at MIT (Cline *et al.*, 1980). In order to achieve smaller temperatures than ≈ 100 mK and limit the effects of three-body recombination at the cryostat walls, H. Hess proposed the implementation of wall-free confinement in a pure magnetic trap and evaporative cooling as the path towards BEC (Hess, 1986). These ideas were successful: spin-polarized hydrogen was confined in a pure magnetic trap in 1987 by the MIT group (Hess *et al.*, 1987) and evaporative cooling was demonstrated in 1988 (Masuhara *et al.*, 1988), several years before its application to alkali gases. Eventually, Bose–Einstein condensation of hydrogen was demonstrated in 1998 (Fried *et al.*, 1998).

The hydrogen BEC is the only Bose–Einstein condensate of atomic gases so far which is not produced by laser cooling. As discussed in Sec. 1.5.2, laser cooling of hydrogen presents a major obstacle with respect to alkalis, owing to the large Lyman-α transition energy which would require vacuum-ultraviolet laser light at a $\lambda = 121.5$ nm wavelength. As a consequence, hydrogen pre-cooling has to rely on traditional cryogenic techniques, which result in higher temperatures than the ones of laser-cooled alkali gases (≈ 100 mK) but, compared to the latter, allow orders of magnitude more atoms to be trapped, resulting in hydrogen BECs that are much larger than the ones achieved in other systems. Differently from the imaging techniques discussed previously in this chapter, in Fried *et al.* (1998) the detection of the BEC was performed spectroscopically, by investigating the two-photon spectrum of the $1s - 2s$ transition at 243 nm (which was extensively discussed in Chapter 1 for its spectroscopic implications). The spectrum, shown in Fig. 3.26a, shows that the BEC signal (solid circles) is shifted and asymmetrically broadened compared to the signal of the normal gas (empty circles) owing to the collisional redshift in the condensed part of the hydrogen cloud, which has a much larger density than the normal component. The analysis of this bimodal spectrum provides a measurement of the number of condensed atoms $N_C \sim 10^9$ and of the condensate fraction, which is limited to $\sim 5\%$ because of the large dipolar-relaxation rate which provides a source of heating in the trap. From the same spectra it is also possible to extract the density of the normal component. The plot in Fig. 3.26b shows that, lowering the trap depth during the evaporative cooling process, at a critical value corresponding to a critical temperature of ≈ 50 μK, the normal component density

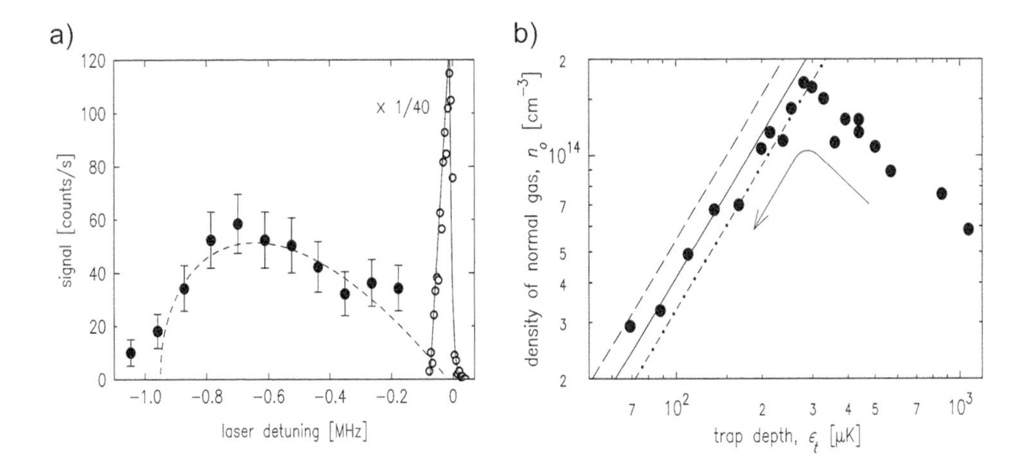

Fig. 3.26 Hydrogen BEC. a) $1s - 2s$ two-photon spectrum showing the collisionally shifted and broadened spectrum of the high-density BEC (filled circles) compared to the narrower and symmetric spectrum of the normal component (empty circles). b) Density of the normal component, showing saturation below the critical trap depth for the onset of Bose–Einstein condensation (central line: predicted saturation density). Reprinted with permission from Fried *et al.* (1998). © American Physical Society.

saturates at the value predicted by Bose–Einstein statistics (central line): below this critical trap depth the normal gas density remains saturated and the condensate forms.

3.4.2 Two-electron atoms

Metastable helium. Although helium gases have not been condensed yet in their $^1S_0$ ground state, BECs of ^{4}He and Fermi gases of ^{3}He in the metastable $^3S_1$ electronic state have been produced since 2001 (Robert *et al.*, 2001; Pereira Dos Santos *et al.*, 2001) and are currently used for quantum metrology and for quantum atom-optics experiments. Their production and applications will be discussed extensively in Sec. 4.3 within the chapter dedicated to helium.

Alkaline-earth atoms. In atomic physics experiments alkaline-earth atoms are mostly used for high-precision spectroscopic measurements, since they exhibit ultra-narrow intercombination transitions which makes them appealing for the realization of optical atomic clocks, as will be discussed in Chapter 5. Recently, however, a growing interest has been put into the realization of alkaline-earth quantum gases. Among the peculiar properties of these atoms, the rich electronic structure and the absence of electronic angular momentum in the ground state (causing the emergence of a purely nuclear spin) make them attractive for applications ranging from quantum information processing to quantum simulation of novel condensed-matter states. The first BEC of this kind was realized with alkaline-earth-like ytterbium (Yb) (Takasu *et al.*, 2003), which was also used to achieve two-electron Fermi degeneracy (Fukuhara *et al.*, 2007) and different kinds of quantum mixtures. More recently, quantum degenerate gases of alkaline-earth atoms have been realized also in bosonic calcium (Ca) (Kraft *et al.*, 2009) and bosonic and fermionic strontium (Sr) gases (Stellmer *et al.*, 2009; Martinez de Escobar *et al.*, 2009; DeSalvo *et al.*, 2010; Tey *et al.*, 2010).

3.4.3 Dipolar atoms

Recent experiments have obtained Bose–Einstein condensation of transition and rare-earth metals which are characterized by a complex electronic structure with open d or f electronic shells. The partial filling of these shells is responsible for the very large magnetic dipole moment of these elements: $6\mu_B$ for chromium (Cr), $7\mu_B$ for erbium (Er), and $10\mu_B$ for dysprosium (Dy). The goal of these experiments is the investigation of new physics arising from the *dipole–dipole interaction*,[24] which differs from the usual s-wave contact interaction discussed in Sec. 3.1.4 by its long range and its strongly anisotropic character. These properties can be used to investigate novel strongly correlated quantum states involving quantum magnetism, spontaneous spatial symmetry breaking, and exotic superfluidity. A detailed discussion of the properties of the dipole–dipole interaction will be presented in Sec. 3.5 devoted to the physics of ultracold polar molecules, in which strong dipole–dipole interactions arise from an *electric*, instead of a magnetic, dipole moment. Although the dipole moment of magnetic atoms is weaker than the electric dipole moment of molecules, quantum degenerate gases of magnetic atoms can be obtained in a straightforward manner by extending the techniques previously discussed for alkali atoms.

The first achieved dipolar BEC was obtained with ^{52}Cr atoms by the group of T. Pfau at Stuttgart (Griesmaier *et al.*, 2005) and later in Beaufils *et al.* (2008). Although in ^{52}Cr the dipole–dipole interaction is weaker than contact interactions, the Stuttgart group demonstrated the possibility of cancelling the latter with a Feshbach resonance, thus realizing quantum gases interacting with a pure dipole–dipole interaction (Lahaye *et al.*, 2007; Lahaye *et al.*, 2008).[25] In very recent experiments lanthanide dysprosium (Lu *et al.*, 2011) and erbium (Aikawa *et al.*, 2012) have been Bose-condensed. For these two elements fermionic isotopes exist as well, which enlarge the possible physics that can be explored with dipolar gases, and in the case of dysprosium they have been already brought to quantum degeneracy (Lu *et al.*, 2012). As a major consequence of the unusually large magnetic moment, the dipole–dipole interaction in lanthanides can be comparable or even much stronger than the background ("bare") contact interaction. In the case of dysprosium strong effects of the dipole–dipole interaction have been observed even in the absence of Feshbach resonances cancelling the contact interaction (Lu *et al.*, 2011).

3.5 Cold molecules

We conclude this chapter with a brief introduction to the ongoing research on ultracold molecules. Compared to atoms, molecules offer an enriched range of possibilities,

[24] For comparison, the magnetic moment μ of rubidium is only $1\mu_B$. Since the intensity of the dipole–dipole interaction scales as μ^2 (see Sec. 3.5.2), dipolar effects in dysprosium are expected to be a factor 100 greater than in rubidium.

[25] A weak effect of dipole–dipole interaction in alkali atoms can be evidenced as well when the contact interactions are suppressed, as discussed in Sec. 6.4.2.

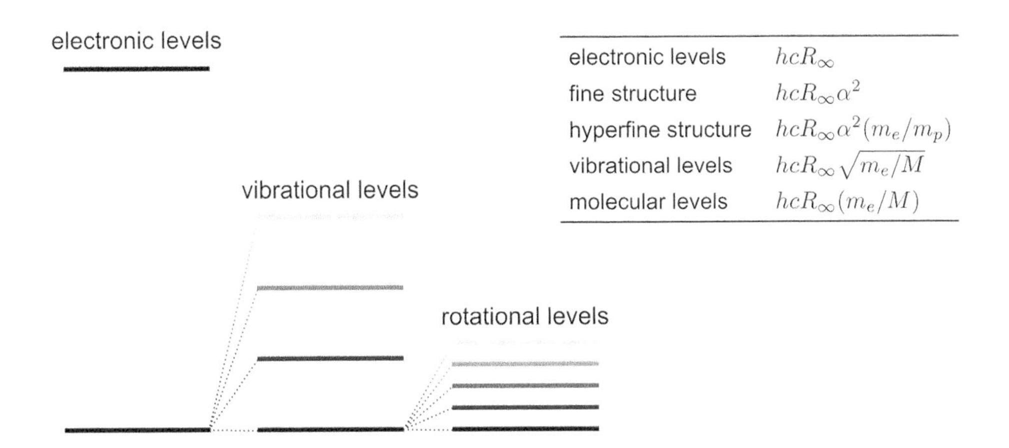

electronic levels	hcR_∞
fine structure	$hcR_\infty\alpha^2$
hyperfine structure	$hcR_\infty\alpha^2(m_e/m_p)$
vibrational levels	$hcR_\infty\sqrt{m_e/M}$
molecular levels	$hcR_\infty(m_e/M)$

Fig. 3.27 Simplified scheme of the energy levels in a molecule, with the typical energy scales for atomic and molecular level separations (R_∞ is the Rydberg constant, α is the fine structure constant, m_e is the electron mass, m_p is the proton mass and M is the molecular mass).

since they are characterized by additional internal degrees of freedom. Besides their electronic structure, molecules possess vibrational and rotational levels (see Fig. 3.27 for a simplified scheme of the rovibrational molecular structure, with the typical energy scales which are associated with the level splittings). Furthermore, heteronuclear molecules, i.e. molecules formed by two or more atoms of different chemical species, typically have a permanent electric dipole moment, which is responsible for the large electric polarizability of the molecule and for long-range dipolar interactions with a very different character from the contact interactions between neutral atoms discussed in Sec. 3.1.4.

The realization of cold molecules is a long-pursued goal that could have important applications in many different fields, ranging from high-precision molecular spectroscopy for tests of fundamental symmetries, to the quantum control of chemical reactions, to the production of degenerate quantum gases of molecules with dipolar interactions, to applications of polar molecules in quantum computation. In this section we will mention only some applications, leaving the reader to more specialized reviews, e.g. Krems *et al.* (2009) and Carr *et al.* (2009).

3.5.1 Cooling molecules

Unfortunately, the laser-cooling techniques developed for atoms cannot be directly extended to molecules: the rich internal structure of molecules becomes a major drawback for laser cooling, since it determines the lack of those closed, cycling transitions that can be used for an efficient cooling of atoms. As a matter of fact, a molecule in an electronically excited state can decay into a multitude of vibrational and rotational levels (see Fig. 3.27) which are generally uncoupled to the cooling light. Therefore, different strategies have been implemented to produce cold molecules (see Fig. 3.28 for a scheme of the different approaches).

direct methods

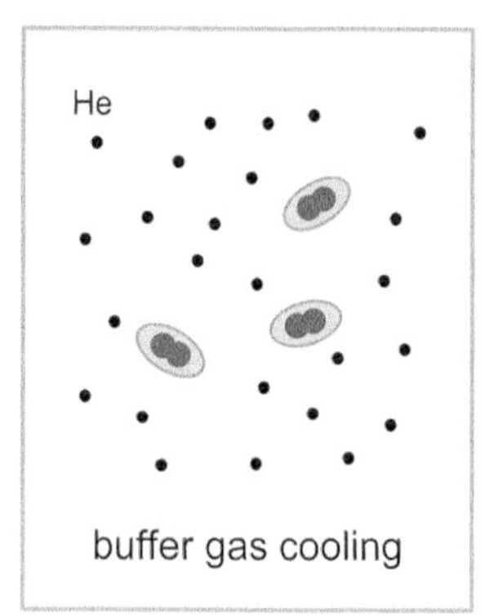

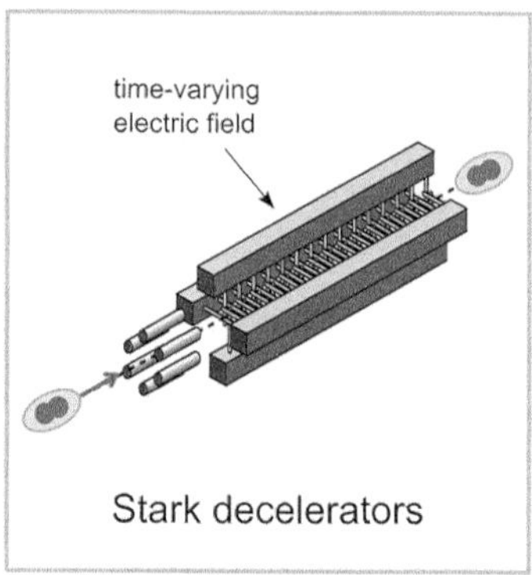

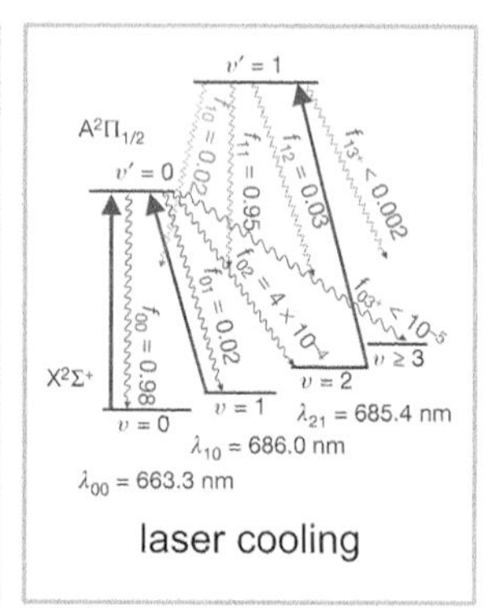

indirect methods

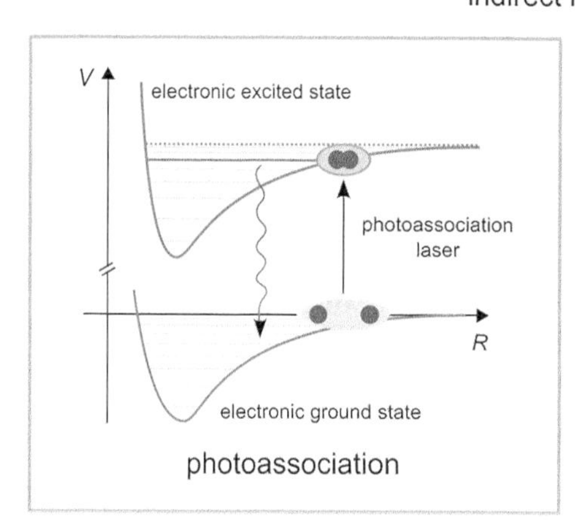

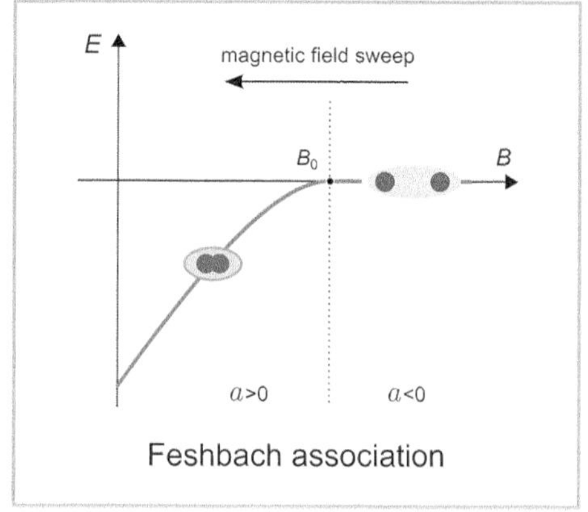

Fig. 3.28 Scheme of the different approaches to the production of ultracold molecules. The Stark decelerator figure is reprinted with permission from Bethlem *et al.* (1999). © American Physical Society. The laser-cooling scheme is reprinted with permission from Shuman *et al.* (2010). © Macmillan Publishers Ltd.

Direct cooling of stable molecules. A first strategy to achieve cold molecular samples deals with *direct cooling* methods, where the starting point is a hot gas of stable molecules.

A quite general technique, which is applicable to a large variety of molecular species, is based on *buffer gas cooling*. This method, successfully demonstrated for CaH molecules in Weinstein *et al.* (1998), relies on elastic collisions with a cold buffer gas of helium, which is mantained cryogenically at a temperature of ≈ 100 mK. The buffer gas both protects the molecules from sticking to the cryostat walls and makes them thermalize at low temperature by sympathetic cooling in such a way to allow them to be confined, e.g. in magnetic traps.

A second technique, which is applicable to polar molecules with first-order Stark effect, consists in the deceleration of molecular beams by means of time-varying electric fields (van de Meerakker *et al.*, 2008). This technique was demonstrated in Bethlem *et al.* (1999) with the operation of a *Stark decelerator*, in which an array of electrodes produces a time-varying electric field which forces the polar molecules to acquire potential Stark energy at the cost of a reduction of their kinetic energy.[26] In this way a pulsed beam of polar molecules can be decelerated near standstill and the translationally cold molecules ($\approx$ 100 mK) can be trapped, e.g. in electrostatic traps (Bethlem *et al.*, 2000).

While *laser cooling* of molecules seemed a totally prohibitive task until a few years ago, very recent experimental advances have shown the possibility of following this route. The first laser cooling of molecules has been demonstrated in Shuman *et al.* (2010) on a cryogenic buffer-gas beam of strontium monofluoride (SrF). This was possible thanks to a favourable structure of the SrF molecule, in which the decay from the first electronically excited state causes the population of only a few vibrational levels. As a consequence, the molecular beam could be cooled with only three lasers coupling the three lowest vibrational levels to the electronically excited state, resulting in a measured transverse temperature of only a few millikelvin.

A novel and possibly more general approach to laser cooling has been demonstrated in Zeppenfeld *et al.* (2012) with the realization of Sisyphus cooling for electrostatically trapped polar molecules. Similarly to the Sisyphus effect in sub-Doppler cooling (see Sec. 2.3.4), the molecules acquire Stark potential energy (reducing their kinetic energy) as they move towards the higher fields at the edge of the trap: here spatially selective radiofrequency transitions drive them to a different rotational level which is less sensitive to the electric field,[27] which allows the lost kinetic energy not to be reconverted into potential energy (as the optical pumping event in the atomic Sisyphus effect). In this technique, demonstrated in Zeppenfeld *et al.* (2012) for polar CH_3F molecules, a large fraction of the kinetic energy is removed at each step and only a few repumping cycles (with an infrared laser exciting a vibrational transition) are required to re-fill the initially populated rotational levels. As a result, high-density samples of molecules at a temperature of a few tens of mK could be produced, with a more-than-one order of magnitude increase in phase-space density.

Ultracold association of molecules. The cold molecules produced with the techniques described above already offer important advantages for many applications, e.g. for molecular spectroscopy experiments, and are attractive since they can be generalized to large classes of molecules. However, the phase-space density demonstrated to date is still much below the one needed for quantum degeneracy. A second category of experiments, which relies on *indirect cooling* methods, is aimed at the creation of quantum degenerate gases of specific diatomic molecules starting from trapped ultracold atomic samples at a large phase-space density.

[26] The electrodes switching is synchronized with the molecule longitudinal motion in such a way that the molecule continues to climb one potential hill after the other, similarly to the Sisyphus cooling mechanism discussed in Sec. 2.3.4.

[27] This approach is similar to the RF evaporative cooling discussed in Sec. 3.1.2, but here the molecules are not lost.

Ultracold atoms can be associated to form diatomic molecules via different schemes. A first strategy is *photoassociation*, in which the molecule formation is induced by a laser beam, red-detuned with respect to an atomic resonance (see sketch in Fig. 3.28). If the laser detuning matches the energy of a bound state in the electronically excited molecular potential, a molecule forms and then decays in the electronic ground state (see e.g. Sage *et al.* (2005), Viteau *et al.* (2008), and Aikawa *et al.* (2010)). In a different approach, molecules can be formed by sweeping the intensity of an external magnetic field across a *Feshbach resonance* (see Sec. 3.1.4), starting from the $a < 0$ side of the resonance (which corresponds to the molecular bound state lying above the dissociation threshold energy) and stopping on the $a > 0$ side (which corresponds to the molecular bound state lying below the dissociation threshold energy) (Chin *et al.*, 2010) (see also Sec. 3.3), or by using a resonantly modulated magnetic field close to a Feshbach resonance (Weber *et al.*, 2008).

In both cases molecules are produced in a highly excited vibrational state. Stimulated Raman adiabatic passage (STIRAP) techniques, in which a Raman transition couples the excited vibrational state to a lower-lying state (Danzl *et al.*, 2008), can be used to coherently transfer the molecules down to their rovibrational ground state (Lang *et al.*, 2008; Ni *et al.*, 2008; Aikawa *et al.*, 2010) with negligible temperature increase and conservation of the phase-space density. This method is very promising for producing quantum degenerate gases of polar molecules, as shown in the JILA experiments with fermionic KRb ground-state molecules (Ni *et al.*, 2008). Recent experiments have studied the collisional stability of molecules in their rovibrational ground state, evidencing the activation of ultracold chemical reactions between KRb molecules and studying the effect of quantum statistics and of the internal molecular quantum state on the reaction dynamics (Ospelkaus *et al.*, 2010). Optical lattices offer the possibility of preventing inelastic collisions or impeding the activation of chemical reactions, increasing the lifetime of the molecules to several seconds, as demonstrated in experiments with homonuclear Cs_2 (Danzl *et al.*, 2010) and polar KRb molecules (de Miranda *et al.*, 2011; Chotia *et al.*, 2012).

3.5.2 Quantum gases with dipolar interaction

Polar molecules, formed by two or more atoms of different chemical species, are characterized by a comparatively large electric dipole moment, which can be externally tuned via strong polarizing electric fields (Carr *et al.*, 2009). Values of the electric dipole moment for such molecules in a low-lying vibrational state typically range from 0.1 to a few debye (Sage *et al.*, 2005; Ni *et al.*, 2008; Deiglmayr *et al.*, 2008; Weinstein *et al.*, 1998). Because of the permanent electric dipole, ultracold polar molecules are characterized by a strong *dipole–dipole interaction* (DDI).

This interaction, acting between particles with a permanent electric $\mathbf{d}$ or magnetic dipole moment $\boldsymbol{\mu}$, is intrinsically different from the contact interaction discussed in Sec. 3.1.4 and promises the realization of a large variety of novel many-body quantum effects (Baranov, 2008; Carr *et al.*, 2009; Lahaye *et al.*, 2009; Baranov *et al.*, 2012; Krems *et al.*, 2009). As prominent aspects, such an interaction has a *long-range* and *anisotropic* character. For dipoles aligned along a polarization axis (fixed e.g. by an external electric/magnetic field), the DDI potential can be written as

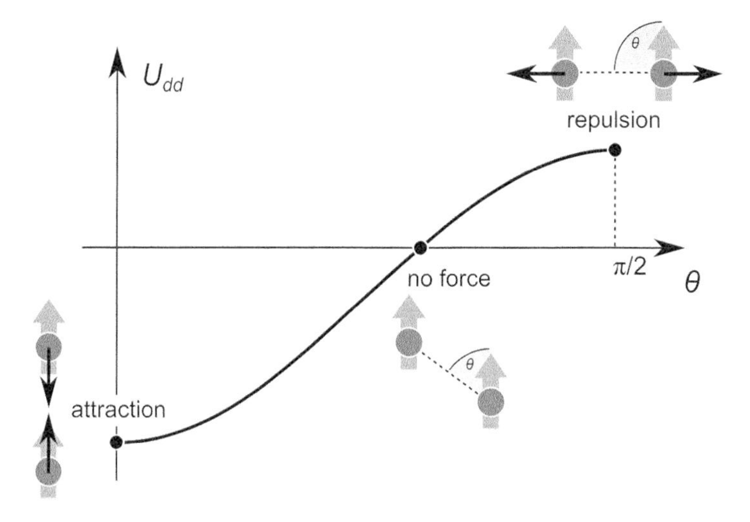

Fig. 3.29 Angular dependence of the dipole–dipole interaction in eqn (3.26). The interaction sign changes from attractive to repulsive as the angle θ between the atom–atom separation and the dipole orientation changes from 0 to 2π.

$$V(\mathbf{r}) = \frac{C_{\rm dd}}{4\pi} \frac{1 - 3\cos^2\theta}{r^3}, \tag{3.26}$$

where r is the distance between two dipoles and θ is the angle between the polarization and the interatomic axis. The coupling constant $C_{\rm dd}$ is d^2/ϵ_0 $(\mu_0\mu^2)$ for particles having a permanent electric (magnetic) dipole moment d (μ), where ϵ_0 (μ_0) is the electric (magnetic) constant. Remarkably, the DDI changes from being attractive to repulsive as the angle θ varies between 0 (head-to-tail arrangement of the dipoles) and $\pi/2$ (side-to-side arrangement), as shown in Fig. 3.29. A paradigm of this anisotropy is the d-wave collapse observed with dipolar atomic BECs of Cr (Lahaye *et al.*, 2008) and Er (Aikawa *et al.*, 2012) magnetic atoms. The long-range nature of the potential lies in its $1/r^3$ behaviour and has a profound impact on both the scattering properties (Lahaye *et al.*, 2009; Bohn *et al.*, 2009; Ospelkaus *et al.*, 2010; de Miranda *et al.*, 2011) and on the behaviour of dipoles confined in periodic potentials (Baranov *et al.*, 2012; Krems *et al.*, 2009; Capogrosso-Sansone *et al.*, 2010; Góral *et al.*, 2002).

In order to explore the physics originated from the DDI, it is necessary to produce large ensembles of particles with permanent electric or magnetic dipole moments, at a large value of the phase-space density. The most promising candidates for dipolar physics are polar molecules (Carr *et al.*, 2009) and strongly magnetic atoms (Lahaye *et al.*, 2009).[28] While dipolar quantum gases have already been realized with open-shell transition metals and lanthanide atoms (see Sec. 3.4.3), polar molecules could offer the advantage of being $\sim 10^3$ times more dipolar than those magnetic atoms (and $\sim 10^6$ more dipolar than alkali atoms). The recent experimental advances discussed

[28] Other possible candidates are Rydberg atoms and light-induced dipoles (Lahaye *et al.*, 2009).

in Sec. 3.5 indicate promising directions in cooling and manipulating polar molecules, pointing to the ultimate goal of quantum degeneracy.

3.5.3 Tests of fundamental physics

Molecular spectroscopy represents a precious resource for precision measurements, in particular for testing the stability of fundamental constants and for tests of fundamental symmetries of Nature (see Tarbutt *et al.* (2009) for a recent review). Although for most of these applications laser cooling is not strictly required, the realization of cold molecular beams or trapped samples of molecules might allow a better control of systematic effects and could considerably increase the resolution of spectroscopic measurements.

Stability of fundamental constants. A first application of precision molecular spectroscopy is connected with the possibility of detecting temporal variations in the fundamental constants. Molecular spectra depend on several fundamental constants, including the fine structure constant and the electron-to-proton mass ratio, which can be probed by comparing astronomical molecular spectra (coming from remote times in the past) with laboratory spectra. We will discuss this possibility in Sec. 5.5.2, in connection with related experiments of high-resolution spectroscopy of forbidden atomic transitions in optical atomic clocks.

Electron electric dipole moment. Molecules can be used as efficient probes of fundamental laws and symmetries of Nature. Among these, the search for the electric dipole moment (EDM) of the electron has a particular significance. The existence of a permanent EDM for an elementary particle implies a violation of the time-reversal (T) discrete symmetry: assuming the EDM to be aligned with the particle spin, a time-reversal operation inverts the spin, but it keeps the EDM unchanged. The Standard Model allows for the breaking of this symmetry (which is equivalent to a violation of charge conjugation and parity (CP) symmetries, according to the CPT theorem) and predicts an exceedingly small value for the electron EDM $< 10^{-38} e$ cm (with e the elementary charge), which is far below the current resolution of experiments. However, the amount of symmetry breaking in the Standard Model seems not to be compatible with the matter–antimatter asymmetry of our Universe. Extensions of the Standard Model, introducing new sources of CP violation, predict much larger values of the electron EDM from $10^{-30} e$ cm up to $10^{-26} e$ cm, which can be tested in laboratory experiments.

Upper limits to the electron EDM d_e have been determined in several experiments performed in different physical systems, including atomic physics systems. Among these, a Ramsey-type magnetic resonance experiment performed on atomic beams of Thallium resulted in an upper limit $|d_e| < 1.6 \times 10^{-27} e$ cm for the electron EDM (Regan *et al.*, 2002). Polar molecules offer much larger sensitivites than atoms for the measurement of the electron EDM, as a consequence of their larger polarizability when they are placed in an external electric field. Because of strong relativistic effects, in polar molecules formed by heavy atoms the external electrons experience an effective electric field which can be several orders of magnitude larger than the external applied field. This is the case of the YbF molecule investigated in the experiments at the Imperial College in

London (Hudson *et al.*, 2011), in which the ground-state molecules of a molecular YbF beam are prepared in a superposition of two hyperfine states $|F=1, m_F=1\rangle$ and $|F=1, m_F=-1\rangle$ with opposite orientation of the electron spin. In the presence of a static electric field, the energy of these two states is expected to be different because of the Stark interaction $-\mathbf{d_e}\cdot\mathbf{E}$ between the electron EDM $\mathbf{d_e}$ and the electric field $\mathbf{E}$. This difference was measured in Hudson *et al.* (2011) in a Ramsey experiment with two radiofrequency excitations, yielding a result consistent with zero energy separation and setting an upper boundary to the electron EDM $|d_e| < 1.0 \times 10^{-27} e$ cm, essentially limited by statistics. The implementation of a cryogenic buffer gas source and possibly direct laser cooling (capable of providing higher fluxes and slower molecules) could increase the precision of the experiment, with a target resolution of $10^{-30} e$ cm (Kara *et al.*, 2012), in such a way as to accurately test the predictions of many extensions of the Standard Model.

4 Helium

After the discussion on alkali atoms and on the revolutionary applications opened by laser cooling in the fields of metrology, atom interferometry, and quantum degenerate gases, this chapter starts with the presentation of the helium spectrum as the next step of complexity in atomic structures. Alkali atoms have basically the same single-electron structure as hydrogen, since the electrons in the closed inner shells are characterized by very large excitation energies and do not contribute to the spectrum in the visible–near infrared regions. The helium atom, instead, has a qualitatively different and richer electronic structure, which is determined by the coupling between the two external electrons.

Helium spectroscopy is presented in this chapter as a resource for precision measurements, in particular for performing tests of quantum electrodynamics and nuclear models and for the determination of fundamental quantities. We will also discuss laser cooling and quantum degeneracy of metastable helium gases, which allowed the development of new detection techniques and the measurement of spatio-temporal correlations in quantum atom optics experiments.

High-precision spectroscopy of the helium fine structure is connected with the possibility of measuring the fine structure constant. Given the importance of this fundamental constant of Physics, the last section of this chapter is devoted to a discussion of the different experimental techniques for its determination. These include the approach based on the atomic recoil, which is one of the most relevant applications of ultracold atoms for precise measurements.

4.1 The helium spectrum

In the early days of spectroscopy, it was believed that helium was actually composed of two different varieties exhibiting spectral lines belonging to two separated series: *ortho-helium* and *para-helium*. The reason for this old belief is the existence of two independent sets of helium states which are uncoupled by the radiation. These two sets correspond to two different ways in which the spins of the two helium electrons can combine: they can form either a singlet state with total spin $S = 0$ (para-helium) or triplet states with total spin $S = 1$ (ortho-helium). Transitions between these two classes of states are forbidden by selection rules.

Singlet and triplet states are not degenerate. This is caused by the electron–electron interaction and by the symmetrization postulate of quantum mechanics, which requires the wavefunction of a system of identical particles to have a well-defined symmetry with respect to the exchange of two of them. Fermions have globally antisymmetric wavefunctions, hence the wavefunction $\Psi(1,2)$ describing the two identical electrons in helium must change sign if the two particles are swapped:

$$\Psi(1,2) = -\Psi(2,1) . \tag{4.1}$$

The global wavefunction $\Psi(1,2)$ includes both the spatial and spin degrees of freedom, described by a spatial wavefunction $\psi(\mathbf{r}_1, \mathbf{r}_2)$ and a spin wavefunction $\chi(s_1, s_2)$:

$$\Psi(1,2) = \psi^{\pm}(\mathbf{r}_1, \mathbf{r}_2)\chi^{\mp}(s_1, s_2) . \tag{4.2}$$

The plus and minus signs in the above equation refer to the symmetry of the wavefunction, e.g. ψ^+ indicates a symmetric spatial wavefunction and ψ^- indicates an antisymmetric spatial wavefunction with respect to the exchange of the two electrons. The spatial wavefunction $\psi^{\pm}$ is associated with a spin wavefunction $\chi^{\mp}$ with opposite symmetry in order to make the global wavefunction Ψ antisymmetric. By neglecting the electrostatic repulsion between the electrons, the symmetrized spatial wavefunction can be written as

$$\psi^{\pm}_{nlm_l}(\mathbf{r}_1, \mathbf{r}_2) = \frac{1}{\sqrt{2}}\left[\phi_{100}(\mathbf{r}_1)\phi_{nlm_l}(\mathbf{r}_2) \pm \phi_{100}(\mathbf{r}_2)\phi_{nlm_l}(\mathbf{r}_1)\right] , \tag{4.3}$$

where $\phi_{nlm_l}(\mathbf{r})$ denote the single-electron hydrogen-like wavefunctions corresponding to principal quantum number n and orbital angular momentum quantum numbers l and m_l.[1] In terms of the single electron spin states $|s_1 s_2\rangle$, the four possible spin wavefunctions are:

$$\begin{cases} \chi^+_{1,1} = |\uparrow\uparrow\rangle \\ \chi^+_{1,0} = \frac{1}{\sqrt{2}}\left(|\uparrow\downarrow\rangle + |\downarrow\uparrow\rangle\right) \\ \chi^+_{1,-1} = |\downarrow\downarrow\rangle \\ \chi^-_{0,0} = \frac{1}{\sqrt{2}}\left(|\uparrow\downarrow\rangle - |\downarrow\uparrow\rangle\right) , \end{cases} \tag{4.4}$$

where the subscripts in χ_{S,m_S} indicate the total electron spin quantum number S and its component m_S. The symmetry of the spin state is determined by the value of S: the singlet state with $S=0$ is described by the antisymmetric combination of spins, the triplet states with $S=1$ are described by symmetric combinations.

When the repulsion between the electrons is introduced as a perturbation, the energy of the states in eqn (4.3) becomes, at first order in perturbation theory,[2]

$$E^{\pm}_{nlm_l} = E^0 + J \pm K , \tag{4.5}$$

where the sign $+$ refers to the symmetric spatial wavefunction (singlet state $S=0$) and the sign $-$ refers to the antisymmetric spatial wavefunction (triplet states $S=1$). The

[1] In eqn (4.3) one electron is assumed to occupy the ground state orbital $n=1$, $l=0$, $m_l=0$. Helium configurations with both electrons in excited orbitals are autoionizing, since their energy is always higher than the ionization energy.

[2] Although first-order perturbation theory does not give very accurate quantitative results, it describes the qualitative features of the spectrum well.

different energy terms in eqn (4.5) are the unperturbed hydrogen-like Bohr energies for the two electrons independently interacting with a nucleus with charge $2e$

$$E^0 = -2hcR_\infty \left(1 + \frac{1}{n^2}\right) , \tag{4.6}$$

the classical electrostatic repulsion energy between the two electron densities

$$J = \int \mathrm{d}\mathbf{r}_1 \mathrm{d}\mathbf{r}_2 \left|\phi_{100}(\mathbf{r}_1)\right|^2 \left|\phi_{nlm_l}(\mathbf{r}_2)\right|^2 \frac{e^2}{4\pi\epsilon_0 |\mathbf{r}_1 - \mathbf{r}_2|} , \tag{4.7}$$

and an *exchange energy*

$$K = \int \mathrm{d}\mathbf{r}_1 \mathrm{d}\mathbf{r}_2 \phi^*_{100}(\mathbf{r}_1)\phi^*_{nlm_l}(\mathbf{r}_2)\phi_{nlm_l}(\mathbf{r}_1)\phi_{100}(\mathbf{r}_2)\frac{e^2}{4\pi\epsilon_0 |\mathbf{r}_1 - \mathbf{r}_2|} , \tag{4.8}$$

with no classical analogue, which gives a positive or negative contribution to the total energy in eqn (4.5) depending on the symmetry of the wavefunction. For this reason triplet states (ortho-helium) have slightly lower energies than singlet states (para-helium) with the same electronic configuration, as shown in the energy level diagram of Fig. 4.1. The helium ground state $\psi^+_{100}(\mathbf{r}_1, \mathbf{r}_2)$ is only present in the spatially symmetric singlet state $S = 0$, since the triplet state $S = 1$ would correspond to an identically null spatial wavefunction, as evident from eqn (4.3).

Transitions between singlet and triplet states are strongly forbidden by selection rules. As a matter of fact, the electromagnetic field cannot change the spin quantum number S, not only for electric dipole transitions, but for higher transition orders as well.[3] This is the reason why the ortho-helium spectrum is separated from the para-helium spectrum, and also explains the existence of helium metastable states. In particular, the lowest-energy triplet state $1s2s\,^3S$ has an extremely long lifetime $\tau \simeq 7870(510)$ s, the largest of all known atomic metastable states: its decay to the absolute ground state is doubly forbidden by both the electric dipole selection rule for the orbital angular momentum ($\Delta l = \pm 1$) and by the spin conservation selection rule ($\Delta S = 0$). Although it can be considered a ground state, this level has a large internal energy of ≈ 20 eV, which allows the implementation of efficient detection techniques that will be discussed in Sec. 4.3 in the context of quantum atom optics experiments with ultracold metastable helium atoms. More information on the properties of metastable helium and its applications to atomic physics experiments can be found in the review article Vassen *et al.* (2012) on cold and trapped metastable noble gases.

4.1.1 Helium laser spectroscopy

In this section we present some experiments on high-resolution laser spectroscopy of atomic helium. More experiments will be discussed in Sec. 4.2, devoted to the helium fine structure, and in Sec. 4.4.1, devoted to recent experimental efforts for the determination of helium nuclear charge radii.

[3] As far as the total electronic spin S is a well-defined quantum number. The approximation in which S is well defined (L–S coupling, or Russell–Saunders scheme) is very well satisfied in helium, while deviations from this rule can be observed in heavier atoms, as we shall discuss in Chapter 5. Nevertheless, the excitation of ultra-weak triplet–singlet transitions in helium has been recently reported and it will be discussed in Sec 4.4.1.

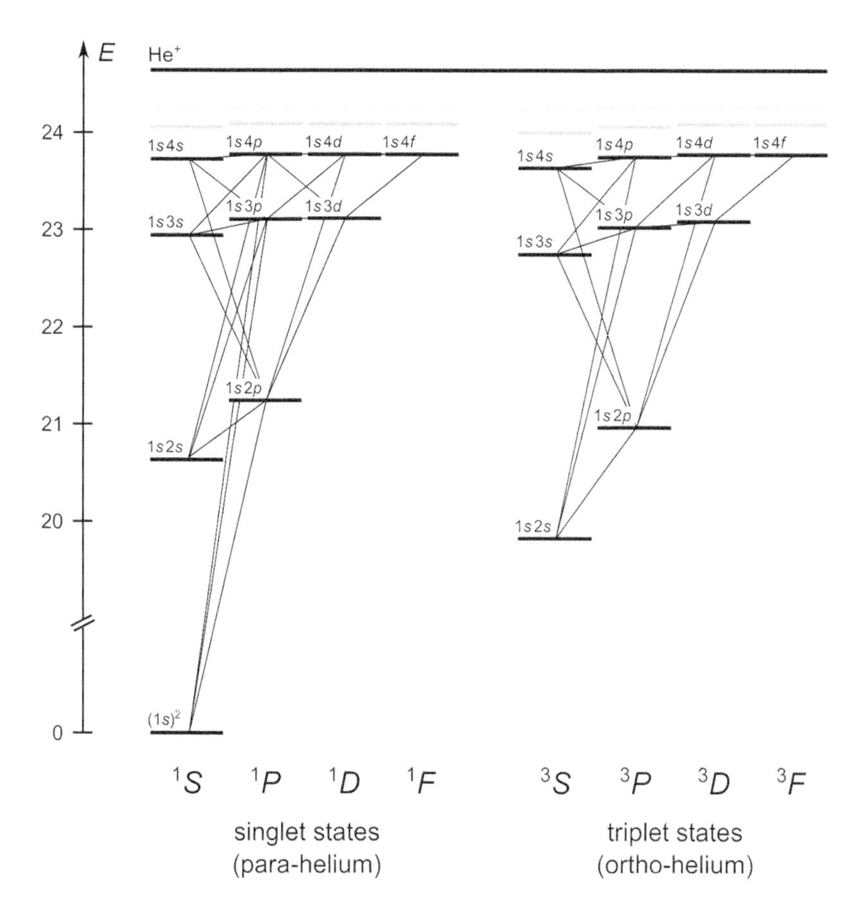

Fig. 4.1 Scheme of helium levels. Dipole-allowed transitions are indicated as thin lines connecting the different levels. Note the absence of transitions between singlet and triplet states and the large energy difference ≈ 20 eV between the ground state and the first excited states.

Para-helium spectroscopy. The spectroscopic investigation of the helium ground state $1\,^1S$ is important since this state is the most affected by Lamb shift corrections, the experimental determination of which can be used to test QED for three-body bound states and for the determination of the helium nuclear charge radius (see also Sec. 4.4.1). However, high-resolution spectroscopy of para-helium transitions from the ground state is complicated by the difficulty of producing narrow-band, tunable radiation in the extreme ultraviolet (XUV) wavelength range below 60 nm.

The first laser excitation from the $1\,^1S$ state was reported by the Amsterdam group with the measurement of the frequency of the $1\,^1S \to 2\,^1P$ transition at $\lambda = 58.4$ nm in an atomic beam of ground-state helium (Eikema *et al.*, 1993). The light for the excitation of this transition was produced by coherent fifth-harmonic generation of the UV light from a frequency-doubled pulsed dye laser at $\lambda = 584$ nm (see Appendix B.3 for an introduction to nonlinear optics techniques) by nonlinear effects in an expanding pulsed jet of C_2H_2. As a reference for the frequency measurement (performed before the advent of the frequency-comb era) the rovibronic spectrum of molecular iodine I_2 was chosen, which has been widely used for metrological calibration and represents a secondary wavelength standard recommended by the Comité International des Poids et Mesures (CIPM). From an experimental point of view, the I_2 spectrum is characterized by a multitude of strong absorption lines extending from about 900 nm in the near infrared (NIR) to about 500 nm in the blue-green visible spectrum.

Combining the measured helium transition energy with the accurately known energy of the upper $2\,^1P$ state measured in previous experiments, the authors of Eikema *et al.* (1993) obtained the energy of the helium ground state, which, combined with the non-QED theoretical value, yields a measurement of the helium ground state Lamb shift. In following experiments the measurement was refined, providing a precise value of the ^{4}He Lamb shift 41 224(45) MHz (Eikema *et al.*, 1997) in good agreement with the QED prediction 41 233(45) MHz, and a determination of the ^{3}He–^{4}He isotope shift 263 410(7) MHz (Eikema *et al.*, 1996), again in excellent agreement with the theoretical value.

Recently, para-helium spectroscopy has been performed by the Amsterdam group by direct excitation of ground-state helium with radiation from an XUV-converted frequency comb (see Sec. 1.4.2). Quantum interference spectroscopy was performed by amplifying two consecutive laser pulses from the output of an infrared frequency comb and up-converting them to the 15th harmonic by nonlinear high-harmonic generation in a gas jet. This spectroscopy, resembling an optical time-delayed version of the Ramsey interferometer scheme (Witte *et al.*, 2005), was applied to the $1\,^1S \to n\,^1P$ transitions, resulting in a more precise determination of the $1\,^1S$ ^{4}He Lamb shift 41 247(6) MHz (Kandula *et al.*, 2011), in very good agreement with the QED prediction.

Ortho-helium spectroscopy. As discussed above, no visible transitions can be excited from the helium ground state. On the contrary, near-visible transitions can be conveniently excited either in para-helium, starting from the $2\,^1S_1$ state, or in ortho-helium, starting from the lowest-energy $2\,^3S_1$ triplet state. These long-lived states can be populated in an electric discharge by collisions between the helium atoms and the plasma electrons accelerated by the voltage difference between the discharge electrodes.

The first direct frequency measurement of an optical transition in helium was performed at LENS in Pavone *et al.* (1994) for the $2\,^3S_1 \to 3\,^3P_0$ ortho-helium transition at $\lambda = 389$ nm. The frequency reference for this measurement was provided by the near coincidence of the transition energy with the energy difference between the $5\,S_{1/2}$ and $5\,D_{3/2,5/2}$ levels of rubidium, connected by a two-photon transition at $\lambda = 778$ nm. The latter transition provides a well-characterized secondary frequency standard, the absolute frequency of which was first measured in Nez *et al.* (1993) at the kHz level by taking advantage of a near coincidence with the frequency difference between two He–Ne lasers, stabilized respectively to an I_2 line at $\lambda = 633$ nm and to a methane transition at $\lambda = 3.39$ μm. The comparison of the measured frequency of the $2\,^3S_1 \to 3\,^3P_0$ helium transition, with a fractional accuracy of 2.4 parts in 10^{10}, with the non-QED prediction, allowed the determination of the ^{4}He $2\,^3S_1$ Lamb shift value 4 057.6(8) MHz. Using the same experimental approach, the measurement of the ^{3}He–^{4}He isotope shift of the $2\,^3S_1 \to 3\,^3P_0$ transition 42 184.18(14) MHz allowed the determination of the rms nuclear charge radius of ^{3}He as 1.923(36) fm (Marin *et al.*, 1994). An extended discussion on the helium nuclear charge radius, together with new experimental results, will be presented in Sec. 4.4.1.

4.2 Helium fine structure

This section is devoted to the high-resolution spectroscopy of the helium fine structure, which is particularly relevant for precision measurements. When relativistic effects are included, a fine structure appears in the helium spectrum, which causes the triplet levels[4] to be split according to the total electron spin J. The fine structure of helium is notably more complex than the one of hydrogen, owing to the presence of multiple interaction terms, e.g. spin–orbit coupling for the same electron, spin–other-orbit coupling between different electrons, spin–spin electron–electron interaction (Heisenberg, 1926; Bethe and Salpeter, 1957). Among the early contributions to the theory of helium, we recall the pioneering work of G. Breit on the relativistic theory of two-electron atoms (Breit, 1929; Minardi and Inguscio, 2001).

Despite the complexity of its fine structure, helium can still be considered a "simple atom", since the energies of its levels can be calculated numerically with very good precision. Although the problem is not exactly solvable, not even at the level of the three-body Schrödinger problem, its nonrelativistic wavefunctions are known with extremely good accuracy. An important contribution to the theory of helium and to the consequent development of high precision helium spectroscopy has come from G. W. F. Drake, who performed very precise calculations of the nonrelativistic wavefunctions and of the lowest-order relativistic contributions (Drake, 1989; Drake, 2006). The knowledge of the wavefunctions is the basis for the evaluation of higher-order relativistic and QED corrections, which have been recently evaluated down to the $m_ec^2\alpha^7$ level (Pachucki and Yerokhin, 2010) of the expansion of the atomic energies in terms of powers of the fine structure constant α:

$$E = m_ec^2\left[Z + a_2\alpha^2 + a_4\alpha^4 + \cdots\right] . \qquad (4.9)$$

The first term in parentheses is the rest energy of the electrons ($Z = 2$ is the atomic number and m_e is the electron mass). The α^2 term determines the gross structure of the spectrum, where the a_2 coefficient is a function of the quantum numbers $\{n_i, l_i\}$ describing the electron configuration. The α^4 term includes the leading-order contributions to the fine structure, where the a_4 coefficient depends not only on the electron configuration, but also on the total electron angular momentum J.[5] Finally, the remaining $o(\alpha^4)$ terms contain the higher-order QED corrections. Therefore, the energy separation between components J and J' of the same fine-structure manifold is given by

$$\Delta E_{J,J'} = E_J - E_{J'} = m_ec^2\left[(a_{4,J} - a_{4,J'})\,\alpha^4 + \cdots\right] . \qquad (4.10)$$

The lowest-energy helium configuration with a fine structure is the $2\,^3P_J$ state, which is split in three different levels with $J = 0, 1, 2$. In late 1950s sub-MHz spectroscopy of

[4]Only triplet states with $S = 1$ exhibit a fine structure, which is of course absent in singlet $S = 0$ states.

[5]In hydrogen, $Z = 1$, $a_2 = -1/2n^2$, $a_4 = -1/2n^3(1/(j + 1/2) - 3/4n)$, see eqn (1.6).

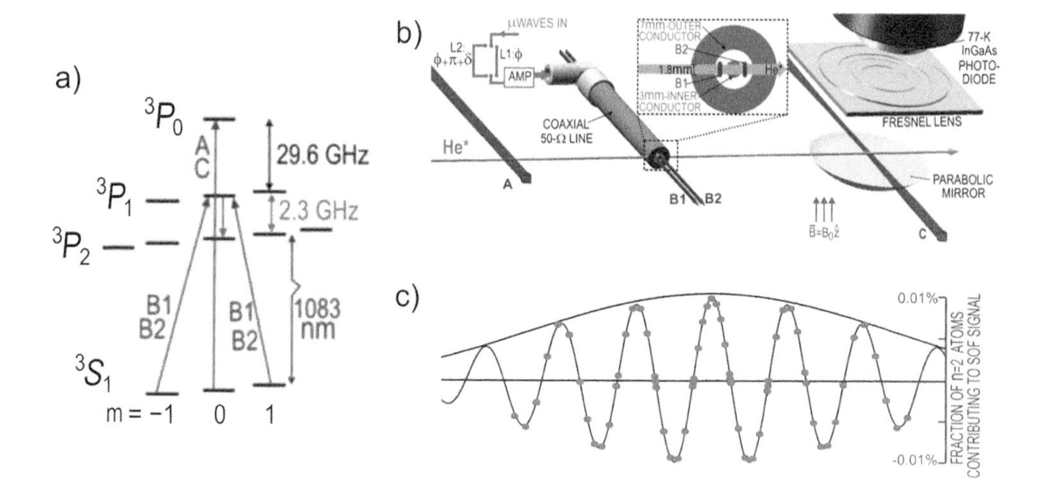

Fig. 4.2 Microwave measurement of the $\Delta\nu_{12}$ helium fine-structure separation between states $2\,^3P_1$ and $2\,^3P_2$. a) Scheme of the energy levels. b) Scheme of the experimental atomic beam setup. c) Ramsey fringes arising from the double microwave excitation. Reprinted with permission from Borbely *et al.* (2009). © American Physical Society.

helium fine structure was started by W. E. Lamb Jr, who proposed and implemented a method based on direct excitation of the $J = 2 \rightarrow J' = 1$ and $J = 1 \rightarrow J' = 0$ transitions in the microwave domain (Lamb Jr, 1957; Wieder and Lamb Jr, 1957). The high-precision measurement of helium fine-structure intervals was early considered as a valuable resource for the spectroscopic determination of α, as recognized by C. Schwartz (Schwartz, 1964) and by V. W. Hughes (Hughes, 1969). The group at Yale also carried out the first experimental high-precision measurements combining the microwave excitation with optical excitation of the $2\,^3S_1 \rightarrow 2\,^3P_J$ transition at 1.08 µm (Pichanick *et al.*, 1968).

Provided that the theoretical coefficients a_n in eqn (4.10) are known with a sufficient level of accuracy, the measurement of the energy separation $\Delta E_{J,J'}$ can be used to determine the value of α. The separation between the helium $2\,^3P_J$ components is larger than the hydrogen $2p$ fine structure ($\approx$ 30 GHz vs $\approx$ 10 GHz), and the uncertainty coming from the inverse lifetime of the $2\,^3P_J$ states is much smaller ($\approx$ 1.6 MHz in helium vs $\approx$ 100 MHz in hydrogen), which allows the experimental determination of helium fine-structure intervals to be more precise than in the hydrogen case. In the following we illustrate two recent experimental approaches based on two different methods: direct microwave excitation within the fine-structure manifold and a heterodyne measurement based on optical excitation to different fine-structure states.

4.2.1 Microwave measurements

A first way to measure helium $2\,^3P_J$ fine-structure intervals $\Delta\nu_{JJ'}$ relies on the direct microwave excitation of transitions between different J states of the manifold. Since the measurement involves quite short-lived states, a combination of optical excitation from the metastable $2\,^3S_1$ state and optical pumping effects following microwave excitation has to be used. In a recent measurement at York University (Borbely *et al.*, 2009) this method allowed a precise measurement of the smallest fine-structure separation $\Delta\nu_{12} = 2\,291\,177.53(35)$ kHz, which is the least sensitive fine-structure interval for the determination of α, but it can be used as a validity test of the helium QED theory.

As shown in Fig. 4.2b, the measurement was performed on a thermal beam of metastable atomic helium, produced in a dc discharge. These metastable atoms are then excited by a 1.08 μm linearly polarized laser resonant with the $^3S_1(m=0) \to {}^3P_0$ transition, which rapidly pumps the atoms into the stretched $^3S_1(m = +1)$ and $^3S_1(m=-1)$ states (see Fig. 4.2a for a scheme of the levels of interest). The atoms then travel through a microwave waveguide, where they are optically excited by circularly polarized laser light to the $^3P_1(m=0)$ level and by a microwave tuned around the $^3P_1(m=0) \to {}^3P_2(m=0)$ transition. In the absence of microwave excitation, the atoms cannot decay from the $^3P_1(m=0)$ state to the $^3S_1(m=0)$, owing to selection rules which prohibit the $m_J = 0 \to m'_J = 0$ for a $J=1 \to J'=1$ transition. Instead, in the case of successful microwave excitation, the atoms can decay to the $^3S_1(m=0)$ level, the population of which is taken as the excitation signal and is detected in a spatially separated region, by exciting the $^3S_1(m=0) \to {}^3P_0$ transition and monitoring the fluorescence. A typical spectrum as a function of the microwave excitation frequency is shown in Fig. 4.2c, where the oscillations represent Ramsey fringes (see Sec. 2.2.2) obtained by double excitation of the atoms in two spatially separated points within the waveguide.

4.2.2 Optical measurements

The second way to measure helium $2\,^3P_J$ fine-structure intervals relies on a heterodyne-type measurement in which the separation between ^3P_J states is obtained by *difference* of optical $^3S_1 \to {}^3P_J$ transition frequencies. This kind of measurement does not require the complex optical pumping schemes of the microwave measurement and can provide the same levels of accuracy. Since it is a differential measurement, it is not necessary to measure the absolute transition frequencies, however a stable frequency reference has to be used in order to avoid systematic effects.

In a LENS experiment (Giusfredi *et al.*, 2005) the $\Delta\nu_{01}$ frequency interval was measured by using a *master/slave* technique in which a 1.08 μm master laser was locked to a stable frequency reference and a twin 1.08 μm slave laser, referenced to the master laser, was used to scan the $^3S_1 \to {}^3P_0$ and $^3S_1 \to {}^3P_1$ helium transitions, as sketched in Fig. 4.3. Also in this case, the source of metastable helium was an atomic beam excited to the metastable $^3S_1$ state by a dc discharge. Frequency-modulated (FM) saturation spectroscopy on the atomic beam by two counterpropagating laser beams at the same frequency was used to determine the line centres. A typical third-derivative spectrum of a 1.08 μm helium resonance is shown in the inset of Fig. 4.3.

As a frequency reference for the master laser, the rovibronic spectrum of molecular iodine I_2 was used. In order to take advantage of this frequency standard, extending in the visible region, the helium 1.08 μm master laser was frequency-doubled to a wavelength of 541 nm with second-harmonic generation techniques (see Appendix B.3)

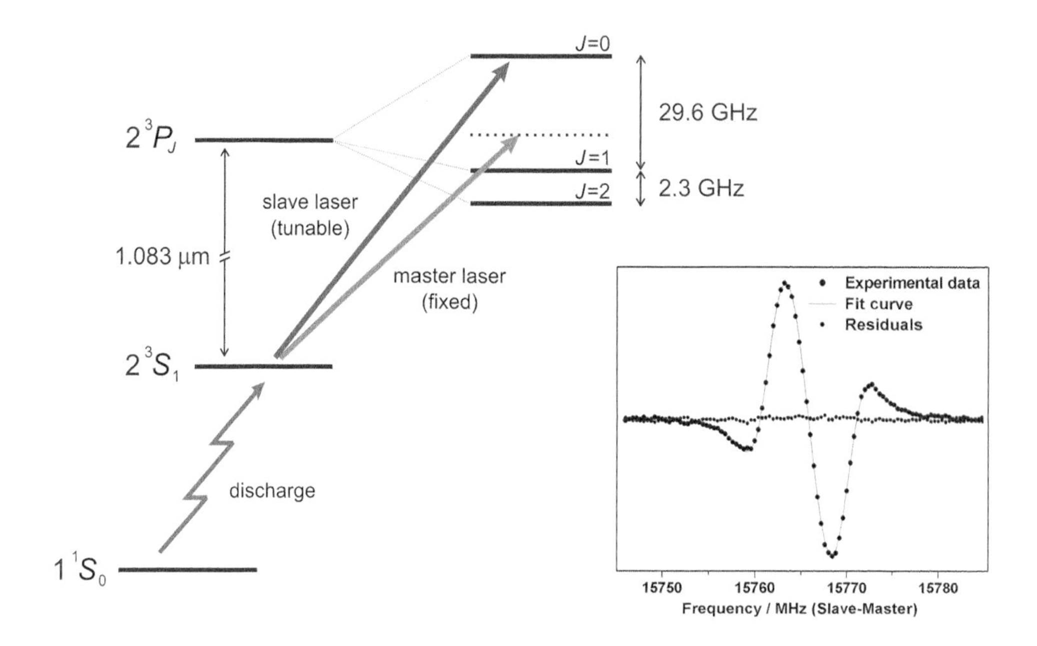

Fig. 4.3 Optical heterodyne measurement of the $\Delta\nu_{01}$ helium fine-structure separation between states $2\,^3P_0$ and $2\,^3P_1$. The inset shows a typical third-derivative spectrum of the saturation dip of one of the optical 1.08 μm helium resonances. Adapted from Giusfredi *et al.* (2005).

and it was locked to a hyperfine component of the I_2 rovibronic spectrum detected with FM saturation spectroscopy in a cell. The relative frequency of the slave laser with respect to the master laser was controlled by monitoring the beat-note of the two lasers with a fast photodiode. A feedback loop acting on the slave laser was used to keep the two lasers phase-locked with a controllable frequency offset in the microwave range. The master laser provides a constant optical frequency shift on top of a tunable microwave fine frequency control: this makes it possible to scan the different transitions of the fine-structure manifold, preserving microwave accuracy in the frequency difference between two transitions. In Giusfredi *et al.* (2005) this method allowed a measurement of the fine-structure separation $\Delta\nu_{01} = 29\,616\,952.7(1.0)$ kHz.

On the same experimental setup, an early extension of the frequency comb to the near infrared allowed the absolute frequency measurements of helium transitions around 1.08 μm. In Cancio Pastor *et al.* (2004) and Cancio Pastor *et al.* (2006) the frequency of the I_2-stabilized master laser was measured, which resulted in the absolute frequency measurement $\nu = 276\,736\,495\,653.1(2.4)$ kHz for the centroid of the $^3S_1 \rightarrow^3 P_J$ transitions. This measurement, combined with the theoretical knowledge of the non-QED level energies, allowed the determination of the Lamb shift both for the $^3S_1$ and the ^3P_J states, which constitutes an important test for the validity of QED calculations for the simplest three-body atomic system.

4.3 Quantum degenerate metastable helium

Bose–Einstein condensation of helium does not require laser cooling, since it naturally occurs as liquid helium is cooled below the λ-point (2.17 K) where it starts to show superfluidity. Reaching Bose–Einstein condensation in dilute helium gases, however, is more difficult than in alkali atoms. As a matter of fact, laser cooling of ground-state helium is a prohibitive task, owing to the large $\sim$ 20 eV energy interval between the ground state and the first dipole-allowed transition, which would require lasers to be operated in the extreme ultraviolet region. However, it turned out that the extremely long lifetime of the $2\,^3S_1$ state makes the standard cooling techniques originally developed for alkali atoms to be applicable also in the case of helium. Metastable atomic helium can be laser-cooled by taking advantage of the 1.08 μm transition connecting the lowest-energy triplet state $2\,^3S_1$ to the upper $2\,^3P_J$ fine-structure manifold. After laser cooling, atoms in the metastable state can be polarized in low-field-seeking Zeeman states ($m_J = +1$ for bosonic metastable ^{4}He), magnetically trapped and evaporatively cooled using the techniques described in Secs. 3.1.1 and 3.1.2.

The first Bose–Einstein condensate of metastable ^{4}He was produced in 2001 by the group of A. Aspect at Institut d'Optique (Orsay) (Robert *et al.*, 2001), shortly followed by the group of C. Cohen-Tannoudji et École Normale Supérieure (Paris) (Pereira Dos Santos *et al.*, 2001). By using heteronuclear sympathetic cooling with metastable ^{4}He, in 2006 the experimental group at Vrije University (Amsterdam) was able to cool a gas of metastable ^{3}He (prepared in the same $2\,^3S_1$ state) down to Fermi degeneracy (McNamara *et al.*, 2006).

Although the atoms in the metastable state are very long-lived (lifetime >10 minutes, as discussed in Sec. 4.1), a large energy of $\sim$ 20 eV is stored in their internal state. In principle, this might cause a problem for the stability of the ultracold gas, since atoms can be lost from the trap owing to collisional Penning ionization processes $\mathrm{He}^* + \mathrm{He}^* \rightarrow \mathrm{He} + \mathrm{He}^+ + e^-$. However this lossy collisional channel is suppressed by several orders of magnitude for samples which are spin-polarized in states with maximal angular momentum projection ($m_J = +1$ for bosonic ^{4}He* and $m_F = +3/2$ for fermionic ^{3}He*) since it would violate the spin conservation rule, as earlier predicted by G. Shlyapnikov, J. Walraven, and coworkers in Shlyapnikov *et al.* (1994). The suppression of these inelastic collisions allows evaporative cooling and sympathetic cooling to work flawlessly and a lifetime of several seconds for high-density helium BECs was reported (Robert *et al.*, 2001; Pereira Dos Santos *et al.*, 2001).

The large internal energy of the metastable He* atoms turned out to be a powerful instrument which enables new detection possibilities for ultracold quantum gases. As a matter of fact, when a metastable helium atom interacts with a matter surface, it is very likely that it releases its internal energy causing the extraction of one electron from the material.[6] In microchannel plate detectors (MCP) the electrons emitted by the material after the collisional de-excitation of the He* atoms are amplified in a cascade process (similar to the one in a photomultiplier), resulting in short measurable current pulses for each incident atom. By using suitable electronic delay lines, it is

[6] The work function for electron extraction in a large class of materials is in the 3–5 eV range.

possible to achieve single-atom detection with high spatial and temporal resolution. This detection technique, developed by A. Aspect, C. Westbrook and coworkers, has been used in beautiful experiments (Schellekens *et al.*, 2005; Jeltes *et al.*, 2007) where the quantum statistical properties of bosonic and fermionic ultracold helium gases have been investigated through the measurement of two-body density–density correlations.

4.3.1 Detecting atom–atom correlations

The coherence properties of light are intimately connected with the spatial and temporal correlations that exist between pairs of photons. In 1956 R. Hanbury Brown and R. Q. Twiss clearly demonstrated the existence of temporal correlations between the intensity fluctuations recorded by two separated detectors observing the light emitted by the same incoherent classical source (Hanbury Brown and Twiss, 1956*b*). The experimental results showed an increase in the joint detection probability when the time delay between the two detection events was smaller than the coherence time of the source: this means that the photons emitted by a classical source exhibit a *bunching* behaviour, i.e. they have the tendency to travel in bunches and arrive together at the detectors. This behaviour, initially demonstrated in a laboratory experiment, allowed the same authors to propose and demonstrate a method for the accurate measurement of the diameter of stars via *intensity* correlation measurements, which were able to outperform the traditional interferometers based on *phase* correlations (which are easily ruined by atmosphere fluctuations and other systematic effects) (Hanbury Brown and Twiss, 1956*a*). Shortly following the invention of lasers, in 1966 F. T. Arecchi et al. measured the statistical distribution of the arrival times of laser photons onto a photodetector (Arecchi *et al.*, 1966). By performing correlation measurements on single-photon detection events they were able to show that, differently from classical "thermal" light, laser light does not show any Hanbury Brown and Twiss correlation and laser photons do not bunch together. These and other seminal experiments represented a fundamental contribution for understanding the differences between classical (thermal and laser) and nonclassical light. These contributions then found an elegant and complete explanation in the theoretical framework of *quantum optics*, the quantum theory of light developed starting from the 1950s with fundamental contributions by R. Glauber (Glauber, 1963), who was awarded the Nobel Prize in Physics in 2005 (Glauber, 2006). For an introduction to the quantum optics theory the reader can refer e.g. to Mandel and Wolf (1995), Scully and Zubairy (1997), and Gerry and Knight (2005).

The tendency of photons to bunch together comes purely from quantum statistics and it is not an exclusive property of photons: it should be observable also for a system of identical bosonic massive particles. In order to quantify the coherence properties of a matter-wave field, it is possible to define the first- and second-order spatial correlation functions as

$$G^{(1)}(\mathbf{r},\mathbf{r}') = \langle \hat{\Psi}^\dagger(\mathbf{r}')\hat{\Psi}(\mathbf{r})\rangle \tag{4.11}$$

$$G^{(2)}(\mathbf{r},\mathbf{r}') = \langle \hat{\Psi}^\dagger(\mathbf{r})\hat{\Psi}^\dagger(\mathbf{r}')\hat{\Psi}(\mathbf{r}')\hat{\Psi}(\mathbf{r})\rangle\ , \tag{4.12}$$

where $\hat{\Psi}(\mathbf{r})$ ($\hat{\Psi}^\dagger(\mathbf{r})$) is the second-quantization field operator which annihilates (creates) a particle at position $\mathbf{r}$. The first-order correlation function gives information on the *phase* correlations in the system, the second-order correlation function describes *density* correlations. These functions, introduced here for the matter field operator $\hat{\Psi}(\mathbf{r})$, have the same form in quantum optics, where the matter field operator is substituted with the electric field operator $\hat{E}(\mathbf{r})$. Very often, one refers to the normalized versions of the above functions

$$g^{(1)}(\mathbf{r},\mathbf{r}') = \frac{G^{(1)}(\mathbf{r},\mathbf{r}')}{\sqrt{G^{(1)}(\mathbf{r},\mathbf{r})G^{(1)}(\mathbf{r}',\mathbf{r}')}} \tag{4.13}$$

$$g^{(2)}(\mathbf{r},\mathbf{r}') = \frac{G^{(2)}(\mathbf{r},\mathbf{r}')}{G^{(1)}(\mathbf{r},\mathbf{r})G^{(1)}(\mathbf{r}',\mathbf{r}')}\,. \tag{4.14}$$

Considering the last equation, it is possible to show that $g^{(2)} = 1$ corresponds to the absence of density correlations, $g^{(2)} > 1$ indicates the presence of bunching (positive correlations), while $g^{(2)} < 1$ indicates the presence of *antibunching* (negative correlations). The first observation of quantum-statistical two-atom density correlations was reported by M. Yasuda and F. Shimizu, who detected a bunching effect in a laser-cooled atomic beam of metastable Ne* by measuring the correlation of the atom arrival times onto a detector (Yasuda and Shimizu, 1996).

The normalized second-order correlation function $g^{(2)}$ was measured in Schellekens *et al.* (2005) by A. Aspect, C. Westbrook, and coworkers for a gas of ultracold bosonic ^{4}He* across the BEC transition. In this experiment the gas was released from the trap and, after a free fall, the atoms were detected on a delay-line microchannel plate (a sketch of the experiment is shown in Fig. 4.4a). Then, the second-order correlation function was determined with a statistical analysis of all the detection events recorded for a single experiment. The result is shown in Fig. 4.4b. At temperatures larger than the critical temperature for Bose–Einstein condensation (upper three plots) and small spatial separations the $g^{(2)}$ function is larger than 1, i.e. there is an increased probability of detecting pairs of atoms at the same position, which represents a direct observation of the same bunching effect which is observed with the photons of a classical (incoherent) light. Lowering the temperature of the gas below the BEC critical temperature (lowest plot) this bunching effect disappears and the $g^{(2)}$ function is always 1, i.e. atoms lose their tendency to be detected at the same position. This behaviour is the atomic counterpart of the suppression of photon bunching which can be observed in the case of coherent laser light.

Differently from photons, massive particles can be fermions and fermions exhibit a different and even more intriguing behaviour. In a collaboration between the A. Aspect and W. Vassen groups (Jeltes *et al.*, 2007), similar correlation measurements were performed on ultracold samples of fermionic ^{3}He*. The result of this experiment, shown by the black points in Fig. 4.4c, shows a clear antibunching effect: differently from bosons (either thermal or Bose-condensed) fermions are characterized by a suppression of $g^{(2)}$ at short distance, which is intimately connected with the Pauli exclusion principle and with the impossibility for them to occupy the same position in space at the same time. This antibunching effect is particularly interesting since, differently from bunching, which can be explained also in classical terms, it can be explained only as a consequence of quantum statistics.

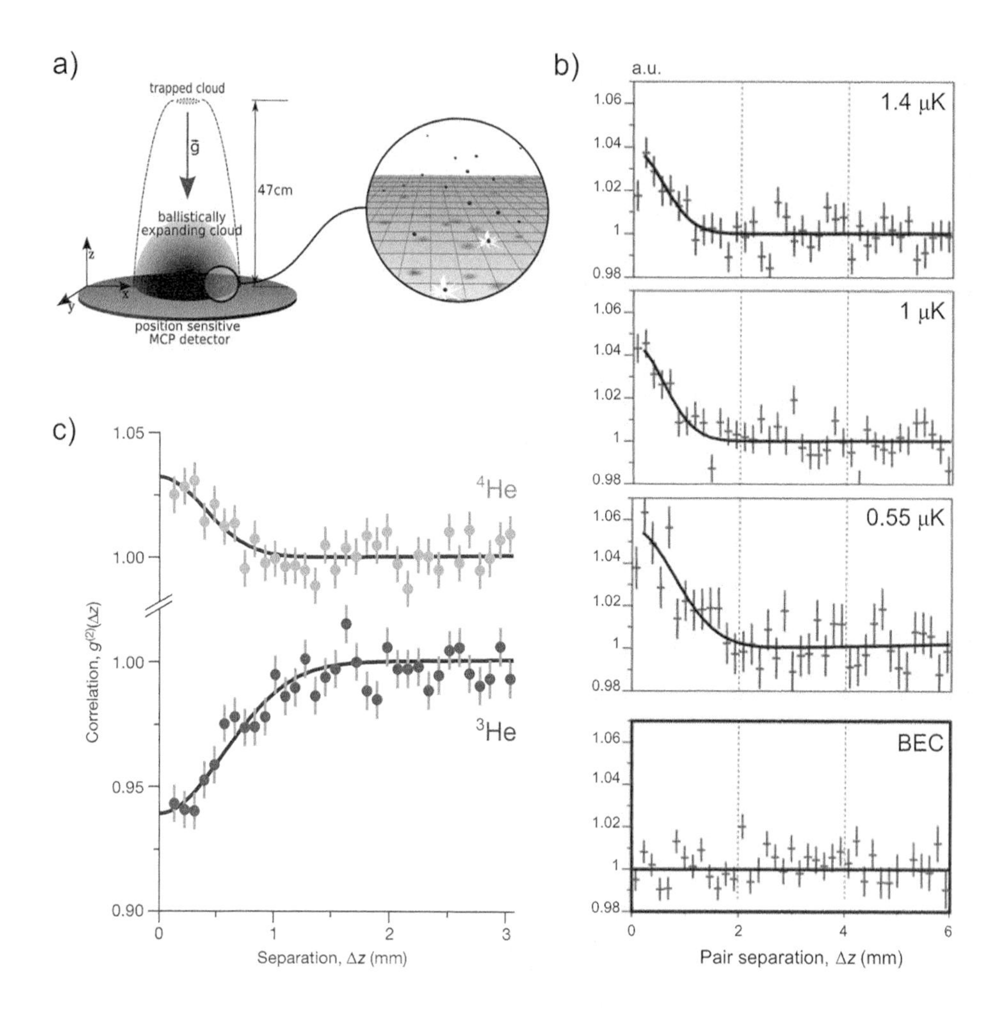

Fig. 4.4 Hanbury Brown and Twiss correlations in an ultracold metastable helium gas. a) Scheme of the experimental setup. b) Correlation function $g^{(2)}(z)$ for bosonic ^{4}He* across the BEC transition. Reprinted with permission from Schellekens *et al.* (2005). © AAAS. c) Comparison of the correlation function for thermal bosonic ^{4}He* (grey points) and fermionic ^{3}He* (black points). Reprinted with permission from Jeltes *et al.* (2007). © Macmillan Publishers Ltd.

Higher-order coherence. Exploring correlation functions in ultracold bosonic gases is important because, differently from photons, atoms interact (as discussed in Sec. 3.1.4). As a consequence, it is interesting to measure if interactions cause a deviation in the correlation functions of an atomic BEC from the values expected for a fully coherent state. Higher nth-order correlation functions (corresponding to the simultaneous arrival of n particles at the detector position) can provide additional and more sensitive tests of coherence. As a matter of fact, the quantum theory of coherence predicts that the difference between a fully incoherent (thermal) state, for which $g^n(0) = n!$ and a fully coherent state, for which $g^n(0) = 1 \ \forall n$, increases as n factorial (Mandel and Wolf, 1995).[7] The third-order g^3 correlation function has been recently measured in Hodgman *et al.* (2011), again recording the arrival times and positions of metastable $^4\mathrm{He}^*$ atoms (both below and above T_C) onto a microchannel plate detector. The results showed that for a BEC the condition $g^3(0) = 1$ is verified, which means that a BEC remains coherent also at third order, while a thermal cloud exhibits nonzero $g^3(0) - 1$ and $g^2(0) - 1$ in agreement with the predictions of the quantum theory of boson statistics.

Related results were obtained in Burt *et al.* (1997) by measuring the three-body decay rate of the number of atoms in a cold trapped ^{87}Rb Bose gas, both above and below the critical temperature for condensation. Confirming the early predictions of Kagan *et al.* (1985), the three-body decay rate for a BEC was found to be 3! smaller than the one for a thermal cloud, in agreement with the expected reduction of $g^3(0)$ for a coherent state. A further reduction of the three-body decay rate was observed in Laburthe Tolra *et al.* (2004) for a system of strongly repulsively interacting 1D Bose gases: in this regime the repulsion between the atoms mimics the effect of the Pauli exclusion principle for fermions, further reducing the probability of finding three particles at the same time in the same position.

More on bunching and antibunching. Related experimental work has been performed in the context of ultracold atoms in optical lattices. In bosonic Mott insulators or fermionic band insulators (described in Chapter 7) single isolated atoms are trapped in a periodic array of microscopic optical traps, separated by a constant lattice spacing d. If the trap is removed, the particles expand and, after a time-of-flight expansion, an absorption image (see Sec. 3.1.5) like the one in Fig. 4.5a is taken. This image is characterized by atomic shot noise, i.e. quantum fluctuations (on the order of $\sqrt{N}$) of the number of particles N detected at each pixel of the CCD camera. Following the proposal of Altman *et al.* (2004), by analysing the density–density correlations of the atomic shot noise (defined in analogy to eqns (4.12) and (4.14)), it is possible to extract information on the spatial ordering in the trap and on the quantum statistics of the atoms. These Hanbury Brown and Twiss correlations arise from the quantum interference between different detection paths, as earlier recognized by U. Fano in Fano (1961*b*): if two identical particles are released from two sites of the lattice, the joint probability of detecting them in two separated positions (e.g. imaging them on two separate pixels of the CCD) depends on the phase acquired during the time-of-flight expansion, hence on the distance between the detection points. These correlations, first observed for ^{87}Rb bosons in Fölling *et al.* (2005) and then also for ^{40}K fermions in Rom *et al.* (2006), are shown in Fig. 4.5b–c: the position of the correlation peaks in the top figures reflects the spatial ordering of the atoms in the optical lattice, while the sign of the correlations (shown in the lower cross sections) depends on the quantum statistics: while bosons show positive correlations, due to their tendency to arrive together at the detector, fermions exhibit negative correlations, i.e. antibunching, as a consequence of the Pauli exclusion principle.

[7] This result holds for an ideal detector with unity quantum efficiency and infinite spatial resolution. Because of the finite resolution of the detection system, the zero-distance peak in the correlation functions is smaller than the expected one (see Fig. 4.4).

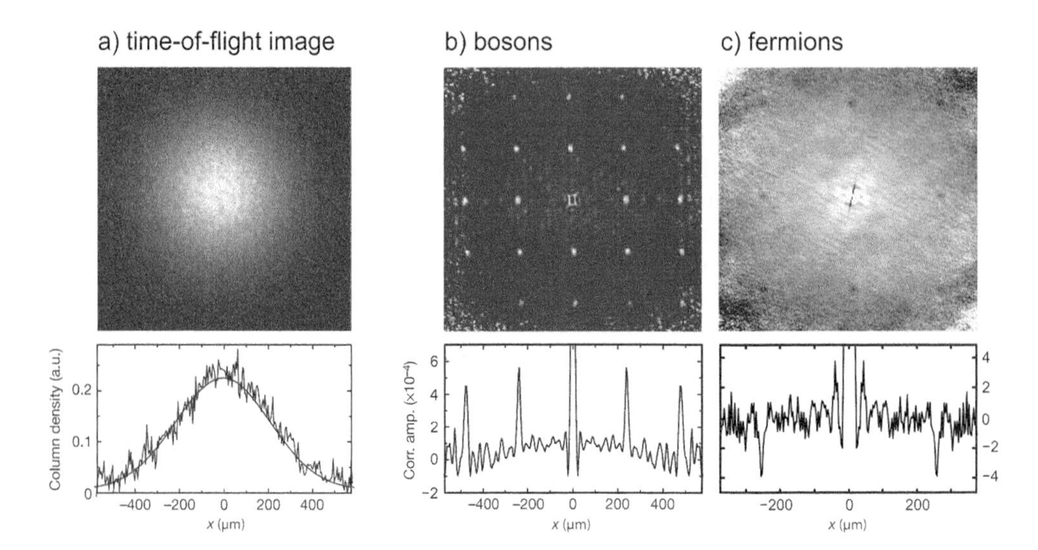

Fig. 4.5 a) Time-of-flight absorption image of ultracold atoms released from an optical lattice. b–c) Density–density correlation of the atomic shot noise for systems of identical bosons or fermions. The upper images show the 2D density–density correlation function evaluated on the time-of-flight images (the centre corresponds to correlation at zero distance). The lower plots show cross sections of the above images along the horizontal median line: in the case of fermions the correlation peaks have opposite sign (antibunching) with respect to bosons (bunching). Adapted with permission from Fölling *et al.* (2005) and Rom *et al.* (2006). © Macmillan Publishers Ltd.

4.4 More on helium spectroscopy

4.4.1 Helium nuclear charge radius

The possibility of performing accurate QED calculations on the "simple" helium atom allows the determination of important physical quantities. Recently, isotope shift measurements, performed on different helium transitions and different experimental arrangements (quantum degenerate ultracold gases, thermal atomic beams, or magneto-optically trapped unstable isotopes), have been used to obtain precise spectroscopic determinations of the helium nuclear charge radius. Indeed, besides the dominant "mass" term, the isotope shift includes a "volume" contribution which depends on the rms radius of the charge distribution within the nucleus.

^{4}He–^{3}He isotope shift. The group of W. Vassen at Amsterdam used ultracold degenerate samples of ^{4}He and ^{3}He to study the isotope shift of an ultra-weak intercombination transition at a wavelength 1557 nm connecting the ortho-helium $2\,^3S_1$ state to the para-helium $2\,^1S_0$ state (see Fig. 4.6a). This magnetic dipole transition, doubly forbidden by both dipole and spin selection rules, has a linewidth of only 8 Hz, which is limited by the two-photon decay channel of the $2\,^1S_0$ state to the $1\,^1S_0$ ground

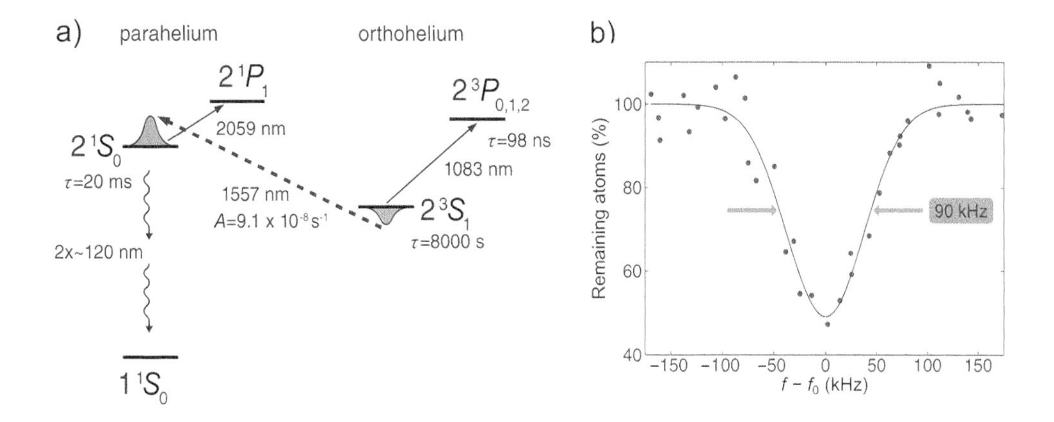

Fig. 4.6 Frequency metrology with quantum degenerate helium gases. a) Level scheme evidencing the ultra-weak intercombination transition $2\,^3S_1 \rightarrow 2\,^1S_0$ observed in van Rooij *et al.* (2011). b) Remaining atoms in the trapped metastable state $2\,^3S_1$ as a function of the excitation laser frequency f (f_0 is the fitted center of the resonance). Reprinted with permission from van Rooij *et al.* (2011). © AAAS.

state. The doubly forbidden nature of the $2\,^3S_1 \rightarrow 2\,^1S_0$ transition causes its strength to be 14 orders of magnitude smaller than the strength of the dipole-allowed 1.08 μm transitions discussed in Sec. 4.2! In the experiment reported in van Rooij *et al.* (2011) high-precision spectroscopy of this transition was performed on ultracold Bose–Einstein condensates and Fermi gases of metastable helium. The long lifetime of the metastable atoms in the trap allowed very long excitation times of several seconds, enabling an efficient detection of the ultra-weak transition. Figure 4.6b shows a plot of the excitation efficiency, detected by observing the decrease in the trapped atom number as a function of the laser detuning. The resonance frequency was determined by measuring the frequency of the excitation laser with an optical frequency comb, the results being 192 510 702 145.6(1.8) kHz for ^{4}He and 192 504 914 426.4(1.5) kHz for ^{3}He ($F = 3/2 \rightarrow F = 1/2$).[8]

The difference between these two values provides a measurement of the ^{4}He–^{3}He isotope shift for this transition, which includes a contribution coming from the finite nuclear size. The comparison of the result with high-precision QED theory (in which many mass-independent terms cancel to give an accurate calculation of the shift) resulted in a value for the difference of the ^{3}He and ^{4}He nuclear charge radii squared $\delta r^2 = r^2(^3\text{He}) - r^2(^4\text{He}) = 1.028(11)$ fm^2 (re-evaluated according to a more recent theoretical analysis presented in Cancio Pastor *et al.* (2012)).

[8]An accurate analysis of the systematic effects was carried out in van Rooij *et al.* (2011) to take into account the ac-Stark shift induced by the optical trap confining the atoms (see Sec. 5.3.1 for the discussion of this aspect in connection with alkaline-earth optical clocks) and the mean-field shift induced by the collisions between the atoms in a high-density inhomogeneous BEC, which is a relevant source of perturbation in high-precision spectroscopic or interferometric measurements.

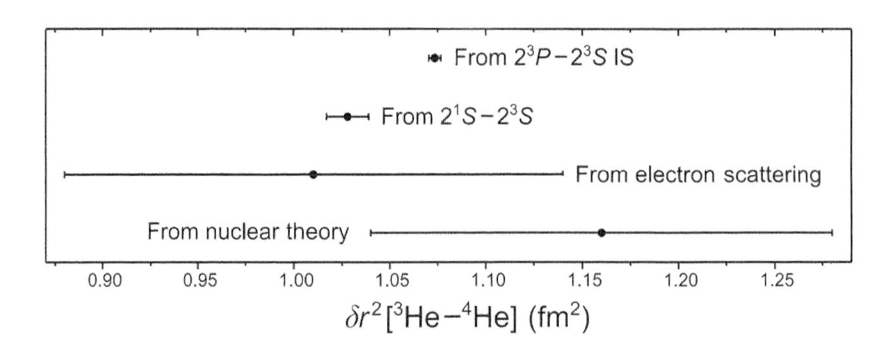

Fig. 4.7 Different measurements of the difference in the square of the nuclear charge radius for ^{4}He and ^{3}He. The spectroscopic determination on top refers to the isotope shift of the $2\,^3S_1 \to 2\,^3P_J$ transitions (Cancio Pastor *et al.*, 2012), while the measurement below refers to the $2\,^3S_1 \to 2\,^1S_0$ transitions (van Rooij *et al.*, 2011), both analysed with the refined QED theory of Cancio Pastor *et al.* (2012). Adapted from Cancio Pastor *et al.* (2012).

The result of van Rooij *et al.* (2011) with ultracold helium can be compared with the measurements performed on helium atoms at room temperature. In addition to the early measurement discussed in Sec. 4.1.1 on the ortho-helium 389 nm $2\,^3S_1 \to 3\,^3P_J$ transition (Marin *et al.*, 1994), more recent isotope shift measurements have been performed at LENS on the 1.08 μm $2\,^3S_1 \to 2\,^3P_J$ transition (already discussed in Sec. 4.2.2) on a thermal beam of metastable helium (Cancio Pastor *et al.*, 2012). The frequency-comb-assisted measurement of the 1.08 μm ^{4}He–^{3}He isotope shift with an uncertainty below 10^{-11}, combined with the same QED theory used to analyse the results of van Rooij *et al.* (2011), provided a very precise value $\delta r^2 = 1.074(3)$ fm^2 for the difference in the nuclear charge radii squared.

Figure 4.7 shows a plot with the two isotope shift determinations of δr^2, together with the values obtained from electron-scattering experiments and from nuclear theory. While the spectroscopic measurements are both in agreement with the much less precise nuclear values, they are inconsistent with each other. In particular, a 4σ discrepancy is present between the value of van Rooij *et al.* (2011) and that of Cancio Pastor *et al.* (2012), for which a satisfactory explanation is currently missing. The observed discrepancy may be in principle explained by some hidden systematics in experiments or by yet unknown effects in the electron–nucleus interaction. The possibility that some additional effects beyond the standard QED may exist has been discussed also in the context of the muonic hydrogen experiment (Pohl *et al.*, 2010) presented in Sec. 1.5.1, which raised what is now known as the "proton charge radius puzzle". One of the ways for solving this puzzle is to investigate similar systems, aiming to confirm or to disprove the disagreement observed for hydrogen. Important hints might come from the measurement of the Lamb shift in muonic helium $\mu^4\text{He}^+$ and $\mu^3\text{He}^+$ (exotic atoms in which a muon takes the place of one helium electron): after the measurements performed as early as the 1970s in Carboni *et al.* (1978), new high-precision experiments are now under preparation and are aimed at the measurement of ^{4}He and ^{3}He nuclear charge radii with a fractional accuracy of 3×10^{-4} (Antognini *et al.*, 2011).

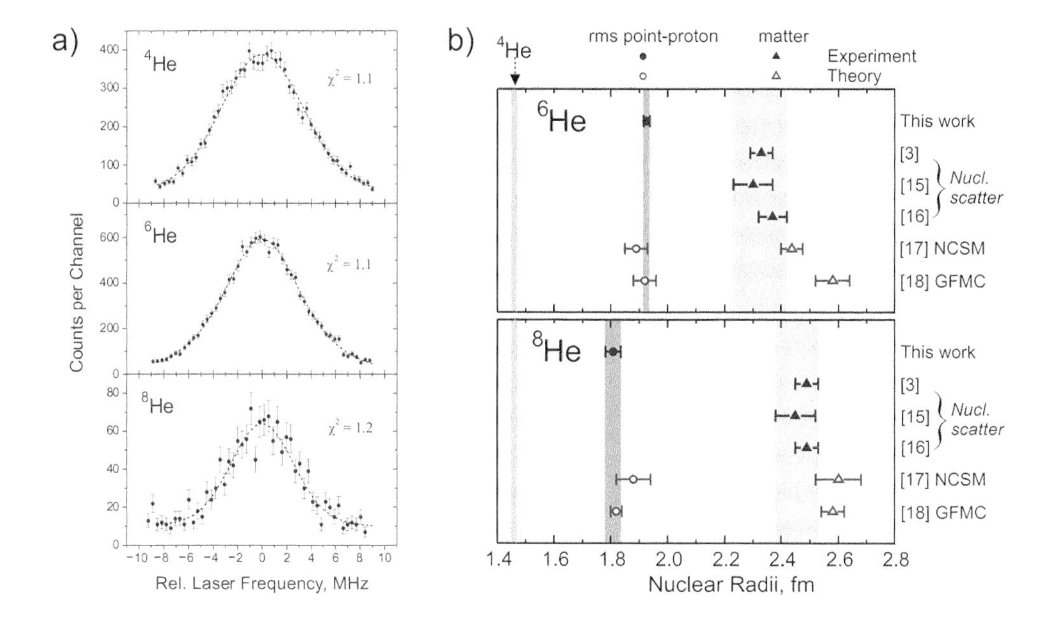

Fig. 4.8 Measurement of the isotope shift of the 389 nm $2^3S_1 \rightarrow 3^3P_J$ transition in unstable helium isotopes. a) Line profiles for different helium isotopes (the isotope shift has been subtracted). b) Comparison between the experimentally determined rms proton distance (filled circles) with two different nuclear theories (empty circles). Triangles refer to the comparison between the matter radius (reflecting the mass distribution of nucleons inside the nucleus) extracted from nuclear scattering experiments and the predictions of the models. Reprinted with permission from Mueller *et al.* (2007). © American Physical Society.

Unstable helium isotopes. High-resolution spectroscopy of unstable isotopes can provide important model-independent tests of nuclear theories. Neutron-rich nuclei are particularly interesting in this respect. Helium has two neutron-rich isotopes ^{6}He and ^{8}He, the lifetimes of which are long enough (807 ms the first, 119 ms the second) for them to be laser-cooled and confined in a magneto-optical trap, as experimentally realized in Wang *et al.* (2004*b*) and Mueller *et al.* (2007). In the nuclei of these isotopes the additional neutrons form a large "halo" with respect to the compact α-particle configuration of the ^{4}He nucleus. The modification in the nuclear charge radius comes from the effective relative motion of the protons (tightly bound in the α-core) around the weakly bound neutrons.

In Mueller *et al.* (2007) unstable helium isotopes, produced in nuclear reactions at the GANIL cyclotron facility in Caen (France), were guided in an atomic beam setup and excited to the metastable 2^3S_1 state in a discharge. Then selected isotopes were laser-cooled on the 1.08 μm transition and were captured in a magneto-optical trap. Here the atoms were excited on the near-UV 389 nm $2^3S_1 \rightarrow 3^3P_J$ transition in order to measure the ^{8}He–^{4}He and ^{6}He–^{4}He isotope shifts. Because of the relatively small number of unstable atoms, the excitation and detection system was optimized in such a way as to perform a complete scan of the resonance on single trapped atoms. In order to continuously cool the atoms, without perturbing them with ac-Stark shifts from the MOT beams, the trapping laser and the spectroscopy laser were alternatively chopped and the fluorescence was detected only synchronously with the excitation phase.

Figure 4.8a shows the measured line profiles for three different helium isotopes. The results of the isotope shift measurements were compared with precise atomic theory calculations including relativistic and QED corrections (Drake, 2004), from which the nuclear charge radius was determined. Figure 4.8b shows a comparison between the measured values (filled circles), which are independent of any assumption on the nuclear structure, and the predictions of two different nuclear models (empty circles). The result is in good agreement with both the nuclear models and confirms the peculiar non-monotonic behaviour of the charge radius in these neutron-rich nuclei.[9]

4.4.2 Antiprotonic helium

Spectroscopy of exotic simple atoms is a valuable tool for measuring fundamental quantities and testing paramount symmetries of Nature, as we have already discussed in the case of muonic hydrogen and antihydrogen in Sec. 1.5. Exotic helium offers very interesting possibilities as well. Recent results come from spectroscopy of *antiprotonic helium* $\bar{p}\mathrm{He}^+$, which is a two-electron atom in which a normal electron and an antiproton are bound to an ordinary helium nucleus. This atom can be synthesized spontaneously from the interaction of a low-energy antiproton $\bar{p}$ beam with a target of gaseous ^{4}He or ^{3}He: an antiproton replaces one of the helium electrons and antiprotonic helium forms. The lifetime of antiprotonic atoms is usually extremely small ($\approx 10^{-12}$ s) because of annihilation of the antiproton with one of the protons in the nucleus. However, the lifetime is remarkably increased (up to the μs scale) if the $\bar{p}\mathrm{He}^+$ electron is in the ground state and the antiproton occupies a circular (or almost-circular) Rydberg state with very large principal quantum number $n \approx 35$ and orbital angular momentum number $l \simeq n - 1$. The reason for this suppression is the smaller and smaller overlap between the antiproton wavefunction and the nucleus as n and l increase and the orbital takes a "torus" shape with large radius. The electron, although in its ground state, is lighter than the antiproton and occupies an orbital which extends at distances larger than the antiproton orbit size,[10] which effectively protects the latter from annihilation during collisions with other helium atoms (as sketched in Fig. 4.9a). Finally, antiprotonic helium in this highly excited state is metastable, with a typical radiative lifetime of several μs: an atom in the circular Rydberg state $|n, l\rangle$, with $l = n - 1$, can only decay radiatively into the $|n-1, l-1\rangle$ state (because of the electric dipole selection rule $\Delta l = \pm 1$ applied to circular Rydberg states), which is suppressed by the small energy spacing for large n. For more information on the properties of antiprotonic helium and on the early investigations of its formation and its spectroscopy, see Yamazaki *et al.* (2002) and Hayano *et al.* (2007).

Antiprotonic helium is an excellent system for tests of CPT invariance (see also Sec. 1.5.2), since its spectroscopy can provide a measurement of the antiproton-to-

[9] The charge radius increases going from ^{4}He to ^{6}He because the α-core recoils according to the motion of the two "halo" neutrons. The charge radius decreases going from ^{6}He to ^{8}He because the four "halo" neutrons move in a more spherically symmetric way, decreasing the motion of the α-core.

[10] For a hydrogenoid atom the orbital size scales as $\sim n^2/m_e$, where m_e is the electron mass.

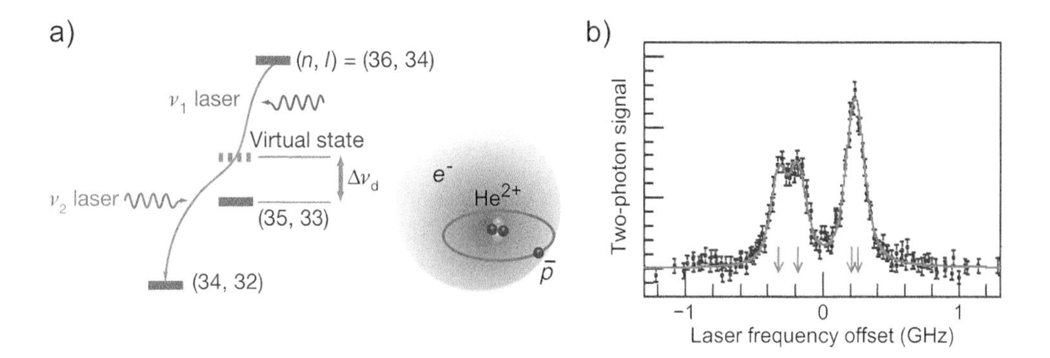

Fig. 4.9 a) Level scheme of antiprotonic helium $\bar{p}He^+$ and sketch of the orbitals: the antiproton $\bar{p}$ occupies a near-circular Rydberg state while the electron is in the ground state. b) Two-photon spectrum of the $\bar{p}He^+$ $|n = 36, l = 34\rangle \rightarrow |n = 34, l = 32\rangle$ two-photon transition. Adapted with permission from Hori *et al.* (2011). © Macmillan Publishers Ltd.

electron mass ratio and put constraints on the equality between antiproton and proton charges and masses. Recently, sub-Doppler two-photon spectroscopy of antiprotonic helium has been performed at the CERN Antiproton Decelerator facility (Hori *et al.*, 2011). Two-photon transitions between levels $|n, l\rangle$ and $|n - 2, l - 2\rangle$ were induced by using two counterpropagating near-UV lasers at different frequencies ν_1 and ν_2: although this choice does not result in a complete suppression of the first-order Doppler broadening,[11] the authors could take advantage of the near resonance with the intermediate $|n - 1, l - 1\rangle$ level in order to enhance by a factor 10^5 the two-photon transition probability. The level scheme and the two-photon sub-Doppler spectra for the $\bar{p}He^+$ transition $|n = 36, l = 34\rangle \rightarrow |n = 34, l = 32\rangle$ is shown in Fig. 4.9 (the fine structure of the spectrum arises from spin–orbit and hyperfine interactions).

Comparing the experimental results with the three-body QED theory of antiprotonic helium, the authors derived an antiproton-to-electron mass ratio

$$\frac{m_{\bar{p}}}{m_e} = 1836.1526736\,(23)\,, \tag{4.15}$$

which well agrees with previous measurements of the proton-to-electron mass ratio m_p/m_e, obtained in Penning trap experiments, which is known with similar precision (Hori *et al.*, 2011). This agreement represents a test of CPT invariance of the proton mass at the level of 1.2×10^{-9}.

4.5 The fine structure constant α

We conclude this chapter with a discussion of the experimental methods for the determination of the fine structure constant α, introduced earlier in this book and

[11] If the frequencies ν_1 and ν_2 of the two counterpropagating beams are different, the first-order Doppler broadening is reduced by a factor $|\nu_1 - \nu_2|/(\nu_1 + \nu_2)$ (< 0.1 for the frequencies used in the experiment) with respect to a single-photon transition.

discussed in connection with the helium fine structure (see Sec. 4.2). In the International System of Units the fine structure constant is defined as

$$\alpha = \frac{e^2}{4\pi\epsilon_0 \hbar c} , \tag{4.16}$$

where e is the electron charge, ϵ_0 is the electric constant, and c is the speed of light. This quantity is particularly important since it is a dimensionless number, therefore it has the same value independently of the chosen system of units. This importance is enforced by the fact that the fine structure constant is the fundamental constant which quantifies the strength of the interaction between photons and charged particles. In fact, it is the fundamental coupling constant of QED, and its small value $\alpha \approx 1/137$ makes it possible to use it as a small parameter for complex and lengthy QED perturbation calculations. However, its precise value has to be experimentally measured, since QED does not make any prediction about it. Regarding the role of the fine structure constant in Physics, Feynman wrote (Feynman, 1988):

It's one of the greatest damn mysteries of physics: a magic number that comes to us with no understanding by man ... We know what kind of dance to do experimentally to measure this number very accurately, but we don't know what kind of dance to do on a computer to make this number come out without putting it in secretly!

What are the techniques for measuring α? There are several experimental methods, which are all based on quite different approaches. As an example of the universal role that the fine structure constant α has in Physics and of the different possibilities to access it, recently it was found that the transmission of visible light by a single-atom-thick sheet of graphene (the revolutionary material for which the Nobel Prize in Physics was awarded in 2010) is exactly $\pi\alpha$ (Nair *et al.*, 2008). Even if α can be measured by such a simple measurement as taking a photograph of a graphene sheet, high-precision measurements require larger efforts.

Figure 4.10 shows the results of all the different measurements of α considered by CODATA in the last 2010 adjustment of the fundamental constants (Mohr *et al.*, 2012). The fitted value

$$\alpha^{-1} = 137.035\,999\,074\,(44) , \tag{4.17}$$

with a precision of 0.32 ppb (parts per billion), is mostly determined by the precise measurement of the electron gyromagnetic anomaly and by the measurement of atomic h/m factors (shaded in grey in the figure).

In the following we discuss these different experimental techniques, as well as other possible approaches. Measuring the same quantity with a variety of different methods is indeed a precious resource, since it provides a sensitive test of the consistency of Physics across a range of energy scales and physical phenomena.

4.5.1 Electron gyromagnetic anomaly

The most precise determination of the fine structure constant is based on the measurement of the electron gyromagnetic anomaly. The magnetic dipole moment of the electron can be written as

$$\boldsymbol{\mu}_e = -g_e \mu_B \frac{\boldsymbol{s}}{\hbar} , \tag{4.18}$$

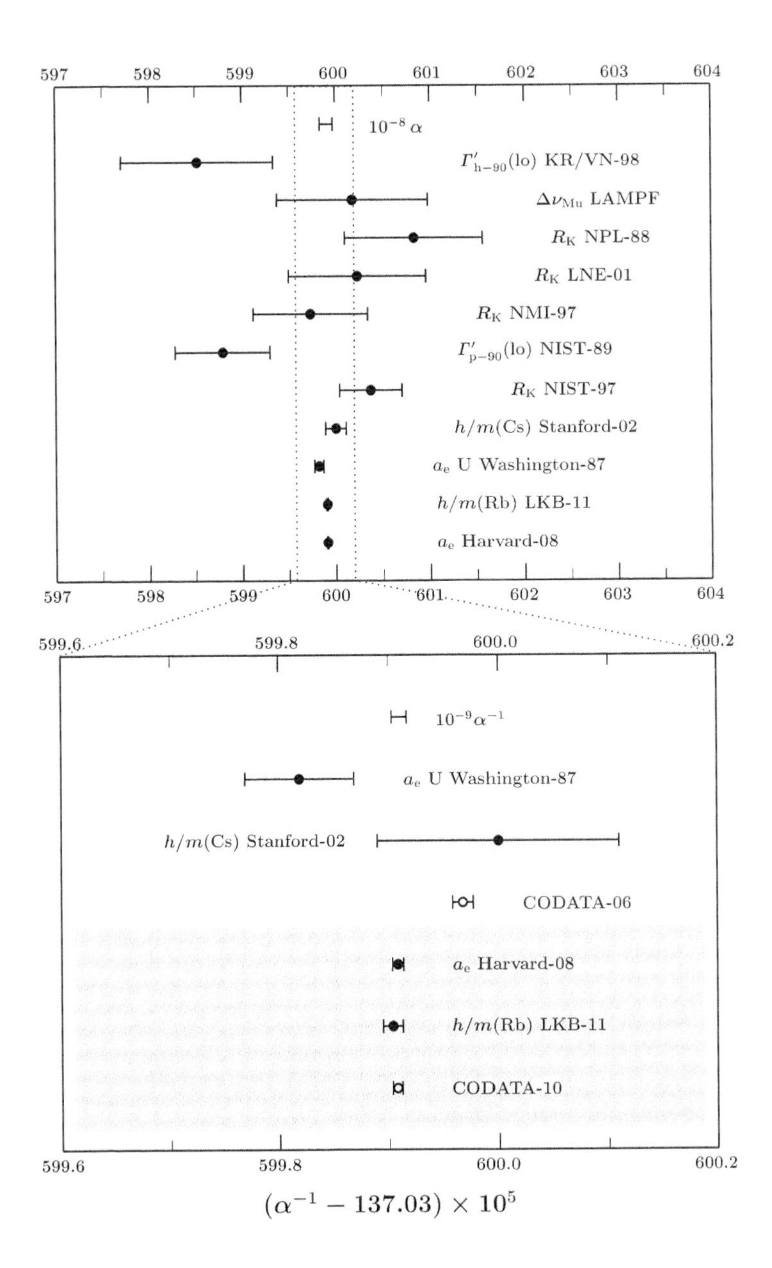

Fig. 4.10 Plot of the different measurements of the fine structure constant α considered by CODATA in the last 2010 adjustment of the fundamental constants. Note that the latest results shaded in grey led to a 2010 determination of α which is more precise and shifted from the 2006 value by more than six times the uncertainty. Reprinted with permission from Mohr *et al.* (2012). © American Physical Society.

where $\boldsymbol{s}$ is the electron spin, $\mu_B = e\hbar/2m_e$ is the Bohr magneton, and g_e is the electron gyromagnetic factor.[12] According to the Dirac quantum relativistic theory, which applies to all the elementary spin-1/2 particles, $g_e = 2$ exactly. However, it was found experimentally (Kusch and Foley, 1948) that the electron gyromagnetic factor has a slightly larger value, nowadays measured as

$$g_e = 2.0023193\ldots \tag{4.19}$$

This deviation from the Dirac theory, well accounted for by quantum electrodynamics, is quantified by the *electron gyromagnetic anomaly* a_e, which can be expressed by a series in α as

$$a_e = \frac{g_e - 2}{2} = a_2\alpha + a_4\alpha^2 + a_6\alpha^3 + \cdots \tag{4.20}$$

with the first-order term known analytically as $a_2 = 1/2\pi$. Extensive calculations of the QED theory for a free electron have been performed by T. Kinoshita, M. Nio, and coworkers, resulting in the full evaluation of the QED contributions up to the a_{10} coefficient, corresponding to the α^5 term in the above expansion (Aoyama *et al.*, 2012).[13] This theoretical knowledge, combined with a precise measurement of the electron magnetic moment, allows the determination of the fine structure constant.

The electron gyromagnetic anomaly was measured in 1987 in a celebrated experiment by H. G. Dehmelt (then awarded with the Nobel Prize for Physics in 1989) and coworkers (van Dyck *et al.*, 1987; Dehmelt, 1990), the result of which has constituted for many years the most precise determination of a_e, yielding a value for the fine structure constant with a fractional uncertainty of 20 ppb (included in Fig. 4.10 as "a_e U Washington-87"). More recently, this quantity has been measured at amazing levels of precision by the group of G. Gabrielse at Harvard University in a series of experiments, the most recent of which has been reported in Hanneke *et al.* (2008). In these experiments single electrons were confined in a *Penning trap*, as first demonstrated in Wineland *et al.* (1973). A Penning trap is an electrostatic trap composed by a strong uniform magnetic field $\mathbf{B}$ and a quadrupole electric field $\mathbf{E}$ produced by a positive ring electrode with the same axis as $\mathbf{B}$ and two negative end-caps, as sketched in Fig. 4.11a. The quadrupole electric potential has a saddle point in the trap centre and is capable of trapping the electron in the direction of the trap axis $\hat{z}$. The magnetic field, which is oriented along $\hat{z}$, dynamically traps the electron in the $\hat{x}\hat{y}$ plane by constraining it to a cyclotron motion with angular frequency

$$\omega_C = \frac{e}{m_e} B\,. \tag{4.21}$$

For typical values of the parameters used in the Harvard experiment (Hanneke *et al.*, 2008), the trap frequency along $\hat{z}$ direction is $\omega_z/2\pi \approx 50$ MHz, while the cyclotron

[12]Here the gyromagnetic factor is taken to be positive and a minus sign has been put in front of the right-hand side of eqn (4.18). Sometimes a different choice on the sign of g_e is made and no minus sign appears in eqn (4.18).

[13]Additional non-QED corrections, coming from hadronic and electroweak effects, have been evaluated as well, but they are very small and well understood in the context of the Standard Model.

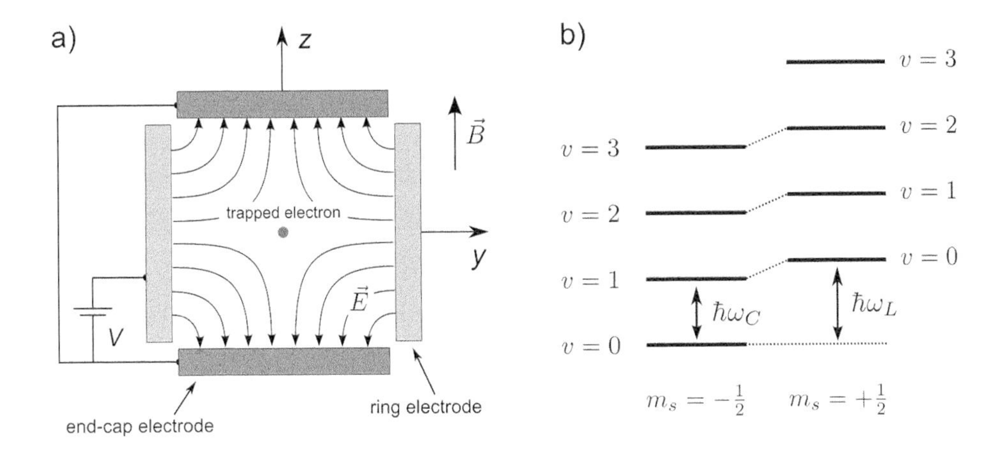

Fig. 4.11 a) Scheme of an electrostatic Penning trap for trapping negatively charged particles. b) Simplified level scheme for the cyclotron motion + spin state of an electron in a Penning trap (relativistic effects and other corrections are not included).

frequency is $\omega_C/2\pi \approx 150$ GHz. The cyclotron motion is quantized to energy levels labelled by the cyclotron quantum number v

$$E_v = \hbar\omega_C \left(v + \frac{1}{2}\right) \tag{4.22}$$

and the large value of ω_C makes the thermal occupation of excited cyclotron states with $v > 0$ negligible at the cryogenic temperature $T \approx 100$ mK of the experiment.

When the electron spin is considered, the presence of a magnetic field along $\hat{z}$ removes the degeneracy between the two possible spin orientations. The differency in energy between the two electron spin states $m_s = +1/2$ and $m_s = -1/2$ is $\hbar\omega_L$, where

$$\omega_L = \frac{g_e}{2}\frac{e}{m_e}B \tag{4.23}$$

is the *Larmor frequency*, which has the classical analogue of the precession frequency of the electron spin around the magnetic field axis. Since g_e is close to 2, this frequency is only slightly different from the cyclotron frequency, as can be seen in Fig. 4.11b. This simple system constituted by an electron bounded to the centre of a Penning trap is often referred to as *geonium* (a bound state between an electron and the Earth!) and extensive literature exists on the properties of this artificial atom (see Brown and Gabrielse (1986) for a review).[14]

[14]Actually the electron motion and the associated geonium spectrum is more complex than what has been discussed so far. An important effect comes from the forces exerted by the quadrupole electric field in the plane $\hat{x}\hat{y}$ as soon as the electron deviates from the trap axis. The result of these forces is a *magnetron* motion in which the axis of the fast cyclotron motion slowly precesses around the trap axis in a cycloid-type fashion. Other relativistic corrections have to be considered as well.

By measuring the cyclotron frequency and the Larmor frequency (van Dyck *et al.*, 1987) it is possible to determine the electron gyromagnetic anomaly as[15]

$$a_e = \frac{g_e - 2}{2} = \frac{\omega_L - \omega_C}{\omega_C} \,. \tag{4.24}$$

The most recent measurement of a_e by the Gabrielse group (Hanneke *et al.*, 2008) provided the up-to-now most precise value of the electron gyromagnetic anomaly

$$a_e = 1.159\,652\,180\,73\,(28) \times 10^{-3} \tag{4.25}$$

with a relative accuracy of $2.4 \times 10^{-10} = 0.24$ ppb (parts per billion). Comparing this result with the theory summarized in Gabrielse *et al.* (2006) yielded an extremely precise determination of α (included in Fig. 4.10 as "a_e Harvard-08")

$$\alpha^{-1} = 137.035\,999\,084\,(51) \tag{4.26}$$

with a relative accuracy of $3.7 \times 10^{-10} = 0.37$ ppb. Very recent improvements in the QED theory (Aoyama *et al.*, 2012), consisting in a full calculation of the α^5 contributions (obtained by the evaluation of 12 672 different Feynman diagrams!), resulted in the most precise determination of the fine structure constant so far, not yet included in the CODATA adjustment of the fundamental constants

$$\alpha^{-1} = 137.035\,999\,173\,(35) \tag{4.27}$$

with a fractional uncertainty of only 0.25 ppb.

4.5.2 h/m ratio

A different method of determining α involves the measurement of masses, more precisely the measurement of the ratio h/m_X between the Planck constant and an atomic mass m_X (Weiss *et al.*, 1993). Differently from the measurement of the electron gyromagnetic anomaly (and high-precision spectroscopy of helium as well), this method is not dependent on QED calculations. This is a very important feature of h/m_X experiments: since they are based on a completely different approach to the measurement of α, the comparison of their outcomes with the results coming from the electron gyromagnetic anomaly can be used as a test of QED.

In Sec. 1.2.1 we have already pointed out that α can be rewritten in terms of the Rydberg constant R_∞ as

$$\alpha^2 = \frac{2hR_\infty}{m_e c} \,. \tag{4.28}$$

By introducing the mass of a generic atom X this equation can be recast as

$$\alpha = \left[\frac{2R_\infty}{c}\frac{m_X}{m_e}\frac{h}{m_X}\right]^{1/2} \,. \tag{4.29}$$

In the above expression all the quantities multiplying h/m_X are known with extremely high precision. The Rydberg constant R_∞ is the most accurately measured physical

[15] The real expression is more complex. As a matter of fact, the magnetron motion is coupled to the cyclotron motion, resulting in a slightly different value of the cyclotron frequency from the one reported in eqn (4.21). This correction to the frequency has to be considered in eqn (4.24), together with other corrections, but it does not compromise the general scheme of the measurement.

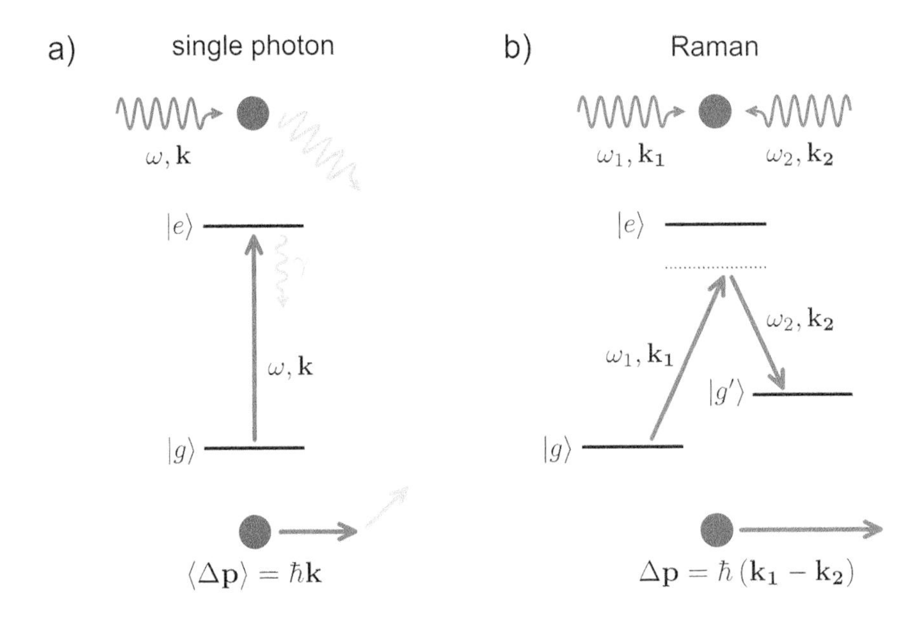

Fig. 4.12 a) Scheme of a single-photon transition between the ground state $|g\rangle$ and an excited state $|e\rangle$, including the spontaneously emitted photon (in grey). b) Scheme of a two-photon Raman transition (absorption + stimulated emission) between two long-lived states $|g\rangle$ and $|g'\rangle$. The momentum transfer in these processes is shown in the bottom part of the figure.

constant with a fractional uncertainty of 5.0×10^{-12} (see Sec. 1.4.3). The speed of light has an exact value $c = 299\ 792\ 458$ m/s. Mass ratios can be measured very accurately in Penning trap experiments by comparing the cyclotron frequencies for different ions (see Sec. 4.5.1): the electron mass in atomic mass units m_e/u is known with a fractional uncertainty of 4.0×10^{-10} (Mohr *et al.*, 2012), while the masses of alkali atoms in atomic mass units m_X/u have been measured very accurately by the group of D. Pritchard with uncertainties down to 1.2×10^{-10} (Bradley *et al.*, 1999). Therefore the measurement of α reduces to a measurement of the ratio h/m_X between the Planck constant and the atomic mass m_X.

Experimentally, it is actually easier to measure the ratio h/m_X than the mass alone, since the former quantity is directly connected with the recoil velocity that an atom acquires after absorbing one photon of frequency ν, thereby changing its momentum by $\Delta p = h\nu/c$:

$$v_R = \frac{h\nu}{m_X c} . \tag{4.30}$$

The simplified principle of the h/m_X measurement therefore involves making an atom absorb a precisely known number of photons with precisely known frequency ν (which can be measured e.g. with an optical frequency comb) and then measuring its velocity change (e.g. by means of a spectroscopic measurement of the Doppler shift).

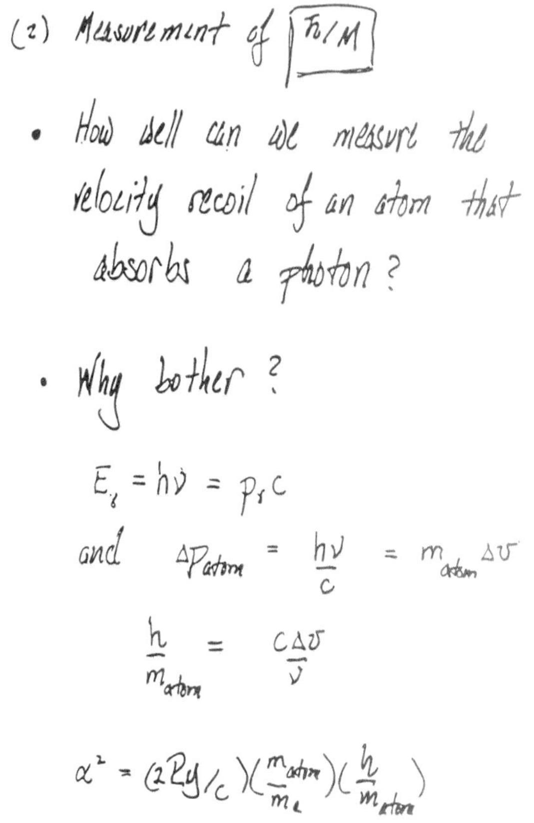

(2) Measurement of $\hbar/M$

- How well can we measure the velocity recoil of an atom that absorbs a photon?

- Why bother?

$$E_\gamma = h\nu = p_\gamma c$$

and $$\Delta p_{atom} = \frac{h\nu}{c} = m_{atom}\,\Delta v$$

$$\frac{h}{m_{atom}} = \frac{c\,\Delta v}{\nu}$$

$$\alpha^2 = (2Ry/c)\left(\frac{m_{atom}}{m_e}\right)\left(\frac{h}{m_{atom}}\right)$$

Fig. 4.13 Handwritten slide by S. Chu illustrating the principle of the h/m measurement from the atomic recoil and its importance for the determination of α. Reprinted from Cancio *et al.* (2001) by courtesy of S. Chu.

However, a single-photon transition (see Fig. 4.12a) between different electronic states $|g\rangle$ and $|e\rangle$ is not the best choice: the upper electronic state is not stable, which strongly limits the measurement time to intervals smaller than the $|e\rangle$ lifetime γ^{-1} before a spontaneous emission event occurs changing the atom velocity randomly. A transition between two internal "stable" states has to be preferred. As an example, we consider an alkali atom which has two hyperfine ground states $|g\rangle$ and $|g'\rangle$ (see Fig. 4.12b), corresponding to different values of the total angular momentum F and $F+1$. Instead of using a microwave transition, which would impart extremely small momentum to the atoms, one can use a Raman two-photon transition, which can be interpreted as the combination of absorption and stimulated emission by an excited virtual level. This combines the advantage of working with stable states and microwave energy separations (which can be controlled very accurately) with the large momentum transfer offered by visible photons. The momentum transfer Δp in the Raman process depends on the direction of the laser beams and it is maximum when the two lasers are counterpropagating, for which $\mathbf{\Delta p} = \hbar(\mathbf{k_1} - \mathbf{k_2}) \simeq 2\hbar\mathbf{k}$ is twice the single photon momentum. In this version of the experiment, one can repeat similar excitations many times (in such a way as to increase the absorbed momentum) and then measure the final velocity acquired by the atom by measuring the Doppler shift on the Raman transition. As a matter of fact, transitions between hyperfine states of the same electronic configuration are extremely narrow and their linewidth is typically limited by the finite interrogation time (which can be quite long, in laser-cooled atomic fountains, as discussed in Sec. 2.4).[16]

The idea of the h/m_X measurement as a tool to determine the fine structure constant and the first seminal experiments in this direction were realized by S. Chu (Chu, 1998) (see the handwritten note illustrating the idea of the method in Fig. 4.13). The actual measurement of h/m_X is more complex than the simplified experiment described above. Atom interferometric techniques (see Sec. 2.5) have to be used in order to improve the sensitivity of the velocity measurement, which is limited by the interrogation time. In Fig. 4.14a we show the sketch of the modified double Ramsey interferometer employed in Weiss *et al.* (1993), which is based on two pairs of $\pi/2$ Raman pulses, with the same time-delay from the first to the second pulse of each pair. The number n in the notation $|ng\rangle/|ng'\rangle$ indicates the number of Raman recoil kicks acquired after the interaction with the Raman beams. Measuring the output population of e.g. state $|g\rangle$, one observes Ramsey fringes centred around two different frequencies: one corresponding to the resonance frequency for $n = 0$, the other one corresponding to $n = 2$. The difference between the two resonance frequencies is independent of the initial atomic velocity and gives the recoil shift corresponding to the absorption of $\Delta n = 2$ recoil kicks.

The experimental scheme is sketched in Fig. 4.14b. Atoms are cooled at the bottom of the atomic fountain and their velocity spread is reduced to a fraction of a photon recoil,[17] then they are launched upwards and, close to the turning point of their free-falling motion, the interferometric sequence takes place. In order to increase the sensitivity of the measurement, the number of recoil kicks can be increased by inserting m additional π pulses between the two $\pi/2$ pulse pairs (see Weiss *et al.* (1993) for details) which change the momentum separation of the two output ports from $|0g\rangle$ / $|2g\rangle$ ($\Delta n = 2$) to $|-mg\rangle$ / $|2+m, g\rangle$ ($\Delta n = 2(m+1)$). Figure 4.14c shows experimental Ramsey fringes for a total momentum transfer of $\Delta n = 18$ Raman recoil kicks.

Using this interferometric scheme, the authors of Wicht *et al.* (2002) performed a measurement of the h/m ratio of ^{133}Cs, which resulted in a determination of the fine structure constant

$$\alpha^{-1} = 137.036\,0000(11) \tag{4.31}$$

[16]The reader will recall that the microwave $F = 3 \rightarrow F = 4$ transition in ground state atomic caesium is adopted as the SI definition of the second.

[17]This is possible by taking advantage of the narrow linewidth of Raman transitions with a sub-recoil laser-cooling technique known as *Raman cooling*; see Metcalf and van Der Straten (1999) and Foot (2005) for more details.

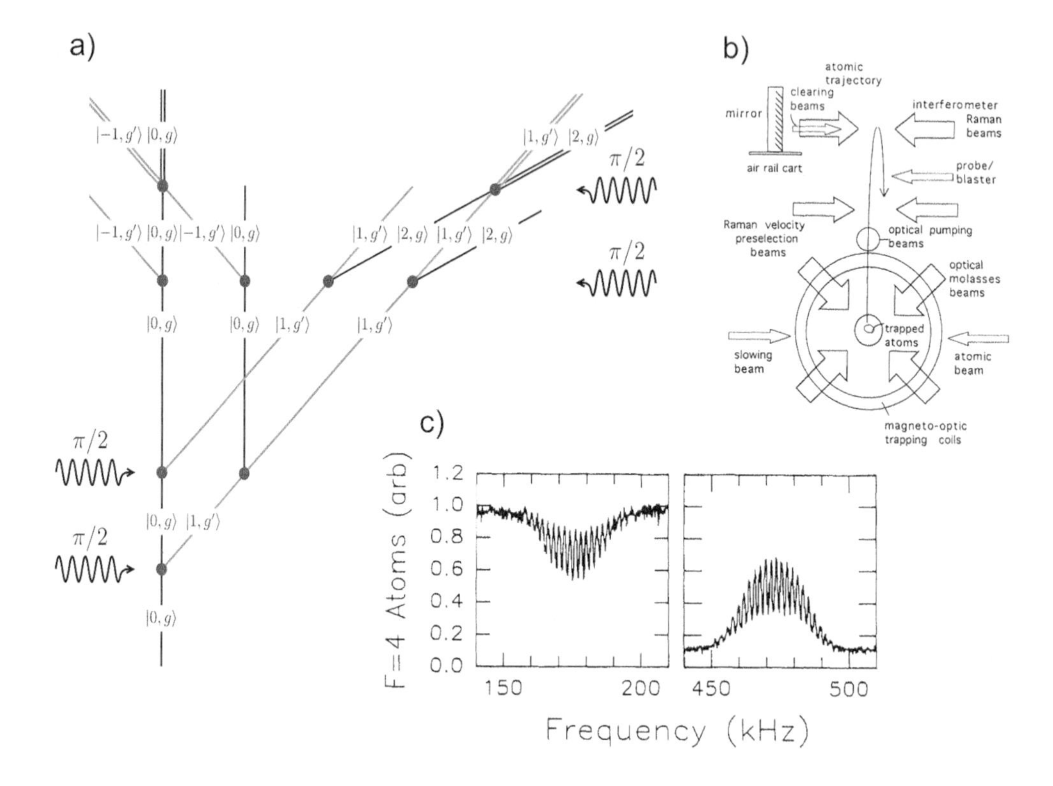

Fig. 4.14 Interferometric measurement of h/m for caesium atoms. a) Simplified scheme of the double Raman interferometer employed in Weiss *et al.* (1993): the number in the state labels indicates the number of Raman velocity recoils the atom has acquired (the direction of the $\pi/2$ pulses refers to the direction of the effective Raman recoil). b) Scheme of the experimental setup. c) Ramsey fringes for atoms in different interferometer output channels separated by $\Delta n = 18$ Raman recoil kicks. b) and c) are reprinted with permission from Weiss *et al.* (1993). © American Physical Society.

with a relative accuracy of 7.4 ppb (the value is indicated with "h/m(Cs) Stanford-02" in Fig. 4.10).[18] Technical improvements of the laser stability and the transmission of a larger atomic recoil by means of large-momentum beam splitters (see Sec. 6.3.3) might improve considerably the precision of this interferometric measurement towards future experiments with a target accuracy of 0.5 ppb (Müller *et al.*, 2006).

We will further discuss the measurement of h/m_X in Chapter 6 devoted to the physics of ultracold atoms in optical lattices. An optical lattice is a very useful tool to impart a large and controlled number of recoil kicks to an atom. The application of this technique to the measurement of h/m_X ratios is very promising: recently, a measurement of the h/m ratio of ^{87}Rb has been performed in Bouchendira *et al.* (2011)

[18]For this measurement the absolute frequency of the caesium transitions had to be known precisely. The authors used the result of one of the first absolute frequency measurements performed with early optical frequency combs (Udem *et al.*, 1999).

("h/m(Rb) LKB-11" in Fig. 4.10) resulting in a determination of α with a fractional uncertainty of 0.66 ppb, less than twice the smallest uncertainty obtained from the electron gyromagnetic anomaly.

4.5.3 Quantum Hall effect

After the discussion of the h/m_X measurement, we mention another approach to the measurement of α which does not rely on QED. This method is based on the *quantum Hall effect* exhibited by 2D degenerate electron gases at low temperatures and strong magnetic fields. In 1980 K. von Klitzing (Nobel Prize in Physics 1985), while performing electric measurements on MOSFETs devices placed in liquid helium cryostats, observed that the Hall conductivity σ was quantized according to the relation $\sigma = n/R_K$, where n is a positive integer and $R_K = h/e^2$ is the von Klitzing constant (von Klitzing *et al.*, 1980). This behaviour can be explained from the quantum treatment of the electron motion in a magnetic field, which is described in terms of quantized *Landau levels* (quantum versions of the classical cyclotron orbits). For strong fields the energy spacing between Landau levels increases and becomes of the same order as the Fermi energy, which causes the density of states to become heavily discretized, which is at the basis of the observed quantization of the conductance.[19]

From the definition of the fine structure constant in eqn (4.16), it follows that a direct measurement of R_K can provide a determination of α through the relation

$$\alpha = \mu_0 c/2R_K \ , \tag{4.32}$$

where both μ_0 and c are exact constants. Since no theoretical calculation is involved in this expression, the determination of α is only limited by the relative accuracy in the R_K measurement.

The main obstacle for this kind of measurement is connected with the experimental difficulties in the realization of accurate standards of resistance. This problem is typically faced by converting the resistance measurement into a capacitance measurement and adopting favourable standard capacitors the value of which can be calculated with relatively large accuracy. The most precise determination of α with this method is

$$\alpha^{-1} = 137.036\,0037(33) \tag{4.33}$$

with a relative accuracy of 24 ppb (Jeffery *et al.*, 1998) (the value is indicated with "R_K NIST-97" in Fig. 4.10).

4.5.4 Helium fine structure and three-body QED

In the 2006 CODATA adjustments of the fundamental constants the α measurement coming from the fine structure of helium was not included. Although the fine-structure interval $\Delta\nu_{01}$ had been measured with an accuracy of 0.5 kHz (combining the results of three different experiments (George *et al.*, 2001; Zelevinsky *et al.*, 2005; Giusfredi *et al.*, 2005) giving consistent results), which was capable of providing a determination

[19] While the *integer* quantum Hall effect can be explained in terms of single-electron physics, the *fractional* quantum Hall effect observed in certain materials has a much more complex explanation, since it is associated with the formation of strongly correlated electronic states.

of α at the 8 ppb level, at the time of the 2006 adjustment the theoretical value of $\Delta\nu_{01}$ based on the best determination of α was found to be not consistent with the experimental value by several standard deviations. For this reason, waiting for the solution of this controversy, the helium measurement was not taken into account.

The 2006 QED calculation of helium fine structure (Pachucki, 2006), performed at the α^7 level in the expansion of eqn (4.10), was later found to contain numerical errors, that were corrected in Pachucki and Yerokhin (2010). The current theoretical determination of $\Delta\nu_{01}$ is in agreement with the experimental value. However, despite the improvements in the theory, the calculated value has a quite large uncertainty of 1.7 kHz, which is limited by a conservative estimate of the following, not yet evaluated, α^8 QED terms in the theoretical expansion. This uncertainty makes a determination of the fine structure constant possible only at the 28 ppb level, which is not competitive with the last measurements of α coming from the electron gyromagnetic anomaly (0.37 ppb) or the h/m ratio (0.66 ppb).

While recent more precise measurements with an uncertainty of 0.3 kHz have been reported (Smiciklas and Shiner, 2010) (capable of providing a measurement of α at the 5 ppb level), major advances in the theory are required for the fine structure of helium to give a competitive contribution to the measurement of α. From an alternative perspective, using the precise value of α obtained e.g. with h/m experiments (which do not depend on QED, as discussed in Sec. 4.5.2), a comparison between the experimental and theoretical values for the helium fine-structure intervals can provide a test of validity of the bound-state QED theory for a three-body problem.

5

Alkaline-earth atoms and ions

In this chapter we will consider the properties and applications of alkaline-earth atoms. The two-electron spectrum of these atoms shares important similarities with the helium spectrum, in particular regarding the existence of spin-singlet and triplet states.

The absence of spin in the ground state makes magnetic trapping impossible, therefore different techniques based on far-off resonant optical trapping (Sec. 5.2) have to be used to confine these atoms. The existence of ultra-narrow transitions connecting singlet and triplet states motivates the use of alkaline-earth atoms for the realization of very accurate optical clocks (Sec. 5.3), which are already outperforming microwave atomic clocks in terms of accuracy.

Optical clocks can also be realized with forbidden transitions in trapped ions, the spectroscopy of which shares similarities and complementarities with respect to the neutral alkaline-earth systems. Trapping and cooling of ions will be discused in Sec. 5.4, together with the realization of extremely accurate ion clocks which can be used to test general relativity or search for temporal variations of the fundamental constants.

5.1 Alkaline-earth atoms

As the electronic structure of alkali atoms is similar to that of hydrogen, alkaline-earth atoms in the second column of the periodic table have an electronic structure which is similar to that of helium. The ground state of an alkaline-earth atom with atomic number Z has the electronic configuration $\{Z-2\}\,(ns)^2$, given by the closed configuration $\{Z-2\}$ of the noble gas with atomic number $Z-2$ and by two external electrons occupying levels with principal quantum number n and angular momentum $l=0$. The spectrum of alkaline-earth atoms in the visible or near-visible region is determined by these two external electrons, which, similarly to helium, can form either singlet $S=0$ or triplet $S=1$ states. The lowest-energy excited states are represented in Fig. 5.1, while Table 5.1 contains the values of some relevant quantities for different atoms of this family, with the addition of lanthanide ytterbium, the electron structure of which closely resembles that of alkaline-earth atoms.[1]

The longest-wavelength allowed transition is between the ground state $(ns)^2\,{}^1S_0$ and the first excited singlet state $ns\,np\,{}^1P_1$, which typically lies in the blue-ultraviolet region of the electromagnetic spectrum and has a linewidth of several tens of MHz (see Table 5.1). The lowest-energy triplet states $ns\,np\ {}^3P_J$ have energies lower than the $ns\,np\,{}^1P_1$ state and exhibit a fine structure with three different terms $J=0,1,2$.

[1]The electronic configuration of ytterbium is [Xe]$4f^{14}6s^2$: since the inner f shell is complete, the electronic spectrum is largely determined by the excitation of the two $6s$ electrons.

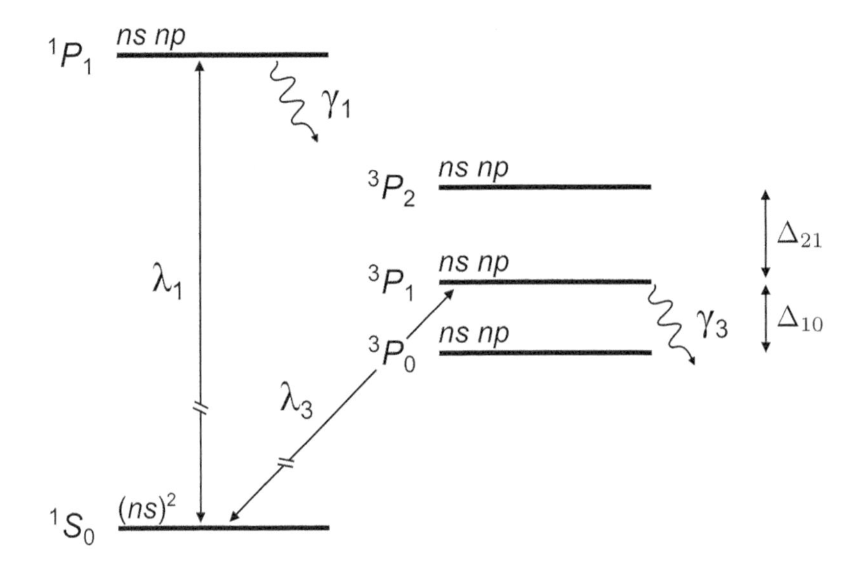

Fig. 5.1 Simplified level structure of an alkaline-earth atom.

Intercombination transitions between singlet and triplet states are forbidden by the same spin selection rule that prohibits transitions between para- and ortho-helium. Among the transitions to the ^3P_J triplet states, the one connecting the ground state to the intermediate $^3P_1$ is forbidden only by the above selection rule, while the ones to the $^3P_0$ and $^3P_2$ states are also forbidden by angular momentum selection rules.[2]

Selection rules, however, are strictly observed when the quantum numbers labelling the states are well defined. As the atomic number increases the $L-S$ coupling scheme becomes less accurate and selection rules are somewhat relaxed. This behaviour can be inferred from the values reported in Table 5.1 if one considers the linewidth Γ_3 of the "less-forbidden" intercombination transition connecting the ground state to the intermediate $^3P_1$ state. The value for this quantity increases considerably moving to heavier atoms, which results from increasing admixtures of the $^1P_1$ state to the state labelled as $^3P_1$, which determines a nonzero transition matrix element and a finite linewidth. There is another hint to the limits of the $L-S$ coupling scheme for heavier atoms. In many-electron atoms the dominant fine-structure correction is given by the spin–orbit coupling proportional to $\langle \mathbf{L}\cdot\mathbf{S}\rangle = \frac{\hbar^2}{2}\left[\,J(J+1)-L(L+1)-S(S+1)\right]$, from which it follows that the energy separation $\Delta_{J,J-1}$ between adjacent fine-structure components has to be proportional to J. While this interval rule is well respected for lighter alkaline-earth atoms such as magnesium (Mg) and calcium (Ca), where $\Delta_{21} \simeq 2\Delta_{10}$ (see Table 5.1), this is not true for heavier atoms such as barium (Ba) or alkaline-earth-like ytterbium (Yb).[3] As the atomic number increases, fine-structure

[2] The $^1S_0 \to {}^3P_0$ transition is forbidden at all the orders of the multipole expansion since it connects two states with $J=0$, while the $^1S_0 \to {}^3P_2$ transition is forbidden since $\Delta J = 2$.

[3] A violation of this interval rule can be observed in beryllium (Be) as well, however this does not depend on a failure of the $L-S$ coupling scheme, which is very well respected for light atoms

Table 5.1 Some values for the stable alkaline-earth atoms, with the inclusion of alkaline-earth-like rare-earth ytterbium (the mass number indicates the most abundant isotope).

	n	Z	λ_1 (nm)	$\gamma_1/2\pi$ (MHz)	λ_3 (nm)	$\gamma_3/2\pi$ (kHz)	Δ_{10} (THz)	Δ_{21} (THz)
^{9}Be	2	4	235	89	455	$\approx 10^{-3}$	0.019	0.070
^{24}Mg	3	12	285	79	457	0.048	0.6	1.2
^{40}Ca	4	20	423	34	657	0.37	1.6	3.2
^{88}Sr	5	38	461	32	689	7.4	5.6	11.8
^{138}Ba	6	56	554	20	791	50	11.1	26.3
^{174}Yb	*	70	399	28	556	180	21.1	51.5

corrections become more important,[4] reducing the validity of the $L-S$ coupling scheme (which is well justified when the fine structure is a perturbation to the level energies) and making intercombination transitions "less forbidden" (since the S quantum number is not exactly defined).

Finally, we note that the ground state of alkaline-earth atoms has electronic angular momentum $J = 0$. All the bosonic alkaline-earth isotopes have nuclear spin $I = 0$ as well, which causes the absence of any internal structure in the ground state. Instead, the fermionic isotopes have nonzero nuclear spin I, which is, however, decoupled from the electronic degree of freedom in the ground state owing to the absence of electron angular momentum, hence of hyperfine interaction.

5.1.1 Laser cooling of alkaline-earth atoms

Laser cooling of two-electron atoms (the most commonly used elements are calcium, strontium, and ytterbium) offers interesting possibilities, owing to the richer electronic spectrum than alkalis. In particular, the existence of both strong dipole-allowed and weak intercombination transitions permits the investigation of different regimes of laser cooling, which have their own advantages and drawbacks. The commonly employed strategy involves two different steps of laser cooling described below.

The atoms, initially heated up in an oven,[5] form an atomic beam which is typically decelerated in a Zeeman slower configuration (see Sec. 2.3.2) operating on the strong allowed $^1S_0 \rightarrow {}^1P_1$ transition. A magneto-optical trap (MOT) operating on the same transition then usually follows (see Sec. 2.3.5). The attainable MOT temperatures are higher than in alkalis: this is a consequence of the larger linewidth of the cooling transition, which determines a Doppler temperature (see Sec. 2.3.3) $T_D^{min} = \hbar\gamma_1/2k_B$

(intercombination transitions are indeed suppressed at a large extent). It is a consequence of other relativistic corrections contributing to the fine structure of light atoms with a similar weight as the spin–orbit coupling. An example is given by the fine structure of the $2\,^3P_J$ energy terms of helium (see Sec. 4.2), in which spin–spin and spin–other-orbit interactions cause a departure from the interval rule and lead to an inversion of the fine structure (see e.g. Fig. 4.3).

[4]Fine-structure separations increase with the atomic number Z approximately as $\sim Z^2$.

[5]All the elements in Table 5.1 are solid at room temperature and have a melting temperature higher than 900 K, which means that at room temperature their vapour pressure is negligible and that larger temperatures are required to have a sizeable flux of atoms.

on the order of 1 mK. Furthermore, the bosonic isotopes do not allow sub-Doppler cooling (see Sec. 2.3.4), because of the lack of internal structure in the ground state.

In order to reach lower temperatures, a second-stage MOT is performed on the $^1S_0 \rightarrow {}^3P_1$ intercombination transition. The reduced linewidth of this transition (see Table 5.1) allows much smaller Doppler temperatures $T_D^{min} = \hbar\gamma_3/2k_B$ (e.g. 4 μK for Yb, 180 nK for Sr), which compensate for the absence of sub-Doppler mechanisms for the bosonic isotopes. These very small temperatures come at a price, since the reduced linewidth also makes the pressure radiation force weaker, as shown by eqn (2.5). This means that the capture velocity is accordingly small, i.e. only atoms which are initially slow enough can be laser-cooled before escaping from the interaction region. A first-stage "collection" MOT working on the allowed $^1S_0 \rightarrow {}^1P_1$ transition is therefore required in order to operate the second-stage "colder" MOT on the $^1S_0 \rightarrow {}^3P_1$ intercombination transition.[6]

Owing to the absence of hyperfine structure in the ground state and to the absence of other decay channels from the $^3P_1$ state, no repumper is needed to operate an intercombination MOT for alkaline-earth atoms. In addition, the lack of sub-Doppler mechanisms for bosonic isotopes allows stringent tests of the Doppler theory of laser cooling.[7]

5.2 Optical traps

Differently from alkalis, two-electron atoms cannot be trapped in magnetic traps. As a matter of fact, their diamagnetic ground state $^1S_0$ implies the absence of magnetic dipole moment originating from the electronic angular momentum. Bosonic alkaline-earth isotopes have zero nuclear spin as well, resulting in a completely magnetic-insensitive ground state. Fermionic isotopes have a nonzero nuclear spin I, resulting in a purely nuclear magnetic moment $\boldsymbol{\mu} = g_I \mu_N \boldsymbol{I}/\hbar$, where g_I is the nuclear Landé factor and $\mu_N = e\hbar/2m_p$ is the nuclear magneton. The latter quantity, depending on the proton mass m_p, is a factor $m_p/m_e \simeq 1836$ times smaller than the Bohr magneton, which instead depends on the electron mass m_e. As a consequence, fermionic alkaline-earth isotopes are $\approx 10^3$ times less sensitive to magnetic fields than alkali atoms and their trapping would require impractical magnetic field values.

Alkaline-earth atoms can be trapped with a different class of conservative traps based on the illumination with far-off resonance light. In this section we discuss the origin of the *optical dipole potential* which is at the basis of the optical trapping techniques.

5.2.1 Optical dipole force

Semiclassical derivation. We start with a semiclassical derivation of the optical dipole potential, following the approach of Grimm *et al.* (2000). We consider an atom interacting with a classical radiation field $\mathbf{E}(\mathbf{r}, t)$ oscillating at frequency ω:

[6] An exception is represented by ytterbium, in which the linewidth of the $^1S_0 \rightarrow {}^3P_1$ transition is large enough to allow the atoms of the decelerated beam to be captured directly in the intercombination MOT.

[7] We note that a MOT operating on the $^1S_0 \rightarrow {}^3P_1$ transition realizes the ideal $J = 0 \rightarrow J' = 1$ scheme discussed in Sec. 2.3.5.

$$\mathbf{E}(\mathbf{r},t) = \hat{\mathbf{e}}\left[E(\mathbf{r})e^{-i\omega t} + h.c.\right]/2\ , \tag{5.1}$$

where $\hat{\mathbf{e}}$ is the polarization vector of the field and $E(\mathbf{r})$ is its amplitude. This field induces an electric polarization of the atom, i.e. a deformation of its charge distribution, which can be described by an electric dipole moment

$$\mathbf{d}(\mathbf{r},t) = \hat{\mathbf{e}}\left[d(\mathbf{r})e^{-i\omega t} + h.c.\right]/2 \tag{5.2}$$

oscillating at the same frequency as the driving field and proportional to the latter through the *atomic polarizability* α:

$$d(\mathbf{r}) = \alpha E(\mathbf{r})\ . \tag{5.3}$$

In a classical model of the atom, the electron is forced to oscillate and the dipole moment arises from the periodic displacement from its equilibrium position.[8] Quantum-mechanically, the dipole moment arises from mixing the ground-state and excited-state orbitals, which have different charge distributions. It is worth noting that the proportionality between d and E stated by eqn (5.3) holds only in the linear regime of weak excitation, when the saturation of the two-level system is negligible (i.e. the atomic population stays mostly in the ground state) and α does not depend on E.

Solving the optical Bloch equations for a two-level system (see Appendix A for a full derivation), one can derive an analytical expression for the atomic polarizability

$$\alpha = -\frac{e^2\mu_{eg}^2}{\hbar}\frac{\delta - i\frac{\gamma}{2}}{\delta^2 + \frac{\gamma^2}{4} + \frac{|\Omega|^2}{2}}\ , \tag{5.4}$$

where $\delta = \omega - \omega_0$ is the detuning, ω_0 is the (angular) resonance frequency, γ is the natural linewidth, μ_{eg} is the transition dipole moment, and Ω is the Rabi frequency. The atomic polarizability α is a complex quantity: the real part, representing the component of $\mathbf{d}$ oscillating in phase with $\mathbf{E}$, is responsible for the dispersive properties of the interaction, while the imaginary part, describing the out-of-phase component of $\mathbf{d}$, is connected to the absorption. Both the real and imaginary components of α strongly depend on the detuning as the laser is scanned across the atomic resonance, as shown in Fig. 5.2. In particular, at resonance $\delta = 0$ the absorption is maximum and α is purely imaginary. Moving away from resonance, the contribution of the real part dominates, since it decays as $\sim 1/\delta$, while the imaginary part decays faster as $\sim 1/\delta^2$: far from the resonance absorption can be neglected and the atomic medium becomes purely dispersive.

What are the effects of this polarizability on the motion of the atoms? In classical electrodynamics, the motion of an electric dipole $\mathbf{d}$ in an inhomogeneous electric field $\mathbf{E}(\mathbf{r})$ can be described with a potential energy $U(\mathbf{r}) = -\mathbf{d}\cdot\mathbf{E}(\mathbf{r})$. Here, in a second-order

[8]Equation (5.3) is very general and can be derived in a very simplified classical model, in which the electron bounded to the atom is treated as a damped harmonic oscillator. The position $x(t)$ of the electron under the action of a uniform oscillating field is described by the differential equation $mx''(t) + \gamma x'(t) + kx(t) = -eE(t)$, where $\omega_0 = \sqrt{k/m}$ is the atomic resonance frequency and γ is a damping constant (describing the radiative decay). Solving this differential equation, it can be demonstrated that the induced electric dipole moment $d(t) = -ex(t)$ can be written as in eqn (5.3).

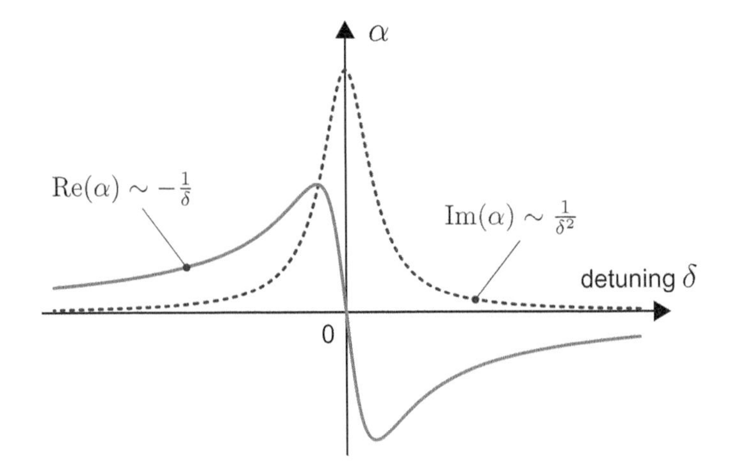

Fig. 5.2 Real and imaginary part of the atomic polarizability near a resonance.

process, the induced dipole moment interacts with the same oscillatory electric field that has created it. Similarly to the classical expression, we define the *optical dipole potential* as the potential energy of the induced electric dipole $\mathbf{d}(\mathbf{r})$ interacting with the driving electric field $\mathbf{E}(\mathbf{r})$ as

$$U_{dip}(\mathbf{r}) = -\frac{1}{2}\langle \mathbf{d} \cdot \mathbf{E} \rangle = -\frac{1}{2\epsilon_0 c}\mathrm{Re}(\alpha) I(\mathbf{r}) , \tag{5.5}$$

where the angle brackets indicate the time average over the fast optical oscillations and $I(\mathbf{r}) = \epsilon_0 c |E(\mathbf{r})|^2/2$ is the average field intensity.[9] We observe that the dipole potential only depends on the real part of α, i.e. only on the dispersive part of the interaction. By differentiating eqn (5.5) one can calculate the *optical dipole force* exerted on the atom:

$$\mathbf{F}_{dip}(\mathbf{r}) = -\nabla U_{dip}(\mathbf{r}) = \frac{1}{2\epsilon_0 c}\mathrm{Re}(\alpha)\nabla I(\mathbf{r}) . \tag{5.6}$$

This *conservative* force is proportional to the intensity gradient ∇I, so it vanishes when the field intensity is uniform, as in the case of a plane wave.

The imaginary part of α describes the *non-conservative* component of the interaction, which is associated with the absorption of photons from the incident light field and with the excitation of the atom to higher-energy states. This process generally determines a heating of the trapped sample[10] and can be quantified by the average number of photons which are scattered in the unit of time through cycles of absorption and subsequent

[9] This definition of the interaction energy differs from the classical expression by a factor 1/2, accounting for the induced nature of the dipole $\mathbf{d}$ and for the electrostatic energy needed to create it.

[10] Note that this mechanism does not always cause heating. The same non-conservative component of the interaction is responsible for the radiation pressure force on which most of the existing laser-cooling techniques are based. There, clever laser arrangements ensure that the non-conservative character is used in such a way as to *reduce* the temperature of the gas. However, at low temperatures, the stochastic nature of radiation pressure (involving random spontaneous emission events) becomes important, as it determines a residual heating due to fluctuations of the cooling force.

spontaneous emission (see also Appendix A.3.1). It is possible to demonstrate that the photon scattering rate is

$$\Gamma_{sc}(\mathbf{r}) = \frac{1}{\hbar\epsilon_0 c}\mathrm{Im}(\alpha)I(\mathbf{r})\,. \tag{5.7}$$

By substituting the expression for the polarizability in eqn (5.4) (after some manipulation of the constants) it is possible to derive the explicit form of eqns (5.5) and (5.7) in the far-off resonant regime, in which the detuning δ is much larger than both the radiative linewidth γ and of the Rabi frequency Ω (Grimm *et al.*, 2000):

$$U_{dip}(\mathbf{r}) = \frac{3\pi c^2}{2\omega_0^3}\left(\frac{\gamma}{\delta}\right)I(\mathbf{r}) \tag{5.8}$$

$$\Gamma_{sc}(\mathbf{r}) = \frac{3\pi c^2}{2\hbar\omega_0^3}\left(\frac{\gamma}{\delta}\right)^2 I(\mathbf{r})\,. \tag{5.9}$$

From eqn (5.8) we note that the sign of the dipole potential depends on the sign of the detuning $\delta = \omega - \omega_0$. More precisely, if the light is red-detuned ($\delta < 0$) the dipole potential is negative, hence maxima of intensity correspond to minima of the potential: as a consequence, the atoms tend to localize in regions of high field intensity. On the contrary, if the light is blue-detuned ($\delta > 0$) the dipole potential is positive, hence maxima of intensity correspond to maxima of the potential and the atoms tend to localize in regions of low field intensity.

From eqns (5.8) and (5.9) we also note that the dependence on the detuning is different in the two cases, as expected from the previous discussion: while the dipole potential U_{dip} scales as $1/\delta$, the scattering rate Γ_{sc} scales as $1/\delta^2$. As a consequence, moving out of resonance, the contribution of the dispersive dipole force dominates over the dissipative processes. In other words, for a fixed intensity of the dipole potential, the spontaneous scattering of photons can be suppressed increasing δ while increasing I in order to keep the ratio I/δ constant. When Γ_{sc} can be neglected, what is left is a purely conservative force that can be used to design spatially dependent potentials for the atoms.

Dressed-state approach. The semiclassical model presented above is not the only possible approach to the derivation of the optical dipole force. Following a more quantum-mechanical treatment, the dipole force can be obtained from the energy shift introduced by the radiation field to the unperturbed atomic levels (Metcalf and van Der Straten, 1999). We consider a two-level system with a ground state $|g\rangle$ and an excited state $|e\rangle$, separated by an energy difference $\hbar\omega_0$. In the presence of coupling with the radiation field, which introduces off-diagonal terms in the Hamiltonian, these two states cease to be eigenstates of the system. Under the rotating-wave approximation introduced in Appendix A.1.2, the Hamiltonian can be written as

$$\hat{H} = \begin{bmatrix} 0 & \hbar\frac{\Omega^*}{2} \\ \hbar\frac{\Omega}{2} & -\hbar\delta \end{bmatrix}, \tag{5.10}$$

where Ω is the Rabi frequency and $\delta = \omega - \omega_0$ is the detuning. Instead of considering the time evolution induced by the off-diagonal terms, we can study the stationary

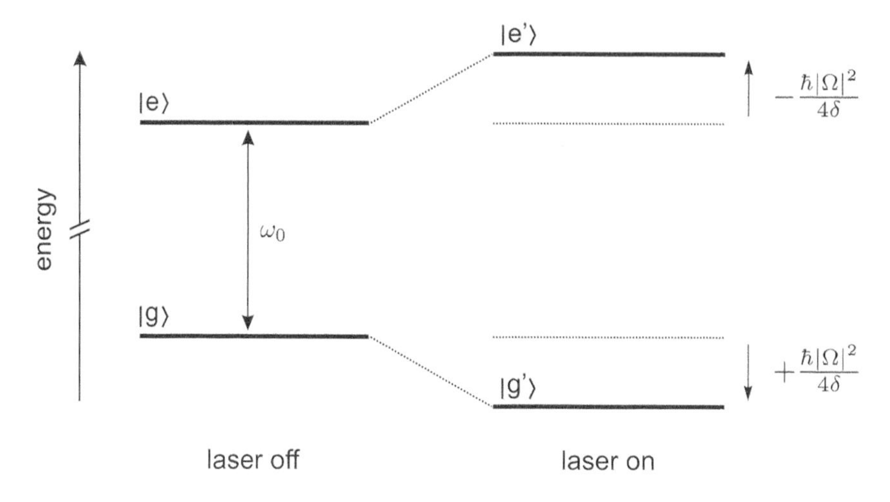

Fig. 5.3 Light shift in a two-level system. The energy shifts depicted in figure refer to the experimentally relevant case of red-detuned laser light ($\delta < 0$).

solution of the Schrödinger problem by diagonalizing the total Hamiltonian including the interaction term. We find new eigenstates $|g'\rangle$ and $|e'\rangle$, that can be written as mixtures of the initial states $|g\rangle$ and $|e\rangle$, according to the linear transformation

$$\begin{bmatrix} |g'\rangle \\ |e'\rangle \end{bmatrix} = \begin{bmatrix} \cos\beta & \sin\beta \\ -\sin\beta & \cos\beta \end{bmatrix} \begin{bmatrix} |g\rangle \\ |e\rangle \end{bmatrix}, \tag{5.11}$$

where the rotation angle $\beta \in [0, \pi/4]$ is defined such that $\cos 2\beta = 1/\sqrt{1 + |\Omega|^2/\delta^2}$. These new stationary states are called *dressed states.*[11] We can verify that, when the coupling is turned off and $\Omega = 0$, these states coincide with the unperturbed states $|g\rangle$ and $|e\rangle$.

In this rotating-wave frame the new eigenstates have energies $\hbar(-\delta \pm \sqrt{\delta^2 + |\Omega|^2})/2$. The energy shift introduced by the coupling with the dressing field is called *light shift*, or *ac-Stark shift*. In the far-detuned regime, when $|\delta| \gg \Omega$, the two states $|g'\rangle$ and $|e'\rangle$ are almost identical to the initial states $|g\rangle$ and $|e\rangle$ and the energy shifts reduce to $+\hbar|\Omega|^2/4\delta$ and $-\hbar|\Omega|^2/4\delta$, respectively, as shown in Fig. 5.3. In this limit the diagonal Hamiltonian on the new basis can be written as

$$\hat{H}' = \begin{bmatrix} \frac{\hbar|\Omega|^2}{4\delta} & 0 \\ 0 & -\hbar\delta - \frac{\hbar|\Omega|^2}{4\delta} \end{bmatrix}. \tag{5.12}$$

In high-precision spectroscopy the light shift generally represents an unwanted effect, since it leads to a shift of the transition frequency which depends on the intensity (its implications will be discussed e.g. in Sec. 5.3.1 devoted to optical clocks). However, it

[11] The term *dressed states* usually refers to the combined atom + photon states introduced when also the radiation field is treated quantum-mechanically (Cohen-Tannoudji, 1992). Here we have adopted a classical description of the field and, similarly, we use the *dressed state* expression to denote the different eigenstates of the atom in the presence of coupling with the field.

turned out to be an important resource for atomic physics experiments as well. In Sec. 2.3.4 we have illustrated the role of the spatially dependent light shift in a polarization gradient as a key element for sub-Doppler Sisyphus cooling. The light shift is also a resource for quantum optics experiments, e.g. for the nondestructive detection of photons by measuring the light shift on "probe" atoms in the cavity-QED experiments that led S. Haroche to the award of the Nobel Prize in Physics 2012.

Here we focus on the applications of the light shift to atom trapping. As a matter of fact, in the case of inhomogeneous light fields, in which the intensity $I(\mathbf{r}) \propto |\Omega(\mathbf{r})|^2$ depends on the position $\mathbf{r}$, also the light shift of the ground state turns out to be a function of the position. The optical dipole potential $U_{dip}(\mathbf{r})$ can thus be defined as the spatially dependent light shift of the ground state

$$U_{dip}(\mathbf{r}) = \frac{\hbar|\Omega(\mathbf{r})|^2}{4\delta} = \frac{e^2\mu_{eg}^2}{4\hbar\delta}E^2(\mathbf{r}) \ . \tag{5.13}$$

By using the definition of the transition dipole moment μ_{eg} (see Appendix A) and of the average field intensity $I(\mathbf{r}) = \epsilon_0 c E^2(\mathbf{r})/2$ it is easy to demonstrate that this definition reduces to the one given in eqn (5.8).

The same results, here obtained through exact diagonalization of the rotating-wave Hamiltonian in eqn (5.10), could be derived by using the standard quantum-mechanical perturbation theory at the second order in the coupling strength. Following this approach, which generalizes the results for the two-level atom, the energy shift of the ground state $|g\rangle$ induced by the interaction with the electromagnetic field can be written, within the rotating-wave approximation, as

$$\Delta E_g = \sum_{i \neq g} \frac{|\langle i|\hat{H}_{int}|g\rangle|^2}{E_i - E_g} = \sum_{i \neq g} \frac{\hbar|\Omega_i|^2}{4(\omega - \omega_{ig})} \ , \tag{5.14}$$

where $|i\rangle$ is a generic state of the unperturbed atomic Hamiltonian with energy E_i, $\hat{H}_{int}$ is the interaction Hamiltonian, Ω_i are the Rabi frequencies for a transition connecting states $|g\rangle$ and $|i\rangle$, and $\omega_{ig} = (E_i - E_g)/\hbar$. When the light is far-detuned from an atomic resonance the two-level approximation for the atom may fail and the light shift can be evaluated with eqn (5.14) taking into account the contribution of multiple excited levels coupled to the ground state by the radiation field.[12]

5.2.2 Applications of optical trapping

When an atom experiences an inhomogeneous light intensity $I(\mathbf{r})$, the optical dipole potential $U_{dip}(\mathbf{r})$ derived in the previous section is a function of the position $\mathbf{r}$, which results in an optical dipole force

$$\mathbf{F}_{dip}(\mathbf{r}) = -\nabla U_{dip}(\mathbf{r}) \ . \tag{5.15}$$

The possibility of using the optical dipole force as a tool to manipulate the atomic motion dates back to the late 1960s, when V. S. Letokhov proposed to use an optical

[12]When the radiation frequency is far from any atomic resonance the rotating-wave approximation may also fail and *counter-rotating-wave* terms in the interaction Hamiltonian have to be considered as well (Grimm *et al.*, 2000).

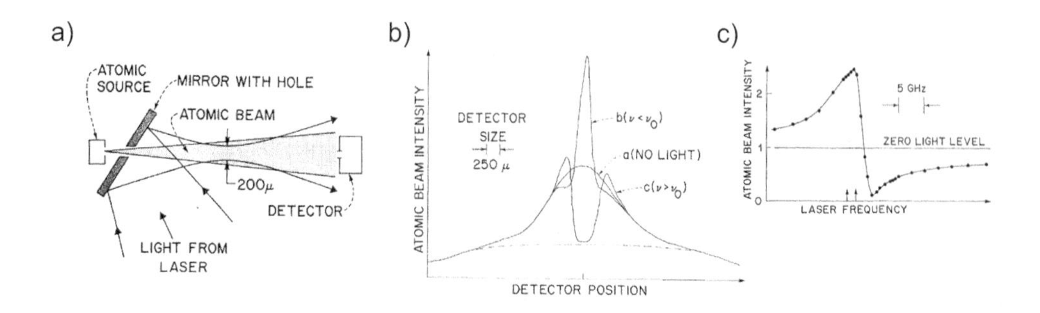

Fig. 5.4 Experimental observation of the optical dipole force. a) An atomic sodium beam moves along an intense focused laser beam and its profile is measured with a detector. b) Effect on the atomic beam profile for red-detuning and blue-detuning. c) Central intensity of the atomic beam as a function of the laser detuning. Reprinted with permission from Bjorkholm *et al.* (1978). © American Physical Society.

standing wave to affect the velocity distribution of atoms in order to increase the Doppler resolution of spectroscopic measurements (Letokhov and Lebedev, 1968).

The first clear effect of the dipole force on neutral atoms was evidenced in 1978 by J. E. Bjorkholm et al. in an experiment (Bjorkholm *et al.*, 1978) where a strong laser beam was directed collinearly to an atomic beam of sodium (see Fig. 5.4a for the experimental setup). The effect of the laser beam on the velocity distribution of the atoms was revealed by measuring the atomic beam intensity along a direction orthogonal to the beam axis. The results, shown in Fig. 5.4b, were extremely clear: when the laser was red-detuned ($\delta = \omega - \omega_0 < 0$) a strong peak appeared in the beam profile, owing to the focusing effect of the dipole force attracting the atoms towards the axis (where the light intensity was maximal); instead, in the case of blue detuning ($\delta > 0$) the opposite effect was detected and atoms were expelled from the centre of the beam towards regions of smaller light intensity. A plot of the atomic beam intensity as a function of the detuning (see Fig. 5.4c) showed exactly the dispersive behaviour expected for the optical dipole force.

An important application of the optical dipole force is the possibility of creating *far-off resonance traps (FORT)* in which cold atoms can be stored for long times (several seconds) with negligible heating. The first demonstration of optical trapping by the optical dipole force was obtained by S. Chu and coworkers in 1986 (Chu *et al.*, 1986) in an experiment in which ≈ 500 sodium atoms were captured in a strongly focused laser beam.

Red-detuned traps. As discussed in Sec. 5.2.1, in the case of red detuning $\delta < 0$, maxima of intensity correspond to minima of the potential. Local intensity maxima can be experimentally realized in quite an easy way: actually, the simplest trap configuration is just given by a focused red-detuned laser beam, the focus of which represents a trap where the atoms can be captured. According to Gaussian beam optics (see Appendix B.1), the spatial intensity distribution of a Gaussian laser beam propagating along the $\hat{z}$ direction is given by

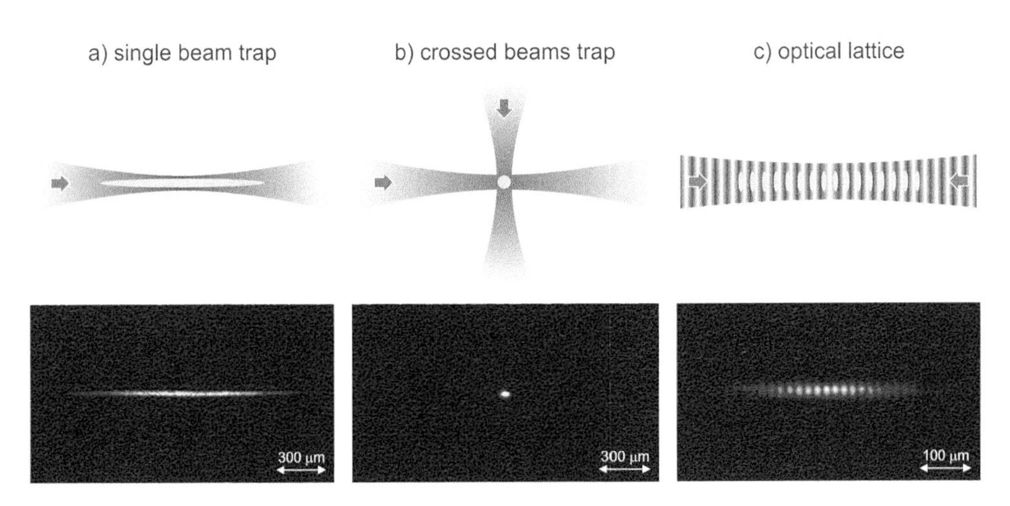

Fig. 5.5 Three different geometries of red-detuned optical dipole traps: a) a single focused beam; b) two crossed focused beams with orthogonal polarization; c) an optical lattice formed by two intersecting beams with the same frequency and polarization. The images in the bottom are pictures of ^{87}Rb ultracold samples confined in the traps sketched above (LENS, Florence).

$$I(z,r) = \frac{2P}{\pi w^2(z)} e^{-2r^2/w^2(z)}\,, \tag{5.16}$$

where P is the total optical power carried by the beam and

$$w(z) = w_0\sqrt{1 + (z/z_R)^2} \tag{5.17}$$

is the distance from the axis at which the intensity drops by a factor $1/e^2$, where w_0 is the beam waist size, $z_R = \pi w_0^2/\lambda$ is the Rayleigh length, and λ is the wavelength. The minimum of the optical dipole potential corresponds to the beam waist at $z = 0$ and the trap depth is

$$U_0 = \frac{3c^2}{\omega_0^3 w_0^2}\left(\frac{\gamma}{\delta}\right) P \tag{5.18}$$

according to eqn (5.8). We note that the trapping forces generated by intense focused laser beams are rather weak. Attainable trap depths are at most around 1 mK (for ^{87}Rb atoms trapped in a 1 W laser beam with $\lambda = 1064$ nm focused to 10 μm), which is orders of magnitude smaller than the thermal energy of room-temperature atoms. Therefore, as in the case of magnetic traps, laser cooling is a fundamental step to reduce the temperature of the atoms below the trap depth.

Using eqn (5.8) and expanding eqn (5.16) to the second order in z and r (harmonic approximation) one can derive the following expressions for the trapping frequencies experienced by an atom in the beam focus:

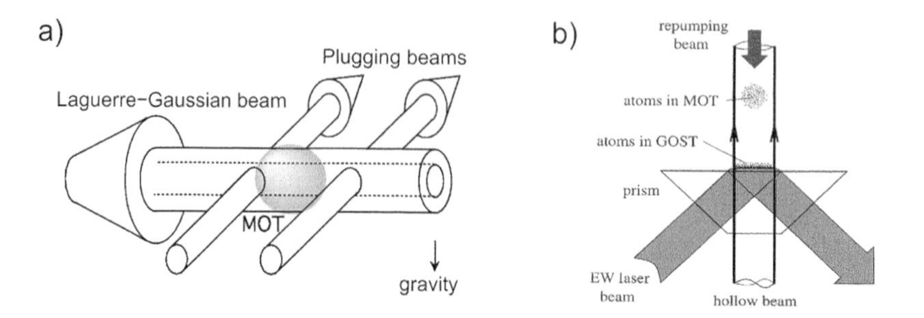

Fig. 5.6 Two different geometries of blue-detuned optical dipole traps using hollow Laguerre–Gaussian beams. Reprinted with permission from Kuga *et al.* (1997) (a) and Ovchinnikov *et al.* (1997) (b). © American Physical Society.

$$\omega_z = \sqrt{\frac{6c^2}{m z_R^2 w_0^2 \omega_0^3} \frac{\gamma}{\delta} P} \tag{5.19}$$

$$\omega_r = \sqrt{\frac{12c^2}{m w_0^4 \omega_0^3} \frac{\gamma}{\delta} P} \,. \tag{5.20}$$

The aspect ratio $\omega_r/\omega_z = \sqrt{2}\pi w_0/\lambda$ obtained with this simple trap geometry is generally large, since the typical waist size w_0 used in the experiments is several tens or hundreds of optical wavelengths.[13] For this reason this kind of geometry is particularly suited to producing elongated traps or waveguides with weak axial confinement.

In Fig. 5.5 we show absorption images of ultracold ^{87}Rb atoms trapped by Gaussian beams in different geometries. Figure 5.5a refers to the case of a single red-detuned focused laser beam, oriented along the horizontal direction: here the trapping anisotropy of single-beam traps is evident. Different trap geometries can be obtained by adding additional laser beams. For example, in order to obtain similar trapping frequencies in all the directions, one can use two independent red-detuned laser beams intersecting at right angles, thus producing a quasi-spherical trap. An example of such a crossed-beam dipole trap is shown in Fig. 5.5b.

Blue-detuned traps. Blue-detuned traps can offer important advantages, coming from the possibility of trapping atoms in a local minimum of light intensity. This can result in a highly suppressed heating rate (according to eqn (5.7)) and in the suppression of unwanted perturbations (as the ac-Stark shift itself, affecting precise spectroscopic measurements). However, since the trap centre corresponds to an intensity minimum, single focused laser beams are not sufficient and more complicated trapping configurations have to be implemented.

One class of blue-detuned traps is provided by hollow beams, i.e. laser beams with one node in the intensity profile. An example is given by Laguerre–Gaussian beams

[13] Both for the experimental necessity of having a large trapping volume $V \sim w_0^2 z_R$ and for the technical difficulty to focus the beam to a size w_0 smaller than several optical wavelengths.

(see Appendix B.1), which can be produced by different techniques, for example by phase-imprinting with a holographic mask. Laguerre–Gaussian beams $LG_{0,l\neq 0}$ are also called *"doughnut" beams* because in their cross section a dark spot is enclosed by a bright ring,[14] which keeps the atoms confined. The first experimental investigation of cold atoms in a "doughnut" hollow-beam trap has been reported in Kuga *et al.* (1997), where the atoms were also axially confined by two blue-detuned plugging beams placed orthogonally at the two ends of the trap, as shown in Fig. 5.6a.

A different possibility involves the use of *evanescent waves*, which are produced at a dielectric–vacuum interface when a laser beam coming from the dielectric side undergoes total reflection. In this process, no travelling beam is transmitted into the vacuum side because of the impossibility of satisfying the conservation of the (in-plane) photon momentum at the interface. However, continuity boundary conditions for the electric and magnetic fields forbid the electromagnetic intensity from vanishing immediately on the vacuum side: instead of vanishing, it falls down to zero exponentially with a very strong gradient on a typical distance $\sim \lambda/2\pi$.[15] If blue-detuned light is used, atoms close to the dielectric surface can be repelled by the rapidly vanishing evanescent wave field. This strong repulsion was first evidenced in experiments with atomic beams (Balykin *et al.*, 1987) and then used to build atom traps. Figure 5.6b shows a scheme of the Gravito-Optic Surface Trap (GOST) developed in Ovchinnikov *et al.* (1997), where cold atoms were vertically trapped above a horizontal dielectric surface by the combined effect of gravity and of the evanescent wave repulsive force, while the horizontal confinement was provided by a hollow beam created with conically shaped lenses (axicons).

Optical lattices. New possibilities arise when one uses the interference of coherent laser beams to create trapping potentials. The simplest of such configurations is the one obtained from the interference of two coherent laser beams with parallel polarization intersecting at an angle θ, as shown in Fig. 5.7. The interference between the two beams results in a periodic modulation of the intensity, giving rise to a periodic potential varying along the difference of the laser wavevectors with a spatial period

$$d = \frac{\lambda}{2\sin(\theta/2)} \,. \tag{5.21}$$

In Fig. 5.5c we show an image of ultracold atoms trapped at the intersection of two laser beams propagating with a small relative angle $\theta \simeq 40$ mrad. If the two laser beams are counterpropagating, as in the case of just one laser beam retroreflected by a mirror, $\theta = \pi$ and the resulting standing wave has a spatial periodicity $d = \lambda/2$.

[14] Laguerre–Gaussian beams are also characterized by a nonzero phase circulation around the axis, corresponding to an *orbital angular momentum* of the photons. This feature can be used for transferring angular momentum to clouds of trapped atoms, as reported e.g. in Andersen *et al.* (2006), where the transfer of orbital angular momentum in quantized units of $\hbar$ from a Laguerre–Gaussian beam to a Bose–Einstein condensate was demonstrated.

[15] Mathematically, this phenomenon corresponds to the propagation of the e.m. field in the vacuum side with an imaginary wavevector (i.e. exponential decrease of the intensity, as in the case of absorption). The existence of evanescent waves is ubiquitous in physics, as a result of continuity boundary conditions which are applied at the interface between "allowed" and "forbidden" propagation regions. A notable example of this effect is the quantum-mechanical tunnelling of a particle wavefunction across a potential barrier which is higher than the particle energy.

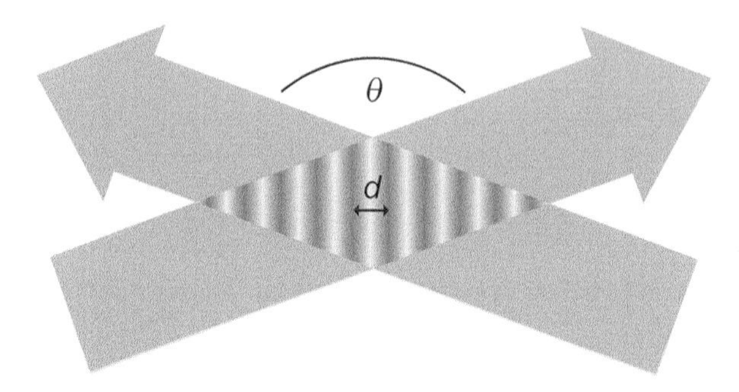

Fig. 5.7 Schematics of the periodically modulated intensity resulting from the intersection of two coherent laser beams propagating at an angle θ.

This trap configuration, which is known as an *optical lattice*, turned out to be particularly interesting both as a tool for atom trapping and for its relevance in the study of fundamental effects of quantum mechanics. Atoms trapped in a standing wave experience very high trapping frequencies along the lattice axis, owing to the steep potential gradient generated by the intensity modulation pattern. If the lattice is red-detuned atoms are trapped in the anti-nodes of the standing wave, while they are trapped in the nodes if the lattice is blue-detuned.[16] A harmonic approximation of the potential close to the minima can easily yield trapping frequencies on the order of $\sim 10^4$ Hz (for alkalis), compared to $\sim 10^2$ Hz of single-beam and crossed-beam traps. This also means that, when ultracold atoms are placed into optical lattices, their motion cannot be described in classical terms, and a quantum-mechanical description involving occupation of the few lowest vibrational states should be invoked.[17] This feature finds quite an important application in high-precision spectroscopic measurements, that will be discussed in the next sections of this chapter. In addition, atoms trapped in laser standing waves can be used to study the physics of quantum transport in periodic structures, which is ultimately connected to the realization of precise interferometric measurements and to the implementation of *quantum simulators* for ideal solid-state models: this research direction will be deeply discussed in Chapters 6 and 7, which are entirely devoted to the quantum physics of ultracold atoms in periodic potentials.

5.3 Optical clocks

In Sec. 2.2.1 we have discussed the importance that the hyperfine structure of alkali atoms had in the development of microwave spectroscopy and in the realization of microwave atomic clocks, on which the current definition of the SI second is based. Figure 5.8 shows how the precision of microwave atomic clocks has steadily increased starting from the first caesium atomic clock in 1955 to contemporary laser-cooled atomic

[16] A 1D red-detuned lattice can provide three-dimensional optical trapping in the beam waist. Instead, a 1D blue-detuned lattice cannot provide radial trapping, owing to the repulsive optical potential pushing the atoms away from the trap along the radial direction.

[17] A typical trapping frequency $\nu = 50$ kHz, easily achievable for ^{87}Rb, corresponds to an energy difference $\Delta E = h\nu \approx k_B \times 2$ µK between the harmonic oscillator levels.

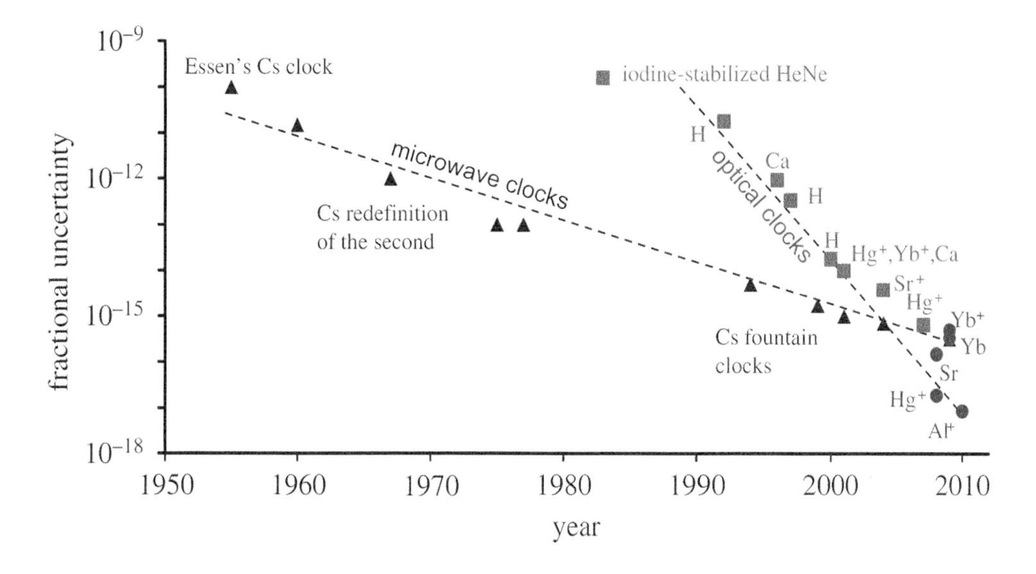

Fig. 5.8 The graph summarizes the impressive historical evolution of the accuracy of atomic clocks, including microwave atomic clocks (triangles), neutral and ion-based optical atomic clocks (squares), optical atomic clocks with estimated uncertainties below the Cs primary standard uncertainty (circles). Reprinted from Gill (2011) by permission of the Royal Society.

fountain clocks, which achieve fractional uncertainties below 10^{-15}. In Sec. 2.4 we have presented the state of the art in laser-cooled microwave atomic clocks, discussing the effects which currently limit their performances and possible strategies to improve them.

In the history of time-keeping (from sundials, to pendulum clocks, to quartz clocks, to microwave atomic clocks), better accuracies and precision had been possible by redefining the standard of time on the basis of physical phenomena characterized by increasing oscillation frequencies. On a very qualitative basis, the precision that a clock can reach is related to the number of oscillations that may be counted per unit of time: if the number of oscillations increases, the precision of the clock increases as well. More quantitatively, it is possible to show that, for an atomic clock operating at the quantum projection noise limit (see Sec. 2.4.1), the fractional frequency instability σ scales with the clock parameters as

$$\sigma \propto \frac{\delta\nu}{\nu_0}\frac{1}{\sqrt{N\tau}}\,, \tag{5.22}$$

where ν_0 is the frequency of the clock transition, $\delta\nu$ is its linewidth (limited e.g. by the interrogation time or, ultimately, by the radiative linewidth of the transition), N is the number of atoms, and τ is the integration time, i.e. the time over which the operation of the clock is averaged (Hollberg *et al.*, 2001). From the above equation it is clear that increasing the quality factor $Q = \nu_0/\delta\nu$ of the oscillator results in a smaller clock uncertainty. While $\delta\nu$ is limited to ≈ 1 Hz by the finite time of interaction in a Ramsey experiment, ν_0 could be significantly increased by using *optical* transitions between different electronic levels ($\nu_0 \approx 10^{15}$ Hz) instead of *microwave* transitions between hyperfine states ($\nu_0 \approx 10^{10}$ Hz).

Starting from the 1990s, the possibility of realizing atomic clocks based on optical transitions has become concrete, thanks to a combination of technological advances, including the development of spectroscopic tools for the excitation of very narrow optical transitions (extreme narrow-band lasers, see Sec. 5.3.2) and techniques to "count" the number of laser oscillations (optical frequency combs, discussed in Sec. 1.4.2). Figure 5.8 shows how optical clocks are already outperforming microwave atomic clocks in terms of reproducibility and stability (the best optical clock reported so far has a fractional uncertainty of 8.6×10^{-18}, see Sec. 5.5), which could eventually result in the near future in a new definition of the second based on the resonance frequency of an optical transition. For an introductory and historical review on the development of optical frequency standards see e.g. Hollberg *et al.* (2005).

Currently, there are two different approaches to the realization of optical atomic clocks. Both are based on the interrogation of forbidden atomic transitions of laser-cooled trapped particles:

1. A first approach is based on ultra-narrow intercombination transitions in neutral alkaline-earth atoms. In this approach, discussed in Sec. 5.3.1, optical lattices are used as a tool to strongly confine large ensembles of laser-cooled atoms in microscopic traps where their external degrees of freedom are frozen out: as a result the Doppler shift is suppressed and long interrogation times can be achieved.
2. A second approach, that will be discussed in Sec. 5.5, is based on forbidden optical transitions in trapped ions. Single ions can be laser-cooled in tightly confining electrodynamic traps, where they can be stored and interrogated for extremely long times.

5.3.1 Neutral atoms lattice clocks

As discussed in Sec. 5.1, the spectrum of alkaline-earth atoms (and lanthanide ytterbium as well) is characterized by the existence of narrow intercombination transitions which are strongly forbidden by the spin selection rule. Laser spectroscopy of these transitions has been marked by impressive progress in the last twenty years. As an example, Fig. 5.9 illustrates the spectroscopy of the intercombination transitions of strontium, from the early application of a semiconductor laser to sub-Doppler spectroscopy in a hollow-cathode discharge (a) to the absolute frequency measurement in a thermal atomic beam (b), to the current detection of the clock transition with ultracold atoms in optical lattices (c).

In this section we focus on the possibility of realizing optical clocks based on the doubly forbidden $^1S_0 \rightarrow {}^3P_0$ transition (see Fig. 5.1), which has a particular significance, since it has a very small radiative linewidth and connects states with zero electronic angular momentum. In fermionic isotopes this transition has a linewidth of ≈ 10 mHz, which is induced by the hyperfine interaction admixing the $^3P_0$ state with the $^3P_1$ and $^1P_1$ states, which can decay to the ground state via electric dipole transitions (Porsev and Derevianko, 2004). In all the bosonic isotopes, instead, the nuclear spin is zero and the hyperfine interaction is absent, which results in the absence of single-photon

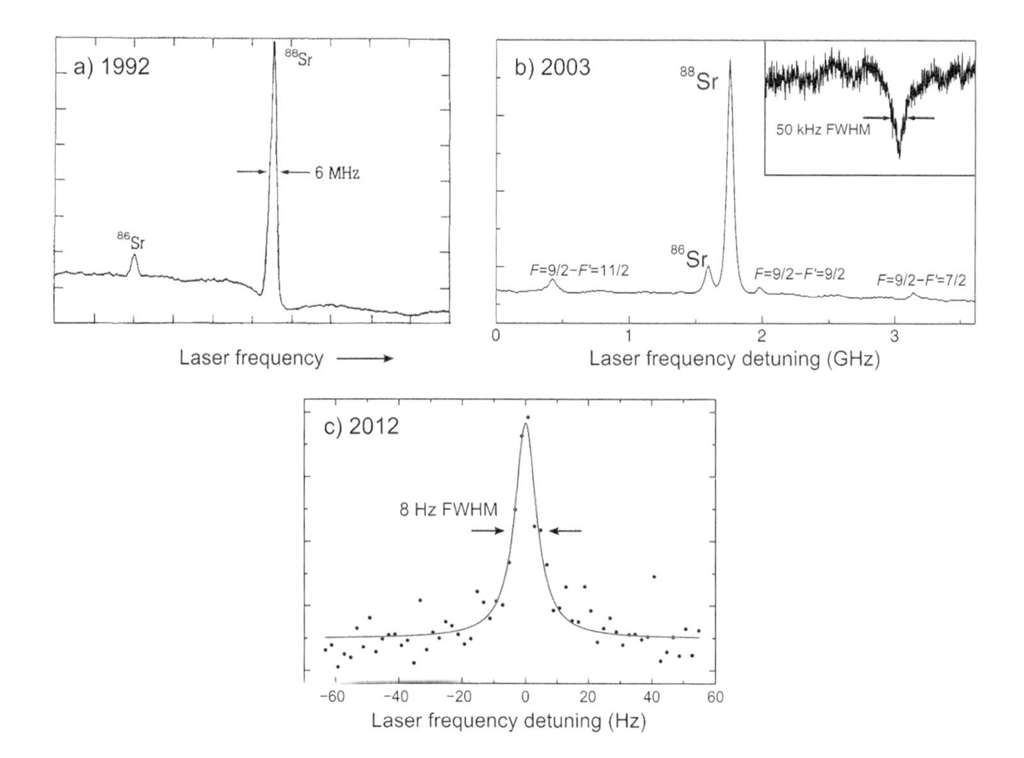

Fig. 5.9 Semiconductor laser spectroscopy of intercombination transitions in atomic strontium. a) Early detection of the $^1S_0 \rightarrow {}^3P_1$ transition in a hollow-cathode discharge (Tino *et al.*, 1992). b) Frequency comb-assisted measurement of the same transition using a thermal atomic beam. Reprinted with permission from Ferrari *et al.* (2003). © American Physical Society. c) Spectrum of the $^1S_0 \rightarrow {}^3P_0$ transition recorded with ultracold strontium atoms in the LENS strontium optical lattice clock (courtesy of G. M. Tino). Twenty years of progress in laser spectroscopy, including laser cooling, have produced a line narrowing of nearly six orders of magnitude.

processes connecting the $^3P_0$ state with the $^1S_0$ state. However, the clock transition can be observed by applying a static external magnetic field, which admixes the $^3P_1$ state with the $^3P_0$ state, resulting in a nonzero transition dipole moment (which can be adjusted by changing the applied magnetic field), as proposed and demonstrated by the group of L. Hollberg at NIST (Taichenachev *et al.*, 2006; Barber *et al.*, 2006).

In the following we discuss the experimental realiation of optical clocks with ultracold alkaline-earth atoms trapped in optical lattices. More advanced information on this topic can be found in very good review articles (Ye *et al.*, 2008; Derevianko and Katori, 2011; Katori, 2011).

Freezing the atomic motion. In order to take full advantage of the very large quality factor of these resonances, the atoms have to be laser-cooled and trapped, in order to suppress perturbing Doppler and recoil shifts and achieve long interrogation times. The

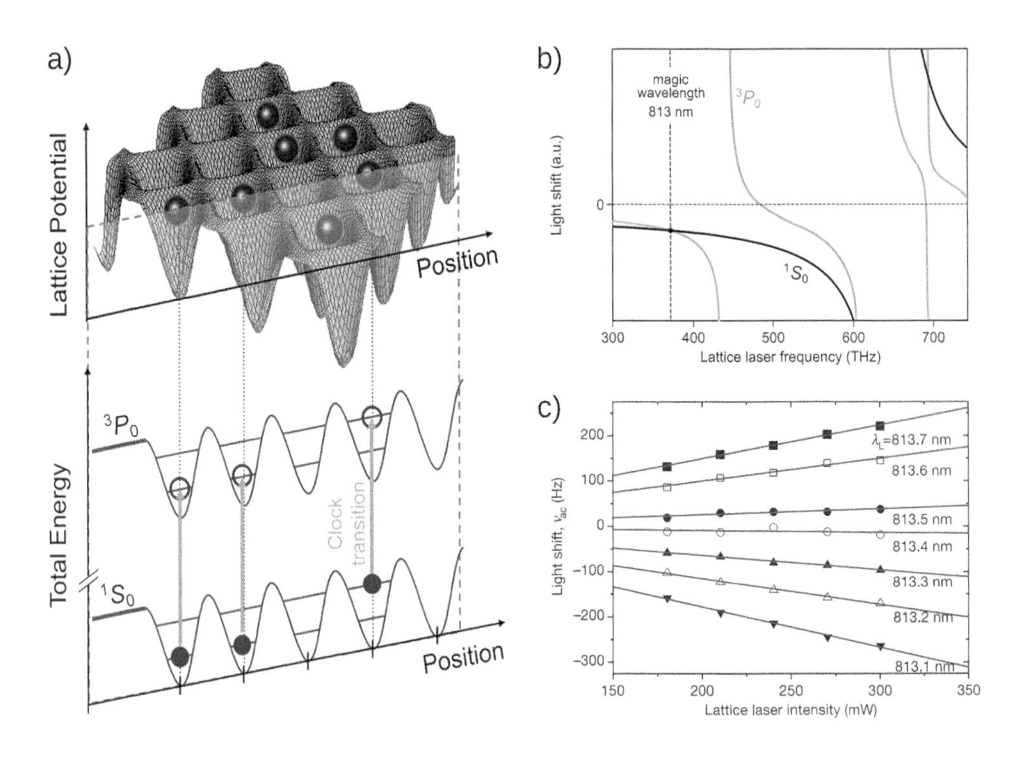

Fig. 5.10 a) Scheme of an optical lattice clock working on the $^1S_0 \rightarrow {}^3P_0$ transition of atomic ^{87}Sr. Reprinted with permission from Derevianko and Katori (2011). © American Physical Society. b) ac-Stark shift of the $^1S_0$ and $^3P_0$ states as a function of the lattice frequency. c) ac-Stark shift of the clock transition for different lattice wavelengths: at the magic wavelength $\lambda = 813.4$ nm the ac-Stark shift of the $^1S_0$ and $^3P_0$ levels are the same and the observed transition frequency does not depend on the trapping light intensity. Reprinted with permission from Takamoto *et al.* (2005). © Macmillan Publishers Ltd.

ultimate step in cooling consists in reaching the regime in which the atomic motion is quantized and the atoms populate the ground motional state, or at most a few discrete states of the confining potential. This situation can be obtained by trapping the atoms in tightly confining traps in which the energy separation between different bound states is much larger than the linewidth of the spectroscopic signals; therefore, occupation of different motional states can be detected and resolved spectroscopically. This strategy, initially applied to single ions trapped in the Lamb–Dicke regime (Dicke, 1953) (see Sec. 5.4.1) turned out to be very successful and allowed the realization of amazingly precise optical ion clocks (Diddams *et al.*, 2001; Margolis *et al.*, 2004; Rosenband *et al.*, 2008).

A similar approach was proposed by H. Katori and coworkers for neutral atoms (Katori *et al.*, 2003), which interact very weakly one with another (contrary to ions) and can be trapped in massive numbers, offering the possibility of a much larger signal-to-noise ratio than ions and a lower quantum projection noise limit for the

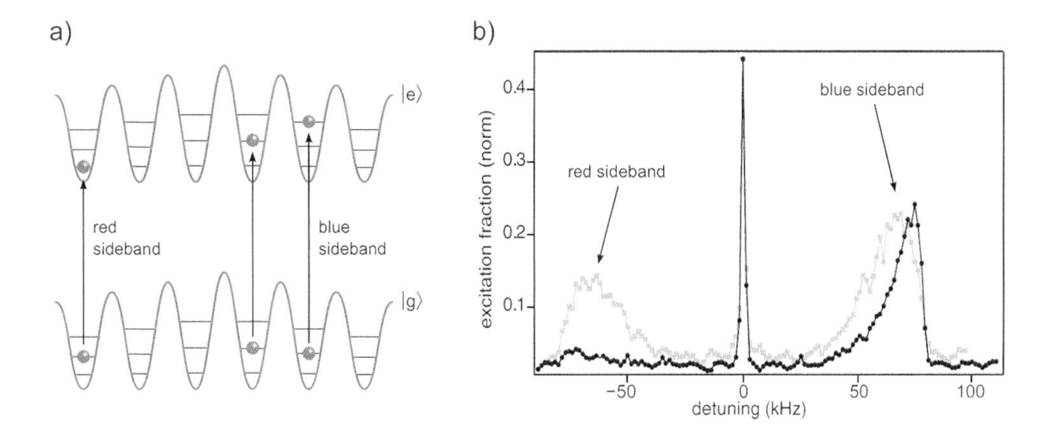

Fig. 5.11 a) Optical transitions can couple different vibrational levels of the optical lattice: $v \to v-1$ (red sideband), $v \to v$ (central band), $v \to v+1$ (blue sideband). b) Spectrum of resolved sidebands of the intercombination ${}^1S_0 \to {}^3P_0$ transition of atomic strontium. The central band is clearly visible as a sharp peak, since it does not suffer from inhomogenous broadening due to lattice anharmonicities and inhomogeneities. The lighter points correspond to a temperature of 3 μK, while the darker points correspond to a temperature of 1 μK (for the colder sample the red sideband disappears because the atoms mostly populate the ground vibrational state $v = 0$). Reprinted with permission from Campbell *et al.* (2009). © AAAS.

same integration time (see eqn (5.22)). The tightly confining traps are provided by the sites of an optical lattice (see Sec. 5.2.2), where the atoms are trapped after being laser-cooled in magneto-optical traps and optical molasses. In the optical lattice the atomic motion is quantized: the atoms occupy discrete vibrational states of the lattice potential wells (see Fig. 5.10a), which are typically separated one from another by an energy $\Delta E \approx h \times 100$ kHz, much larger than the linewidth of the ${}^1S_0 \to {}^3P_0$ clock transition. For this reason optical transitions connecting different motional states can be easily resolved as sidebands around a strong Doppler-free (and recoil-free) peak. As an example, we show in Fig. 5.11b a spectrum of the clock transition for cold strontium atoms in optical lattices (Campbell *et al.*, 2009). The figure shows a narrow central peak, which comes from transitions $v \to v$ in which the vibrational state v of the atom in the lattice potential well does not change. Additional sidebands are present, coming from the excitation towards different vibrational levels: a red sideband at lower frequency for transitions $v \to v-1$ and a blue sideband at higher frequency for transitions $v \to v+1$, as sketched in Fig. 5.11a. Both sidebands are broadened by lattice inhomogeneities, which lead to a change in the vibrational spacing as the lattice intensity is spatially changing, and, at high temperatures, by the anharmonicity of the lattice potential wells.

The tight confinement in the optical lattice allows long interrogation times and the cancellation of the Doppler effect. However, a problem immediately arises, since atoms trapped in an optical lattice experience a strong ac-Stark shift induced by the trapping light, which affects the clock transition frequency.[18] As a matter of fact, the ground and excited states typically have different polarizabilities, as shown in Fig. 5.10b, which lead to a nonzero differential shift of the two levels. The solution for this problem relies on using a well-chosen *magic wavelength* for which the polarizabilities are the same and the two clock states are shifted in the same direction by the same amount. Even if the magic wavelength can be estimated theoretically (knowing position and linewidth of the most important resonances of the spectrum, from eqn (5.14)), the precise value has to be determined experimentally, as shown in Fig. 5.10c for the $^1S_0 \to {}^3P_0$ transition in ^{87}Sr (Takamoto *et al.*, 2005). Far from the magic wavelength the frequency of the clock transition changes with the lattice intensity, but at the magic wavelength (approximately 813.4 nm in figure) no dependence on the trapping light intensity can be observed.

Experimental realizations. Different two-electron atomic candidates are being explored for the determination of the most advantageous atoms for the realization of optical lattice clocks. The first experiments have been performed with Sr (clock wavelength $\lambda = 698$ nm) (Takamoto *et al.*, 2005; Ludlow *et al.*, 2006; Le Targat *et al.*, 2006) and alkaline-earth-like Yb (Barber *et al.*, 2006) ($\lambda = 578$ nm), and efforts are underway to build optical lattice clocks with Mg (Friebe *et al.*, 2008) ($\lambda = 458$ nm) and alkaline-earth-like Hg (Petersen *et al.*, 2008) ($\lambda = 266$ nm).[19] Experiments can be run with either bosonic or fermionic isotopes and with either 1D or higher-dimensional optical lattices. There are several practical and fundamental issues to be considered:

- Alkaline-earth bosons have the advantage of being spinless (nuclear spin $I = 0$), which makes the clock transition insensitive to magnetic fields and immune to vector light shifts,[20] but they generally interact collisionally, which determines unwanted collisional shifts. For identical spin-polarized fermions s-wave collisions are suppressed, but the nonzero nuclear spin I leads to a sensitivity of the clock transition to magnetic fields and vector light shift.
- The 1D lattice geometry is more beneficial since the polarization of the lattice light is uniform, which simplifies the evaluation of the vector light shift, however many atoms ($\approx 10^2$) are trapped in the same lattice sites, which favours the choice of fermions in order to reduce the collisional shift. The 3D geometry is more beneficial for the collisional shift, since the higher density of lattice sites allows single atoms to be trapped in separated potential wells eliminating collisions, however the presence of unavoidable polarization gradients complicates the evaluation of the vector light shift, favouring in this case the choice of bosons.

A direct comparison between two strontium optical lattice clocks operating in different configurations, spin-polarized ^{87}Sr fermions in a 1D lattice and ^{88}Sr bosons in a 3D

[18] In Sec. 5.2.1 we have seen that the ac-Stark shift is indeed the quantum-mechanical equivalent of the optical dipole potential.

[19] Similarly to ytterbium, mercury has an electronic structure [Xe]$4f^{14}5d^{10}6s^2$, which is characterized by completely filled f and d shells, resulting in a spectrum dominated by the excitation of the two $6s$ electrons.

[20] The vector light shift arises due to the dependence of the ac-Stark shift on the Zeeman state m_F of an atom with nonzero spin F.

lattice, was reported in Akatsuka *et al.* (2008). The analysis of their frequency difference showed a fractional instability of 5×10^{-16} after an interrogation time of 2000 s, yielding a precise determination of the ^{88}Sr–^{87}Sr isotope shift at the Hz level.

Collisional shift in ultracold fermions. Increasing the number of atoms N is a good strategy to improve the signal-to-noise ratio and reduce the clock uncertainty, as shown by eqn (5.22). On the other hand, high-density samples have the drawback of larger collisional shifts, that can in principle be suppressed by working with identical fermions, which at low temperature cannot interact by s-wave collisions (see Sec. 3.1.4). However, even using fermionic isotopes, atom–atom interactions can affect the operation of an optical lattice clock, as evidenced at JILA in Campbell *et al.* (2009) for a system of ^{87}Sr spin-polarized fermions trapped in a 1D optical lattice. The collisions arise from the fact that atoms excited by the clock laser into the $^3P_0$ state become distinguishable from atoms in their ground state $^1S_0$ and can undergo collisions. If the excitation Rabi frequency (see Appendix A.1.3) is not homogeneous, e.g. because of a misalignment between the clock laser beam and the optical lattice direction, the dynamical evolution of two fermions occupying the same lattice site is different, which makes them (partially) distinguishable and leads to a reduction of the Fermi collisional suppression. As a consequence of this mechanism, a shift of the clock resonance at the 10^{-15} level was observed and characterized by changing the temperature of the atoms, the alignment of the beam, and the fraction of excited atoms.

Counterintuitively, collisional shifts in a fermionic lattice clock can be suppressed by *increasing* the interactions between the atoms, as recently demonstrated in Swallows *et al.* (2011) by using a more tightly confining 2D optical lattice. In this configuration the local density is increased, leading to an enhancement of the atomic interactions, which can be described by an interaction energy U of the atom pair in distinguishable states. However, if U becomes much larger than the energy scale $\Delta\Omega$ associated with the inhomogeneous excitation (the difference in Rabi frequencies), a "blockade" mechanism sets in: the collisional shift U is so strong that it leads to a well-resolved interaction band the occupation of which is energetically suppressed.[21]

Comparisons between clocks. The first experimental demonstration of an optical lattice clock was realized at the University of Tokyo in Takamoto *et al.* (2005) for fermionic ^{87}Sr atoms trapped in a 1D optical lattice: the operation of the clock was compared to a realization of the SI second via an optical frequency comb, resulting in a determination of the ^{87}Sr clock frequency with a fractional uncertainty 3.5×10^{-14}. Figure 5.12 shows the result of a remote comparison between the strontium lattice clocks operating at JILA, SYRTE, and University of Tokyo over a period of 3 years (Blatt *et al.*, 2008). Independent measurements of the clock frequency performed in the three laboratories showed agreement at the 4.0×10^{-15} level, approaching the limit of the current Cs standard of time. The most accurate frequency measurement of the ^{87}Sr clock transition has been reported in Campbell *et al.* (2008) with a fractional

[21]This process shares some similarities with the suppression of recoil shift in the Lamb–Dicke regime, where the excitation of atoms towards higher vibrational states of the optical lattice can be spectrally resolved.

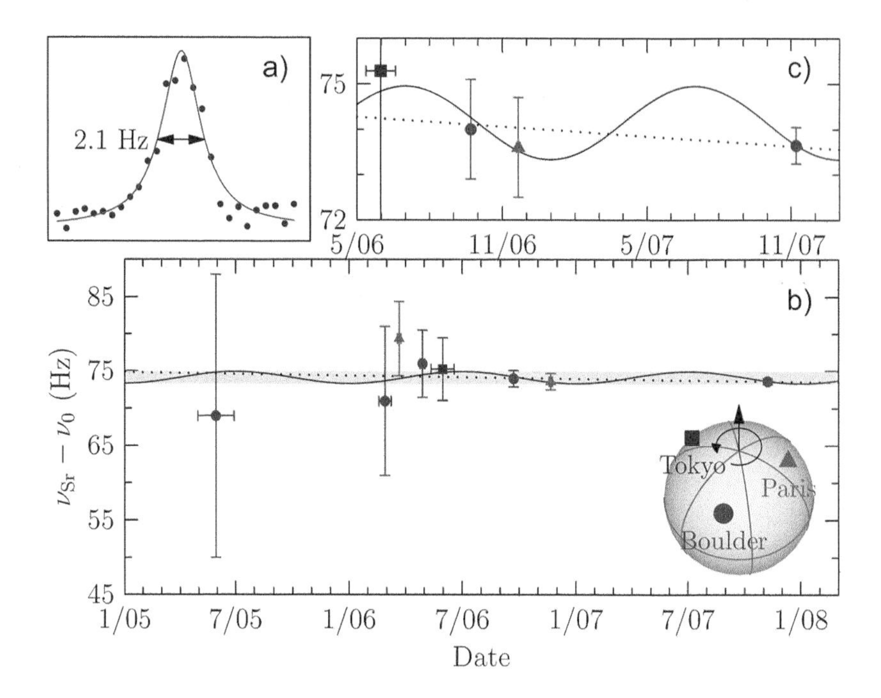

Fig. 5.12 Remote comparison between optical lattice clocks. a) Spectrum of the $^1S_0 \rightarrow {}^3P_0$ clock transition in ^{87}Sr. b) Measurements of the absolute clock frequency from JILA (circles), SYRTE (triangles), and University of Tokyo (squares) over 3 years (the vertical axis has been offset by $\nu_0 = 429\,228\,004\,229\,800$ Hz). The lines are fits to the data with a straight line (dotted) or a sinusoid with 1-year period (straight), which can be used to constrain possible variations of the fundamental constants (see Sec. 5.5.2). c) Zoom into the last four measurements, showing agreement within 1.7 Hz (4.0×10^{-15}). Reprinted with permission from Blatt *et al.* (2008). © American Physical Society.

uncertainty of 8.6×10^{-16}, limited so far by the performance of an intermediate H-maser used to compare the Sr clock with the Cs standard.

As optical lattice clocks become more accurate and precise than microwave clocks, their stability characteristics can be evaluated only by a direct comparison between two different optical clocks, in order not to be limited by the stability of the current standard of time. As an example, a direct comparison of this kind was performed in Boulder between the Sr lattice clock operating at JILA and a Ca optical clock operating at NIST, thanks to a 4-km optical fiber link between the two laboratories (Ludlow *et al.*, 2008). The NIST calcium clock was based on the interrogation of freely expanding laser-cooled ^{40}Ca atoms on the $^1S_0 \rightarrow {}^3P_1$ intercombination transition, providing a robust frequency standard with good short-term stability (Wilpers *et al.*, 2007). This comparison, made possible by a frequency comb bridging the optical gap between the two clock wavelengths, resulted in the accurate evaluation of systematic effects in the JILA Sr lattice clock at the 1.5×10^{-16} level of fractional uncertainty, mostly limited by the determination of the blackbody radiation shift (see Sec. 2.4.1).

On the side of the stability, current optical lattice clocks are still operating above the quantum projection noise limit set by eqn (5.22) (typically $< 10^{-17}$ in 1 s). This is mostly caused by the clock laser frequency noise and by the "dead" time occurring between one measurement and the next due to the preparation of a new laser-cooled atomic ensemble. During this time interval (which can be ten times longer than the laser interrogation time) the frequency of the clock laser may fluctuate without being corrected by the lock to the atomic resonance: this periodic sampling of the laser noise has a negative consequence on the long-term clock stability, according to the so-called *Dick effect* (Santarelli *et al.*, 1999). While nondestructive detection techniques could allow "recycling" the atoms and significantly reducing the dead time (Lodewyck *et al.*, 2009), an improvement in the spectral properties of the clock lasers can help in improving the clock stability, as in such a way to reduce the integration time needed to achieve the desired precision.

Recently, the comparison of two independent strontium optical clocks resulted in a fractional instability of only 1×10^{-17} in 10^3 s of averaging time, which is within a factor of 2 of the quantum projection noise limit (Nicholson *et al.*, 2012). The field is still developing and very recent measurements (Hinkley *et al.*, 2013) suggest that ytterbium clocks are very likely to play an important role for the realization of even more precise clocks. These new developments were possible also thanks to crucial improvements in the stability of the clock lasers, which are discussed in the next section.

5.3.2 Sub-Hz lasers

In order to probe extremely narrow clock transitions it is crucially important to use lasers with a very narrow spectral width. Good-quality single-frequency commercial lasers may have linewidths down to the kHz range (in the case of fibre lasers or solid state, e.g. Nd-YAG lasers), while free-running tunable lasers (e.g. titanium-sapphire lasers, dye lasers, or extended-cavity diode lasers) typically have linewidths on the order of 100 kHz. Since the intercombination clock transitions of alkaline-earth atoms have sub-Hz linewidth, this means that the spectroscopic resolution would be limited by the poor monochromaticity of the radiation source. In the last decade, spectacular progress in the field of laser technology has permitted the realization of ultra-stable lasers with linewidths which are now approaching the mHz range! For a very good review of this subject the reader may refer to Martin and Ye (2011).

The linewidth of a laser is typically limited by the mechanical quality of its cavity.[22] Mirror vibrations or instabilities in the intra-cavity processes induce fast variations in the instantaneous cavity resonance frequency, which, averaged in time, determines the laser linewidth. In order to narrow the emission of a laser, very good-quality reference optical cavities can be employed, in combination with active stabilization techniques which are capable of correcting the laser frequency in order to keep it locked to the cavity resonance frequency, primarily the FM-based Pound–Drever–Hall locking scheme developed in Drever *et al.* (1983).

[22]There is a more fundamental physical limit which is set by the spontaneous emission into the laser mode. This limit is described by the Schawlow–Townes linewidth, which is typically negligible with respect to the linewidth induced by the technical noise.

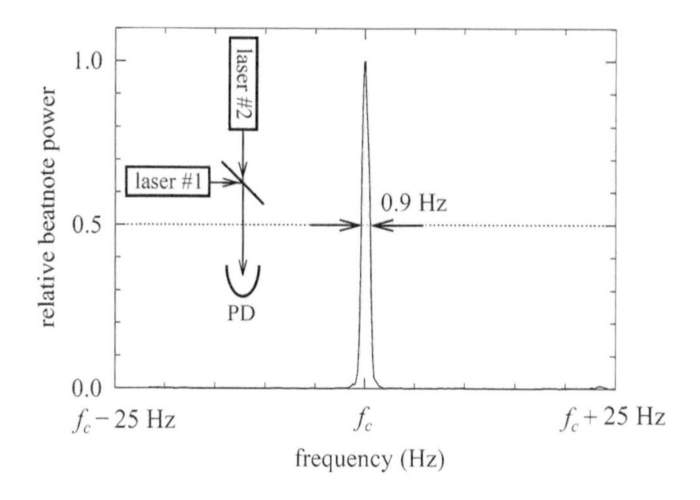

Fig. 5.13 Power spectrum of the beat note between two laser beams stabilized to two independent cavities (with an averaging time of 32 s). Reprinted with permission from Young *et al.* (1999). © American Physical Society.

Achieving sub-Hz linewidth poses very strict requirements on the quality of the optical cavity. In a Fabry–Perot cavity (see Appendix B.2 for a discussion of its main properties) the frequency ν of the n-th allowed mode is given by

$$\nu = n\frac{c}{2L} , \tag{5.23}$$

where L is the length of the cavity and we have assumed an intracavity index of refraction equal to 1. Therefore a fractional variation $\Delta L/L$ on the cavity length determines a fractional variation

$$\frac{\Delta\nu}{\nu} = \frac{\Delta L}{L} \tag{5.24}$$

on the cavity resonance frequency. A maximum allowed change of 1 Hz in the frequency of a visible laser means a fractional change of 10^{-15}. For typical cavity lengths $L \approx 10$ cm, this means that the optical path in the reference cavity has to change by less than one tenth of a femtometer, which is less than the size of a proton!

Given these strict requirements, the reference cavity has to be accurately evacuated (in order to neglect fluctuations in the intracavity refraction index that would result in a change of the effective cavity length) and mechanically decoupled from the environment (in order to neglect fluctuations in L). Vibrations have to be eliminated by carefully choosing the geometry of the cavity, its suspension points, and implementing systems for damping (passively and/or actively) vibrations from the laboratory. Long-term stability can be achieved by using stabilization cavities in which the mirrors are fixed to a spacer made up of materials with low thermal expansion coefficient. In general, the fractional change in cavity length can be written as

$$\Delta L/L = \alpha(T)\Delta T + \alpha'(T)\Delta T^2/2 + O(\Delta T^3) , \tag{5.25}$$

where $\alpha(T)$ is the thermal expansion coefficient of the cavity material, T is the temperature, and ΔT is a small change in temperature around the operating point. The most stable cavities are made of ultra-low expansion (ULE) glasses, in which the thermal expansion coefficient crosses zero at an operating temperature T_C near room temperature, which allows the minimization of the frequency resonance dependence on temperature. Plugging in typical numbers for ULE glasses, it turns out that around the operating point T_C a temperature stability of 1 mK provides a cavity length fractional stability at the level of 10^{-16} (Martin and Ye, 2011). Therefore a very careful thermal insulation and active temperature stabilization are primary requirements for achieving a long-term stable operation of the reference cavity.

When all these requirements are taken into account, sub-Hz linewidths of the laser (on timescales of several tens of seconds) can be achieved.[23] Figure 5.13 from the J. Bergquist group at NIST shows the beat note between two identical lasers locked to two different reference cavities: the measured 0.9 Hz linewidth in the beating signal provided the first demonstration of laser frequency locking to reference cavities with sub-Hz stability (Young *et al.*, 1999). Very recently, new stabilization cavity designs have allowed the reduction of the cavity thermal noise, resulting in clock lasers with a fractional frequency instability of only $\simeq 2 \times 10^{-16}$ over 10 seconds and $\simeq 250$ mHz linewidth, as recently demonstrated by the C. Oates group at NIST (Jiang *et al.*, 2011).

A different approach to the realization of stable reference cavities has been recently followed in Kessler *et al.* (2012), in which the demonstration of a monolithic reference cavity formed by a single silicon crystal is reported. A crystalline material such as silicon has important advantages over glasses, including a superior Young's modulus, which suppresses the sensitivity to environmental vibrations, and the absence of the aging effects typical of amorphous materials as glasses. Operating the silicon reference cavity at 124 K (where the thermal expansion coefficient of silicon has a zero crossing) the authors of Kessler *et al.* (2012) were able to achieve a laser linewidth < 40 mHz and short-term stability of 1×10^{-16} Hz.

5.4 Trapped ions

This section is devoted to the application of ions to precision measurements. We will initially discuss commonly used techniques for trapping and cooling ions (the reader can refer to Leibfried *et al.* (2003*a*) for an excellent review on this field), then we will show how cold trapped ions can be used to realize the most accurate clocks to date.

5.4.1 Ion traps

Charged particles can be trapped by electric fields. However, a static electric field alone is not sufficient to ensure a stable three-dimensional trapping, since Maxwell's equations prevent the electrostatic potential from having a minimum in a region of space free from source charges.[24] In Sec. 4.5.1 we have already discussed the Penning trap, which

[23] Extremely small frequency drifts in the cavity resonance frequency can still be measured and they are ultimately determined by aging effects due to the deformation of the glass itself, which thermodinamically is not a solid, but just a fluid with enormous viscosity!

[24] The Poisson equation for the electrostatic potential $\phi(\mathbf{r})$ in free space reads $\nabla^2\phi = \partial_x^2\phi + \partial_y^2\phi + \partial_z^2\phi = 0$, which means that the potential cannot have $\partial_i^2\phi > 0$ along all directions.

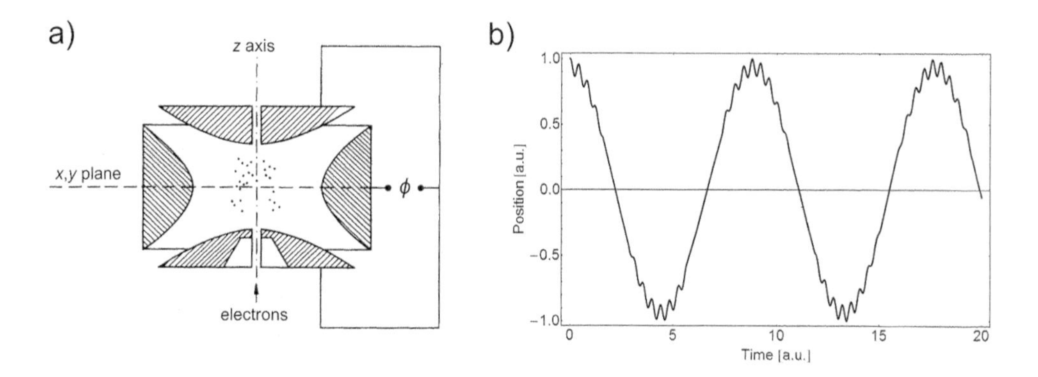

Fig. 5.14 a) Scheme of a 3D Paul trap: a sinusoidally varying voltage ϕ is applied between a ring electrode with $\hat{z}$ symmetry axis and two end-cap electrodes, which produce an oscillating quadrupole field at the centre of the trap. Reprinted with permission from Paul (1990). © American Physical Society. b) Stable solution of the equation of motion of a charged particle in the Paul trap. Note the micromotion impressed on top of the slower secular oscillation.

uses a dc magnetic field to provide confinement along the directions left untrapped by a quadrupole electric field. Although very efficient and simple to implement, this trap is not particularly appropriate for high-precision spectroscopic measurements, since fluctuations in the large magnetic field can lead to shifts of the ion transitions.

A different kind of trap which has turned out to be the most successful design for ion experiments is the *Paul trap*, which circumvents the problem described above by relying on time-dependent electric fields. The idea comes from W. Paul, who was awarded the Nobel Prize in Physics in 1989 (Paul, 1990). In a 3D Paul trap a sinusoidally-varying voltage is applied between two electrodes which produce a quadrupolar electric potential of the form

$$\phi(\mathbf{r}, t) = (U + V \cos \Omega t) \left(x^2 + y^2 - 2z^2\right) , \tag{5.26}$$

where U is a dc constant voltage and V is the amplitude of the ac voltage oscillating at frequency Ω (typically in the tens of MHz range for ion clocks experiments). Figure 5.14a shows a typical electrode arrangement, with a ring electrode with symmetry axis along $\hat{z}$ and two end-cap electrodes (similarly to the Penning trap discussed in Sec. 4.5.1). While for $V = 0$ this trap has a saddle point at $x = y = z = 0$, it is possible to demonstrate that for $V \neq 0$ there is a range of parameters which allows a dynamic trapping of the ion along all the directions.[25] The stable solutions of the equation of motion for a charged particle in the potential of eqn (5.26) show a slow oscillation with well-defined secular frequency $\omega_s = \beta\Omega/2$ (where β is a parameter bounded between 0 and 1, dependent on the potential parameters U, V, Ω, and on the particle mass). This oscillatory motion, typically in the MHz range, can be described in terms of an effective *pseudopotential* which can be approximated by a three-dimensional harmonic well. On

[25] Mathematically, the problem of a charged particle moving in a time-modulated quadrupole field is equivalent to the problem of a quantum particle moving in a static sinusoidal potential (this problem will be extensively discussed in Sec. 6.1), since they are both described by second-order differential Mathieu equations.

top of this secular motion, the ion dynamics shows an additional fast *micromotion*[26] at the modulation frequency Ω. A typical solution for a classical stable trajectory is represented in Fig. 5.14b.

Ion traps are typically very deep, if compared to magnetic or optical traps for neutral atoms. The depth of the pseudopotential can easily exceed 1 eV, which allows trapping of ions even at room temperature and storage times reaching several weeks. The trap frequencies are also much higher than the ones for neutral atoms: at low temperatures the motion of the trapped ion is quantized to discrete vibrational energy levels according to a quantum harmonic oscillator energy

$$E_v = \hbar\omega_S\left(v + \frac{1}{2}\right), \tag{5.27}$$

where v is the quantum vibrational number.[27]

In addition to the 3D Paul trap discussed above, different geometries of radiofrequency ion traps exist: the linear Paul trap, currently used in many experiments, is realized with a quadrupole electric potential of the form $\phi(\mathbf{r}, t) = (U + V\cos\Omega t)\left(x^2 - y^2\right)$, which provides confinement in the plane $\hat{x}\hat{y}$, plus additional dc electrodes which trap along the $\hat{z}$ direction.

Spectrum of a trapped ion. Spectroscopy of trapped ions offers interesting possibilities when forbidden transitions are considered. A common choice for ion experiments is given by alkaline-earth(like) atoms, the photoionization of which results in positively charged ions with the same single-electron structure of alkalis. These ions (e.g. Ca^+, Sr^+, Yb^+, Hg^+) exhibit $S \to D$ electric quadrupole transitions which are forbidden in dipole approximation and have a natural linewidth smaller than 1 Hz. Other choices of forbidden transitions include $S \to F$ electric octupole transitions (Yb^+) or intercombination transitions ${}^1S_0 \to {}^3P_0$ in a different class of group-III ions with two-electron structure (Al^+ and In^+). In all of these cases the transition linewidth γ is much smaller than the frequency ω_s of the secular motion. In this *tight-binding regime* the fluorescence spectrum of the trapped ion, when excited on the narrow optical transition, shows well-resolved sidebands corresponding to transitions between different quantum motional states in the ion trap (similarly to the optical lattice clock spectrum of Fig. 5.11). The ion sideband spectrum contains precious information on the motional state of the ion, since from the measured intensities of the sideband peaks it is possible to extract information on the population of the vibrational modes and on the temperature of the ion (see Fig. 5.15 for an example).

The spectrum is particularly simple in the *Lamb–Dicke regime*, already introduced in Sec. 5.3.1. The Lamb–Dicke parameter

$$\eta = ka_s \tag{5.28}$$

[26]The micromotion represents an unwanted perturbation, since it complicates the control of the ion dynamics in the trap. Although the micromotion is null at the centre of an ideal quadrupole trap, in a real experimental setting the centre of the time-averaged potential can be slightly displaced from the quadrupole centre; therefore, dc fields corrections are needed in order to decrease the micromotion amplitude.

[27]For a typical value $\omega_S = 2\pi \times 2$ MHz, the quantum of oscillation energy $\hbar\omega$ corresponds to an energy $k_B \times 0.1$ mK.

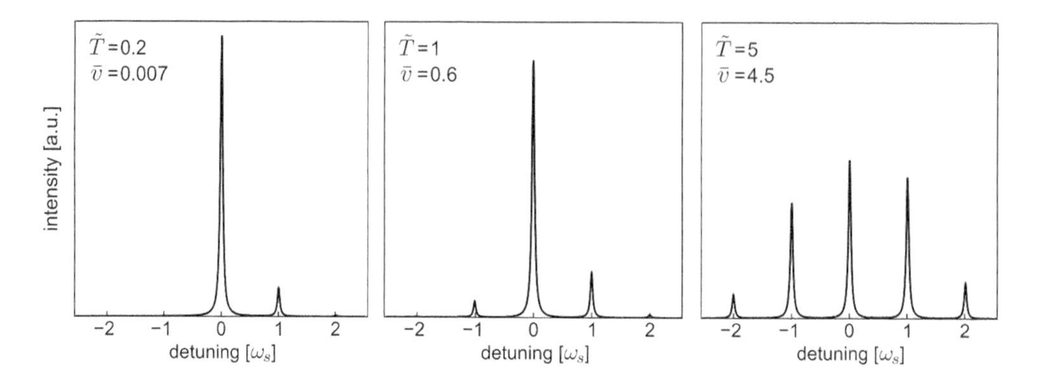

Fig. 5.15 Calculated fluorescence spectrum of an ion in a Paul trap for a Lamb–Dicke parameter $\eta = 0.3$. The spectra correspond to three different reduced temperatures $\tilde{T} = k_B T/\hbar\omega_s$, determining three different values of the average vibrational quantum number $\bar{v}$.

is defined as the product of the laser wavenumber k by the ion trap oscillator length $a_s = \sqrt{\hbar/m\omega_s}$, where m is the ion mass. Equivalently, it can be written as

$$\eta = \frac{2\omega_R}{\omega_s}\,, \tag{5.29}$$

where $\hbar\omega_R = \hbar^2 k^2/2m$ is the recoil kinetic energy that an atom at rest acquires after absorbing one photon, typically on the order of $h \times 10$ kHz for optical transitions. The Lamb–Dicke regime sets in when

$$\eta \ll 1/\sqrt{\bar{v}}\,, \tag{5.30}$$

where $\bar{v}$ is the expectation value of the quantum vibrational number, which depends on the ion temperature in the case of a thermal distribution:

$$\bar{v} = \frac{\sum_{v=0}^{\infty} v e^{-E_n/k_b T}}{\sum_{v=0}^{\infty} e^{-E_n/k_b T}} = \left(e^{\hbar\omega_S/k_b T} - 1\right)^{-1}\,. \tag{5.31}$$

In the Lamb–Dicke regime the excitation wavelength is much larger than the length scale associated with the trap dynamics, so the radiation can barely "distinguish" the ion motion: as a result, light does not couple efficiently motional states with different v and only transitions with $v \to v-1, v, v+1$ can be observed (see Fig. 5.15). This regime can be achieved by decreasing the oscillator length a_s, which can be obtained by making tightly confining traps with small electrodes driven by large RF frequencies, or by using long excitation wavelengths (this is the case e.g. of transitions between different hyperfine states driven by Raman or microwave excitations with small k).

5.4.2 Ion cooling

Laser cooling of ions works quite similarly to the laser cooling of neutral atoms described in Sec. 2.3, as far as single trapped ions and dipole-allowed transitions with large linewidth are considered. Laser cooling of trapped ions was demonstrated

Table 5.2 Different frequency scales for the trap dynamics and light–matter interaction in experiments using trapped ions for atomic clocks or quantum computation applications.

Secular motion frequency in the trap	$\omega_s/2\pi$	$\sim 10^6$ Hz
Micro-motion frequency in the trap	$\Omega/2\pi$	$\sim 10^7$ Hz
Linewidth of the spectroscopy narrow transition	$\gamma/2\pi$	$\lesssim 1$ Hz
Linewidth of the cooling allowed transition	$\gamma_c/2\pi$	$\sim 10^7$ Hz
Recoil frequency	$\omega_R/2\pi$	$\sim 10^4$ Hz

in 1978 independently in Boulder (USA) (Wineland *et al.*, 1978), where Mg^+ ions were cooled to temperatures lower than 40 K by the interaction with a red-detuned laser beam, and in Heidelberg (Germany) (Neuhauser *et al.*, 1978), where a similar technique was used to cool Ba^+ ions trapped in a radiofrequency Paul trap. By using $S \to P$ dipole-allowed transitions Doppler cooling can be operated without any major restriction[28] and temperatures very close to the Doppler limit $T_D^{min} = \hbar\gamma_c/2k_B$ can be achieved. At the Doppler temperature the number of populated vibrational modes is still quite large, since the laser linewidth γ_c is typically one order of magnitude larger than the secular trap frequency ω_s. The average vibrational number after Doppler cooling can be easily calculated from eqn (5.31) as $\bar{n} = \left(e^{2\omega_s/\gamma_c} - 1\right)^{-1} \approx 10$.

Different cooling possibilities emerge when narrow transitions are used in combination with ion trapping, enabling the possibility of controlling the motion of the ions at the quantum level (see e.g. Eschner *et al.* (2003) for a review). As a reference, the typical frequency scales associated with the laser–ion interaction and the ion trap dynamics are reported in Table 5.2. Dipole-forbidden transitions (e.g. electric quadrupole $S \to D$ in group-II ions) can be used to perform additional laser cooling in the tight-binding regime in which the laser linewidth γ is much smaller than the secular frequency ω_s and individual sidebands can be resolved. This allows the possibility of performing *resolved sideband cooling* in which the laser is tuned to the frequency of the red sideband $\omega_0 - \omega_s$ (where ω_0 is the transition frequency for the free ion). After the absorption the ion may re-emit the fluorescent photon on different sidebands, but, in the Lamb–Dicke regime, the probability of emission is by far the largest for the central resonance centred at ω_0, which means that in the net absorption/emission process the ion loses a quantum of vibrational energy $\hbar\omega_s$. This process is very efficient and can be iterated to cool the motion of the ion down to the vibrational quantum state ($\bar{v} \ll 1$), as demonstrated in 1989 by the D. Wineland group (Diedrich *et al.*, 1989).

Ion crystals. When more than one ion is loaded in a trap, this simple picture changes considerably, since ions are strongly interacting with each other via the Coulomb repulsion. This strong repulsion is responsible for the formation of ordered crystalline structures such as the ones shown in Fig. 5.16. In linear Paul traps, where the confinement in the radial directions is much stronger than along the trap axis, the ions arrange in linear chains in which their motion is strongly coupled. These chains exhibit

[28] Even with some advantages, since shining the cooling laser along one direction only is typically sufficient for reducing the temperature of the ion in the trap.

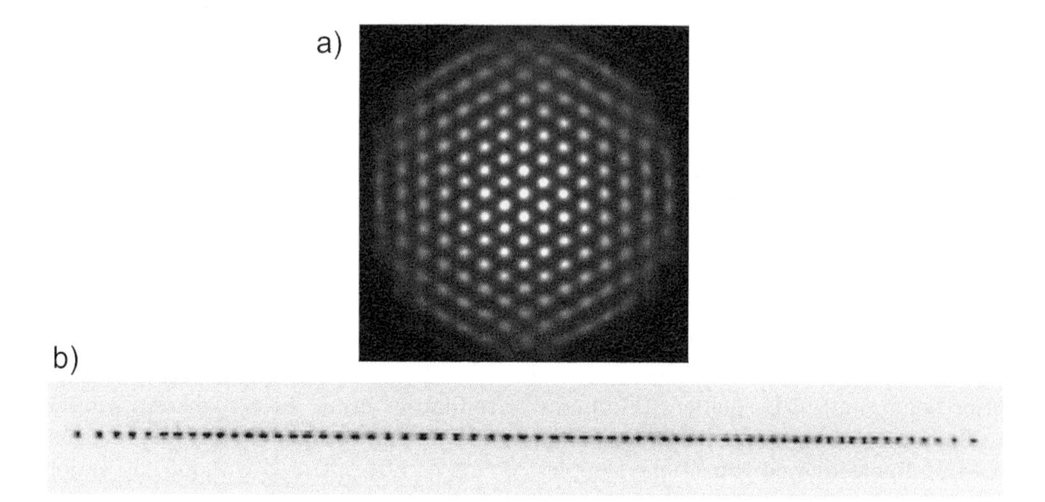

Fig. 5.16 Examples of laser-cooled ion crystals. a) 2D triangular array of Be^+ ions in a Penning trap realized at NIST for quantum simulation of Ising models (see Sec. 7.3). Reprinted with permission from Britton *et al.* (2012). © Macmillan Publishers Ltd. b) Linear 1D chain of Ca^+ ions in a linear Paul trap. Entangled states of up to 14 ion qubits have been recently produced in the group of R. Blatt (Monz *et al.*, 2011). © University of Innsbruck.

collective modes in the form of phonon excitations, in which the excitation is shared by all the ions. At the end of the 1990s resolved sideband cooling allowed small linear ion chains to be cooled down to the ground state of their collective motion (King *et al.*, 1998; Rohde *et al.*, 2001). The coupling between internal state and collective motion of a system of ions is an extremely powerful resource for applications, since it can be used to entangle different ions together. This entanglement is at the basis of a number of breakthroughs in quantum information science, including the demonstration of high-fidelity quantum gates (Cirac and Zoller, 1995; Monroe *et al.*, 1995; Schmidt-Kaler *et al.*, 2003; Leibfried *et al.*, 2003*b*) and the production of entangled states of a large number of ions (Leibfried *et al.*, 2005; Häffner *et al.*, 2005; Monz *et al.*, 2011).

5.5 Ion clocks

Single trapped ions are among the best candidates for the realization of optical clocks, since they can be trapped and interrogated continuously for very long times (even days and weeks!) and the systematic effects can be kept under control with very good precision. After the first trapping of single electrons in Wineland *et al.* (1973), trapping of single ions was reported for the first time in Neuhauser *et al.* (1980), and later in Wineland and Itano (1981), with the observation of discrete steps in the fluorescence signal due to the trapping of a few countable number of ions in a regime of small trap-loading efficiency. Single trapped ions offer a complementary system to the ensembles of neutral atoms discussed in Sec. 5.3.1, where the large number of absorbers ($\approx 10^5$ clocks working in parallel) gives a larger signal-to-noise ratio (thus permitting shorter integration times), but systematic effects can be more difficult to evaluate.

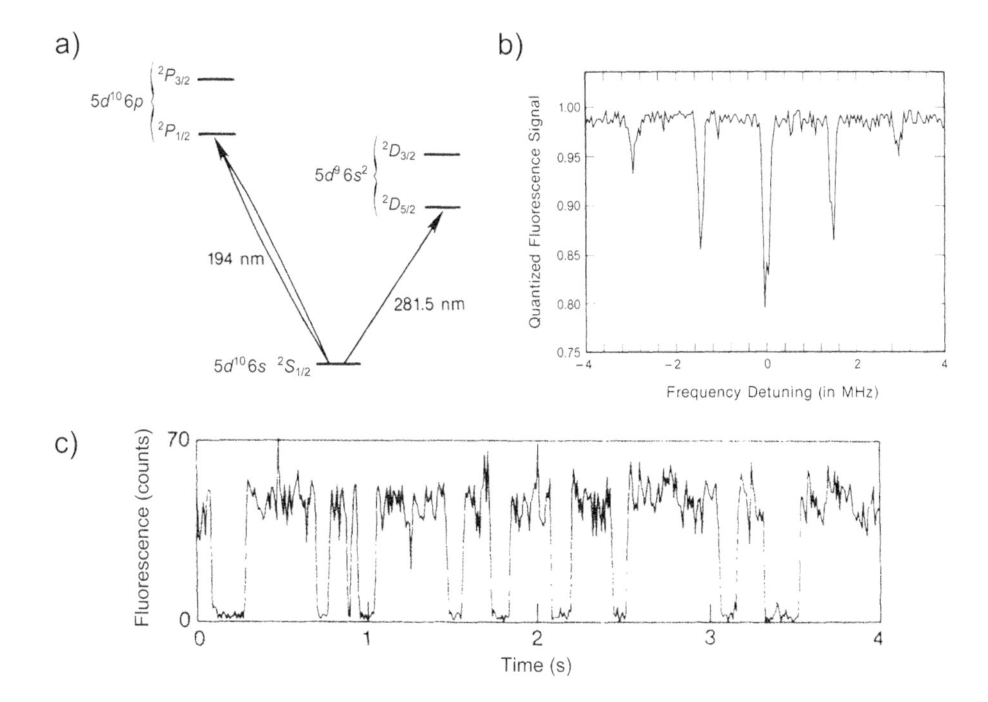

Fig. 5.17 a) Level scheme for the Hg^+ ion showing the cooling/detection transition at 194 nm and the forbidden clock transition at 281.5 nm. b) Resolved sideband spectrum for the 281.5 nm transition in Hg^+. c) Electron shelving technique showing quantum jumps of a single ion: the presence of 194 nm fluorescence light indicates that the ion is in the $^2S_{1/2}$ state, while the absence of fluorescence indicates that it is in the metastable $^2D_{5/2}$ state. Reprinted with permission from Bergquist *et al.* (1986) and Bergquist *et al.* (1987). © American Physical Society.

Detection of fluorescence light from a single ion requires a strong dipole-allowed transition capable of guaranteeing a sufficiently intense flux of photons to be detected. However, the spectroscopic transitions used in ion clocks are dipole-forbidden and the scattering rate is typically limited to less than one photon per second. In order to detect the excitation of the clock transition a very efficient technique, known as *electron shelving detection*, was proposed by H. G. Dehmelt in 1975 (Dehmelt, 1975) and then experimentally realized in 1986 (Nagourney *et al.*, 1986; Bergquist *et al.*, 1986; Sauter *et al.*, 1986). In this technique the ion is illuminated on both the cooling and the spectroscopy transitions and state information is acquired by detecting the fluorescence light emitted on the strong cooling transition.[29] As an example, we consider the case of the Hg^+ ion illustrated in Fig. 5.17a. If the ion has been excited by the 281.5 nm clock laser to the metastable $|m\rangle = {}^2D_{5/2}$ state (the *shelf* state), no photons at the detection wavelength 194 nm can be observed until the atom has decayed to the ground state

[29]Actually, the two lasers are applied in sequence and not simultaneously, to avoid ac-Stark shifts of the clock transition induced by the cooling light.

$|g\rangle = {}^2S_{1/2}$. If the ion has not been excited to the $|m\rangle$ state, it can be coupled to the $|e\rangle = {}^2P_{1/2}$ state by the detection laser and a large amount of fluorescence 194 nm photons is detected.

It is interesting to note how this detection technique is intrinsically "digital" and therefore largely immune from technical noise. As far as a single ion is considered, the amount of fluorescence can only assume two values - "yes" or "no" - as clearly shown in Fig. 5.17c from the group at NIST (Bergquist *et al.*, 1986), which shows characteristic *quantum jumps* in the amount of detected light as a function of time. The time intervals during which the ion is in the metastable state $|m\rangle$ (and no photons are detected) have different durations, reflecting the statistical distribution of decay times: from the average duration of these dark periods it is therefore possible to infer the lifetime of the metastable state. Of course, from a quantum-mechanical point of view, the ion can be found in any superposition state $\alpha|g\rangle + \beta|m\rangle$, depending on the detuning, intensity, and duration of the spectroscopy pulse. The measurement of fluorescence on the allowed transition projects the state of the ion on the two possible measurement outcomes: the ground state $|g\rangle$ with probability $|\alpha|^2$ and the metastable state $|m\rangle$ with probability $|\beta|^2$. Making a statistical analysis on the outcomes of several repeated measurements it is possible to measure the average excitation probability and plot it as a function of the clock laser frequency, resulting in a graph such as the one reported in Fig. 5.17b, clearly showing the sideband spectrum discussed in the previous section.[30]

Quantum logic Al^+ clock. The most precise ion clocks demonstrated so far are based on group-III Al^+ ions, which have the same two-electron structure as neutral alkaline-earth atoms. In particular, they exhibit the same ultra-narrow intercombination transition ${}^1S_0 \rightarrow {}^3P_0$ already discussed in Sec. 5.3.1 in the context of optical lattice clocks. The choice of this transition (originally suggested by H. Dehmelt) is particularly convenient because it is insensitive to external fields and, in the case of Al^+, it is affected by the smallest known room-temperature blackbody radiation shift. However, the choice of Al^+ is problematic from the point of view of ion preparation and detection, since strong transitions for cooling and state detection are not experimentally accessible.

This problem has been solved by the group of the 2012 Nobel Prize winner D. J. Wineland at NIST by trapping two different ions in the same linear Paul trap: the Al^+ "spectroscopy" ion which is interrogated on its clock transition and a Be^+ "logic" ion which is used both for sympathetic cooling and for state detection of the Al^+ ion (Schmidt *et al.*, 2005). The level scheme for the two ions is plotted in Fig. 5.18a. The two ions are initially prepared in their ground states ${}^1S_0$ (Al^+) and ${}^2S_{1/2}(F=2)$ (Be^+). After the application of the clock laser, a succession of laser pulses (indicated as 4, 5, 6 in the figure) maps the initial Al^+ state to the state of the Be^+ ion using well-established protocols demonstrated in the context of quantum information experiments with ions (Häffner *et al.*, 2008). This detection procedure, which can be considered an extension of the electron shelving method, takes advantage of the common vibrational mode of Al^+/ Be^+ as transfer state of the excitation.

[30] The uncertainty related to this statistical reconstruction of the excitation probability is the *quantum projection noise*, already discussed in Sec. 2.4.1. It typically scales as $1/\sqrt{N}$, where N is the number of measurements.

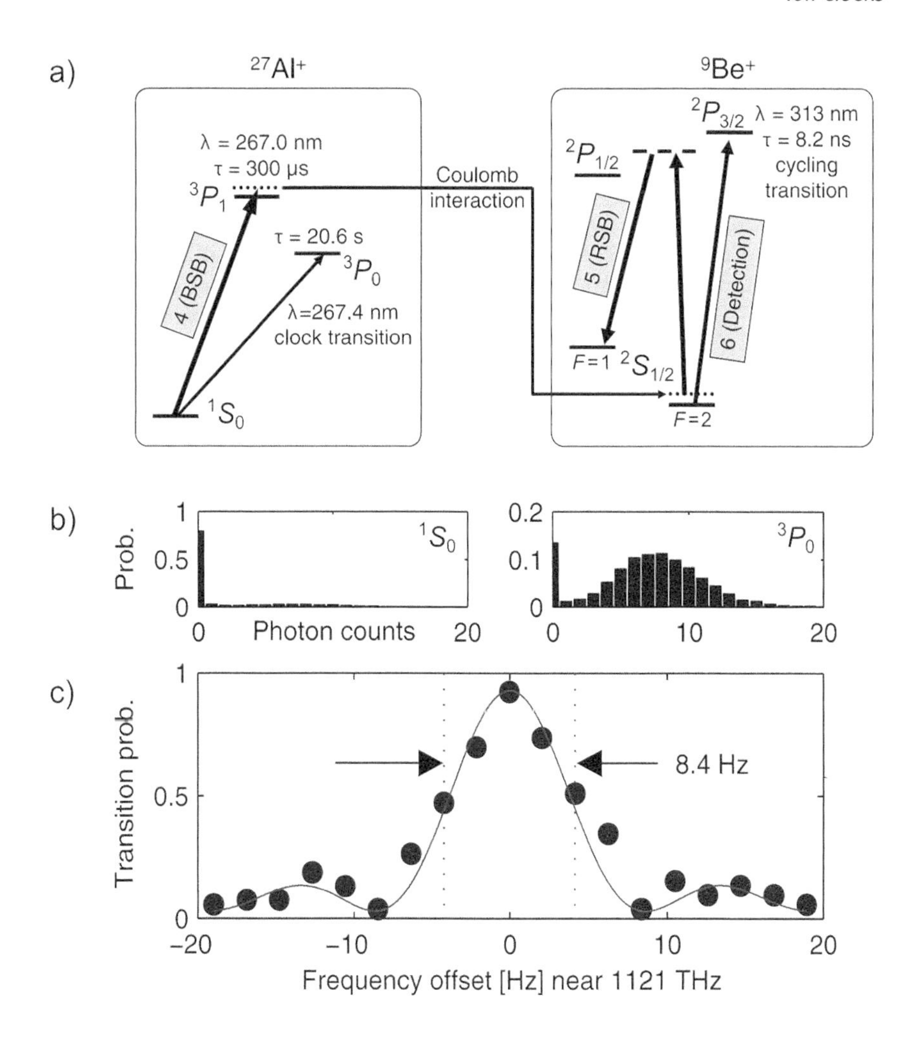

Fig. 5.18 a) Level scheme of $^{27}Al^+$ and $^9Be^+$ investigated in Rosenband *et al.* (2007). The electronic state of the "spectroscopy" ion $^{27}Al^+$ is transferred to the electronic state of the "logic" ion $^9Be^+$ through the intermediate excitation of a vibrational normal mode determined by the Coulomb interaction between the ions. b) Distribution of detected $^9Be^+$ photons as a function of the internal $^{27}Al^+$ state. c) Spectrum of the $^{27}Al^+$ $^1S_0 \rightarrow {}^3P_0$ clock transition. Reprinted with permission from Rosenband *et al.* (2007). © American Physical Society.

More in detail, in the case of absence of excitation to the Al^+ $^3P_0$ state, laser 4 (on the blue sideband of Al^+ $^1S_0 \rightarrow {}^3P_1$ transition) excites one quantum of the Al^+/Be^+ normal vibrational mode; then Raman lasers 5 (on the red sideband of Be^+ $^2S_{1/2}(F = 2) \rightarrow {}^2S_{1/2}(F = 1)$ transition) transfer the excitation of the vibrational mode to the $^2S_{1/2}(F = 1)$ state; then no fluorescence photons are detected on the Be^+ $^2S_{1/2}(F = 2) \rightarrow {}^2P_{3/2}$ transition driven by laser 6. This can be evidenced in the left panel of Fig. 5.18b, showing the absence of fluorescence photons at the detection wavelength. Instead, if the Al^+ is excited to the $^3P_0$ state, the state transfer steps described before do not take place and the Be^+ ion stays in the $^2S_{1/2}(F=2)$ state, where it is able to scatter photons from the detection laser. This can be shown in the right panel of Fig. 5.18b, evidencing the distribution of fluorescence photons (with average value ≈ 8).

Using this *quantum logic* technique, the frequency of the Al^+ clock transition has been evaluated in Chou *et al.* (2010*a*) with a fractional inaccuracy of 8.6×10^{-18}, which is currently the best performance for an atomic clock of any kind. The comparison of two similar Al^+ quantum-logic based clocks has resulted in a fractional frequency difference $(-1.8\pm0.7)\times10^{-17}$, which is consistent with the inaccuracy of the least accurate of the two Al^+ clocks.

Towards a new generation of ion clocks. We conclude this section by mentioning two recent theoretical works, which have proposed new classes of "exotic" optical ion clocks which might break the level of accuracy of the best optical clocks realized to date. A first proposal (Campbell *et al.*, 2012) relies on the investigation of a transition between the nuclear ground state and a nuclear metastable isomer configuration of a single $^{229}Th^{3+}$ ion. This nuclear magnetic dipole transition, predicted to occur at a vacuum-UV wavelength, has the peculiarity of connecting states with the same electronic wavefunction, which strongly suppresses the dependence of the transition frequency on external perturbations (by e.g. magnetic or electric fields) due to common-mode rejection. A second proposal (Derevianko *et al.*, 2012) is based on the investigation of highly forbidden near-visible transitions connecting different states within the open-shell $4f^{12}$ configuration of highly charged ions such as $^{209}Bi^{25+}$. Also in this system the transition frequency is largely independent of external perturbations, owing to the reduced size of the electron wavefunctions (experiencing a larger effective nuclear charge) with respect to ordinary singly charged ions. In both cases the extremely large quality factor of the transitions and the excellent level of rejection to external perturbations could make a fractional accuracy of 10^{-19} potentially achievable for a future generation of optical ion clocks.

5.5.1 General relativity tests

Accurate frequency measurements or comparisons between different atomic clocks are valuable tools for quantitative tests of Einstein's theory of relativity. Second-order Doppler shifts, induced by time dilation in a reference frame moving with respect to the observer, have already been discussed in the context of hydrogen spectroscopy in Sec. 1.3.1. Not only is the motion responsible for a shift in the frequency of the electromagnetic radiation emitted or absorbed by an atom. According to the theory of general relativity a clock that is operating in the presence of a gravitational field runs slower than a clock operating in free space, even if the two clocks are not in relative motion. Both of these effects have to be taken into account daily by the tens of millions GPS receivers which calculate navigation information from the time counted by the atomic clocks operating onboard the GPS satellites (Ashby, 2002).

The performances reached by ion clocks (described in Sec. 5.5) are so spectacular in terms of accuracy that it is possible to use them to detect relativistic effects on

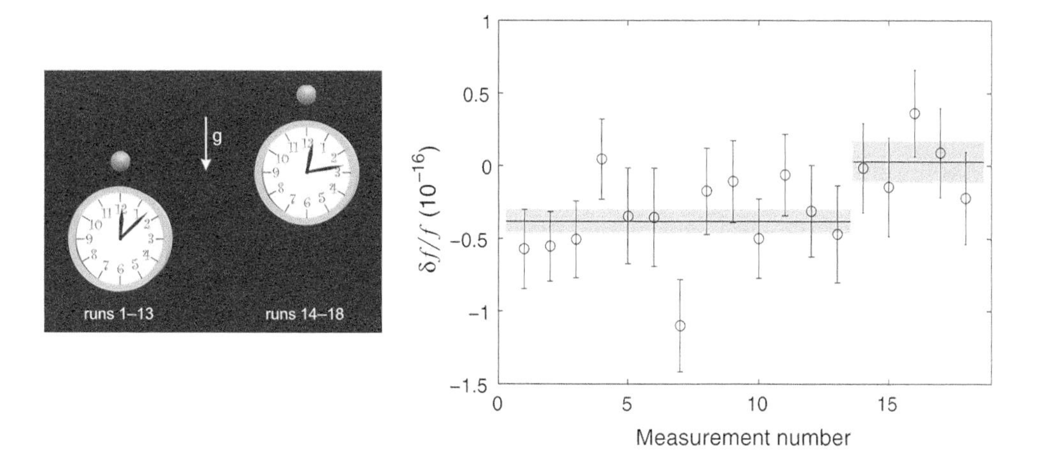

Fig. 5.19 According to general relativity, a clock operating at a smaller gravitational potential (which means closer to the source mass for the gravity field) runs more slowly than a clock at a larger potential. The plot shows the fractional change in the frequency of an Al^+ optical ion clock. Measurement runs 1–13 have been taken when the clock was operated at 33 cm lower elevation than runs 14–18 and show a change in the measured frequency of $(4.1 \pm 1.6) \times 10^{-17}$ with respect to measurement runs 14–18. Reprinted with permission from Chou *et al.* (2010*b*). © AAAS.

daily-life velocity and length scales. In Chou *et al.* (2010*b*) an optical clock based on the Al^+ ion has been used to detect and measure the time dilation effect induced by tiny differences of the gravitational field. Two clocks operating on the Earth at two different elevations are expected to run at different rates, their frequencies differing by an amount

$$\delta\nu = \nu_0 \frac{g\delta h}{c^2} , \tag{5.32}$$

where ν_0 is the clock resonance frequency, g is the local gravity acceleration, δh is the difference in height between the two positions, and c is the speed of light. The fractional frequency shift $\delta\nu/\nu_0$, corresponding to the ratio between the difference in gravitational potential energy $mg\delta h$ and the rest energy of the ion mc^2, is extremely small, amounting to only 1.1×10^{-16} per metre. Altough very small to be practically unimportant for our daily life, the effect is large enough to be detected by ion clocks even for distances below 1 metre. In the experiment reported in Chou *et al.* (2010*b*) the authors have measured the frequency of an Al^+ ion clock at two different elevations differing only by 33 cm. The results are shown in Fig. 5.19, which evidences that the measurements at lower height (runs 1–13 in figure) are redshifted with respect to the measurements at larger height (runs 14–18). The difference amounts to a fractional change $(4.1 \pm 1.6) \times 10^{-17}$, that corresponds to the shift induced by a change in height 37 ± 15 cm, which agrees well with the actual elevation difference.

5.5.2 Stability of fundamental constants

A spectacular application of high-resolution frequency measurements regards the possibility of investigating possible variations in the value of fundamental constants. Among these, dimensionless constants are particularly significant, since their value does not depend on the definition of any system of units. The best known example of dimensionless constant is the fine structure constant α (see Sec. 4.5), representing the strength of the electromagnetic interaction. Despite the spectacular success of QED, this theory does not provide a value for α, nor makes any prediction on the fact that this value may change in time: basically, it is a parameter which has to be put in "by hand" in the theory. The discussion on the stability of fundamental constants is strictly connected with Einstein's Equivalence Principle, which states that the results of any non-gravitational experiment must be independent of time and space. Non-metric string theories, which aim at the unification of gravity with other fundamental interactions, allow for a variation of the fundamental coupling constants, in contrast with the Equivalence Principle. These variations would happen on the spatial and temporal scales of the cosmic dynamics, therefore the possible detection of related effects is extremely challenging. For a discussion of the theoretical framework and a review of the main experimental constraints related to the variation of the fundamental constants see e.g. Uzan (2011) and Chiba (2011).

Astronomical molecular spectroscopy. A first approach to the experimental investigation of the stability of fundamental constants relies on the observation of signals coming from the very remote past. This information can be obtained in astronomical observations by analyzing the spectrum of the light absorbed in remote regions of the Universe, billions of years before its detection on Earth. In recent experiments, the study of absorption lines by molecular clouds along quasar sightlines allowed the determination of precise constraints on the variation of fundamental constants. In these experiments the astronomical spectrum is compared with the spectrum of the same molecule in the laboratory searching for quantitative deviations in their structure.[31]

The analysis of the microwave spectrum of NH_3 inversional and rotational lines (Murphy *et al.*, 2008) and optical H_2 absorption lines (King *et al.*, 2008; van Weerdenburg *et al.*, 2011) have been recently used to investigate the stability of the proton-to-electron mass ratio $\mu = m_p/m_e$, an important dimensionless constant which ultimately quantifies the ratio between the strengths of the nuclear and the electromagnetic interactions. The limit on the variation of this quantity, determined in van Weerdenburg *et al.* (2011) as the average of several measurements on different molecules and different astronomical objects (see Fig. 5.20a), was determined to be $\Delta\mu/\mu = (5.2 \pm 2.2) \times 10^{-6}$, which is consistent with no shift within 2.3 standard deviations.

Similar experiments were performed for the investigation of variations of the fine structure constant, obtaining controversial results (see e.g. Murphy *et al.* (2003) and Srianand *et al.* (2004)). A recent work, combining data from different telescopes, has

[31] The largest difference between laboratory spectra and astronomical spectra is given by the cosmological redshift associated with the expansion of the Universe. This effect, capable of shifting the Lyα hydrogen line from the UV to the visible spectrum, can be measured from the spectra themselves and taken into account in the comparison.

reported surprising indications of α variations which depend on the direction of sight along the Universe (Webb *et al.*, 2011) (negative in the northern sky and positive on the southern side) according to a dipole pattern, which has been determined with 4.2 standard deviation significance. Further experiments will be needed to confirm the validity of this unexpected result and exclude possible hidden systematics.

High-resolution atomic spectroscopy. A second strategy for the investigation of possible drifts in the fundamental constants is based on purely laboratory-based experiments. Even if the observation timescales are much smaller, the precision is much higher and the investigation is not restricted to the limited number of atoms and molecules that are present in Space with sufficient concentration to be detected. High-resolution spectroscopy of narrow atomic transitions allows us to reach similar levels of precision to the astronomical observations when the experiment is repeated only after a few years (or months).[32] As an example, we consider the $1s - 2s$ transition in atomic hydrogen (see Sec. 1.3). Repeated measurements performed over a period of 44 months resulted in a fractional drift of the transition frequency $\Delta\nu_{\mathrm{H}} = (1.2 \pm 2.3) \times 10^{-14}$, which is consistent with zero (Fischer *et al.*, 2004). In order to analyse this drift, and similar results coming from optical atomic or ion clocks, one has to consider that absolute frequency measurements are performed by comparing the optical frequency ν to the international frequency standard ν_{Cs} defined by the microwave transition between the two hyperfine ground states of caesium (see Sec. 2.2). It is possible to demonstrate that temporal variations of ν/ν_{Cs} could be due either to a drift of the fine structure constant α or to a drift of the nuclear magnetic moment of caesium μ_{Cs}, which enters the definition of the second.

In order to isolate possible variations of α from variations of μ_{Cs} (coming from contingent drifts in the strong interaction), it is possible to compare the absolute frequency measurements of two or more different transitions, which depend differently on these two quantities. As an example, we show in Fig. 5.20b the comparison made in Kolachevsky *et al.* (2009*b*) between the variations measured in the $1s - 2s$ transition of hydrogen and the variations observed in narrow optical transitions of Yb^+ and Hg^+. The grey regions represent the variations of α and of the nuclear magnetic moment of caesium $\tilde{\mu} = \mu_{\mathrm{Cs}}/\mu_B$ (expressed in units of the Bohr magneton μ_B) which are compatible with the measured frequency drifts: the widths of the zones are determined by the experimental uncertainty, while the different slopes reflect the different dependency on α and $\tilde{mu}$ for the different transitions. Owing to the different slopes, the intersection of the grey zones results in the dark grey ellipse which defines the limiting bounds for a possible variations of these constants, as determined in Kolachevsky *et al.* (2009*b*).

Recently, improved constraints on the variation of α have been determined thanks to the direct comparison of two optical frequencies without referring to the caesium standard, which itself ultimately limits the accuracy of the measurements. The direct measurement of the frequency ratio between Hg^+ and Al^+ transitions carried out at NIST (USA) (Rosenband *et al.*, 2008) resulted in a much stricter limit on the variation of the fine structure constant

[32] Assuming a drift of the constants which is linear in time.

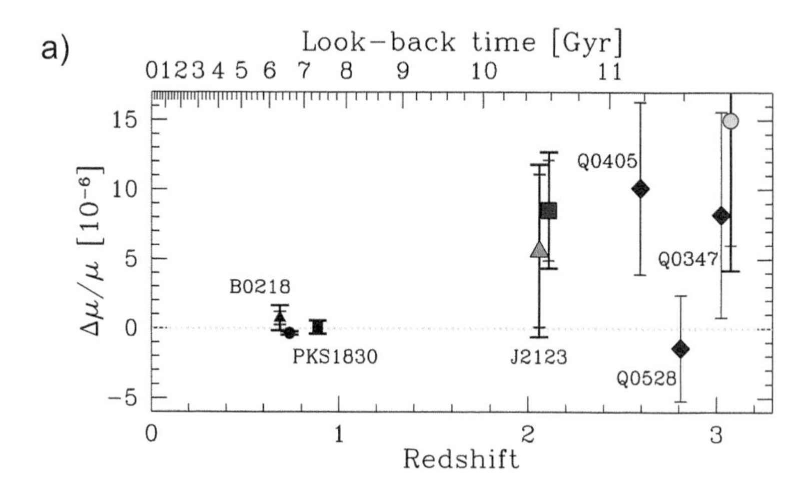

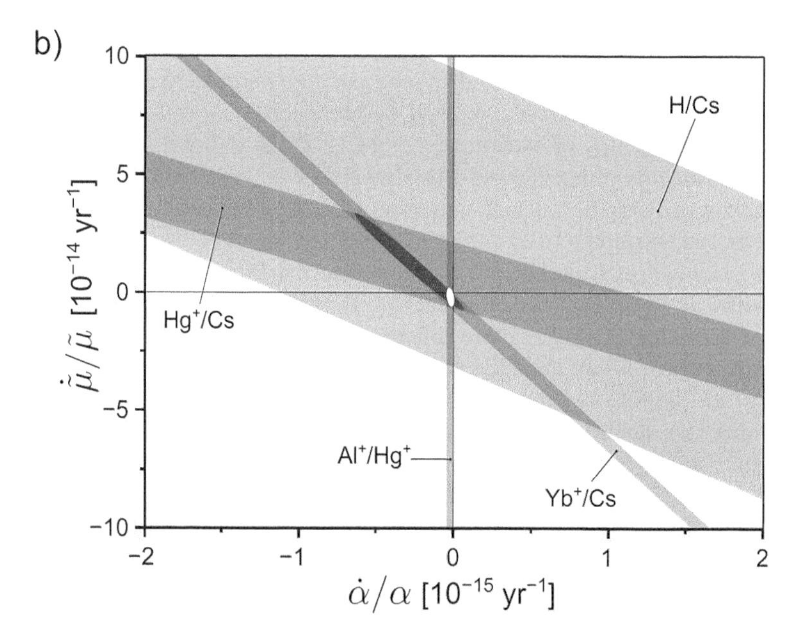

Fig. 5.20 a) Constraints on the variation of the proton-to-electron mass ratio $\mu = m_p/m_e$ obtained by the investigation of astronomical molecular spectra. Reprinted with permission from van Weerdenburg *et al.* (2011). © American Physical Society. b) By combining the measured drifts in the transition frequencies of different atom/ions, this plot sets the boundaries (determined by the intersection regions) for a possible drift of the fine structure constant α and of the caesium nuclear magnetic moment in units of Bohr magnetons $\tilde{\mu} = \mu_{Cs}/\mu_B$. No variations of the fundamental constants with statistical significance has been evidenced so far. The boundaries are taken from Kolachevsky *et al.* (2009*b*) and Rosenband *et al.* (2008).

$$\dot{\alpha}/\alpha = (-1.6 \pm 2.3) \times 10^{-17}\ \mathrm{yr}^{-1}\,, \tag{5.33}$$

independently of the comparison with the caesium standard and of possible variations of $\tilde{\mu}$. This is represented by the narrow vertical stripe in Fig. 5.20b, which intersect the other zones in the small white ellipse at the centre: comparing the result of Rosenband *et al.* (2008) with the previous result yields a determination of the time variation of the caesium nuclear magnetic moment

$$\dot{\tilde{\mu}}/\tilde{\mu} = (-1.9 \pm 4.0) \times 10^{-16}\ \mathrm{yr}^{-1}\,. \tag{5.34}$$

So far, the measured drifts in the laboratory are consistent with no variation of the constants.

6

Optical lattices and precise measurements

The possibility of using far-detuned laser light to engineer trapping potentials for ultracold atoms allows an unprecedented control of the forces exerted on a quantum system. In this chapter we discuss the application of ultracold atoms in optical lattices to the investigation of fundamental quantum-mechanical effects in strong connection with the physics of solid-state systems.

An atom moves in an optical lattice as an electron moves in an ideal crystalline solid, since both can be described as quantum particles interacting with a periodic potential. The extremely long coherence times and the absence of defects in the optical lattice provide the ideal environment for investigating the transport of quantum particles in periodic structures. The intrinsic periodicity of these "crystals" of light also make them valuable tools for precise measurements: they can be used for the accurate determination of fundamental constants and can provide quantum sensors for the interferometric detection of forces with high sensitivity and spatial resolution.

6.1 Quantum transport in periodic potentials

6.1.1 Bloch theorem and energy bands

Before discussing the implementations with cold atoms, in this section we will review the quantum-mechanical solution to the problem of a particle in a periodic potential (Ashcroft and Mermin, 1976), which serves as a theoretical background for the whole chapter. This problem is central in the description of electric conduction in crystalline solids, where the atoms are arranged in an ordered configuration with well-defined symmetries, giving rise to the *Bravais lattice*. In a solid the valence electrons, instead of forming bound states with individual atoms, experience the combined attraction to all the atoms of the crystal, which can be modelled with a periodic potential $V(x)$ having the same discrete translational symmetries as the lattice.

In the one-dimensional case the periodicity condition for the potential $V(x)$ takes the form

$$V(x) = V(x + d) \; , \tag{6.1}$$

where d is the distance between nearest-neighbouring lattice sites. In solids the potential $V(x)$ does not have a simple analytical form, being the sum of the Coulomb attraction to the nuclei of the lattice atoms, screened by an effective repulsion generated by the more tightly bound electrons in the inner shells. In the following we will consider a particular choice of the potential $V(x)$, which is relevant for its physical implementation with optical lattices for neutral atoms. This is the sinusoidal potential

$$V(x) = V_0 \cos^2(kx) , \tag{6.2}$$

where $k = \pi/d$ is related to the lattice spacing d and V_0 describes the strength of the potential. The time-independent Schrödinger equation for the wavefunction $\Psi(x)$ of a particle of mass m moving in the potential $V(x)$ reads

$$\hat{H}\Psi(x) = \left[-\frac{\hbar^2}{2m}\frac{\partial^2}{\partial x^2} + V(x)\right]\Psi(x) = E\Psi(x) . \tag{6.3}$$

The Bloch theorem (Bloch, 1928) states that, when the discrete translational invariance in eqn (6.1) is considered, the solutions of eqn (6.3) take the generic form

$$\Psi_{n,q}(x) = e^{iqx} u_{n,q}(x) \tag{6.4}$$

$$u_{n,q}(x) = u_{n,q}(x+d) , \tag{6.5}$$

which describes plane waves e^{iqx} modulated by functions $u_{n,q}(x)$ having the same periodicity as the lattice. These stationary solutions are labelled by two quantum numbers: the *band index* n and the *quasimomentum* (or *crystal momentum*) q.

The quantum number q is called *quasimomentum* because (as discussed in Sec. 6.1.2) it presents some analogies with the momentum p, which is a good quantum number to describe the eigenstates of the Schrödinger equation in the absence of any external potential. However, since the potential $V(x)$ does not present a complete translational invariance, the solutions (6.4) are not eigenstates of the momentum operator and $\hbar q$ is *not* the expectation value of the momentum. We note that, because of the discrete invariance of the Hamiltonian under translations $x \to x + nd$ (with n integer), the quasimomentum is defined modulo $2k = 2\pi/d$, that is the period of the reciprocal lattice. As a matter of fact, the periodicity of the problem in real space induces a periodic structure also in momentum space, the elementary cells of which are the so-called *Brillouin zones*.[1]

Energy spectrum. For a given quasimomentum q many different solutions $\Psi_{n,q}(x)$ with different energies $E_n(q)$ exist. These solutions are identified with the band index n. The term *band* refers to the fact that the periodic potential causes a segmentation of the energy spectrum into forbidden zones[2] and allowed zones, the so-called *energy bands*, which in solid-state physics are at the basis of the conduction properties of metals and insulators. Figure 6.1 shows a plot of $E_n(q)$ as a function of the quasimomentum q for the first three energy bands in the first three Brillouin zones. In this example the sinusoidal potential in eqn (6.2) is chosen with $V_0 = 4E_R$, where $E_R = \hbar^2k^2/2m$ is a natural energy scale introduced by the lattice (its meaning will be clear in Sec.

[1]In higher dimensions the periodicity in momentum space is defined by the vectors of the *reciprocal lattice* which can be constructed starting from the base vectors of the Bravais lattice in real space (Ashcroft and Mermin, 1976).

[2]With the term "forbidden zones" we refer to intervals of energies in which the density of states $\rho(E)$ vanishes, while in the "allowed zones" $\rho(E) > 0$.

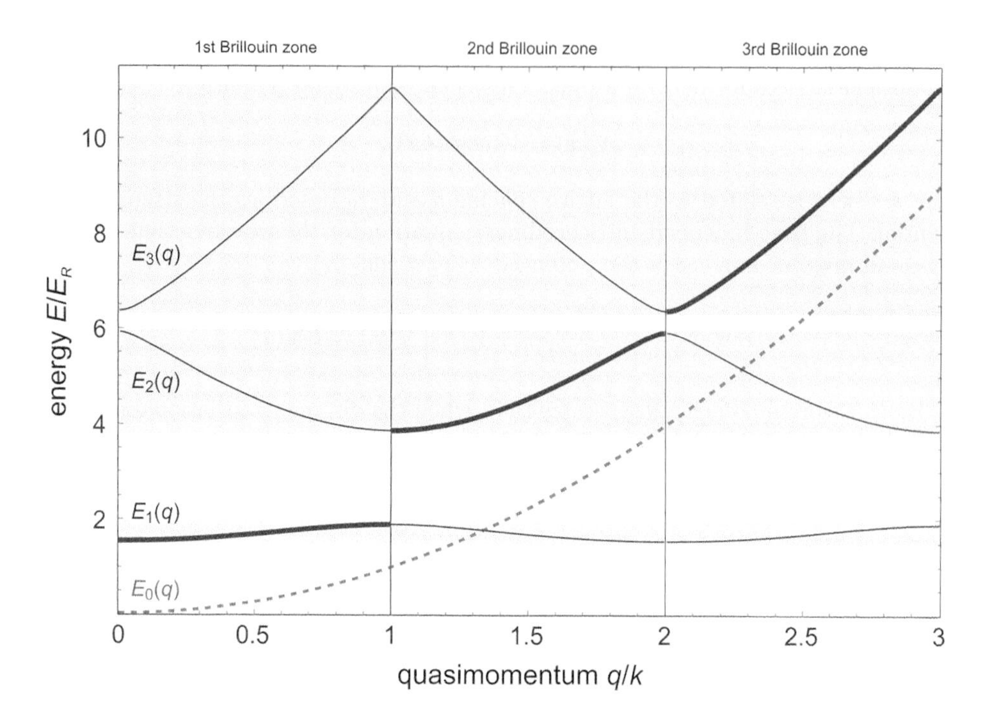

Fig. 6.1 Energy spectrum of a particle in a periodic potential. Solid curves: lowest three energy bands for a particle of mass m in the periodic potential of eqn (6.2) with $V_0 = 4E_R$. Dashed curve: energy spectrum of the free particle. The energies are expressed in natural units $E_R = \hbar^2 k^2/2m$.

6.2 devoted to the experimental realization of this potential). The solid thick lines correspond to a representation of the energy spectrum in the so-called *extended-zone scheme*, in which different energy bands are plotted in different Brillouin zones. However, because the quasimomentum is defined modulo the periodicity $2k$ of the reciprocal lattice, the spectrum $E_n(q)$ can be more generally represented in the *repeated-zone scheme*, in which all the energy bands are plotted in each of the Brillouin zones (both thin and thick solid lines). For comparison, we plot in the same graph also the parabolic spectrum $E_0(q) = \hbar^2 q^2/2m$ of the particle in free space (dashed curve). Some general features of the band structure already emerge from this particular case:

1. At low energies ($E_n \ll V_0$) the bands are almost flat and, for increasing height of the periodic potential, asymptotically tend to the discrete eigenenergies of the harmonic oscillator potential obtained with a parabolic approximation of the single lattice site potential.
2. At high energies ($E_n \gg V_0$) the bands are pretty similar to the free particle spectrum (except for a zero-point energy shift) and differ from the latter only near the boundaries of the Brillouin zones.
3. Near the zone boundaries, in correspondence with the appearance of the energy gap, the energy spectrum $E_n(q)$ has null derivative.

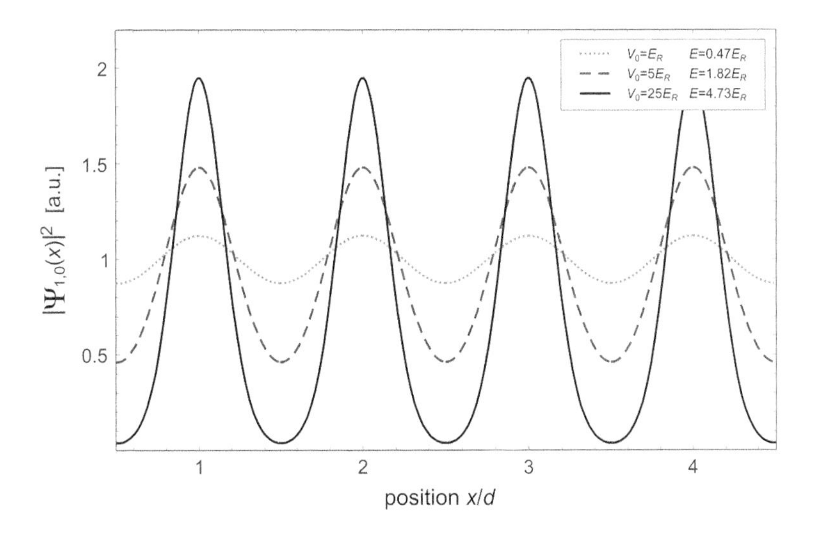

Fig. 6.2 Squared modulus of the ground-state wavefunction $\Psi_{1,0}(x)$ for different heights of the sinusoidal potential in eqn (6.2): $V_0 = E_R$ (dotted), $V_0 = 5E_R$ (dashed), $V_0 = 25E_R$ (solid). Increasing the lattice height, the wavefunction changes from a weakly modulated plane wave to a function that is strongly localized at the lattice sites. The energy of these states is reported in the inset.

Eigenfunctions. We now consider the eigenstates of eqn (6.3) in the case of the sinusoidal potential in eqn (6.2). In Fig. 6.2 we plot the squared modulus of the ground-state wavefunction $\Psi_{1,0}(x)$ for three different heights of the periodic potential. For small potential heights (dotted line) the Bloch states $\Psi_{1,q}(x)$ are similar to the plane waves e^{iqx}, except for a small amplitude modulation at the periodicity of the lattice. Increasing the lattice height (solid line), the energy of the states becomes much smaller than the maximum height of the potential and the Bloch wavefunctions turn out to be strongly modulated. In this *tight-binding* limit $E \ll V_0$ these functions can be more conveniently written as the sum of many wavefunctions located at the lattice sites:

$$\Psi_{n,q}(x) = \sum_{j=-\infty}^{\infty} e^{ijqd}\psi_n(x - jd) \, . \tag{6.6}$$

The functions $\psi_n(x)$ are the so-called *Wannier functions* and e^{ijqd} is a phase factor dependent on the quasimomentum q. Equation (6.6) states that the eigenstates of the system are coherent superpositions of Wannier functions centred at the lattice sites and with a well-defined phase link across the entire lattice. In the tight-binding regime the Wannier functions are strongly localized on individual lattice sites and, for increasing lattice height, asymptotically tend to the wavefunctions describing the eigenstates of the individual lattice wells. However, eqn (6.6) holds for any lattice height, even out of the tight-binding regime: as the lattice height is decreased, the Wannier functions become less localized, significantly spreading over multiple lattice sites.

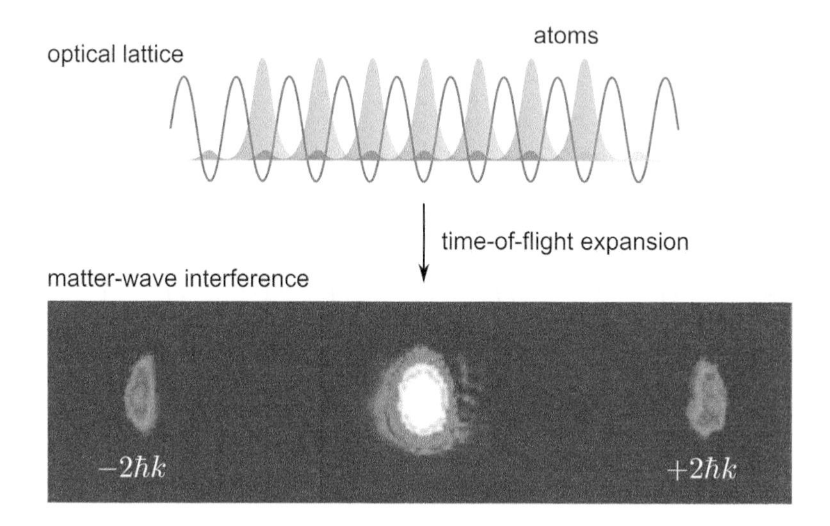

Fig. 6.3 Expansion of a Bose–Einstein condensate released from a 1D optical lattice. The time-of-flight image, corresponding to the atomic momentum distribution, shows a diffraction pattern emerging from the long-range coherence of the atomic wavefunction, which is delocalized across the lattice with a well-defined phase relation between lattice sites. Adapted from Pedri *et al.* (2001).

Using the Wannier function expansion in eqn (6.6), the wavefunction in momentum space can be derived as

$$\begin{aligned}\tilde{\Psi}_{n,q}(p) &= \mathrm{FT}\left[\Psi_{n,q}(x)\right](p/\hbar) = \sum_{j=-\infty}^{\infty} e^{ijqd}\,\mathrm{FT}\left[\psi_n(x-jd)\right](p/\hbar) \\ &= \sum_{j=-\infty}^{\infty} e^{ijqd}e^{-ijpd/\hbar}\,\mathrm{FT}\left[\psi_n(x)\right](p/\hbar) \\ &= \tilde{\psi}_n(p/\hbar)\sum_{n=-\infty}^{\infty}\delta(qd - pd/\hbar + 2n\pi)\ ,\end{aligned} \tag{6.7}$$

which corresponds to discrete components at momenta

$$p_n = \hbar(q + 2n\pi/d) \tag{6.8}$$

separated by multiples of the lattice momentum $2\hbar\pi/d = 2\hbar k$, with an overall envelope given by the Fourier transform of the Wannier function $\tilde{\psi}_n(p/\hbar)$.

Here we anticipate that the momentum distribution of the atom wavefunction in an optical lattice can be directly imaged in a time-of-flight experiment in which the trapping periodic potential is removed and the atoms evolve in free space before being detected (see also Sec. 3.2.1). Figure 6.3 shows the time-of-flight image of a Bose–Einstein condensate prepared in the lowest-energy state of an optical lattice (Pedri *et al.*, 2001). Discrete momentum components at $p = 0, -2\hbar k, +2\hbar k$ are clearly visibile. With an optical analogy, the emergence of this discrete momentum pattern can be interpreted in terms of diffraction of a coherent atomic wavefunction from an "optical" diffraction grating. The long-range atomic coherence, expressed in eqn (6.6) by a well-defined phase link between the wavefunctions $\psi_n(x - jd)$ in each lattice site, is analogous to the spatial coherence of a laser which illuminates a diffraction grating. In the latter case the interference between the light diffracted by different slits builds up in a diffraction pattern in the far field, where only light waves propagating at well-defined angles interfere constructively. Here matter-wave interference is constructive only if atoms propagate at the specific momenta given by eqn (6.8). As further discussed in Sec. 7.1.2, the detection of these diffraction peaks in time-of-flight images provides a measurement of the coherence of the atomic quantum state.

6.1.2 Dynamics of a Bloch wavepacket

Bloch waves represent the stationary flow of a quantum particle in a periodic potential in the absence of any perturbation, similarly to plane waves describing the flow of the particle in free space when no forces are applied. In this section we review the basic concepts describing the dynamics of a Bloch wavepacket in the presence of external fields. We consider a superposition of Bloch states with a mean quasimomentum q and a quasimomentum spread Δq much smaller than the width of the Brillouin zones.[3] It is possible to demonstrate that the group velocity of the wavepacket (also called "Bloch velocity") is

$$v_n(q) = \frac{1}{\hbar}\frac{\partial E_n(q)}{\partial q}\,, \tag{6.9}$$

directly proportional to the first derivative of the energy spectrum (Ashcroft and Mermin, 1976). We observe that, since the energy spectrum is flat at the zone boundaries, in these regions the Bloch velocity vanishes. We also note that eqn (6.9) has an immediate analogy in optics, as the group velocity $\partial\omega/\partial k$ of a packet of electromagnetic waves having a dispersion relation $\omega(k)$. It also reduces to the ordinary velocity $v = \hbar q/m$ in the case of the free-particle spectrum $E(q) = \hbar^2 q^2/2m$.

The simplest model describing the dynamics of a Bloch wavepacket in the presence of external fields is based on a semiclassical approach. We assume that the external forces do not modify the eigenstates and the energy spectrum of the system, and have only the effect of changing the mean position and the quasimomentum of the wavepacket. The main approximation used in this model is the assumption that the force F acting on the system is weak enough, if compared to the lattice forces ($F \ll V_0/d$), not to induce interband transitions (i.e. the band index n can be considered as a constant of the motion). Within these assumptions one can write

$$\dot{x} = \hbar^{-1}\frac{\partial E_n(q)}{\partial q} \tag{6.10}$$

$$\hbar\dot{q} = F\,. \tag{6.11}$$

[3] From the "uncertainty relation" for the Fourier transform, this means that the spatial extension of this wavepacket $\Delta x \sim 1/\Delta q$ is much larger than the lattice spacing, i.e. the wavefunction extends over many lattice sites.

Equation (6.10) is a restatement of eqn (6.9) in which x is now the expectation value of the coordinate operator, i.e. the average position of the wavepacket, while eqn (6.11) describes the change of the mean quasimomentum q as a result of the external fields. We observe that eqn (6.11) is formally analogous to Newton's second law of dynamics, but with two important differences: 1) on the left-hand side the quasimomentum q is present, instead of the true momentum p; 2) on the right-hand side only the external force F is present, and not the total forces acting on the system (actually, the forces due to the lattice potential are implicitly included in the energy spectrum E_n). Nevertheless, the similarity between these two equations is remarkable.

To get more insight into this semiclassical model we can use simple mathematics on eqns (6.10) and (6.11) obtaining

$$\ddot{x} = \frac{d}{dt}\left[\hbar^{-1}\frac{\partial E_n(q)}{\partial q}\right] = \hbar^{-1}\frac{\partial^2 E_n(q)}{\partial q^2}\dot{q} = \hbar^{-2}\frac{\partial^2 E_n(q)}{\partial q^2}F , \qquad (6.12)$$

which is a restatement of Newton's second law of dynamics for a particle subject to an external force F and having an *effective mass*

$$m_n^*(q) = \hbar^2\left[\frac{\partial^2 E_n(q)}{\partial q^2}\right]^{-1} , \qquad (6.13)$$

which is related to the local curvature of the energy bands and reduces to the real mass m in the case of the parabolic free-particle spectrum $E(q) = \hbar^2 q^2/2m$.

In Fig. 6.4 the Bloch velocity in eqn (6.9) and the effective mass in eqn (6.13) are plotted as a function of the quasimomentum q for the first three energy bands in the sinusoidal periodic potential of eqn (6.2) with height $V_0 = 4E_R$. It is worth noting that in the proximity of the zone boundaries the effective mass may become negative: this means that a force F acting on the wavepacket produces a centre-of-mass acceleration in the opposite direction. This is a result of the modification of the energy spectrum $E_n(q)$, which locally changes the sign of its curvature with respect to the free particle case (see Fig. 6.1). For comparison, in the same graphs also the (linear) velocity and the (constant) mass of the free particle (dashed curves) are plotted.

When the approximation of weak forces is relaxed and the acceleration becomes large, *Landau–Zener tunnelling* from the occupied energy band to an adjacent one may take place. The probability for this process to happen is

$$\Gamma \propto e^{-a_c/a} , \qquad (6.14)$$

where $a_c = d(\Delta E)^2/4\hbar^2$ is a critical acceleration dependent on the lattice spacing d and on the energy gap ΔE between the two bands.

6.1.3 Bloch oscillations

A paradigm of quantum transport in periodic potentials is represented by the effect of a constant force. This case, besides constituting the simplest possible example, is also relevant for its applications. As a matter of fact, when a constant voltage is applied to the opposite sides of a metal, the electrons inside the crystal experience a constant electric force in addition to the lattice potential.

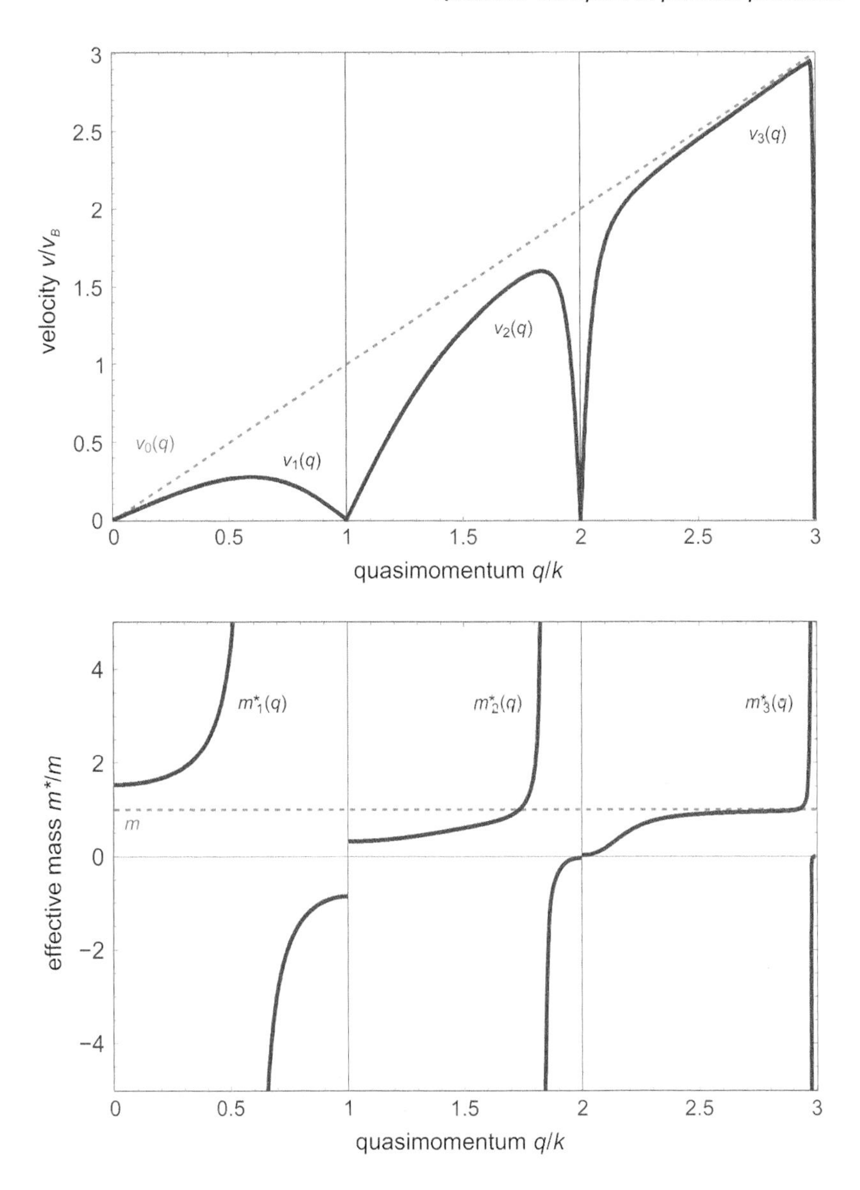

Fig. 6.4 Solid curves: Bloch velocity (top) and effective mass (bottom) in the first three energy bands for a particle of mass m in the periodic potential of eqn (6.2) with $V_0 = 4E_R$. Dashed curves: velocity and mass of the free particle. The velocities are expressed in natural units $v_B = \hbar k/m$ and the effective masses in units of real mass m. Note that the extended-zone scheme is used, in which only one band for each Brillouin zone is plotted.

The dynamics of a wavepacket under a constant external force F can be obtained from the semiclassical approach of Sec. 6.1.2. From eqn (6.11) the quasimomentum evolves linearly in time according to the equation

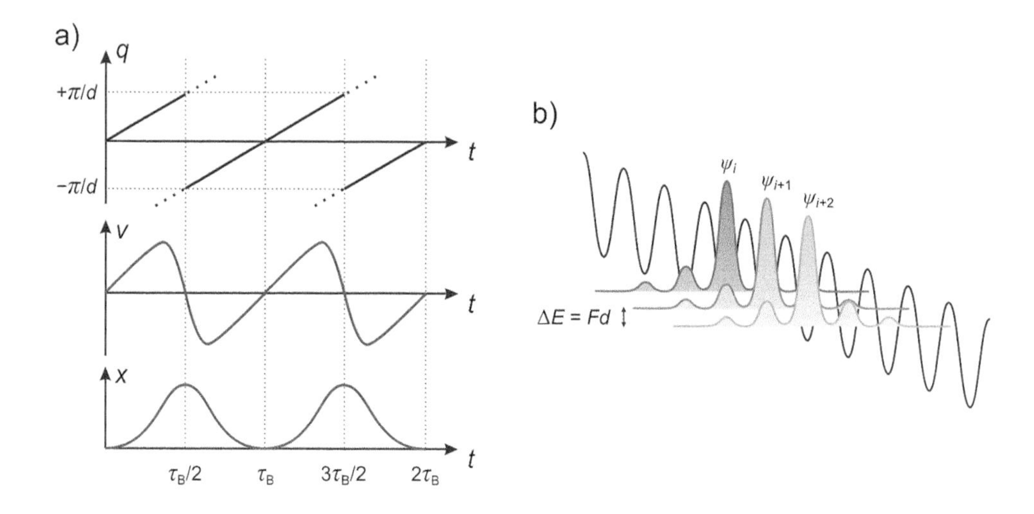

Fig. 6.5 Bloch oscillations in a periodic potential. a) A constant force F induces an uniform motion in quasimomentum space, which repeats periodically after a time τ_B and corresponds to an oscillation in real space. b) Wannier–Stark states in a tilted periodic potential. The energy of the states forms a comb of levels spaced by $\Delta E = Fd$.

$$q(t) = q_0 + \frac{F}{\hbar} t \,. \tag{6.15}$$

Since the quasimomentum is defined modulo a vector of the reciprocal space $2k = 2\pi/d$, this means that after a time

$$\tau_B = \frac{h}{Fd} \tag{6.16}$$

the quasimomentum has spanned a whole Brillouin zone and takes the same starting value q_0 modulo $2\pi/d$. During this time interval both the group velocity and the effective mass change sign, resulting in an oscillatory motion in which the wavepacket moves back and forth with zero average displacement, as shown in Fig. 6.5a. By using eqns (6.15) and (6.9), it is possible to demonstrate that after a time τ_B the average position of the wavepacket has not changed:

$$\Delta x = \int_0^{\tau_B} v(t)dt = \frac{\hbar}{F} \int_{q_0}^{q_0+\frac{2\pi}{d}} v(q)dq = \frac{1}{F} \left[E(q_0 + \frac{2\pi}{d}) - E(q_0) \right] = 0 \,, \tag{6.17}$$

due to the periodicity of the energy spectrum in reciprocal space. This effect, known as *Bloch oscillations*, can be conveniently described in the *reduced-zone scheme* in which the quasimomentum is only defined in the range $[-\pi/d, \pi/d]$. The oscillatory motion has two turning points: one at $q = 0$, where the velocity vanishes, and one at the Brillouin zone edge $q = \pi/d$, where also the velocity vanishes and a Bragg reflection "resets" the dynamics from $-\pi/d$.

There is another way to approach the problem. Instead of using the semiclassical equations for the dynamics, we can revert to a full quantum-mechanical treatment. The presence of a constant force F reduces the problem to that of a quantum particle in a "washboard" potential given by the superposition of the periodic potential and a linear potential $U(x) = -Fx$. The eigenstates of the Hamiltonian for this combined potential are an infinite set of *Wannier–Stark states*, which are localized on a finite number of sites and differ only by a translation by multiples of the lattice period, as sketched in Fig. 6.5b. Their energies $E_j = E_0 - j\Delta E$ differ by a uniform energy separation $\Delta E = Fd$, which amounts to the potential energy difference between nearest-neighbouring lattice minima. A generic wavepacket can be written as a linear combination of Wannier–Stark states. The evolution of the wavepacket is the result of the combined evolution of the Wannier–Stark states, the phases of which evolve according to $e^{-iE_j t/\hbar}$. It is easy to realize that under this time evolution all the Wannier–Stark states rephase after a time $\tau_B = h/\Delta E = h/Fd$, which amounts exactly to the period of the Bloch oscillation derived in eqn (6.16).

Differently from what happens in free space, where a constant force produces a continuous increase in the velocity of the particle (hence a nonzero flux of energy), in the presence of a lattice a constant force produces, on average, no drift. This pure quantum-mechanical effect is also in contrast with the observation of a continuous electric current when a constant electric field is applied within a crystal. In other words, an ideal metal would not support a direct current when a voltage is applied, because the electrons would oscillate very fast around their equilibrium position with period τ_B, instead of drifting from one side of the crystal to the other. An average nonzero conductivity arises in real crystals owing to the presence of dephasing and relaxation effects, which are important in ordinary solids and lead to typical coherence times of the electron dynamics shorter than the Bloch oscillation period.

Bloch oscillations in solid-state devices were reported for artificial semiconductor superlattices (Feldmann *et al.*, 1992; Leo *et al.*, 1992), where the longer spacing of the structures leads to shorter oscillation periods (see eqn (6.16)), which result in the emission of THz radiation by the oscillating electrons (Waschke *et al.*, 1993). More recently Bloch oscillations have been observed also for light propagating in photonic crystals (Sapienza *et al.*, 2003), i.e. dielectric devices in which the index of refraction is periodically modulated, resulting in a photonic bandgap analogous to the energy gap between different bands of the electronic spectrum. None of these systems, however, reaches the impressively long coherence times that have been achieved in experiments with ultracold atoms trapped in optical lattices, where thousands of Bloch oscillations can be observed without decay of the observed signal, as will be discussed in Sec. 6.4.2.

6.1.4 Josephson picture of the tight-binding limit

We conclude this theoretical introduction by presenting an alternative approach, which is based on a different formalism, more closely related to the physics of Josephson junctions. Following the definition of the Wannier functions in eqn (6.6), the time-dependent wavefunction describing a generic state $\Psi(x,t)$ in the lowest-energy band can be written as

$$\Psi(x,t) = \sum_{j=-\infty}^{\infty} a_j(t)\phi_j(x) , \tag{6.18}$$

where $\phi_j(x) = \psi_1(x - jd)$ is the Wannier state of the lowest-energy band associated with the j-th lattice site and

$$a_j(t) = \sqrt{n_j(t)} e^{i\varphi_j(t)} \tag{6.19}$$

is a complex function describing the amplitude $\sqrt{n_j}$ and the phase φ_j associated with the wavefunction centred in the j-th site. We consider the tight-binding limit, in which the Wannier functions are strongly localized at the lattice sites, with a small (but nonzero) overlap between functions $\phi_j(x)$ and $\phi_{j\pm1}(x)$ located at neighbouring sites. Substituting eqn (6.18) into the time-dependent version of the Schrödinger eqn (6.3)

$$i\hbar \frac{\partial \Psi(x,t)}{\partial t} = \left[-\frac{\hbar^2}{2m} \frac{\partial^2}{\partial x^2} + V(x) \right] \Psi(x,t) \tag{6.20}$$

we obtain a set of differential equations for the complex amplitudes $a_j(t)$:

$$i\hbar \frac{da_j}{dt} = -J\left(a_{j-1} + a_{j+1}\right) + E_0 a_j \quad \forall j \ , \tag{6.21}$$

in which we have defined an on-site energy

$$E_0 = \int \phi_j^*(x) \left[-\frac{\hbar^2}{2m} \frac{\partial^2}{\partial x^2} + V(x) \right] \phi_j(x) dx \tag{6.22}$$

and a *tunnelling energy*

$$J = -\int \phi_{j\pm1}^*(x) \left[-\frac{\hbar^2}{2m} \frac{\partial^2}{\partial x^2} + V(x) \right] \phi_j(x) dx \ , \tag{6.23}$$

calculated as an overlap integral between neighbouring Wannier wavefunctions. The latter quantity is related to the quantum-mechanical probability of a transition from a Wannier state to a neighbouring state via quantum tunnelling through the optical potential barrier. The tight-binding approximation is applied in eqn (6.21), where tunnelling is allowed only between nearest-neighbouring sites, neglecting all the other longer-distance couplings. We note that J is a function of the lattice height, since the states $\phi_j(x)$ implicitly depend on the shape of the lattice potential.

Exact solutions of eqn (6.21) are the usual Bloch waves, in which the complex amplitudes $a_j(t)$ take the form of plane waves $a_j(t) = e^{i(jqd - Et/\hbar)}$. The phase difference between neighbouring sites $\Delta\varphi = \varphi_{j+1} - \varphi_j = qd$ is constant across the entire lattice and depends on the quasimomentum state q. Substituting this Bloch ansatz in eqn (6.21) one finds an analytical expression for the shape of the lowest-energy band

$$E(q) = E_0 - 2J\cos(qd) \ , \tag{6.24}$$

which has a cosinusoidal dependence on the quasimomentum and is characterized by an energy width $4J$ depending only on the tunnelling energy.

Again, we can study the dynamics of a wavepacket in the presence of external forces. Following the definitions in eqns (6.9) and (6.13), the group velocity and the effective mass turn out to be

$$v(q) = \frac{2Jd}{\hbar}\sin(qd) \tag{6.25}$$

$$m^*(q) = m\left(\frac{E_R}{J}\right)\frac{1}{\pi^2\cos(qd)}\,, \tag{6.26}$$

both dependent on the tunnelling rate J, which is the relevant quantity for the description of the dynamics.

Since the group velocity of the wavepacket is proportional to the quantum-mechanical current i, eqn (6.25) can be rewritten as

$$i = i_c \sin\Delta\varphi\,, \tag{6.27}$$

which has the form of the current/phase relation holding for Josephson junctions, $i_c \propto 2Jd/\hbar$ being a critical current that is directly proportional to the tunnelling rate J. Indeed, in this tight-binding regime the system forms an array of Josephson junctions, i.e. a chain of quantum systems weakly linked by a tunnelling mechanism which preserves a phase coherence across the entire array. The current flowing in the array depends on the phase difference $\Delta\varphi$ between adjacent sites, as in the case of a weak insulating junction between two superconductors, in which the electronic current due to the quantum tunnelling through the barrier is proportional to the sine of the phase difference between the macroscopic wavefunctions in the two superconductors.[4]

The phase dynamics is sensitive to the presence of forces driving the system out of equilibrium. If we apply a constant force $F = -\Delta U/d$ across the array, with ΔU difference of potential between adjacent sites, eqn (6.11) becomes

$$\hbar\frac{d}{dt}\Delta\varphi = -\Delta U\,. \tag{6.28}$$

This equation shows that a constant ΔU results in a phase difference $\Delta\varphi$ growing linearly in time, hence, using eqn (6.27), an ac-current oscillating at a frequency proportional to ΔU. This phenomenon, at the basis of the *Josephson effect*, is the transposition, in the language of current-phase variables, of the Bloch oscillations discussed in Sec. 6.1.3.

We note that this tight-binding Josephson approach is particularly suitable in the case of large lattice heights, when the total wavefunction can be expanded as the sum of many localized wavefunctions. In the case of small lattice heights, when the wavefunction is just weakly modulated and we cannot neglect the overlap integral of Wannier functions located at more distant sites, the nearest-neighbour tunnelling rate J is not sufficient to describe the system. In this case other energy scales J', J'', etc. describing longer-range couplings should be included in eqn (6.21) and the simple form of the energy spectrum in eqn (6.24) will get more complicated.

[4] We note that eqn (6.27) takes the form of a pendulum equation with the phase difference $\Delta\varphi$ corresponding to the angle of the pendulum with respect to the vertical axis.

6.2 Optical lattices

A periodic potential for cold neutral atoms can be easily produced by using laser light. As discussed in Sec. 5.2, the mechanical effects of the non-resonant interaction between radiation and matter can be described in terms of a conservative optical dipole potential. In Sec. 5.2.2 we have shown that a periodic potential is obtained whenever two phase-locked laser beams with the same optical frequency cross and interfere, generating a periodic modulation of the light intensity. The simplest and most common experimental setting is provided by two counterpropagating beams, which form a standing wave with an intensity modulation of period $d = \lambda/2$. In this counterpropagating configuration, the resulting optical dipole potential can be written in the form

$$V(x) = V_0 \cos^2(kx) , \tag{6.29}$$

already introduced in eqn (6.2), where $k = 2\pi/\lambda$ is the laser wavenumber and $V_0 = 6\pi c^2 \gamma I_0/\omega_0^3 \delta$ is the depth of the periodic potential, I_0 being the intensity of each of the laser beams, ω_0 the atomic resonance frequency, γ the transition linewidth, and $\delta = \omega - \omega_0$ the detuning of the laser frequency ω from the resonance. The laser wavenumber k sets the lattice spacing $d = \pi/k$ and the position of the Brillouin zone edges, which are located at $\pm k, \pm 2k$, ... The height of the optical lattice V_0 is often measured in units of recoil energies $E_R = \hbar^2 k^2/2m = h^2/8md^2$, already introduced in Sec. 6.1.1 and physically corresponding to the kinetic energy an atom at rest acquires after absorption of a photon from one of the lattice beams.

This optical way to produce periodic potentials for neutral atoms offers several unique features. The first is the extreme *tunability* of the potential: the height of the optical barriers V_0 can be manipulated by changing laser frequency and detuning from the atomic resonance, while the lattice spacing d can be tuned by changing the wavelength or the angle formed by the beams. A second important feature is the intrinsic *periodicity* of the pattern: approximating the laser beams with plane monochromatic waves, the intensity modulation turns out to be exactly periodic, with no inhomogeneities or defects. A third unique feature is offered by the possibility of changing the potential parameters in *real-time*, simply by amplitude or frequency modulation of the laser light.

As an example to illustrate this latter possibility, we consider the situation in which the two laser beams have slightly different frequencies ω and $\omega + \delta\omega$. It is easy to demonstrate that the resulting interference pattern (averaged on the fast optical oscillations) is time-dependent, giving a slowly varying potential of the form

$$V(x) = V_0 \cos^2(kx - \frac{\delta\omega}{2} t) , \tag{6.30}$$

representing a standing wave with nodes and antinodes moving in the laboratory frame at a velocity[5]

$$v_L = \frac{\delta\omega}{2k} . \tag{6.31}$$

[5]This can be derived quite easily considering the fact that an atom with velocity $-v_L$ experiences two Doppler-shifted laser frequencies both equal to $\omega + \delta\omega/2$, and in its reference frame their interference is a stationary standing wave.

This means that, by controlling the relative frequency of the two lattice beams, it is possible to produce a periodic potential which can be set in motion and accelerated. Later in this chapter we will show that this is a particularly convenient way to accurately control the relative motion between the atoms and the optical lattice, allowing a precise investigation of band-structure properties.

We note that an important feature of optical periodic potentials is the absence of phonons. In a real crystal the atoms are arranged "on average" according to the lattice structure (because this is the configuration which minimizes the total energy), but at finite temperature they can oscillate around this equilibrium configuration. The coupled oscillations of the atoms in the crystalline lattice result in a vibration of the whole crystal, which can be quantum-mechanically described with phonons. The resulting vibration of the lattice potential then affects the motion of the electrons and this can be modelled with an effective electron–phonon interaction. In an optical lattice, instead, vibrations of the lattice can be neglected (as far as instrumental sources of noise, such as the laser phase noise and mirror vibrations, are suppressed or can be neglected on the timescale of the experiments) and cold atoms can be used to study the physics of a gas of particles in an "ideal" crystal.

6.3 Experiments with cold atoms

In this section we will review some of the first experimental work in which quantum transport of cold atoms in optical lattices has been studied and we will discuss some applications of these systems for precise measurements.

The use of optical lattices for manipulating the motion of atoms dates back to the late 1970s/early 1980s, when first pioneering experiments studied diffraction of an atomic beam from a laser light standing wave (Arimondo *et al.*, 1979; Grinchuk *et al.*, 1981; Moskowitz *et al.*, 1983). Later, the observation of discrete momentum kicks at even multiples of the photon momentum was clearly reported in Gould *et al.* (1986) and Martin *et al.* (1988). This phenomenon is analogous to the diffraction of light from a grating, but the roles of matter and light are swapped: the light forms the periodic structure which diffracts a collimated beam of atoms in multiple beams with different transverse momentum $p_\perp = 2n\hbar k$, where n is an integer number (see also Fig. 6.3 for the diffraction of a BEC from an optical lattice). This diffraction later provided a mechanism for building atom interferometers based on optical splitting and recombination of an atomic beam (Rasel *et al.*, 1995; Giltner *et al.*, 1995). As an example, we show in Fig. 6.6 the scheme of the setup used in Rasel *et al.* (1995) for the realization of a multiple-channel Mach–Zehnder interferometer. A first standing wave of light diffracts a beam of metastable Ar* atoms into three beams, which are again split by a second standing wave and then recombined by a third one. The splitting process is coherent and interference fringes in the atom population are clearly observed at the two output ports of the interferometer as a function of the position of the third grating.

In the early 1990s, different research groups observed first evidence of the quantized motion of laser-cooled atoms in optical standing waves of light (Verkerk *et al.*, 1992; Jessen *et al.*, 1992; Hemmerich and Hänsch, 1993). A few years later the possibility of directly observing quantum transport of cold atoms in optical lattices was clearly

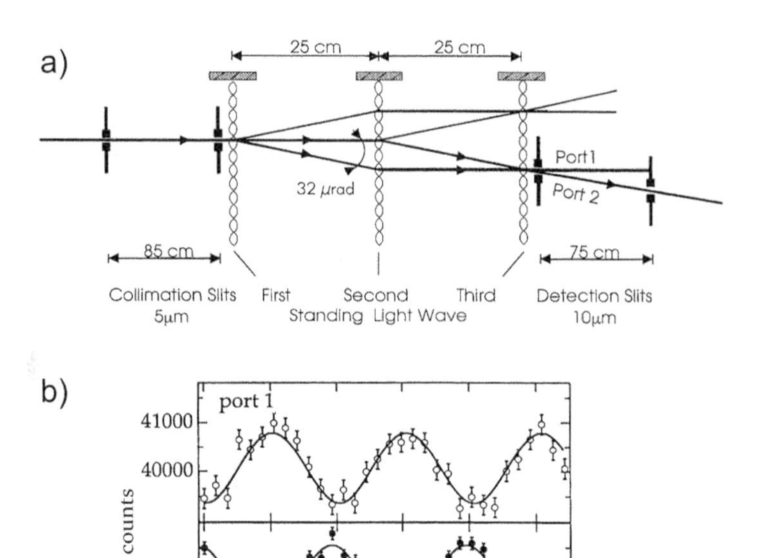

Fig. 6.6 Scheme of a multiple-channel Mach–Zehnder interferometer using optical lattices to diffract and recombine atomic beams. The interference fringes exhibited by the population at the output ports is plotted as a function of the third lattice position. Reprinted with permission from Rasel *et al.* (1995). © American Physical Society.

demonstrated in a series of beautiful experiments (Raizen *et al.*, 1997). The possibility of studying band-structure dynamics for cold atoms in optical lattices is related to the ability of decreasing the temperature of the atomic samples well below the characteristic energy scale E_R set by the lattice. As a matter of fact, in order to resolve the dynamics of the atoms within the energy bands, the Brillouin zone has to be filled by particles only partially. This means that the momentum spread of the atomic cloud Δp along the lattice direction has to be smaller than the (half-)width of the Brillouin zone:

$$\Delta p < \hbar k \, . \tag{6.32}$$

Considering the Maxwellian distribution of velocities for a 1D gas of free particles $\Delta p = \sqrt{k_B T m}$ the above condition reads

$$T < \frac{\hbar^2 k^2}{k_B m} \, , \tag{6.33}$$

which is equivalent to requiring the thermal energy of the sample to be smaller than the recoil energy $E_R = \hbar^2 k^2 / 2m$. The same condition is also equivalent to requiring the thermal de Broglie wavelength of the sample to be larger than the lattice constant, i.e. the range of distances across which the particles exhibit a coherent wave-like behaviour should extend over many lattice sites.

These temperatures are remarkably low ($\sim$ 100 nK for ^{87}Rb) and sophisticated sub-recoil laser-cooling techniques had to be implemented before it was possible to cool the atoms well within the first Brillouin zone of the lattice. Important achievements were obtained with the development of *velocity-selective coherent population trapping*, which allowed the first demonstration of sub-recoil temperatures (Aspect *et al.*, 1988), and *Raman cooling* (Kasevich and Chu, 1992), which can be used to reach ultra-low temperatures by selectively cooling different classes of atomic velocities on a narrow two-photon Raman transition connecting different hyperfine states of an alkali atom.

6.3.1 Observation of Bloch oscillations

As already discussed in Sec. 6.1.3, an important consequence of the energy band structure is the existence of Bloch oscillations. In systems of ultracold atoms in optical lattices the constant force driving the Bloch dynamics could be either provided by a real force (e.g. gravity, as we will consider in the following sections) or by a constant acceleration of the lattice. In the first experiments with cold atoms the latter possibility was used, and the acceleration was induced by linearly chirping the relative detuning $\delta\omega$ of the two lattice beams, according to eqn (6.31). In a reference frame which is co-moving with the lattice, the effect of the acceleration a is indeed replaced by a constant inertial force $F = -ma$.

Bloch oscillations with cold atoms were first observed by the group of C. Salomon at ENS in Paris (Ben Dahan *et al.*, 1996). Clouds of caesium atoms were cooled at a very low temperature around 10 nK by a combination of magneto-optical trapping and Raman cooling, resulting in a momentum spread $\sim \hbar k/4$, i.e. one eighth of the Brillouin zone width. The cloud was then trapped in the optical lattice and the latter was accelerated by linearly increasing the frequency difference $\delta\omega$ between the two lattice beams. The momentum distribution of the atoms in the accelerated frame of the lattice was then measured by exciting the atoms on a narrow velocity-selective Raman transition (see also Sec. 4.5.2) and detecting the excitation Doppler profile. Figure 6.7a shows different velocity distributions measured at intervals $\tau_B/8$, where τ_B is the period of the Bloch oscillations. The data clearly show that the peak at zero velocity starts moving towards higher velocities as the inertial force $F = -ma$ drives the Bloch dynamics within the lowest-energy band according to eqn (6.11). At time $\tau_B/2$ the cloud undergoes a Bragg reflection at the edge of the Brillouin zone $\hbar k$, after which it accelerates again starting from $-\hbar k$. This represents a beautiful direct demonstration of Bloch oscillations, which can be observed for several cycles without decay of the observed signal.

Figure 6.7b shows the average atomic velocity as a function of time, which can be compared with the theoretical prediction shown in the second plot of Fig. 6.5a. The three different plots in Fig. 6.7b refer to Bloch oscillations for the same acceleration, but different heights of the optical lattice: the observation of the same oscillation frequency in the different plots provides experimental evidence that the period of Bloch oscillations does not depend on the height of the periodic potential, but only on the

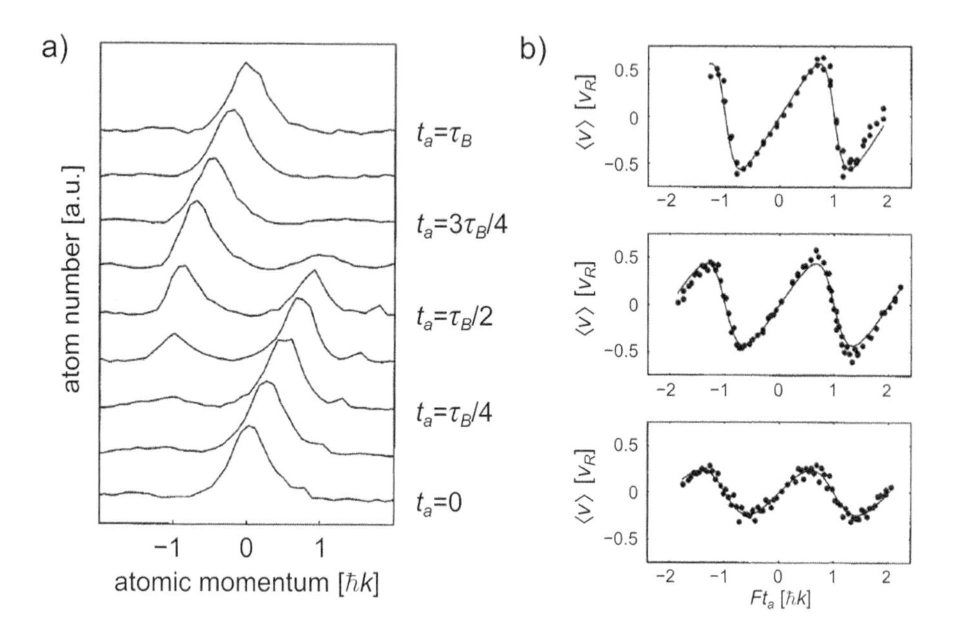

Fig. 6.7 a) Velocity distribution of ultracold caesium atoms performing Bloch oscillations in an accelerated lattice. At time $\tau_B/2$ the atomic cloud undergoes Bragg reflection at the edge of the Brillouin zone. b) Mean velocity in the frame of the lattice for the same acceleration but different lattice depths (increasing from top to bottom). Reprinted with permission from Ben Dahan *et al.* (1996). © American Physical Society.

lattice spacing and on the external force, as expected from eqn (6.16). As we shall see in the following sections, this property has important consequences for using Bloch oscillations for precise measurements of forces and fundamental constants.

6.3.2 Measurement of h/m with optical lattices

When the optical lattice is moved with an acceleration a, the Bloch oscillation period in eqn (6.16) can be written as

$$\tau_B = \left(\frac{h}{m}\right)\frac{2}{a\lambda}\,. \tag{6.34}$$

Since the acceleration and the lattice spacing can be determined with very high precision, in principle a measurement of the Bloch oscillation period τ_B can provide an accurate determination of the ratio h/m. The latter value has a remarkable metrological importance since it can be used to determine the fine structure constant α, as we have already pointed out in Sec. 4.5.2, where we have discussed the measurement of h/m with atomic fountain interferometers. Here we recall that knowing the value of h/m for a specific atom, for example ^{87}Rb, allows a determination of α via the relation

$$\alpha^2 = \frac{2R_\infty}{c}\frac{m(^{87}\mathrm{Rb})}{m_e}\frac{h}{m(^{87}\mathrm{Rb})}\,, \tag{6.35}$$

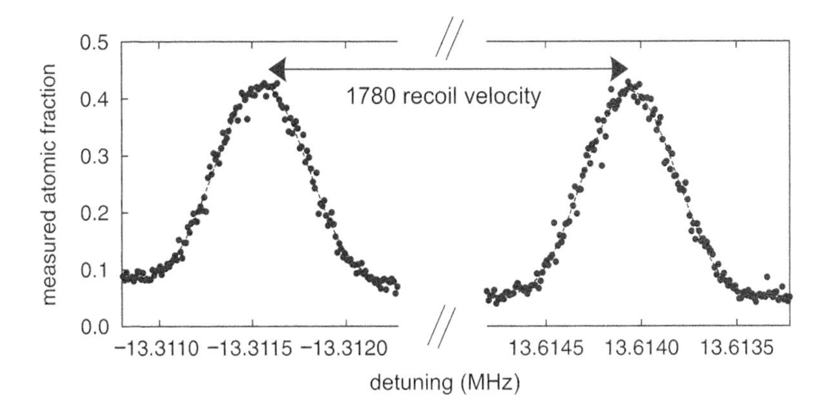

Fig. 6.8 Measurement of h/m with Bloch oscillations. The figure shows the spectrum of the $F = 1 \to F' = 2$ Raman transition in ultracold ^{87}Rb clouds accelerated by an optical lattice. The two Doppler-shifted peaks refer to two different signs of the acceleration and the frequency difference between them provides a measurement of the atomic recoil velocity. Reprinted with permission from Cladé *et al.* (2006). © American Physical Society.

where the Rydberg constant R_∞ (see Sec. 1.4.3) and the mass ratio $m(^{87}\mathrm{Rb})/m_e$ (Bradley *et al.*, 1999; Mount *et al.*, 2010) are known with remarkably small uncertainties.

Very precise measurements of the ^{87}Rb h/m ratio were performed at LKB-ENS in Paris (Cladé *et al.*, 2006; Bouchendira *et al.*, 2011). In these experiments Bloch oscillations in an accelerated lattice were used as a tool to impart a precisely known number of recoil kicks to the atoms (Battesti *et al.*, 2004). While in the frame of the accelerated lattice the atoms exhibit an oscillatory motion at the Bloch period, in the laboratory frame they are dragged by the lattice and follow "in average" its accelerated motion.[6] In the laboratory frame the atoms acquire a momentum $2\hbar k$ per Bloch oscillation, therefore, after an acceleration time $T = N\tau_B$ (corresponding to a well-known number N of Bloch periods), the total momentum transfer is $\Delta p = 2N\hbar k$ and the atom velocity changes by

$$\Delta v = 2N\frac{\hbar k}{m} . \tag{6.36}$$

The advantage of this technique over the measurement presented in Sec. 4.5.2 is the very large momentum transfer efficiency,[7] which allows $N \approx 10^3$ recoil kicks to be transferred to the atoms, with only minor atom losses.

After the acceleration phase, the recoil velocity Δv can be determined with a spectroscopic or an interferometric measurement. Figure 6.8, taken from Cladé *et al.*

[6]For the experimental parameters used in Cladé *et al.* (2006) (very large optical lattice depth $V_0 = 70E_R$ and large acceleration $a = 2000$ m/s^2) the amplitude of the Bloch oscillation in coordinate space is completely negligible, only $\approx 10^{-13}$ m, while the distance travelled by the lattice sites during the acceleration phase is in the cm range.

[7]Provided that the lattice is deep enough, it is possible to suppress Landau–Zener transitions to higher bands, which would limit the efficiency of the momentum transfer protocol.

(2006), shows the spectrum of the ^{87}Rb $F = 1 \rightarrow F' = 2$ two-photon Raman transition for two different signs of the acceleration, hence two different final velocities, that appear as two Doppler-shifted peaks. Measuring the distance between the two peaks allows the determination of the total recoil velocity Δv and, complemented with a precise measurement of the laser wavevector (made possible by a frequency comb), according to eqn (6.36), it provides a value of the h/m ratio. Using this technique, a determination of the fine structure constant α with a fractional uncertainty of 6.7 ppb was obtained (Cladé *et al.*, 2006).

In more recent developments of the Paris experiment, a Ramsey–Bordé interferometric detection has been implemented in order to achieve a higher resolution on the determination of the recoil velocity (Cadoret *et al.*, 2008). This new detection scheme, together with other improvements of the experimental protocol and a better control of the systematic effects, allowed the determination of a new value for the fine structure constant

$$\alpha^{-1} = 137.035\,999\,037\,(91) \tag{6.37}$$

with a fractional uncertainty of only 0.66 ppb (the value is indicated with "h/m(Rb) LKB-11" in Fig. 4.10) (Bouchendira *et al.*, 2011). To date, this measurement constitutes the most precise determination of the fine structure constant with atomic physics techniques. This value is in excellent agreement with the most precise 0.25 ppb α determination obtained from the electron gyromagnetic anomaly measurement (Hanneke *et al.*, 2008; Aoyama *et al.*, 2012) (see Sec. 4.5.1). Since the latter determination is based on advanced QED calculations of the electron magnetic anomaly, while the h/m method does not depend on QED, the agreement between these two results provides a very stringent test of the validity of QED.

6.3.3 Large-area atom interferometers

In atom interferometry, achieving a large separation between the trajectories is a major requirement in order to increase the resolution of the measurements. Increasing the interaction time is one possibility, however limitations arise due to the finite size of the experimental setup and to unwanted systematic effects, such as vibrations or laser phase noise, which could limit the accuracy of the interferometer at longer times. In order to increase the interferometer area it would be much more desirable to keep the same interaction time and devise beam splitters capable of transferring a large momentum to the atoms. Beam splitters based on single Raman pulses (see Secs. 2.5.1 and 4.5.2) allow a momentum transfer of $2\hbar k$. Recently, experimental work has succeeded in building atom interferometers based on more efficient optical splitting techniques capable of transferring a momentum as large as $\simeq 100\hbar k$. This increase in the interferometer area opens the possibility of performing more precise measurements and detecting tiny effects which require an increased sensitivity.

The large momentum that can be transferred to the atoms with Bloch oscillations in an accelerated lattice (discussed in the previous section) has been used in Cladé *et al.* (2009) to realize an interferometer with a separation of $10\hbar k$ between the two interferometer arms. In this beam splitter an initial Raman pulse splits the atom wavefunction into two wavepackets with momentum $p_A = p$ and $p_B = p + 2\hbar k$. When the optical lattice is turned on, the two wavepackets occupy two different lattice bands,

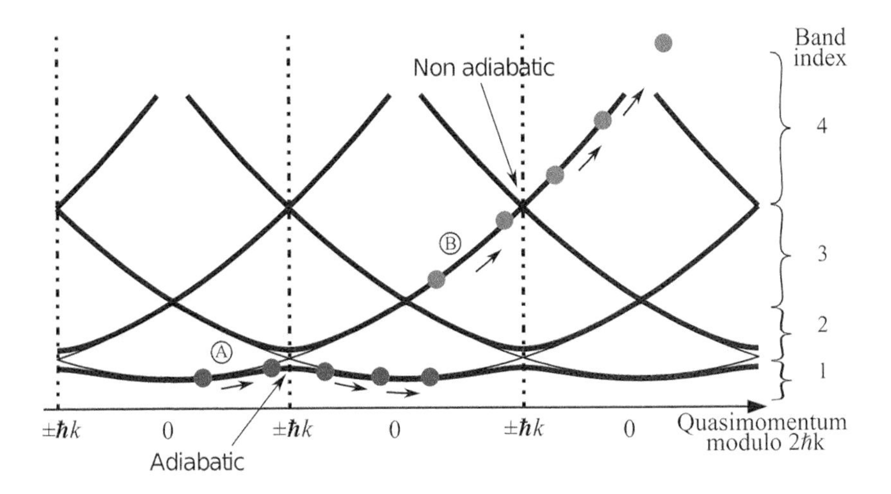

Fig. 6.9 Principle of operation of a Bloch oscillation-based large-momentum beam splitter (see text). Reprinted with permission from Cladé *et al.* (2009). © American Physical Society.

as shown by the A and B states in Fig. 6.9. As the lattice is accelerated, state A performs Bloch oscillations in the lowest lattice band as described in the previous sections, which correspond to an acceleration in the laboratory frame. Since the energy gap between bands decreases with the band index, for an appropriate choice of the lattice depth, it is possible to have a very large probability for state B to perform a non-adiabatic Landau–Zener tunnelling to higher bands (see Sec. 6.1.2), which corresponds to an average absence of motion in the laboratory frame. This mechanism allows the acceleration of only one component of the atomic wavepacket, thus realizing a large-momentum-transfer atomic beam splitter. The main limitation to this technique is represented by the large light shift experienced by the atoms in the lattice (which is dependent on the band index), which produces a systematic effect in the interferometer operation. This effect has been limited in Cladé *et al.* (2009) by designing a Mach–Zehnder interferometer in which the phase shift is applied successively to both interferometer arms, thus making it a common-mode contribution which does not affect the interferometer outcome. The final limitation comes from laser intensity noise and wavefront distortions, which make the cancellation not perfect and limit the usable momentum transfer to $10\hbar k$.

This technique has been used also in Müller *et al.* (2009), where the Bloch oscillation technique was used in combination with Bragg transitions[8] to achieve single beam splitters with $88\hbar k$ and complete interferometers with $24\hbar k$ momentum splittings and 15% fringe contrast. The ac-Stark effect harming the Bloch oscillation technique was further minimized by the simultaneous use of two overlapped optical lattices which were used to accelerate the two atomic wavepackets at the same time in opposite directions.

[8]We recall that Bragg transitions are two-photon Raman transitions in which the internal state of the atom does not change (see also Secs. 3.2.2 and 6.4.1). The resonance condition for the energy difference between the photons is only given by the difference in kinetic energy between the final and initial states.

The largest momentum splitting achieved to date in an atom interferometer has been reported in Chiow *et al.* (2011). Instead of an optical lattice, multiple n-photon Bragg transitions[9] were used to realize a beam splitter with a momentum transfer of $102\hbar k$. This $\sim 50\times$ improvement on the momentum transfer over the traditional $2\hbar k$ interferometers was accompanied by a reduction of fringe contrast (due to the non-ideality of the Bragg pulses) by only a factor ~ 5 with respect to a full-contrast interferogram, thus resulting in a substantial improvement of the interferometer sensitivity.

6.4 Experiments with quantum gases

The achievement of Bose–Einstein condensation (BEC) in 1995 was welcomed as the realization of a coherent source of matter waves with very high brilliance. As we have shown in Chapter 3, this comes from the fact that a BEC can be described by a macroscopic coherent wavefunction, with a momentum spread much smaller than that of a thermal gas at the same temperature. For this reason one can expect that Bloch oscillations and, more generally, all the phenomena of quantum transport in an optical lattice, could be observed in BECs with much higher contrast and much longer coherence times than in the case of dilute thermal gases.

However, if compared to a thermal gas, a BEC is typically much denser. As a consequence, the atoms of the gas interact more strongly and these atom–atom interactions turn the single-particle physics of the Bloch bands into a more complex many-body problem, where novel physical effects can be studied. For not too large atomic densities a mean-field approach can be used, and interactions are taken into account with a nonlinear term in the wave equation for the BEC wavefunction $\Psi(\mathbf{r},t)$, resulting in the Gross–Pitaevskii equation already introduced in Sec. 3.2.1 and further discussed in Appendix C.2

$$i\hbar\frac{\partial}{\partial t}\Psi(\mathbf{r},t) = \left[-\frac{\hbar^2\nabla^2}{2m} + V(\mathbf{r}) + \frac{4\pi\hbar^2 a}{m}|\Psi(\mathbf{r},t)|^2\right]\Psi(\mathbf{r},t)\,, \tag{6.38}$$

where the strength of interactions is described by the scattering length a, positive for repulsive interactions, negative for attractive interactions (see Sec. 3.1.4). The nonlinear term in Ψ, being analogous to the nonlinear Kerr effect for light propagation (discussed in Appendix B.3), is responsible for the self-nonlinear behaviour of matter waves and is at the basis of many intriguing phenomena well known in optics, such as solitonic propagation (Burger *et al.*, 1999; Khaykovich *et al.*, 2002; Al Khawaja *et al.*, 2002) and four-wave mixing (Deng *et al.*, 1999). In the context of optical lattices, the presence of such nonlinearities leads to a number of new effects deviating from the single-particle Bloch theory, including the observation of different kinds of instabilities (Smerzi *et al.*, 2002; Wu and Niu, 2001).

Different possibilities can also be explored by using fermions instead of bosons. The physics of degenerate fermionic gases is dominated by the Pauli exclusion principle, which forbids multiple occupancy of the same quantum state by two or more identical

[9] Higher-order Bragg transitions can be induced by absorption of n photons from one laser beam and stimulated emission of n photons into the second beam. This process is accompanied by a momentum transfer $2n\hbar k$ (when the transition is driven by counterpropagating laser beams).

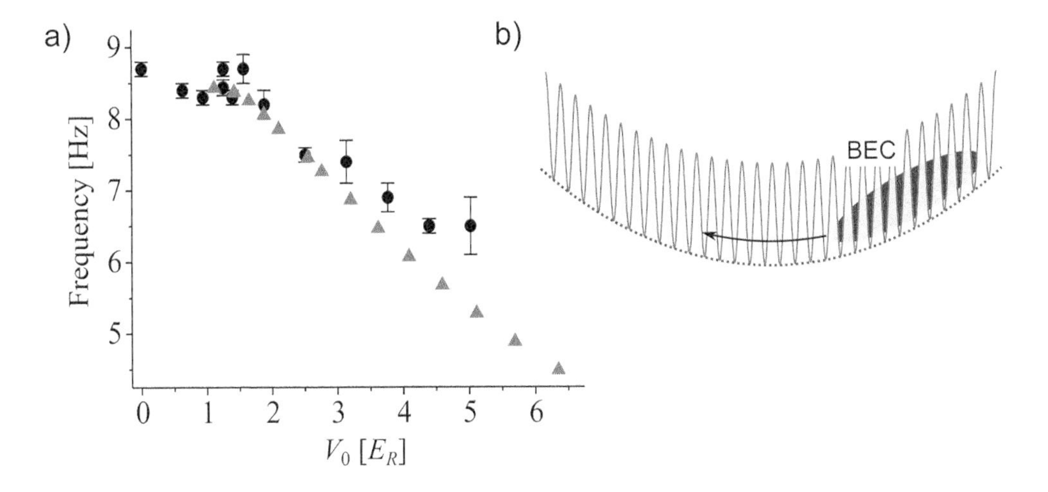

Fig. 6.10 Oscillations of a Bose–Einstein condensate in an optical lattice driven by a harmonic oscillator force. a) Measured oscillation frequency as a function of lattice depth (points), compared with calculated values (triangles). The bare oscillation frequency in the harmonic potential is 8.7 Hz. b) Sketch of the experiment. Adapted from Cataliotti *et al.* (2001).

fermions. An important consequence of the antisymmetrization of the wavefunction, already discussed in Sec 3.1.4, is the absence of ultracold collisions between spin-polarized identical fermions, which thus behave as noninteracting particles. This property has important consequences for interferometric and metrological applications, as will be further discussed in the following sections.

6.4.1 Dynamics of a BEC in a periodic potential

Josephson dynamics. The dynamics of Bose–Einstein condensates in a 1D optical lattice was investigated at LENS in Cataliotti *et al.* (2001). In this experiment a ^{87}Rb BEC was trapped in an optical lattice in the tight-binding regime, where the BEC wavefunction is strongly modulated by the presence of the lattice and can be described as a phase-coherent array of bosonic Josephson junctions, following the approach presented in Sec. 6.1.4. The dynamics of this system was studied by taking advantage of a weak magnetic harmonic potential superimposed on the optical lattice (see Fig. 6.10b). A sudden shift of the harmonic trap centre induced small-amplitude centre-of-mass oscillations of the BEC, that were studied as a function of time. The oscillation was undamped and its frequency was found to be downshifted with respect to the bare oscillation frequency in the harmonic trap: this can be explained as a consequence of the finite tunnelling times through the optical barriers, leading to a higher effective mass, as described by eqn (6.26), and a slower dynamics. Figure 6.10a shows the measured oscillation frequencies as a function of the lattice height V_0 (points) together with the theoretical frequencies (triangles) obtained from a tight-binding description in terms of the current-phase dynamics in a Josephson junction array.

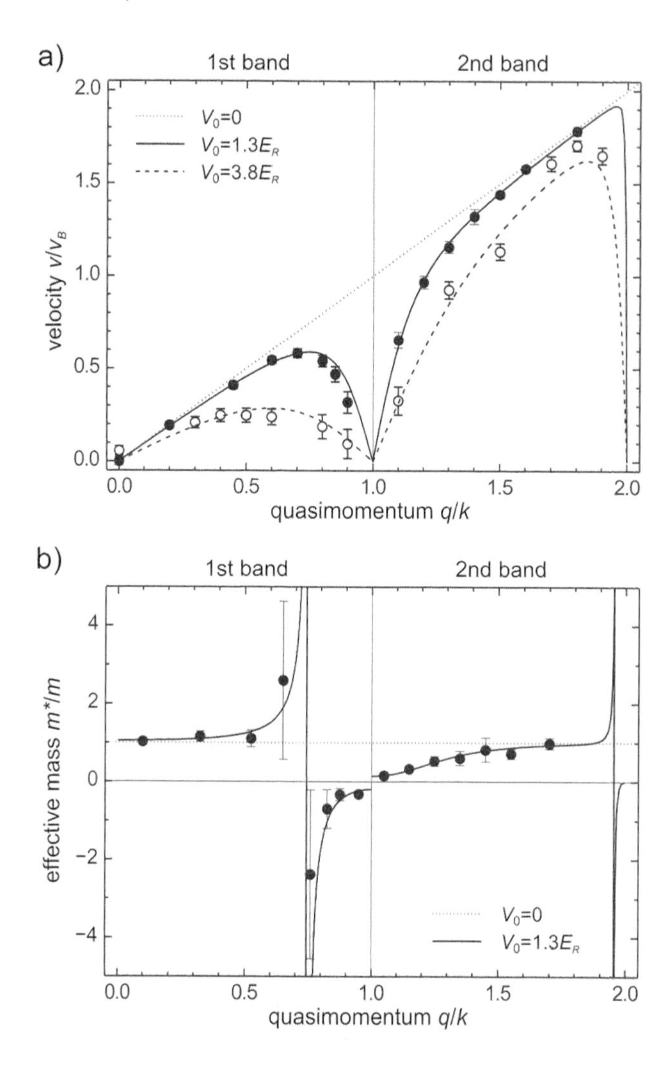

Fig. 6.11 Band spectroscopy of a BEC in an optical lattice. The wavepacket velocity and the effective mass are plotted as a function of the quasimomentum for the first two energy bands. The lines are obtained from a a calculation of the single-particle Bloch spectrum (see also Fig. 6.4). Adapted from Fallani *et al.* (2003).

Probing the band structure. The band structure of a Bose–Einstein condensate in an optical lattice can be probed by addressing single Bloch states and studying their kinematics in the periodic potential (Fallani *et al.*, 2003). In order to selectively load the BEC in a well-defined Bloch state[10] it is convenient to use a moving optical lattice,

[10]Because of its finite size (due to the presence of a trapping potential) and the role of interactions, a trapped BEC can be described as a Bloch wavepacket occupying several Bloch states with a nonzero quasimomentum spread Δq, which however is typically much smaller than the width of the Brillouin zones $2k$.

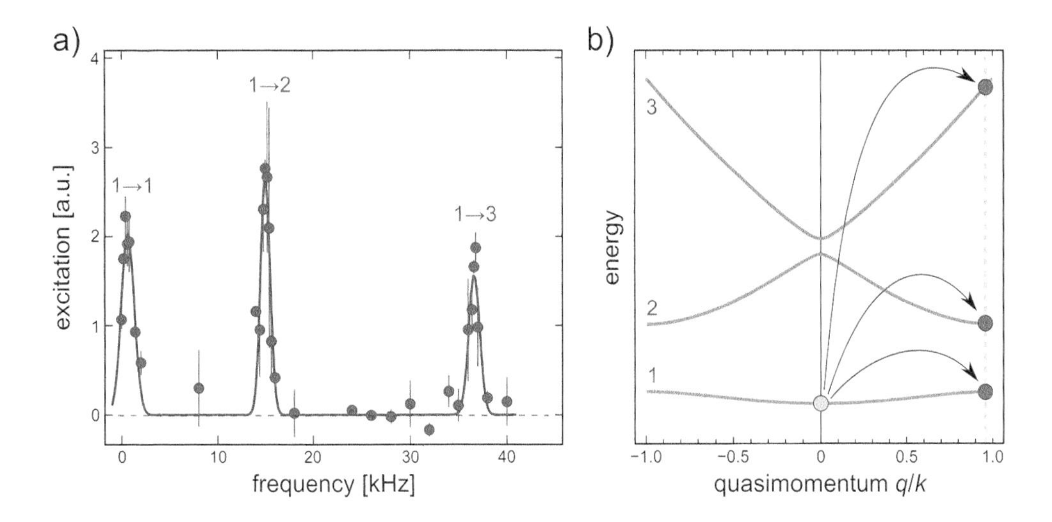

Fig. 6.12 Bragg spectroscopy of a BEC in an optical lattice. a) Energy absorbed by the BEC after a two-photon Bragg excitation with momentum transfer $\hbar\kappa = 0.96\hbar k$, as a function of Bragg excitation frequency. The three peaks refer to excitations in three different bands. b) Scheme of the transitions. Adapted from Clément *et al.* (2009*b*).

which can be produced by detuning the two lattice beams as described in Sec. 6.2. If the lattice beams are switched on slowly enough, the free-space energy spectrum is slowly modified into the band structure and the condensate is adiabatically projected into a Bloch state with band index n and quasimomentum q uniquely defined by v_L in eqn (6.31). When the velocity of the lattice is not zero (i.e. the quasimomentum q is different from zero), the atoms are partially dragged by the lattice in the same direction as v_L. By measuring the displacement Δx after an observation time Δt it is possible to determine the velocity of the condensate in the frame of the periodic potential $v_n(q)$, which is shown in Fig. 6.11a as a function of q for the first two energy bands. The lines correspond to the velocity spectrum calculated from the single-particle Bloch theory with no free parameters, as discussed in Sec. 6.1.2 and shown in Fig. 6.4.[11] A sufficiently dense sampling of the experimental points in quasimomentum space also provides a measurement of the effective mass $m_n^*(q)$, which can be obtained by the numerical evaluation of $dv_n(q)/dq$ from the experimental data, following eqns (6.10) and (6.13), as shown in Fig. 6.11b.

The energy band spectrum $E_n(q)$ can be precisely measured using the Bragg spectroscopy technique that has been introduced in Sec. 3.2.2 to probe the excitation spectrum of a Bose–Einstein condensate. Here we recall that this technique relies on the excitation of the BEC with two laser beams with wavevectors and frequencies $(\mathbf{k}_1, \omega_1)$ and $(\mathbf{k}_2, \omega_2)$ respectively, in order to produce excitations with well-defined momentum $\hbar\boldsymbol{\kappa} = \hbar(\mathbf{k}_1 - \mathbf{k}_2)$ and energy $\hbar\omega = \hbar(\omega_1 - \omega_2)$. At a fixed momentum transfer $\hbar\boldsymbol{\kappa}$,

[11]The experiment is performed with a dilute expanding BEC, in which atom–atom interactions can be neglected.

for noninteracting atoms at rest in free space, excitations can be induced only at a well-defined energy $\hbar\omega = \hbar^2\kappa^2/2m$. For noninteracting atoms in the ground state of an optical lattice, different bands can be excited for the same momentum transfer $\hbar\boldsymbol{\kappa}$, resulting in a spectrum with several peaks at energies $\hbar\omega = E_n(\kappa) - E_0$ (where E_0 is the energy of the ground state in the centre of the lowest band). Bragg spectroscopy of a weakly interacting[12] Bose–Einstein condensate in a 1D optical lattice was performed in Fabbri *et al.* (2009): the spectrum in Fig. 6.12a, taken at a fixed momentum transfer $\hbar\kappa \simeq \hbar k$, features the presence of three well-resolved peaks, which correspond to the energy of the transitions represented in Fig. 6.12b.

6.4.2 Bloch oscillations with quantum gases

The first evidence of Bloch oscillations in atomic quantum gases was obtained at Yale in the first experiment in which Bose–Einstein condensates were trapped in optical lattices (Anderson and Kasevich, 1998). Differently from previously discussed experiments on Bloch oscillations, in this case the periodic potential (aligned along the vertical axis) was static and the force driving the dynamics was a real force, gravity, instead of an inertial force. The small lattice height used in Anderson and Kasevich (1998) resulted in a finite probability of Landau–Zener tunnelling from the lowest-energy band to higher-energy bands. Because of the small lattice height, the upper bands were essentially in the continuum and the atoms could fall under the action of gravity. The result of this Bloch oscilation dynamics was a pulsed atom laser in which bunches of atoms were released from the trapped BEC at a constant rate $1/\tau_B$.

Interaction-induced instability. The effect of repulsive atom–atom interactions on the Bloch oscillations of a Bose–Einstein condensate was studied in Pisa using an accelerated optical lattice. The intriguing observation reported in Morsch *et al.* (2001) was a breakdown of Bloch oscillations when the atomic density was large enough: the BEC stopped performing oscillations and the momentum distribution recorded after time-of-flight rapidly occupied the whole Brillouin zone, signalling the onset of decoherence in the Bose–Einstein condensate, as shown in Fig. 6.13a.

This dephasing behaviour can be interpreted in the frame of a very general class of instabilities which are peculiar of dynamical nonlinear systems (for example, light propagating in nonlinear media characterized by a large Kerr coefficient (Tai *et al.*, 1986)). In the presence of a positive nonlinear term in the wave equation, such as the repulsive mean-field interaction term in the Gross–Pitaevskii eqn (6.38), approaching the edge of the Brillouin zone the Bloch states become dynamically unstable, as a consequence of the interplay between nonlinearity and dispersion: small perturbations induced by thermal or quantum fluctuations can be amplified and grow exponentially in time, eventually destroying the initial state (Smerzi *et al.*, 2002; Wu and Niu, 2001). In the tight-binding regime, this happens in correspondence with the region of the Brillouin zone in which the effective mass becomes negative ($q \gtrsim k/2$), i.e. when the kinetic and nonlinear term of the Hamiltonian have opposite sign.

[12]For weak interactions and large momentum transfer, the BEC excitation spectrum differs from a single-particle spectrum only for an interaction mean-field shift (already discussed in Sec. 3.2.2) (Fabbri *et al.*, 2009).

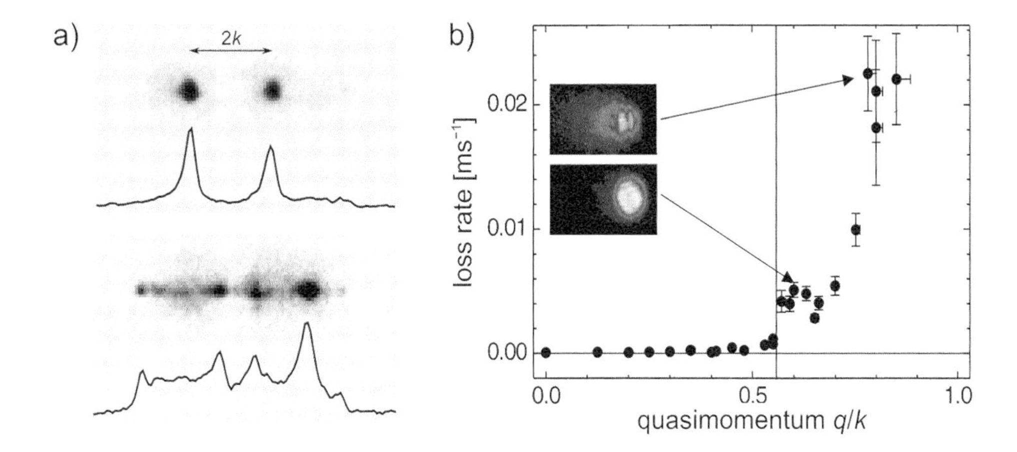

Fig. 6.13 Interaction-induced instability for a Bose–Einstein condensate in an optical lattice. a) Momentum distribution of a ^{87}Rb BEC performing Bloch oscillations in an accelerated optical lattice: at small atomic densities (top) the momentum is well defined (the two peaks represent a Bragg reflection at the edge of the Brillouin zone), at large densities (bottom) the cloud fragments into a complex peaked structure and Bloch oscillations rapidly damp as a consequence of the BEC nonlinearity. Reprinted with permission from Morsch *et al.* (2001). b) Observation of dynamical instability for a ^{87}Rb BEC in an optical lattice: when the quasimomentum is larger than approximately half of the Brillouin zone the BEC fragments and losses of atoms are observed. Adapted from Fallani *et al.* (2004).

The effect of dynamical instability was observed also in centre-of-mass BEC oscillations driven by a harmonic force: while for small amplitudes the BEC performs coherent oscillations without any damping (as discussed in Sec. 6.4.1), for larger amplitudes the coherent dynamics is destroyed as a result of dynamical instability and the BEC stops oscillating in the harmonic trap (Cataliotti *et al.*, 2003). The growth of dynamical instability for a BEC in an optical lattice was then assessed in Fallani *et al.* (2004), where the boundaries of the unstable regions and the timescales for the unstable dynamics were measured. In Fig. 6.5b the inverse lifetime of the BEC in the optical lattice is reported as a function of quasimomentum: clearly, when crossing approximately half of the Brillouin zone, the BEC ceases to be stable and the atom number decays as a consequence of inelastic losses following a complex nonlinear dynamics.

Fermi gases and control of the interactions. Spin-polarized fermionic atoms cannot interact with s-wave collisions (see Sec. 3.1.4), therefore they are expected not to show this behaviour. Early experiments performed in Florence on centre-of-mass oscillations of spin-polarized Fermi gases in optical lattices (Modugno *et al.*, 2003; Ott *et al.*, 2004) already evidenced the remarkable difference between bosons and fermions, which can be attributed to the crucial role of collisions in the quantum transport.

Bloch oscillations of trapped spin-polarized fermionic ^{40}K were then studied at LENS in Roati *et al.* (2004) by observing the dynamics of the quantum gas in a vertical optical lattice under the action of gravity. The number of fermions in the cloud was

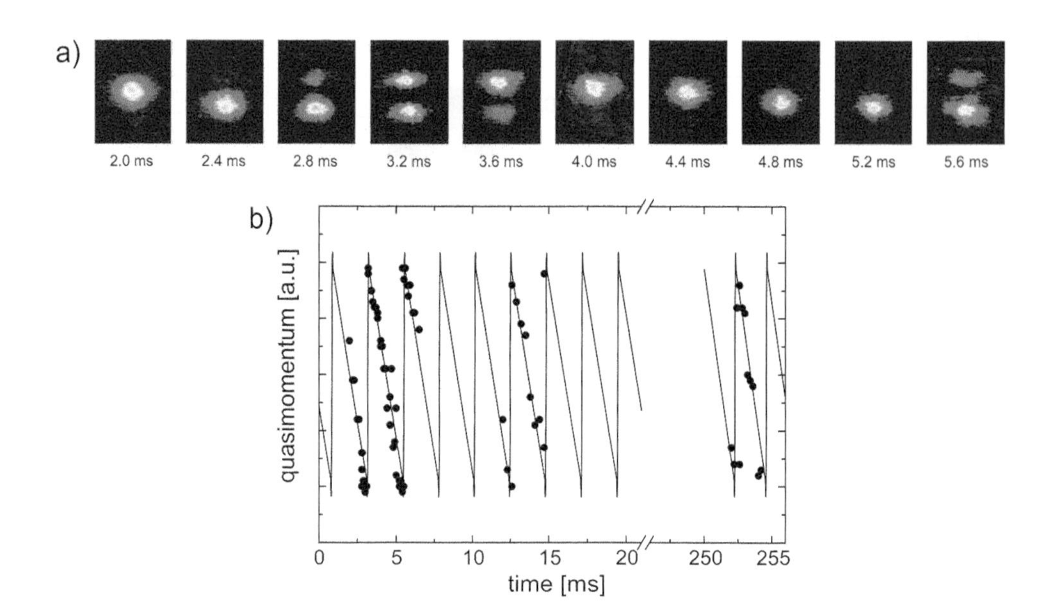

Fig. 6.14 Bloch oscillations of a degenerate spin-polarized ^{40}K Fermi gas in a vertical optical lattice. a) Time-of-flight images for different evolution times. b) Average momentum as a function of time. Adapted from Roati *et al.* (2004).

chosen in such a way to make the Fermi energy lie within the lowest-energy band of the periodic potential, in order to populate only a fraction of the first Brillouin zone. Figure 6.14 shows time-of-flight images of the fermionic cloud for different evolution times in the vertical optical lattice. More than 100 Bloch oscillations could be observed without any loss of contrast, thanks to the absence of collisions between the trapped atoms. The comparison of these results with the behaviour of an interacting Bose–Einstein condensate demonstrated the superiority of spin-polarized fermions for the observation of long-lived quantum coherence effects.

From an interferometric point of view, trapped *noninteracting* Bose–Einstein condensates represent the optimal solution for achieving long coherence times, since they combine a much larger brilliance than fermions with the same absence of collisional dephasing between the atoms as in the case of fermionic gases. Bloch oscillations of almost noninteracting BECs were simultaneously observed at Innsbruck with ^{133}Cs (Gustavsson *et al.*, 2008) and at LENS with ^{39}K (Fattori *et al.*, 2008*a*). In both these experiments the atomic scattering length was changed by using a magnetic Feshbach resonance (see Sec. 3.1.4) and the decay of Bloch oscillation was studied as a function of the strength of interactions. In the proximity of the magnetic field value corresponding to a vanishing scattering length the decoherence was found to be minimum, and up to 20 000 Bloch oscillations were observed in the Innsbruck experiment (Gustavsson *et al.*, 2008), as shown in Fig. 6.15. Similar long-lived Bloch oscillations with bosonic samples were also observed with thermal clouds of ^{88}Sr, thanks to the very small scattering length (Ferrari *et al.*, 2006).

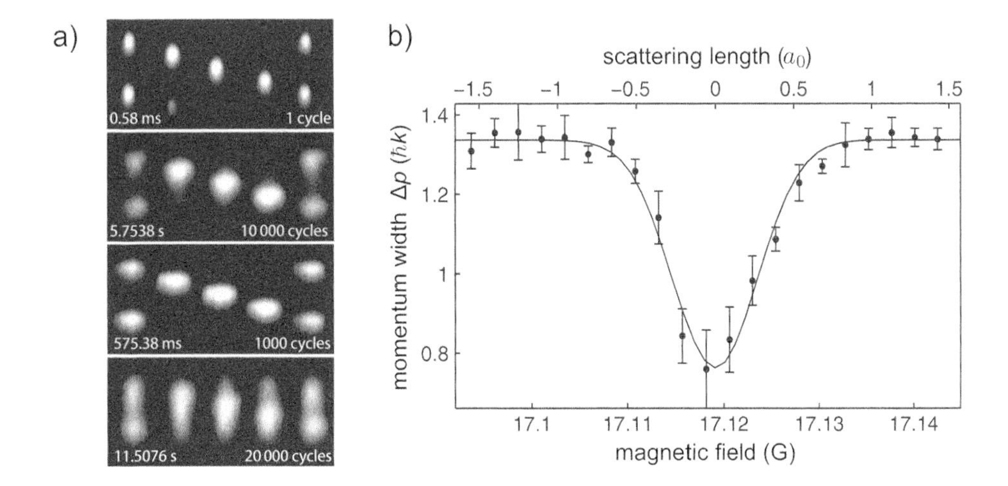

Fig. 6.15 Bloch oscillations of a ^{133}Cs Bose–Einstein condensate in a vertical optical lattice. a) Momentum distribution after an increasing number of Bloch oscillations for an almost noninteracting BEC. b) Momentum spread after ≈ 7000 Bloch oscillations as a function of the magnetic field inducing the Feshbach resonance: a clear minimum is observed at the magnetic field corresponding to zero scattering length. Reprinted with permission from Gustavsson *et al.* (2008). © American Physical Society.

Beyond s-wave scattering. Suppressing s-wave collisional interactions with Feshbach resonances, however, does not eliminate other kinds of weaker interactions between the atoms. At ultralow temperature the next most important interaction mechanism is the dipole–dipole interaction between the atomic magnetic moments, already discussed in Secs. 3.4.3 and 3.5.2, which can be described by a potential of the form

$$V(\mathbf{r}) = \frac{\mu_0 \mu^2}{4\pi} \frac{1 - 3\cos^2\theta}{r^3}, \tag{6.39}$$

where θ is the angle between the direction of the magnetic dipole moment $\boldsymbol{\mu}$ and the direction of the interparticle separation $\mathbf{r}$. A characteristic feature of this dipole–dipole interaction is its anisotropy, since both the strength and sign of the interaction depend on θ. Effects of the dipolar interaction have been evidenced in Fattori *et al.* (2008*b*) in the residual dephasing of Bloch oscillations after cancelling the collisional interactions via Feshbach resonances, as shown in Fig. 6.16. When the dipoles are parallel to the optical lattice (grey circles) it is possible to show that the interaction within each lattice site is mainly repulsive, which leads to a positive mean-field shift which can be minimized by a proper (small) negative value of the scattering length. Similarly, when the dipoles are orthogonal to the lattice (black squares), the resulting negative mean-field shift can be minimized by a positive value of the scattering length. This can be observed in the two distinct minima positions of the decoherence rate shown in Fig. 6.16 as a function of the magnetic field around the zero-crossing of the Feshbach resonance.

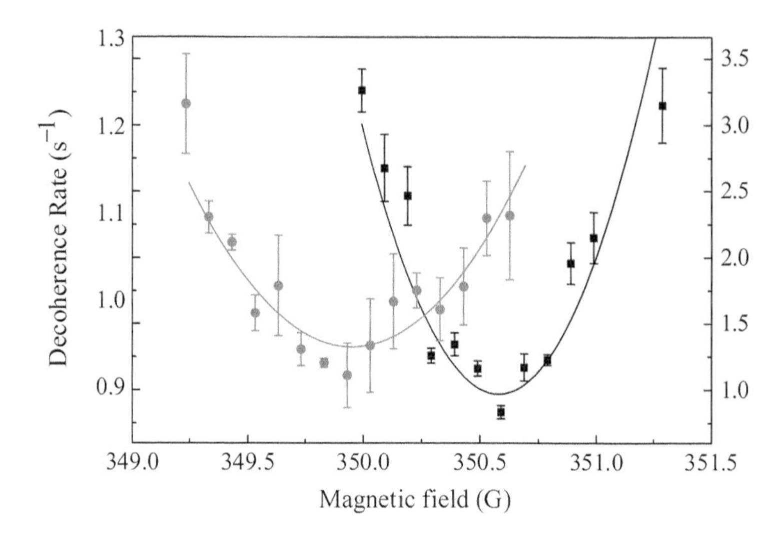

Fig. 6.16 Decoherence rate of Bloch oscillations in bosonic ^{39}K as a function of the external magnetic field for a lattice parallel (grey circles, left vertical scale) and orthogonal (black squares, right vertical scale) to the magnetic field direction. The different minima positions evidence the effect of anisotropic dipole–dipole interactions. Taken from Fattori *et al.* (2008*b*).

6.4.3 High spatial resolution force sensors

In Sec. 6.3.2 we have already presented the application of Bloch oscillations to the high-precision measurement of fundamental constants, thanks to the possibility of imparting a well-controlled momentum to the atoms. Ultracold atoms performing Bloch oscillations can also be used as very accurate sensors of forces. From eqn (6.16), the force F driving the Bloch dynamics is

$$F = \frac{h}{\tau_B d}, \tag{6.40}$$

where τ_B is the Bloch oscillation period. We note that the only lattice parameter entering this expression is the lattice spacing d, which can be very precisely determined via optical interferometric techniques or with frequency combs, while there is no dependence on the lattice depth. Remarkably, with this approach the measurement of a force reduces to the measurement of a frequency, which is the physical quantity which can be measured with the highest precision.

The use of ultracold atoms in optical lattices as quantum sensors of forces combines two advantages: high precision and high spatial resolution. Regarding the first point, the precision of the measurement is essentially given by the uncertainty on the Bloch period τ_B: since in existing experiments several thousands of Bloch oscillations periods can be already tracked, this uncertainty can be made quite small, and a relative uncertainty below 10^{-6} can be achieved. Recent progress in the measurement of Bloch oscillation frequencies has been reported thanks to the use of phase-modulated (Ivanov *et al.*, 2008) or amplitude-modulated (Alberti *et al.*, 2010) optical lattices. Indeed, the "spectroscopic" measurement of harmonics of the Bloch oscillation frequency in an amplitude-modulated vertical lattice has allowed the determination of the local gravity acceleration with a relative uncertainty of only $\approx 10^{-7}$ (Poli *et al.*, 2011).

Regarding the spatial resolution, the typical size of trapped atomic clouds is on the order of 10 μm or less: this means that forces can be sensed at the micrometer scale. These possibilities make ultracold atoms in optical lattices an appealing system for the accurate measurement of forces at small distances. One application regards the measurement of possible deviations from the Newtonian $1/r^2$ dependence of the gravitational force at short distances (Dimopoulos and Geraci, 2003). Another application is the measurement of forces in proximity of surfaces, for instance the Casimir–Polder force already discussed in Sec. 3.2.5: this effect could be precisely investigated by studying the shift of the Bloch period τ_B when the ultracold trapped atoms approach the substrate, as proposed in Carusotto *et al.* (2005).

7 Optical lattices and quantum simulation

In the previous chapter we have discussed the close analogy between the behaviour of ultracold atoms in optical lattices and the physics of an electron gas in an ideal crystalline solid. Then we have discussed the possibility of performing precise measurements of forces and fundamental constants by taking advantage of the quantum transport of ultracold atoms in these ideal periodic potentials.

In this chapter we discuss a different application of optical lattices. Thanks to the techniques developed in the field of ultracold atomic physics, it is possible to precisely engineer and control the system Hamiltonian in such a way as to "simulate" the physics of more complex condensed-matter systems (Bloch *et al.*, 2008). In the spirit of an early conjecture by R. Feynman (Feynman, 1982), the ultimate goal of this approach is the realization of *quantum simulators* (Lloyd, 1996; Buluta and Nori, 2009), i.e. laboratory systems in which quantum models or theories, relevant for different physical systems but hardly solvable with classical computational techniques, can be precisely implemented and directly "solved" by measuring the properties of the resulting quantum state. For a complete and up-to-date illustration of this "quantum simulation" perspective with ultracold atoms in optical lattices, see the recent book by M. Lewenstein et al. (Lewenstein *et al.*, 2012).

As an example of this possibility, in Secs. 7.1 and 7.2 we consider a paradigmatic question in condensed-matter physics, which regards the microscopic distinction between metals and insulators. Moving from the ideal Bloch description given in the previous chapter, we will show that ultracold atoms can be used to study superfluid/metal–insulator quantum phase transitions induced by the presence of *interactions* and *disorder*, which can be experimentally controlled in a very precise way. In Sec. 7.3 we will consider new perspectives opened by novel detection techniques and very recent experimental efforts towards the quantum simulation of different condensed-matter systems.

7.1 Mott insulators

The single-particle Bloch theory reviewed in Sec. 6.1, combined with the Pauli exclusion principle for the electrons, already establishes a criterion for distinguishing metals from insulators: a solid is an insulator, more precisely a *band insulator*, when, owing to the Pauli exclusion principle, there are no allowed states for the electrons to jump into when an external electric field is applied. This is the case, for example, for diamond, in which completely filled energy bands are separated from empty higher-energy bands by a large energy gap (the Fermi energy lies *within the gap*). In a metal, for example copper or gold, the highest-occupied band is only partially filled (the Fermi energy lies *within the band*) and there are free electronic states in which conduction can take place.

This elementary criterion is not universally valid, since there are many materials which do not follow this rule. Furthermore, there are condensed-matter systems which exhibit metal–insulator transitions (Mott, 1990) or superconductor–insulator transitions (Fisher, 1990) from a conducting to an insulating state as some physical parameter is changed. Among the mechanisms responsible for this behaviour, the existence of *interactions* between the particles (electrons) and the presence of *disorder* in the lattice play a very important role. These two effects, which are not included in the simple Bloch theory, may independently turn a conducting system into an insulator. In this section we are going to investigate the effects of interactions, leaving the discussion of disorder to Sec. 7.2.

7.1.1 Bose–Hubbard model

An optical lattice can dramatically increase the effects of interactions between the atoms and change a weakly interacting quantum gas into a strongly correlated many-body state. As a matter of fact, increasing the depth of the optical lattice results in a lower tunnelling rate between the sites, i.e. a lower mobility of the atoms in the lattice, which amplifies the effects of interactions. At the same time, the atoms get more squeezed into the lattice potential wells, thereby increasing their interaction energy.

The physics of a gas of interacting identical bosons in the lowest-energy band of a deep optical lattice is well captured by the Bose–Hubbard model (Jaksch *et al.*, 1998), originally introduced in Fisher *et al.* (1989) to describe the superfluid–insulator transition observed in condensed-matter systems. The Bose–Hubbard Hamiltonian is the second-quantization generalization of the tight-binding description in Sec. 6.1.4 to the case of interacting particles. It takes the form

$$\hat{H} = -J \sum_{\langle j,j' \rangle} \hat{b}_j^\dagger \hat{b}_{j'} + \frac{U}{2} \sum_j \hat{n}_j \left(\hat{n}_j - 1 \right) , \tag{7.1}$$

where $\hat{b}_j$ ($\hat{b}_j^\dagger$) is the annihilation (creation) operator of one particle in the j-th lattice site, $\hat{n}_j = \hat{b}_j^\dagger \hat{b}_j$ is the number operator, and $\langle j, j' \rangle$ indicates the sum on nearest neighbouring sites. The two terms on the right-hand side of eqn (7.1) account for different contributions to the total energy of the system. The first term, proportional to the *tunnelling energy* (or *hopping energy*) J already defined in Sec. 6.1.4, describes the tunnelling of bosons from one site to an adjacent site. The second term, proportional to the *interaction energy* U, arises from atom–atom on-site short-range interactions and gives a nonzero contribution only if more than one particle occupies the same site.

In the presence of repulsive interactions $U > 0$, this model supports a quantum phase transition from a superfluid phase ($U \ll J$), in which the atoms are delocalized occupying extended Bloch states, to an insulating phase ($U \gg J$), where the atom wavefunctions are localized in individual lattice sites as a consequence of the strong interactions.

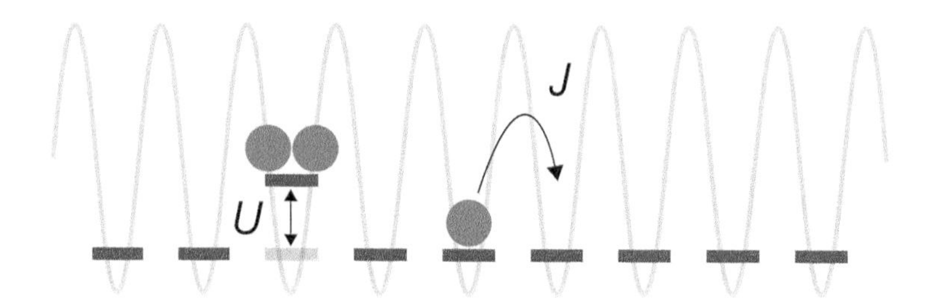

Fig. 7.1 Sketch of interacting bosons in a periodic potential. Hopping of particles from one site to the nearest-neighbours is described by the tunnelling matrix element J, while repulsive short-range two-body interactions are taken into account by the on-site interaction energy U.

Double-well case. The superfluid–insulator transition can be studied in a simplified setting where one considers only two lattice sites.[1] This situation can be described by a potential with two identical wells, which are coupled by a tunnelling probability J. This double-well problem is paradigmatic of a large number of effects where quantum tunnelling is important, e.g. molecular spectra (inversion spectrum of ammonia) and superconducting Josephson junctions. The Hamiltonian describing ultracold interacting bosons in the ground vibrational states of two coupled potential wells is the two-sites reduction of eqn (7.1):

$$\hat{H} = -J\left(\hat{b}_1^\dagger \hat{b}_2 + \hat{b}_2^\dagger \hat{b}_1\right) + \frac{U}{2}\left[\hat{n}_1\left(\hat{n}_1 - 1\right) + \hat{n}_2\left(\hat{n}_2 - 1\right)\right] . \tag{7.2}$$

We can use an occupation-number representation and describe the state of the system with the notation $|\Psi\rangle = |mn\rangle$, which means that m particles are present in the left well and n particles are present in the right well.[2] In the simplest case of two identical bosons only the three basis states shown in Fig. 7.2 have to be considered and a generic state of the system can be written as[3] $|\Psi\rangle = 1/\sqrt{2}\sin\alpha|20\rangle + \cos\alpha|11\rangle + 1/\sqrt{2}\sin\alpha|02\rangle$, which is parametrized by the angle α and automatically satisfies the normalization condition $\langle\Psi|\Psi\rangle = 1$. One can easily work out that the energy of such state is

$$E = \langle\Psi|H|\Psi\rangle = U\sin^2\alpha - 2J\sin(2\alpha) . \tag{7.3}$$

We consider now two interesting limiting cases. When $J \gg U$ the energy can be approximated by $E \simeq -2J\sin(2\alpha)$, which is minimized by $\alpha = \pi/4$, yielding

$$\Psi_{J\gg U} \simeq \frac{1}{2}|20\rangle + \frac{1}{\sqrt{2}}|11\rangle + \frac{1}{2}|02\rangle , \tag{7.4}$$

[1] This simplified model is connected to the double-well BEC interferometers with nonclassical states discussed in Sec. 3.2.6.

[2] We note that this kind of representation automatically takes into account the indistinguishability of the particles.

[3] This parametrization does not describe all the possible states (since superpositions with different relative phases, i.e. with complex coefficients, are possible), but is perfectly adequate to describe the ground state.

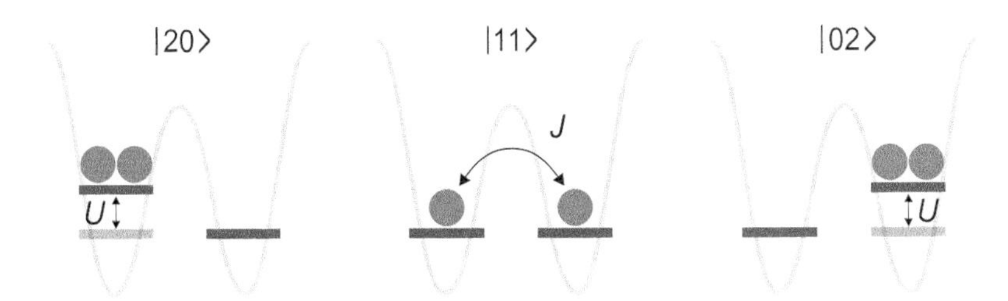

Fig. 7.2 Two interacting identical bosons in a double-well potential. The three possible states of the system are shown.

which means that each particle is delocalized in the two wells and the number of particles in each well has nonzero fluctuations around unity. When instead $U \gg J$ the energy becomes $E \simeq U \sin^2 \alpha$, which is minimized by $\alpha = 0$, yielding

$$\Psi_{U \gg J} \simeq |11\rangle \,, \tag{7.5}$$

which means that no number fluctuations are present: one particle is present in the left well and one particle is present in the right well. This happens because the energy $\sim U$ saved by localizing each particle in a different well is larger than the energy $\sim J$ that would have been gained if the particles were delocalized.

Lattice case. The localization behaviour illustrated above, induced by repulsive interactions between the atoms, is amplified in a lattice, which can be thought of as an array of identical potential wells linked by a tunnelling coupling between nearest-neighbouring sites. In the case of unit filling, i.e. number of sites = number of atoms = N, the two limits are a *superfluid state* for $J \gg U$, with each particle being delocalized in an extended Bloch wave (see Sec. 6.1.1)

$$\Psi^{SF}_{J \gg U} = \frac{1}{\sqrt{N! N^N}} \left(\hat{b}_1^\dagger + \hat{b}_2^\dagger + \cdots + \hat{b}_N^\dagger \right)^N |00\ldots0\rangle \,, \tag{7.6}$$

and a *Mott-insulating* state for $U \gg J$ in which each atom is localized in a single site:

$$\Psi^{MI}_{U \gg J} = \hat{b}_1^\dagger \hat{b}_2^\dagger \ldots \hat{b}_N^\dagger |00\ldots0\rangle = |11\ldots1\rangle \,. \tag{7.7}$$

As we will discuss in the following sections, these two states exhibit remarkably different properties. The transition from a superfluid to a Mott insulator when the Hamiltonian parameters are changed is an example of *quantum phase transition* (Sachdev, 2000). In this kind of transition the control parameter is not temperature, as in the case of classical phase transitions (e.g. the ferromagnetic transition in the Ising model (Ising, 1925)), but a different parameter entering the system Hamiltonian, e.g. the strength of the coupling between particles or the hopping rate. Therefore, the mechanism driving the transition is not temperature fluctuations, but quantum fluctuations.

7.1.2 Superfluid–Mott quantum phase transition

We can interpret the physics of the superfluid–Mott transition in terms of basic concepts of quantum optics. Within this approach we consider the atoms as if they were photons, while the different vibrational levels of the lattice potential wells play the role of the different modes of the electromagnetic field. Within the single-band Bose–Hubbard model, the different occupation of the lattice sites can be interpreted in terms of a single-mode electromagnetic field with different photon statistics. Following this quantum-optics analogy, the transition from a superfluid to a Mott insulator reveals close similarities to the transition from coherent states to Fock states (Scully and Zubairy, 1997).

In the case of the superfluid state in eqn (7.6) it is possible to demonstrate that the on-site number fluctuations $\delta n_i = \sqrt{\langle \hat{n}_i^2 \rangle - \langle \hat{n}_i \rangle^2}$ can be written as

$$\delta n_i = \sqrt{1 - \frac{1}{N}} \, . \tag{7.8}$$

In the limit $N \gg 1$, one recovers the property $\delta n_i = \langle \hat{n}_i \rangle = 1$, which is typical of coherent states with Poisson statistics for the number occupation and average occupation $\langle \hat{n}_i \rangle = 1$. These results can be generalized to the case of arbitrary number of particles per site m and it is possible to demonstrate that the superfluid state can be almost exactly factorized in the tensor product of coherent states in each lattice site according to:

$$\Psi^{SF}_{U \gg J} \simeq \prod_i \sum_n \frac{e^{-m/2} m^{n/2}}{\sqrt{n!}} |n\rangle_i \, , \tag{7.9}$$

where $|n\rangle_i$ represents the occupation of the i-th site by n bosons and $m = \langle \hat{n}_i \rangle$ is the average occupation number (equal to the number standard deviation δn_i). Coherent states represent the most "classical" states, since they correspond to the states of the field with maximally determined phase (Scully and Zubairy, 1997). In a superfluid state there is a well-defined phase in each lattice site and the tunnelling coupling locks the site-to-site phase difference to zero.[4]

In the Mott insulator regime, instead, particles are "frozen" in the lattice with an exact number of atoms per site. This state can thus be described by the tensor product of Fock states as in eqn (7.7), in which number fluctuations δn_i are zero. Correspondingly, owing to the phase-number uncertainty relation $\delta n_i \delta \phi_i \geq 1/2$, the phase in a Mott insulator state is maximally uncertain.[5]

Experimental observation. The first experimental observation of the quantum phase transition from a superfluid to a Mott insulator has been reported in Munich in Greiner *et al.* (2002*a*), following the initial proposal by Jaksch *et al.* (1998). The experiment was performed by trapping a BEC in a three-dimensional cubic optical lattice. The

[4]This is true for the ground state. As we have studied in Sec. 6.1.4, higher-energy states in the lowest band have a nonzero site-to-site phase difference, resulting in a nonzero quasimomentum.

[5]This phase-number uncertainty relation is not formally exact, owing to the difficulties in defining a Hermitian phase operator (see for example Gerry and Knight (2005)), but it turns out to be heuristically correct.

control parameter for the Mott transition, i.e. the ratio U/J between interaction energy and hopping energy, was tuned by adjusting the depth V_0 of the optical lattice. As a matter of fact, in a deep 3D optical lattice, the following relations hold (Zwerger, 2003):

$$U = \sqrt{\tfrac{8}{\pi}} k a E_R \left(\tfrac{V_0}{E_R}\right)^{3/4} \tag{7.10}$$

$$J = \tfrac{4}{\sqrt{\pi}} E_R \left(\tfrac{V_0}{E_R}\right)^{3/4} \exp\left(-2\sqrt{\tfrac{V_0}{E_R}}\right) \tag{7.11}$$

and the ratio U/J increases exponentially with the lattice depth V_0.

The superfluid–Mott transition was investigated by monitoring the interference pattern of the atomic cloud released from the 3D lattice as a function of the lattice height. When the lattice confining the atoms is suddenly switched off, the atom wavefunctions expand in free space and overlap: if the atoms are initially delocalized and long-range phase coherence exists in the system, the interference of the overlapping wavefunctions builds up in a regular diffraction pattern (Greiner *et al.*, 2001; Pedri *et al.*, 2001), as we have already evidenced in the one-dimensional case presented in Sec. 6.1.1. This pattern, resembling the diffraction of light from a grating or the diffraction pattern of X-rays scattered from a crystalline structure, can be taken as a measure of the degree of coherence in the many-body system. The formation of this diffraction pattern is shown in Fig. 7.3a for different times of flight after the lattice switch-off.

The interference patterns shown in Fig. 7.3b correspond to different lattice depths V_0, going from zero to $20E_R$. For small lattice depths the intensity of the diffraction side peaks increases with V_0: the atoms are more tightly confined in the lattice sites and, following the optical analogy, this means that the lattice acts as a grating with narrower slits, resulting in a stronger "diffraction" effect. At a lattice height $V_0 \simeq 15E_R$ the visibility of the interference peaks suddenly decreases, eventually going to zero, reflecting the loss of coherence of the system after entering the Mott insulator state, where the atoms are localized and the number fluctuations δn_i vanish. This coherence loss is reversible: by ramping down the height of the optical lattice one can recover the peaked interference pattern peculiar of the superfluid state, where atoms are delocalized and on-site coherent states are created.

Non-equilibrium dynamics. The "quantum optics" approach illustrated above is useful in unveiling important features of the superfluid–Mott transition. Additionally, it is possible to drive the analogy between atoms and photons one step further and use it to realize experiments of "quantum atom optics" where the time evolution of these states is observed.

A beautiful example of this possibility has been provided by Greiner *et al.* (2002*b*) where the evolution of the atomic states from coherent states to nonclassical states and then back to coherent states was revealed. In the experiment, the system was prepared in a superfluid state at $V_0 = 11E_R$; then the lattice height was suddenly increased at $V_0 = 35E_R$ well above the critical point for the transition, when the ground state was expected to be a Mott insulator. This fast modification of the Hamiltonian parameters was non-adiabatic, i.e. the system did not have the time to reach the new Mott ground state and was left in a non-stationary state (*quantum quench*). Since the final lattice

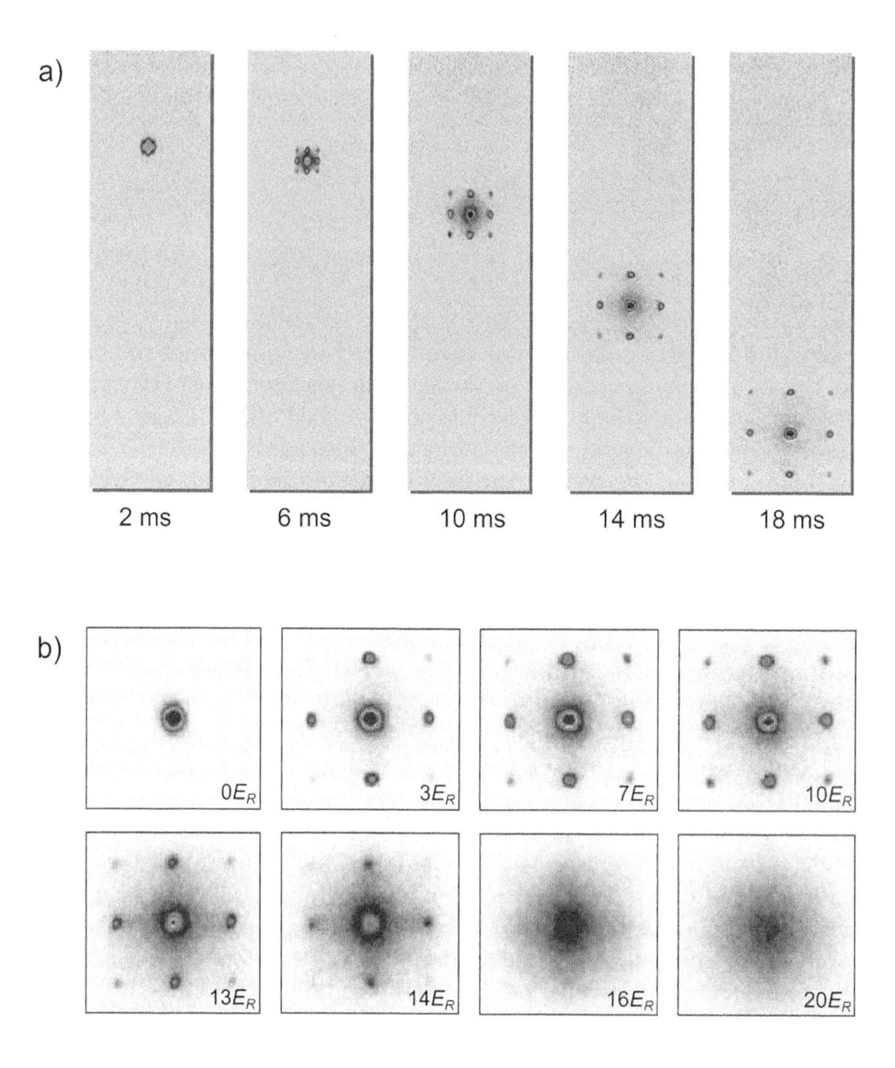

Fig. 7.3 a) Diffraction pattern of a Bose–Einstein released from a 3D optical lattice. The images show the free fall and the evolution of the diffraction pattern for different times of flight. Reprinted from Greiner (2003) by courtesy of the author. b) Superfluid to Mott Insulator transition. Interference pattern of an ensemble of ultracold atoms released from a 3D optical lattice for different lattice heights V_0, expressed in recoil energies E_R. Around $V_0 \simeq 15E_R$ one observes the disappearance of the interference peaks, which indicates entering the Mott Insulator state. Reprinted with permission from Greiner *et al.* (2002*a*). © Macmillan Publishers Ltd.

depth was very large, no tunnelling of particles was allowed and the system evolved according to the independent "internal" evolution of the many identical coherent states $|\phi\rangle_i$ which initially occupied each lattice site i.

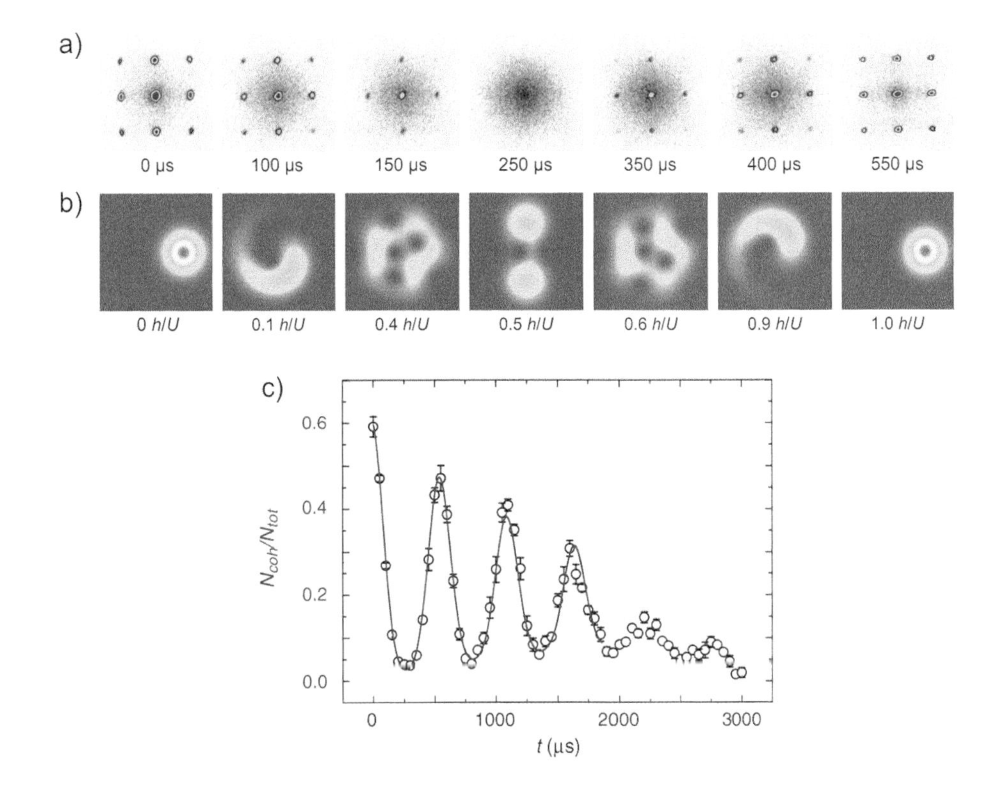

Fig. 7.4 Non-equilibrium dynamics in the Bose–Hubbard model. a) Experimental time-of-flight images showing the collapse and revival of coherent states after a sudden increase of lattice depth. b) Phase-space visualization of the temporal evolution of the states (theory). c) Fraction of atoms in the coherent diffraction pattern as a function of time. Adapted with permission from Greiner *et al.* (2002*b*). © Macmillan Publishers Ltd.

This evolution, monitored by taking the visibility of the diffraction pattern after time-of-flight, is reported in Fig. 7.4a. The visibility of the pattern initially drops, signalling that the coherent states (in each lattice site) are evolving towards nonclassical states with higher phase uncertainty. This happens since the different $|n\rangle_i$ contributions to the initial coherent state $|\phi\rangle_i$ evolve each with a different phase factor $e^{-iUn(n-1)t/2\hbar}$ (since the interaction energy term in eqn (7.1) depends on the occupation number). After a few hundreds μs the different $|n\rangle_i$ contributions are dephased and visibility is completely lost. After this collapse, a revival of visibility is observed periodically at time multiples of $\hbar/U$, in correspondence with the temporary restoration of coherent states when the different $|n\rangle_i$ temporarily rephase, before collapsing again (see Fig. 7.4c, where the fraction of atoms in the coherent diffraction pattern is plotted). This evolution is also represented in the phase-space diagrams in Fig. 7.4b, showing that interesting nonclassical states are created at fractions of the revival time $\hbar/U$, such as the Schrödinger cat state visible at a time $0.5\hbar/U$.

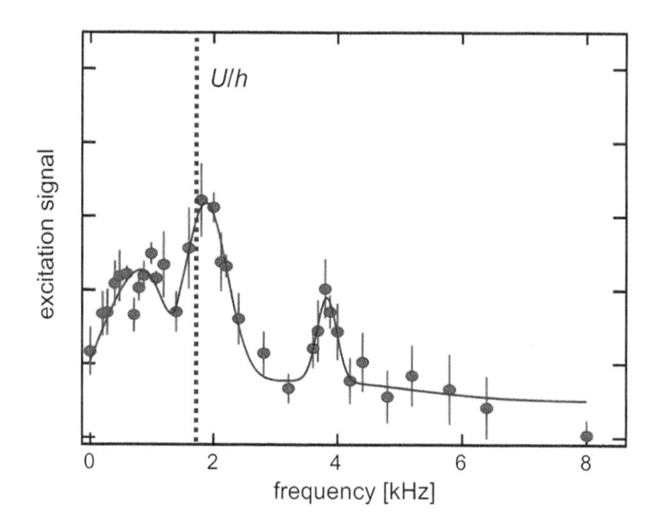

Fig. 7.5 Bragg excitation spectrum of a mixed state composed of a superfluid and a Mott insulating component. The peak at the Mott–Hubbard gap U originates from the Mott insulating state. Taken from Clément *et al.* (2009*a*).

7.1.3 Probing Mott insulators

An important property characterizing Mott insulators is the existence of a gap in the excitation spectrum. A Mott insulator is a "rigid" insulating state, in which no current flow can be sustained and no excitations can be produced, unless an energy gap is broken. This gap coincides with the energy $\epsilon \sim U$ that is required for moving a particle from a site to a nearest-neighbouring already-occupied site, i.e. for the creation of a *particle–hole* excitation. A superfluid, instead, is characterized by a gapless excitation spectrum, with the lowest branch of excitations corresponding to the creation of phonons inside the superfluid: long-wavelength excitations can be created at very low energies ϵ, according to the linear dispersion relation of phonons $\epsilon = c\hbar k$, where c is the sound velocity in the superfluid (as discussed in the context of sound propagation in Bose–Einstein condensates in Sec. 3.2.2).

The existence of a gap was already evidenced in Greiner *et al.* (2002*a*), where the authors applied a magnetic potential gradient in order to produce tunnelling excitations in the Mott state. A marked resonance centred at a site-to-site energy shift $\sim U$ signalled the presence of the Mott gap. Later, the measurement of the excitation spectrum across the superfluid–Mott transition was carried out by using a lattice modulation technique (Stöferle *et al.*, 2004) and inelastic scattering of light (Clément *et al.*, 2009*a*). An example of the spectrum obtained with the latter technique (also called *Bragg spectroscopy*, see Secs. 3.2.2 and 6.4.1) is reported in Fig. 7.5 for a mixed state composed of a superfluid and a Mott insulating component: while the low-energy part of the spectrum (< 1 kHz) describes excitations of the superfluid regions, the peak at ~ 2 kHz (corresponding to U/h), and the smaller peak at twice the energy, account for the gapped excitations of the Mott state.

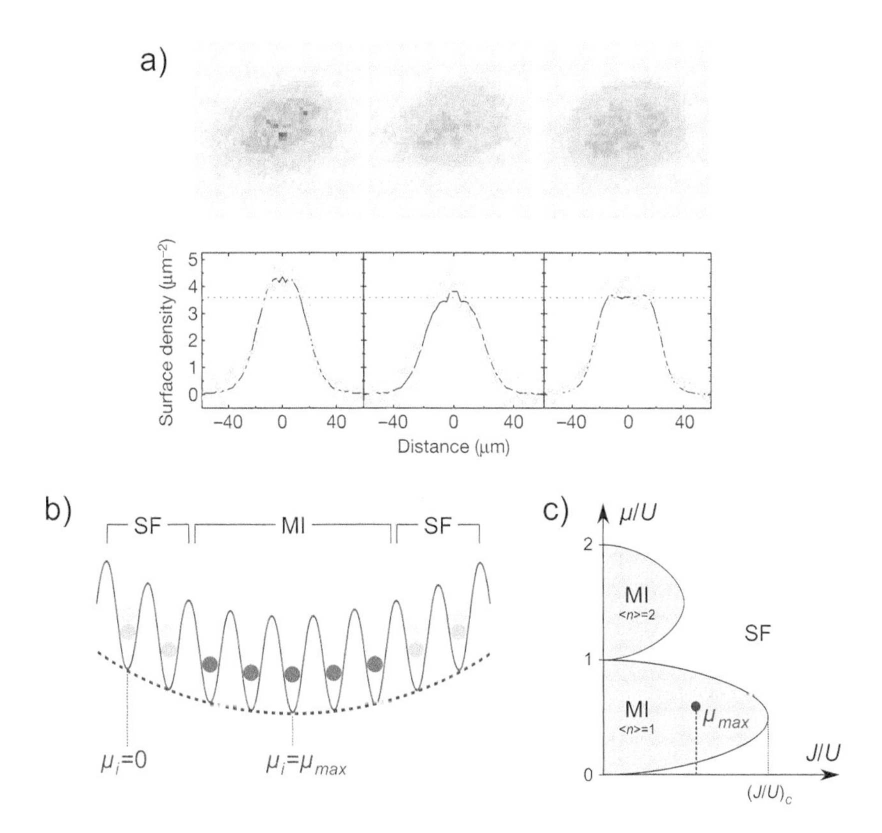

Fig. 7.6 a) In-situ imaging of a 2D bosonic gas across the superfluid–Mott transition (from left to right: superfluid, phase-transition regime, Mott insulator). Taken from Gemelke *et al.* (2009). b) The presence of an external harmonic potential determines a variation of the local chemical potential μ_i. c) Sketch of the phase diagram for interacting bosons on a lattice: the presence of the harmonic trap allows the phase diagram from zero chemical potential up to μ_{max} to be explored.

The presence of a gap in the Mott insulator excitation spectrum is intimately related to its incompressible nature. The *compressibility* $\kappa = n^{-2}\partial n/\partial\mu$ (where n is the mean occupation number) describes the ability of the system to change density and redistribute its constituent particles as the chemical potential μ is changed. The finite compressibility of a superfluid gas is connected with the existence of low-lying excitation modes at $q \to 0$ which can be populated, and its value is determined by the slope of the phonon branch in the excitation spectrum. A Mott insulator is incompressible because of its energy gap, which forbids the rearrangement of particles (unless a variation of chemical potential larger than $\sim U$ is made and particle/hole excitations are created).

The presence of a slowly varying harmonic trapping potential (in addition to the optical lattice) allowed Gemelke et al. (Gemelke *et al.*, 2009) to detect the vanishing compressibility of the Mott insulator by directly imaging domains of the cloud in which the number of particles per site was constant. The slowly varying harmonic potential determines a variation of the local chemical potential μ_i (see Fig. 7.6b), allowing the measurement of compressibility from the spatial density variations measured by in-situ imaging. In Fig. 7.6a it is evident that, as the lattice depth is increased across the Mott transition (from left to right), the density in the central region of the cloud becomes flat: this means that the average site occupation does not change upon variations of the local chemical potential, i.e. the compressibility vanishes. In the external regions of the cloud the density is spatially changing, as expected for a compressible superfluid state. As a matter of fact, the harmonic trap determines a shell structure with the spatial separation of Mott domains and superfluid regions, according to the local chemical potential (changing from zero to μ_{max} in the centre of the trap): this behaviour can be inferred from Fig. 7.6c, where the phase diagram for the Bose–Hubbard model as a function of chemical potential and J/U, originally derived by Fisher and coworkers, is presented (Fisher *et al.*, 1989). More experimental advances on the measurement of the site occupancy statistics across the superfluid–Mott quantum phase transition will be presented in Sec. 7.3.1.

7.1.4 Fermionic Mott insulator

The physics of interacting fermions in optical lattices is even more interesting than the case of bosons. As a matter of fact, the investigation of ultracold fermions in optical lattices allows us to establish a direct connection with condensed-matter physics, where the conducting particles, the electrons, are fermions. With respect to the case of bosons, the physics of fermions is enriched by the Pauli exclusion principle and by the necessity of introducing the spin degree of freedom in order to allow interactions between fermions. The physics of spin-1/2 fermions with repulsive interactions in a single lattice band is described by the *Fermi–Hubbard* model, which represents a key model of condensed matter physics for the understanding of strongly correlated electron systems. The Fermi–Hubbard Hamiltonian can be written as

$$\hat{H} = -J \sum_{\langle j,j' \rangle,\sigma} \hat{c}^{\dagger}_{j,\sigma}\hat{c}_{j',\sigma} + U \sum_{j} \hat{n}_{j,\uparrow}\hat{n}_{j,\downarrow} \,, \tag{7.12}$$

where σ denotes one of the two possible spin states $\uparrow$ and $\downarrow$. Similarly to the Bose–Hubbard Hamiltonian in eqn (7.1), J describes the tunnelling matrix element, U is the interaction energy of a pair of fermions sitting on the same site with opposite spin, $\hat{c}_{j,\sigma}$ ($\hat{c}^{\dagger}_{j,\sigma}$) is the annihilation (creation) operator for a fermion with spin σ on site j, and $\hat{n}_{j,\sigma} = \hat{c}^{\dagger}_{j,\sigma}\hat{c}_{j,\sigma}$ is the number operator.

This Hamiltonian supports different quantum phases depending on the lattice filling and the ratio U/J between interaction energy and hopping energy. A fermionic Mott insulator state exists in the case of "half-filling", i.e. when N sites are occupied by $N/2$ spin-up and $N/2$ spin-down fermions, when the repulsive interaction energy U is much larger than the hopping energy J. The strong repulsion between fermions creates an insulating state with exactly one fermion per site (similarly to the case of bosons), since the simultaneous presence of two fermions with opposite spin on the same site is energetically unfavourable. After the first investigations of ultracold Fermi gases in 3D optical lattices (Köhl *et al.*, 2005), a fermionic Mott insulator was demonstrated in Jördens *et al.* (2008), where a magnetic Feshbach resonance (see Sec. 3.1.4) was used to control the interactions between the fermions. The production of a Mott state was signalled by a suppression of doubly occupied lattice sites and by the appearance of a gapped mode in the excitation spectrum. Around the same time, the evolution of a fermionic system from a metallic state to a Mott insulator and a band insulator was investigated in detail in a related experimental work (Schneider *et al.*, 2008), where the compressibility of the gas (again ^{40}K) was measured via direct *in-situ* imaging. The experimental observations, demonstrating the emergence of an incompressible Mott-insulating core, were compared with the results of a complete theory taking into account also the finite temperature of the gas.

These first investigations of interacting fermions in optical lattices opened the way for the ongoing experimental research on the low-temperature phase diagram of the Fermi–Hubbard model, that is connected to the possibility of studying *quantum magnetism* with ultracold atoms, which will be discussed in Sec. 7.3.3. These studies could also be important for the contemporary research on high-T_C superconductivity, a striking phenomenon discovered in the 1980s and still lacking a complete theoretical understanding (Leggett, 2006). In particular, experiments with ultracold fermionic atoms could highlight the emergence of a superconducting d-wave phase, which is believed to emerge from the 2D Fermi–Hubbard model, and is expected to play a fundamental role in most high-T_C superconductors. A review of the recent developments in the experimental investigation of Fermi–Hubbard physics with ultracold atoms in optical lattices can be found in Esslinger (2010).

7.2 Anderson localization

Not only can interactions force a metal to become an insulator, but also disorder can produce a similar effect. In 1958 P. W. Anderson published a paper (Anderson, 1958) in which he conjectured that the presence of disorder in a crystal could turn a metal into a perfect insulator with zero conductivity. The microscopic mechanism for this insulating behaviour is the localization of the wavefunctions: the eigenstates of the system change from extended to localized. In Sec. 6.1 we have shown that a perfectly periodic crystal supports extended eigenstates (plane-wave-like Bloch waves), which describe stationary currents flowing across the whole crystal. On the other hand, spatially localized eigenstates imply that no current can flow in the system under stationary conditions.

Anderson formulated his localization theory for a simple tight-binding model (see Fig. 7.7) in which each site has a random energy offset $\epsilon_j \in [0, \Delta]$ which models the presence of defects and inhomogeneities in a real solid:

$$\hat{H} = -J \sum_{\langle j,j' \rangle} \hat{b}_j^\dagger \hat{b}_{j'} + \sum_j \epsilon_j \hat{n}_j \,. \tag{7.13}$$

Thanks to a clever mathematical treatment of the problem, Anderson predicted the existence of a transition from extended states to localized states for increasing disorder amplitude Δ. Anderson-localized states are characterized by exponentially decreasing tails, which at large distance can be written as

$$\Psi(x) \sim \exp\left(-x/\xi\right) \,, \tag{7.14}$$

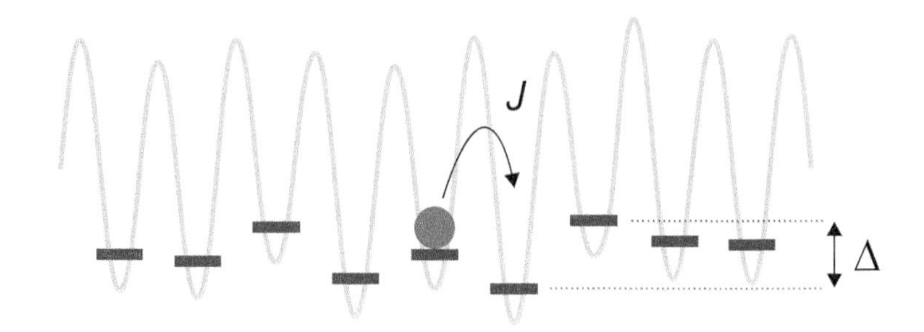

Fig. 7.7 Sketch of the Anderson model for a quantum particle in a disordered lattice. Hopping from one site to the nearest-neighbours is described by the tunnelling matrix element J, while the on-site potential energy changes randomly in an energy interval $[0, \Delta]$.

where ξ is the localization length.[6] Generally speaking, the stronger the disorder the smaller the localization length. This intuition, together with the mathematical tools developed by Anderson to describe the localization transition, led to the award of the Nobel Prize in Physics in 1977 (Anderson, 1978).

In the following decades it was realized that Anderson localization is a much more general phenomenon, holding for propagation of generic linear waves (also classical waves) in disordered media. In the language of wave propagation, Anderson localization arises because of interference effects in the scattering of a wave by disordered defects. In the strong scattering limit $kl \ll 1$, when the wavelength $\lambda = 2\pi/k$ is much larger than the mean free path l between scattering events, these interferences can add up to completely halt the waves inside the random medium, resulting in strong localization, or Anderson localization. This interpretation in terms of interference effects in multiple scattering led to the search for Anderson localization in the propagation of classical waves in random media, where it was eventually demonstrated, e.g. with light (Wiersma *et al.*, 1997; Schwartz *et al.*, 2007) and ultrasound waves (Hu *et al.*, 2008).

A direct observation of Anderson localization for electrons is difficult, because in solid-state systems these subtle interference effects can be spoiled by many other dephasing effects, most notably interactions between the electrons and interactions of the electrons with phonons. Very recently, a direct observation of Anderson localization for matter waves has been obtained with noninteracting atomic Bose–Einstein condensates in disordered optical potentials. Cold atoms offer the advantage of a clean system where it is possible to create disordered potentials in an extremely controlled way, tuning precisely the kind and amount of disorder (Fallani *et al.*, 2008). Furthermore, the localized wavefunctions (more precisely, their squared modulus) can be directly observed by imaging the atomic samples with a CCD camera: in the case of a BEC it is possible to take advantage of $\approx 10^5$ atoms sharing the same single-particle wavefunction (provided that interactions can be neglected). As a result, the typical exponential tails of Anderson-localized states can be observed, allowing the detection of localization.

[6]Stationary localized states cannot exist in a uniform lattice, owing to the discrete translational invariance. If this symmetry is not broken by disorder, all the eigenstates obey the Bloch theorem, which predicts solutions which are extended in the entire space.

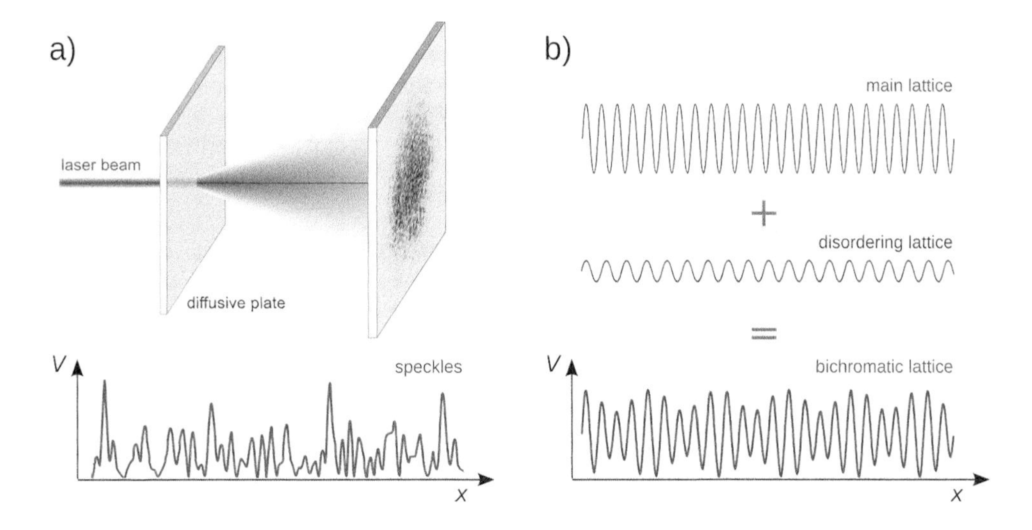

Fig. 7.8 Two different methods for the production of disordered potentials: a) speckle patterns, produced by the scattering of laser light from a diffusive glass plate; b) bichromatic optical lattices, produced by the superposition of two optical standing waves with incommensurate periodicity.

7.2.1 Disordered potentials for ultracold atoms

Different possibilities were proposed for the production of disordered ultracold atomic systems. Static disorder can be created by using inhomogeneous magnetic fields (Wang *et al.*, 2004*a*) or atoms of different species acting as scattering impurities (Gavish and Castin, 2005). However, the most successful implementation has been provided by the optical dipole potential (see Sec. 5.2) produced by inhomogeneous patterns of light intensity, as suggested e.g. in Damski *et al.* (2003) and Roth and Burnett (2003). Bose–Einstein condensates in disordered or quasi-disordered optical potentials were demonstrated in Lye *et al.* (2005) by using speckle patterns and in Fallani *et al.* (2007), respectively, with multi-chromatic incommensurate optical lattices.

Speckle patterns. The first realization of disordered potentials for cold atoms was obtained with *speckle patterns* (Boiron *et al.*, 1999). Speckles are produced whenever light is reflected by a rough surface or transmitted by a diffusive medium made up of many randomly distributed impurities by which the illuminating light is scattered. Since the scattering of laser light is mainly a coherent process, the partial waves emerging from the scattering interfere and produce a strongly inhomogeneous intensity pattern (Goodman, 2008) (see Fig. 7.8a). This disordered distribution of light can be imaged onto the atoms, producing a stationary disordered potential proportional to the local laser intensity $I(\mathbf{r})$ according to eqn (5.8). Speckles patterns are characterized by some degree of spatial correlations, which is reflected in a finite "grain size" of the disordered field. A lower limit to the speckle size is set by the characteristics of the diffuser and, more importantly, by the finite optical resolution of the imaging setup.

Multichromatic lattices. Inhomogeneous lattice potentials with broken translational invariance and small "grain size" can be synthesized by combining several optical standing waves with different non-commensurate spacings. The simplest example is given by a bichromatic lattice in which a main lattice with wavelength λ_1 is perturbed by a weaker secondary lattice with wavelength λ_2, with λ_2/λ_1 irrational (Fallani *et al.*, 2007), as shown in Fig. 7.8b. The resulting quasiperiodic potential can be written in the form

$$V(x) = V_1 \cos^2(k_1 x) + V_2 \cos^2(k_2 x) , \tag{7.15}$$

where $k_1 = 2\pi/\lambda_1$ and $k_2 = 2\pi/\lambda_2$ are the lattice wavenumbers and V_1 and V_2 indicate the heights of the two lattices. In the limit $V_2 \ll V_1$ the height of the optical barriers is roughly constant across the whole lattice and it is possible to define a tunnelling rate J which only depends on the main lattice height V_1. In this limit, the effect of the secondary lattice reduces to an inhomogeneous and non-periodic shift of the potential energy at the bottom of the lattice wells, characterized by an amplitude $\Delta = V_2$. Because of the lack of any translational invariance,[7] a bichromatic incommensurate lattice can be used to investigate the emergence of Anderson localization. This configuration implements the quasiperiodic version of the tight-binding model in eqn (7.13), which exhibits a transition from extended to Anderson-localized states for increasing quasi-periodic modulation strength (Aubry and André, 1980).

7.2.2 Anderson localization of noninteracting BECs

Anderson localization is a single-particle phenomenon, that holds for the wavefunction of a single atom placed in a disordered potential. Interactions between the particles, even the weak ones characterizing the atoms of a BEC, can introduce dephasing and easily destroy localization.[8] As a matter of fact, after early investigation with interacting Bose–Einstein condensates (Fallani *et al.*, 2008), Anderson localization was eventually demonstrated with BECs in the regime of vanishing interactions, obtained with different strategies (Aspect and Inguscio, 2009).

In the Palaiseau experiment reported in Billy *et al.* (2008) an initially trapped ^{87}Rb Bose–Einstein condensate was left free to expand in a disordered 1D waveguide produced by combining a weakly focused red-detuned laser beam with a 1D speckle potential. Interactions could be neglected by studying the BEC expansion only after the initial stage, when the density rapidly decreases and the atomic cloud becomes highly dilute. For weak disorder the BEC expansion was only slowed down, but, for a sufficiently large amount of disorder, the atomic wavefunction soon stopped expanding. In this regime the analysis of the in-situ density profiles showed a clear indication of exponentially decreasing tails, which is a signature of Anderson localization (see the inset of Fig. 7.9a).

[7]Incommensurate lattices have many interesting properties, as they are the simplest examples of *quasicrystals*, i.e. structures showing long-range order but no periodicity (Steinhardt, 1987), the physics of which interpolates between that of periodic systems and that of disorder. A review of the properties of incommensurate lattices, focused on their implementation with ultracold atoms, can be found in Fallani and Inguscio (2011).

[8]This behaviour is reminiscent of the interaction-induced dephasing of Bloch oscillations studied in Sec 6.4.2. In that case, as in the case of Anderson localization, single-particle coherence is essential for the observation of the phenomenon.

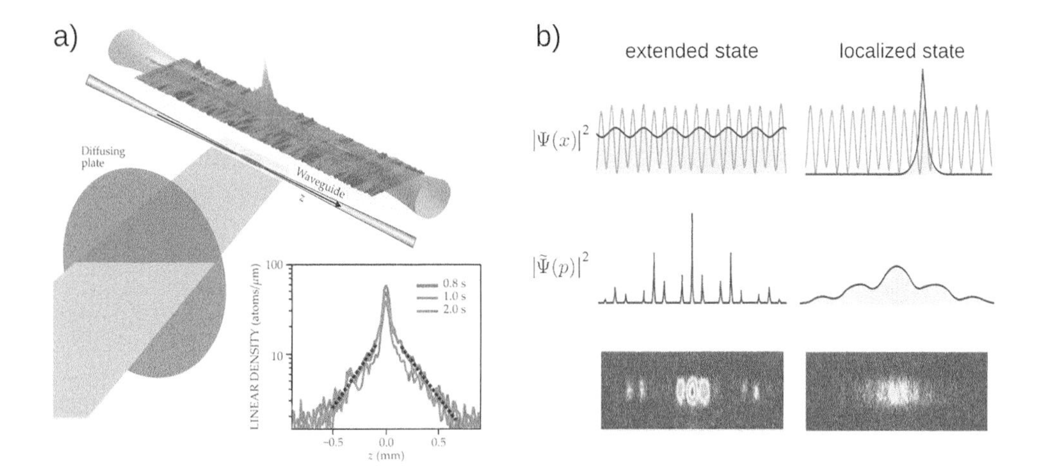

Fig. 7.9 Evidence of Anderson localization with ultracold atoms. a) The wavefunction of a weakly interacting ^{87}Rb BEC expanding in a speckle potential shows exponentially decreasing tails. b) The momentum distribution $|\tilde{\Psi}(p)|^2$ of a noninteracting ^{39}K BEC in a bichromatic lattice becomes broad as the wavefunction $\Psi(x)$ changes from extended to localized. Adapted from Aspect and Inguscio (2009).

In the Florence experiment reported in Roati *et al.* (2008), Anderson localization was observed for a noninteracting ^{39}K BEC in an incommensurate 1D bichromatic lattice. Here the strategy to exclude the effect of interactions was different: instead of working with dilute samples, interactions were cancelled by tuning a static magnetic field in the proximity of a Feshbach resonance (see Sec. 3.1.4) to set the scattering length to zero. Localization in the bichromatic lattice was studied in detail by looking at the expansion of the BEC, similarly to what was done in Billy *et al.* (2008). The localization transition was also characterized by imaging the atomic eigenstates in momentum space, which can be done after a time-of-flight expansion in which the wavefunctions undergo an evolution in free space (see Sec. 3.2.1). Extended states are characterized by a momentum distribution with narrow peaks, while spatially localized states have a broad momentum distribution, as expected from the Heisenberg uncertainty relation. The transition between these two regimes is exactly what was observed in the experiment when the bichromatic perturbation strength of the lattice was increased above a critical value (see Fig. 7.9b).

More recently, three-dimensional localization induced by disorder has been observed with noninteracting spin-polarized ^{40}K fermions in Kondov *et al.* (2011) and with dilute ^{87}Rb Bose–Einstein condensates in Jendrzejewski *et al.* (2012). In both the experiments the disordered potential was provided by 3D speckle patterns with short correlation length. The expansion of the ultracold atomic cloud in the speckle potential resulted in a two-component density distribution, with an expanding mobile component and a localized part that could be explained in terms of 3D Anderson localization.

Finally, a very interesting direction of research is represented by the investigation of the role of atom–atom interactions in the localization. Recently, the effects of weak repulsive interactions in the 1D localization of a ^{39}K Bose–Einstein condensate were investigated in Deissler *et al.* (2010) by tuning the scattering length with a magnetic Feshbach resonance. The experimental data showed that repulsive interactions counteract localization, leading to an increase of the quasidisorder strength necessary to enter the localized regime. Additional topics of interest include the observation of subdiffusive dynamics for the expansion in a disordered lattice in the presence of controlled repulsive interactions (Lucioni *et al.*, 2011) and the effects of disorder in the superfluid-insulator transition of strongly interacting Bose–Hubbard systems (Fallani *et al.*, 2007; Pasienski *et al.*, 2010) (see also Sec. 7.1).

7.3 New frontiers of quantum simulation

In this last section we will review some of the latest developments in the field of quantum simulation with ultracold atoms in optical lattices.

7.3.1 Single-site detection

An important technological advance is represented by the recent demonstration of high-resolution *in-situ* imaging, which allowed the possibility of detecting single atoms in single sites of an optical lattice. From a technical point of view the problem is challenging, since it requires an optical resolution on the same length scale as the wavelength of the light used in the imaging process.[9] For this reason, high-numerical aperture optics have to be used, which are generally difficult to integrate in an existing experimental setup, and accurately designed objectives have to be realized in order to reduce geometric aberrations and reach the diffraction limit.[10]

Recently, high-resolution imaging of a 2D gas of ^{87}Rb atoms trapped in optical lattices has been demonstrated by the group of M. Greiner at Harvard (Bakr *et al.*, 2009). In this experiment a high-numerical hemispherical lens was placed inside the vacuum cell in order to optimize light detection and reach an effective numerical aperture $NA = 0.8$ thanks to a "solid-immersion" effect, yielding an optical resolution $d \simeq 500$ nm. The atoms were detected with fluorescence imaging (see Sec. 3.1.5) by using red-detuned laser light which was simultaneously exciting the atoms and cooling them by optical molasses in the sites of the optical lattice.[11] The scheme of the imaging setup and an image of single atoms detected in the optical lattice is shown in Fig. 7.10. The same optical setup for high-resolution imaging was also used in Bakr *et al.* (2009), in reversed direction, to produce the optical lattice in which the atoms

[9] The distance between sites in an optical lattices is on the order of 0.5 μm, while imaging is generally performed on the first allowed transition in alkaline atoms, which lies in the visible or near infrared (0.5–0.8 μm). Increasing the spacing between sites (e.g. with optical lattices at small angle) would allow an easier optical detection, but at the price of an exponentially reduced tunnelling.

[10] In an optical system, the diffraction-limited resolution is given by $d = 0.61\lambda/NA$, where NA is the numerical aperture of the objective lens, given by $NA = n \sin\theta$, where n is the index of refraction of the medium between sample and objective lens and θ is the maximum angle of the rays with respect to the axis of the optical system (Born and Wolf, 1997).

[11] In this regime fluorescence imaging is much more efficient than absorption imaging, since the optical density of a single atom is too small to produce a detectable absorption in a probe beam owing to technical noise and photon shot noise in the probe beam. Fluorescence imaging is not affected by this problem since it is an imaging technique with ideally zero background (see Sec. 3.1.5).

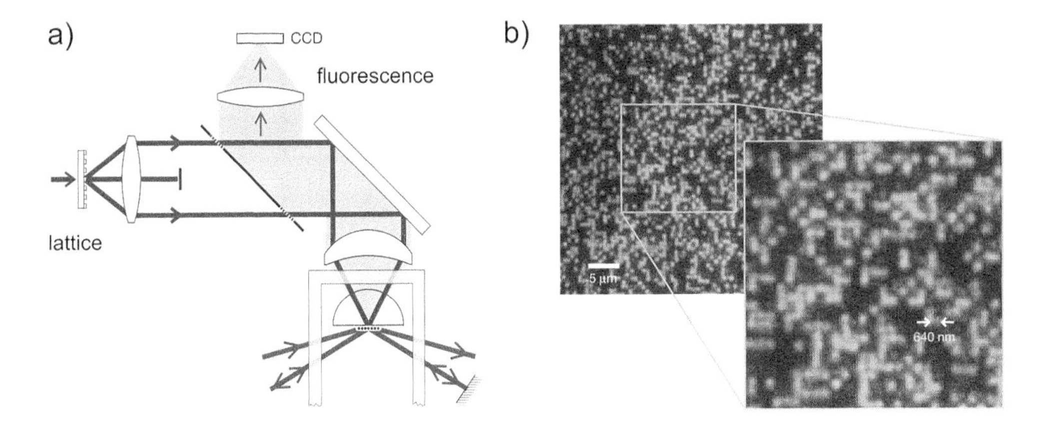

Fig. 7.10 a) Optical setup for single-site imaging of ultracold atoms in optical lattices. b) Fluorescence images from single trapped atoms. Reprinted with permission from Bakr *et al.* (2009). © Macmillan Publishers Ltd.

are trapped. This alternative approach to the production of optical lattices offers important possibilites for the realization of lattices with exotic topologies (such as the triangular and hexagonal lattices recently implemented with optical standing waves e.g. in Becker *et al.* (2010), Soltan-Panahi *et al.* (2011), and Tarruell *et al.* (2012)) or for the manipulation of the atom internal state on individually addressed lattice sites, as realized at MPQ in Weitenberg *et al.* (2011).

The superfluid–Mott transition was investigated with single-site resolution at MPQ (Sherson *et al.*, 2010) and Harvard (Bakr *et al.*, 2010). Figure 7.11a shows images of a 2D gas of ^{87}Rb bosons in the superfluid-BEC state, characterized by strong site-to-site fluctuations in the number of atoms. Figure 7.11b–f, instead, shows images of Mott insulating states with increasing atom number. It is evident how the perfect unit-filling of the Mott insulating state evolves, in the presence of the underlying harmonic trap, towards the expected shell structure at larger atom numbers (see Sec. 7.1.3). We observe that the fluorescent imaging technique implemented in Bakr *et al.* (2009), Sherson *et al.* (2010), and Bakr *et al.* (2010) provides a measurement of the *parity* of the atom number instead of the atom number itself. As a matter of fact, the strong illumination of the atoms during the imaging (together with their tight trapping in the lattice) induces strong light-assisted collisions that cause pairs of atoms to acquire a large kinetic energy and quickly escape from the trap: therefore only sites with an odd number of atoms contribute to the fluorescence signal. This is the reason why sites with two atoms appear as empty sites (central part of Fig. 7.11e), while sites with three atoms appear as bright as sites with one atom (central part of Fig. 7.11g).

7.3.2 Synthetic vector potentials

In the previous sections we have discussed the profound analogies between ultracold atoms in optical lattices and the electrons of a crystalline solid. Nevertheless, there is a major difference arising from the electrical neutrality of ultracold atoms vs. the negative

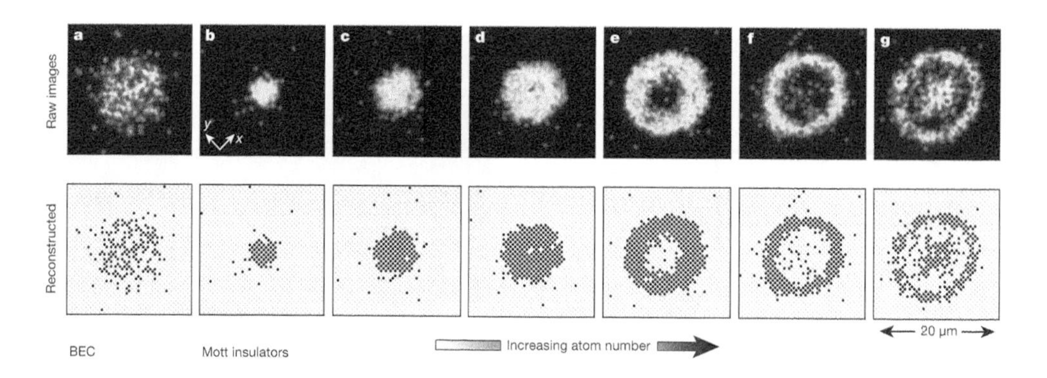

Fig. 7.11 Single-site fluorescence image of ultracold bosons trapped in a two-dimensional optical lattice. Bright spots correspond to sites with an odd number of bosons, dark spots to sites with even (or zero) number of bosons. a) Weakly interacting Bose–Einstein condensate. b–f) Mott insulators for increasing atom number. Reprinted with permission from Sherson *et al.* (2010). © Macmillan Publishers Ltd.

charge of electrons. A number of interesting condensed-matter effects is connected with the Lorentz force $\mathbf{F} = -e\mathbf{v} \times \mathbf{B}$ experienced by an electron moving with velocity $\mathbf{v}$ in the presence of a magnetic field $\mathbf{B}$. The quantum-mechanical Hamiltonian description of this interaction is given in terms of the electromagnetic potential by writing the kinetic energy term as $(\mathbf{p} + e\mathbf{A})^2/2m$, where $\mathbf{p}$ is the canonical momentum and $\mathbf{A}$ is the magnetic vector potential, being $\mathbf{B} = \nabla \times \mathbf{A}$.

The realization of artificial magnetic fields for neutral atoms could enable the simulation of the *quantum Hall effect* (QHE) and other topological effects in systems of ultracold atoms. The simplest example is given by the integer QHE in a 2D electron gas, already discussed in Sec. 4.5.2 for its implications for the measurement of the fine structure constant. Interactions between the particles can lead to the emergence of the fractional quantum Hall effect (FQHE), which is connected to the formation of a strongly correlated 2D electron liquid, as conjectured by R. B. Laughlin (Laughlin, 1999), with excitations characterized by fractional elementary charge and possibly by fractional (anyonic) quantum statistics. The possibility of investigating the FQHE in clean atomic systems is particularly significant and could lead to new insight into the rich physics of strongly correlated 2D systems.

According to quantum mechanics, when an electron moves in a region with a magnetic vector potential $\mathbf{A}$, it acquires a geometric phase given by the path integral $\frac{e}{\hbar}\int \mathbf{A} \cdot d\mathbf{x}$: this is the so-called *Aharonov–Bohm effect*. Recent theoretical proposals (see Dalibard *et al.* (2011) for an excellent review and introduction to this topic) have outlined the possibility of using phase imprinting by laser light for the implementation of artificial magnetic fields for neutral atoms.

Adiabatic potentials for dressed states. The first demonstration of synthetic magnetic fields was realized at NIST and reported in Lin *et al.* (2009). The geometric phase was imprinted by coupling the $F = 1$ Zeeman states of a ^{87}Rb BEC with a pair

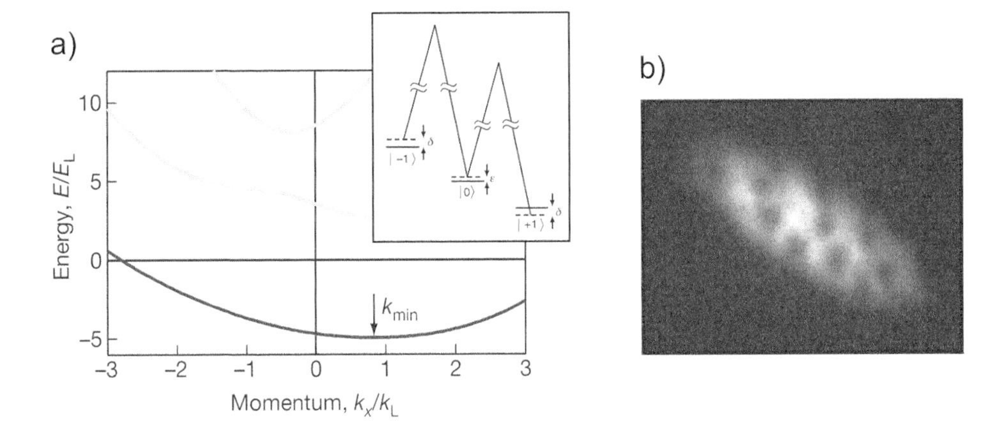

Fig. 7.12 An artificial magnetic field for artificially charged ultracold atoms. a) Two laser beams induce Raman couplings between Zeeman levels of the $F = 1$ ^{87}Rb ground state, resulting in dressed states with modified dispersion relation. b) BEC vortices induced by the artificial magnetic field. Adapted with permission from Lin *et al.* (2009). © Macmillan Publishers Ltd.

of Raman laser beams. In presence of this dressing, for appropriate parameters, the lowest-energy state was characterized by a dispersion relation with a minimum at a nonzero momentum k_{min} (see Fig. 7.12a), resembling the effect of the magnetic vector potential in shifting the minimum of the kinetic energy $(\mathbf{p} + e\mathbf{A})^2/2m$. By making this vector potential spatially dependent, an artificial magnetic field for artificially charged particles was simulated and its effects were detected by observing the appearance of vortices in the Bose–Einstein condensate (see Fig. 7.12b), in agreement with the expected theoretical behaviour.

Further developments of this technique allowed the same experimental group to achieve other important results, such as the demonstration of synthetic electric fields for neutral atoms (Lin *et al.*, 2009) and the realization of spin–orbit coupling for BECs (Lin *et al.*, 2011). The latter achievement is particularly significant, since spin–orbit coupling is an important mechanism in many intriguing condensed-matter effects, as in spin–Hall systems and in other topological insulators.

Laser-assisted tunnelling. An alternative approach, valid for atoms trapped in optical lattices, was proposed in Jaksch and Zoller (2003). This scheme is based on *laser-assisted tunnelling*, a process in which appropriate laser fields can induce tunnelling between the sites of an optical lattice in which ordinary tunnelling is suppressed. This laser-assisted tunnelling is characterized by a complex tunnelling amplitude $Je^{i\theta}$, where the geometric phase θ, dependent on the laser wavevector, is imprinted onto the atom wavefunction during the tunnelling process. By suitable laser arrangements, a nonzero phase ϕ (equivalent to the Aharanov–Bohm effect) can be induced on a closed loop, realizing the equivalent of a strong artificial magnetic field, as recently demonstrated in Aidelsburger *et al.* (2011).

7.3.3 Quantum magnetism with atoms and ions

A very interesting topic in quantum simulation with ultracold atoms in optical lattices regards the possibility of studying *quantum magnetism*. Spin models have been introduced in condensed-matter physics, in order to explain the magnetic properties of strongly correlated materials, and in statistical physics, where they are often used as paradigmatic systems for the description of quantum phase transitions. Quantum simulation with ultracold atoms could improve our understanding of many aspects of quantum magnetism for which known analytical approaches or numerical techniques are not adequate. Among these are the behaviour of magnetic systems in the presence of frustration, spin liquids, and systems with disordered spin couplings (spin glasses).

Magnetic phases in Hubbard models. A first approach to quantum spin models with ultracold atoms in optical lattices is given by the realization of effective spin Hamiltonians such as the low-energy limit of Bose– and Fermi–Hubbard models in the Mott insulating state. Spins can be either real atomic spins or pseudo-spins which effectively describe a mixture of two atomic species (sites occupied by different atoms correspond to different pseudo-spin projections). In order to distinguish the strength of the interaction between spins, usually denoted with J in solid-state physics, from the single-particle tunnelling energy, in this section we will indicate the latter with t.

We consider the Bose– or Fermi–Hubbard model for a system of (pseudo)spin-1/2 in the case of one atom per lattice site. In the atomic limit of a Mott insulator $U/t \to \infty$ number fluctuations are completely frozen out, as we have discussed in the previous sections. At finite t the effect of atom tunnelling can be evaluated with second-order perturbation theory, describing virtual processes in which a particle hops to the nearest-neighbouring site and, after interacting with the atom in that site, comes back to the initial site, as pictorially sketched in Fig. 7.13a. The effects of this virtual tunnelling are equivalent to those of an effective Heisenberg Hamiltonian with a *superexchange* coupling between nearest-neighbouring quantum spins $\mathbf{s}_i$:

$$H_{eff} = J \sum_{<i,j>} \mathbf{s}_i \cdot \mathbf{s}_j \ . \tag{7.16}$$

The spin–spin coupling is given by the superexchange energy

$$J = \pm \frac{4t^2}{U} \ , \tag{7.17}$$

where the sign + corresponds to the case of fermions, which exhibit an antiferromagnetic spin coupling, while the sign − corresponds to the case of bosons, for which the coupling is ferromagnetic.[12]

Experiments are currently on the way for the observation of magnetic ordering in atomic two-component Mott insulators, with a particular interest for fermionic antiferromagnetism. The spin ground state of the Fermi–Hubbard model in a 2D square lattice is expected to be the Néel state, in which a checker-board phase of

[12] The different sign is a consequence of quantum statistics, which forbids second-order tunnelling processes for identical fermionic spins, while they are possible for bosons.

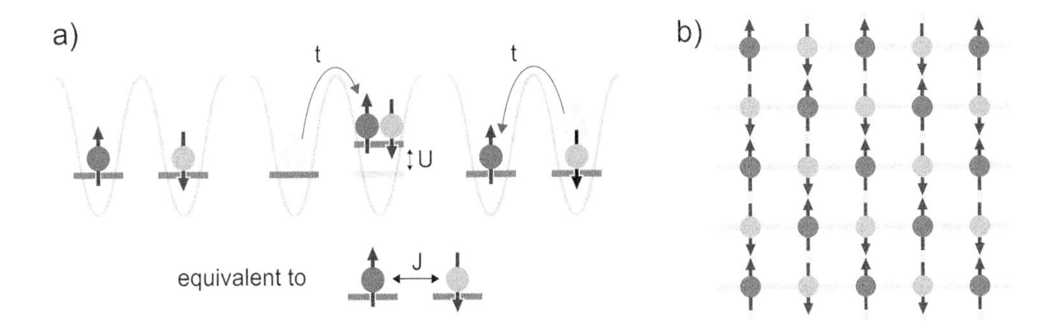

Fig. 7.13 a) Representation of second-order tunnelling processes in a Mott state giving rise to a superexchange interaction between spin-1/2 atoms. b) Magnetic ordering in an antiferromagnetic Néel state.

spin-up/spin-down is formed, exhibiting a staggered magnetization, as shown in Fig. 7.13b.[13] The investigation of the Fermi–Hubbard model away from half-filling is also particularly significant, since in the presence of doping the Fermi–Hubbard model is a strong candidate for describing the origin of high-T_C superconductivity. However, despite a strong theoretical effort in the last decades, no consensus on this point has been reached. This problem is indeed one of the most important open questions in condensed-matter physics, and its quantum simulation with ultracold atoms could bring important elements of understanding.

Fermionic Mott insulators have been experimentally realized (Jördens *et al.*, 2008; Schneider *et al.*, 2008) and the control of the superexchange interaction has already been demonstrated in Trotzky *et al.* (2008) for ultracold bosons in optical superlattices. Binary mixtures of atomic species, realizing pseudospin-1/2 systems, have also been realized in the strongly interacting Hubbard regime, including Fermi–Bose mixtures (Günter *et al.*, 2006; Ospelkaus *et al.*, 2006; Best *et al.*, 2009) and heteronuclear Bose–Bose mixtures (Catani *et al.*, 2008). In the bosonic case different tunnelling rates and different interaction strengths can be engineered for the two atomic species, resulting in more rich Heisenberg models exhibiting different phases, including ferromagnetic and antiferromagnetic Néel states (Duan *et al.*, 2003; Kuklov and Svistunov, 2003; Altman *et al.*, 2003).

However, no magnetic ordering arising from the superexchange interaction has been observed so far in experiments. A major problem which has not yet been solved is related to the critical temperature for the observation of magnetic ordering. As a matter of fact, spin correlations are expected to be rapidly lost above a critical temperature on the same order as the superexchange energy scale $J \sim t^2/U$, which is smaller than the lowest temperature observed in experiments. Novel cooling schemes have been recently proposed and in the near future could lead to the achievement of colder Mott

[13]Antiferromagnetic spin models on different topologies may be characterized by frustration. This happens in the case of a triangular lattice, where the condition of having opposite spins in nearest-neighbouring sites cannot be fulfilled for all the bonds. In this case a highly degenerate ground state is expected.

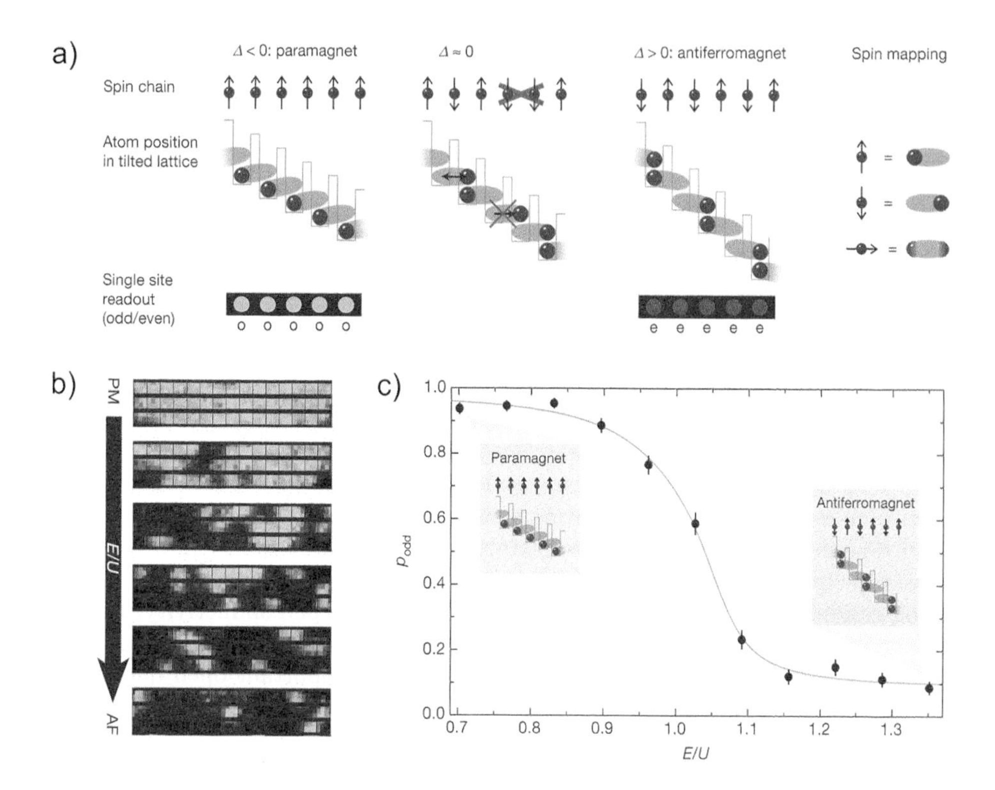

Fig. 7.14 Investigation of the 1D antiferromagnetic quantum Ising model with ultracold bosons in a tilted optical lattice. a) The atom position in the tilted lattice (left/right well) is mapped onto the orientation of an effective spin-1/2. The parity fluorescence detection maps bright or dark rows of atoms onto paramagnetic or antiferromagnetic regions, respectively. b) Transition from a paramagnetic state to an antiferromagnetic state as the strength of the effective longitudinal magnetic field is changed. c) Measured fraction of spins in "paramagnetic regions" around the phase transition at $E/U = 1$. Reprinted with permission from Simon *et al.* (2011). © Macmillan Publishers Ltd.

insulators for the observation of magnetic ordering (see McKay and DeMarco (2011) for a state-of-the-art review on the proposed experimental techniques to lower and measure the temperature of ultracold atomic systems in optical lattices).

1D antiferromagnetism with pseudospins. Recently, a very different approach to quantum magnetism in optical lattices has been experimentally realized in Simon *et al.* (2011) with a 1D Mott insulator of spinless bosons, on the basis of an earlier proposal (Sachdev *et al.*, 2002). An external magnetic field gradient is used to introduce a site-to-site energy difference E in the lattice. If E is smaller than the Mott gap U no excitation occurs, as shown in Fig. 7.14a, and the lattice is occupied by one atom per site. For $E > U$ the atoms can tunnel to the nearest-neighbouring sites, unless the atoms in those sites have not yet tunnelled: this produces a "staggered" state with a

$0-2-0-2-\ldots$ density wave. It is possible to map the state of each particle, whether it has tunnelled or not, onto the state of a pseudospin-1/2: $s_{iz}=1/2$ for absence of tunnelling, $s_{iz}=-1/2$ for a tunnelling event, as shown in the right of Fig. 7.14a. With this mapping it is possible to derive an effective Hamiltonian, which describes a 1D antiferromagnetic Ising spin chain in a homogeneous effective field:

$$H_{eff}=J\sum_{<i,j>}s_{iz}s_{jz}-\sum_{i}\left(h_z s_{iz}+h_x s_{ix}\right)\,, \tag{7.18}$$

where $J\simeq U$ and the effective magnetic field is given by $h_z=1-\frac{E-U}{J}$ ($h_x\simeq 0$ deep in the Mott insulator regime). This model exhibits a phase transition between a paramagnetic state and an antiferromagnetic state as the effective magnetic field h_z decreases below a critical value $h_z\simeq 1$. The phase transition can be observed with the single-site imaging technique described in Sec. 7.3.1: owing to the parity detection, the paramagnetic state with 1 atom per site for $E/U<1$ corresponds to the detection of a bright line of sites, while the antiferromagnetic state with $0-2-0-2-\ldots$ atoms for $E/U>1$ corresponds to a sequence of dark spots, as shown in Fig. 7.14a. This transition can be observed in the experimental images shown in Fig. 7.14b and in the measured fraction of atoms in paramagnetic domains shown in Fig. 7.14c.

One of the advantages of this approach, in addition to the possibility of observing *in situ* the effective "magnetization" domains, resides in the looser temperature requirements, owing to the stronger effective spin–spin coupling: in this case $J\simeq U$ (approximately $h\times 1$ kHz) instead of $J\sim t^2/U$ in the superexchange case (approximately $h\times$ few tens of Hz for typical experimental values in the Mott state).

Simulating quantum magnets with trapped ions. We conclude this section by mentioning the recent application of trapped ions (see Sec. 5.4) to the investigation of quantum spin models (see Blatt and Roos (2012) for a review on quantum simulation with trapped ions). In these novel quantum simulation experiments the internal state of the ion is mapped onto the orientation of an effective spin-1/2. External laser beams couple the ion state to collective modes of motion (induced by the Coulomb interaction) which mediate effective couplings between the spins. The result is an effective quantum Ising Hamiltonian in a transverse field:

$$H_{eff}=\sum_{i<j}J_{ij}s_{iz}s_{jz}-\sum_{i}\left(h_x s_{ix}\right)\,, \tag{7.19}$$

where, differently from experiments with neutral atoms, the couplings J_{ij} can be long-ranged (since they are mediated by collective modes of the ion crystal).

This technique was experimentally demonstrated in Friedenauer *et al.* (2008) on a system of two $^{25}Mg^+$ ions with the realization of tunable ferromagnetic couplings between the spins and the observation of a phase transition from a paramagnetic to a ferromagnetic state. Spin–spin interactions mediated by transverse collective modes were also realized in Kim *et al.* (2009) and Kim *et al.* (2010) on small chains of $^{171}Yb^+$ ions, coupled by either ferromagnetic or antiferromagnetic interactions. The scaling of this approach to much larger two-dimensional systems (where quantum simulation could provide advantages over classical computation) has been demonstated in Britton *et al.*

(2012) on a triangular Wigner crystal formed by hundreds of $^9Be^+$ ions in a Penning trap, where the character of the Ising interaction could be tuned from short-range dipole–dipole interactions, through Coulomb-range interactions, up to infinite-range couplings.

A very different approach to quantum simulations with ions has been reported in Lanyon *et al.* (2011) with the demonstration of a *digital quantum simulation*. Instead of performing an analog quantum simulation by engineering Hamiltonian couplings between the spins, the dynamics of an initial quantum state was "computed" digitally by implementing quantum gates on the ions which approximate the operator $\hat{U}$ governing the time evolution of an initial state $|\Psi_0\rangle$ according to $|\Psi(t)\rangle = \hat{U}|\Psi_0\rangle$. This digital simulator, which is intrinsically "reprogrammable" by changing the sequence of gates, was implemented in Lanyon *et al.* (2011) to study different quantum Ising models describing up to six spins in an external magnetic field with different kinds of interactions. The state of the spins was mapped onto the electronic state of trapped $^{40}Ca^+$ ions and an appropriate sequence of quantum gates was applied to the ions to simulate the time evolution under the Ising Hamiltonian. The results of the digital quantum simulation were then compared with the results of an exact calculation, demonstrating a very good fidelity of the operation of the quantum simulator. This new digital approach introduces a new paradigm in quantum simulation and represents the first step towards the realization of universal quantum simulators capable of simulating the time evolution of larger quantum systems in various Hamiltonian models.

7.3.4 Simulation of relativistic quantum mechanics

We conclude this chapter by mentioning recent experiments in which some aspects of relativistic quantum mechanics have been quantum-simulated in ultralow-energy atomic physics systems.

The first example is given by the investigation of particles with relativistic energy–momentum dispersion relation (Tarruell *et al.*, 2012). This feature arises naturally for a particle moving in a hexagonal lattice, as happens for electrons moving in two-dimensional graphene (Novoselov, 2011). Single-particle band calculations show that the hexagonal lattice structure of graphene determines the existence of points q^* in the reciprocal space around which the dispersion relation $E(q)$ varies linearly with the momentum according to

$$E(q) = \pm\hbar c(q - q^*) , \tag{7.20}$$

as in the case of a massless relativistic particle, instead of quadratically, as in the case of a classical massive particle with energy–momentum dispersion relation $E(q) = \hbar^2 q^2/2m$. A system of spin-polarized fermionic ^{40}K atoms in honeycomb optical lattices was recently realized in Tarruell *et al.* (2012), where the presence of Dirac points in the energy spectrum was demonstrated by a momentum-resolved detection of the energy gap between bands: a controlled deformation of the hexagonal lattice was used to control the position of the Dirac points inside the Brillouin zone and to eventually merge them. This work represents a first step towards the quantum simulation of graphene physics and other relativistic effects with ultracold atoms in optical lattices.

A second example is related to the quantum simulation of the relativistic Dirac equation with trapped ions. In Gerritsma *et al.* (2010) the positive and negative energy components of a Dirac spinor were mapped onto two electronic states of a single trapped $^{40}\text{Ca}^{+}$ ion connected by a narrow electric quadrupole transition. An appropriate laser coupling between the two states resulted in the realization of an effective one-dimensional Dirac equation, which allowed the authors of Gerritsma *et al.* (2010) to demonstrate the existence of the long-searched-for *Zitterbewegung*. This effect is the fast oscillatory motion that is predicted by the Dirac theory for a massive particle as a result of the interference between the positive and negative energy components of the Dirac spinor. This effect, not detectable for electrons because of the too small amplitude oscillation and too large frequency, was clearly observed in the trapped ion experiment thanks to a mapping onto an ion dynamics which could be observed on a time and length scale accessible to the experiment. Extension of this approach with trapped ions allowed the same group to investigate relativistic scattering dynamics, observing the effects of the famous *Klein paradox*, consisting in the nearly total transmission of a relativistic particle by a very high potential barrier, impenetrable for both classical and nonrelativistic quantum mechanics (Gerritsma *et al.*, 2011).

Appendix A
Atom–light interaction

In this appendix we will review the basic concepts regarding the interaction of an atom with classical coherent radiation. More information can be found e.g. in Foot (2005), Grynberg *et al.* (2010), Metcalf and van Der Straten (1999), and Allen and Eberly (1987).

A.1 Interaction with a coherent field

We consider the case of near-resonant interaction, in which the atom can be approximated as a system with only two energetic levels, which are coupled by the oscillating field. A generic state can be expressed as

$$|\psi\rangle = a_g|g\rangle + a_e|e\rangle \, , \tag{A.1}$$

where $|g\rangle$ and $|e\rangle$ indicate the electronic ground state and the excited state, with energies $\hbar\omega_g$ and $\hbar\omega_e$ respectively. The energy separation between these two states defines the resonance (angular) frequency $\omega_0 = \omega_e - \omega_g$. The Hamiltonian of the unperturbed system can be written as

$$\hat{H}_0 = \hbar\omega_g|g\rangle\langle g| + \hbar\omega_e|e\rangle\langle e| \, . \tag{A.2}$$

A.1.1 Interaction Hamiltonian

We will now derive an expression for the Hamiltonian describing the interaction of the atomic electron with an external electromagnetic field. The most general expression for this Hamiltonian is

$$\hat{H} = \frac{[\mathbf{p} + e\mathbf{A}(\mathbf{r},t)]^2}{2m} - eV(\mathbf{r},t) + U(\mathbf{r}) \, , \tag{A.3}$$

where $\mathbf{A}(\mathbf{r},t)$ and $V(\mathbf{r},t)$ are, respectively, the vector and scalar potentials describing the electromagnetic field, which are connected to the electric field $\mathbf{E}$ and to the magnetic field $\mathbf{B}$ by the relations

$$\mathbf{B} = \nabla \times \mathbf{A} \tag{A.4}$$

$$\mathbf{E} = -\nabla V - \frac{\partial A}{\partial t} \, , \tag{A.5}$$

while $U(\mathbf{r})$ describes the interaction of the electron with the atomic nucleus (in the case of a multi-electron atom, with the other electrons as well). We consider the case of classical radiation, in which the field propagates along $\hat{z}$ as a plane wave with electric and magnetic fields

$$\begin{cases} \mathbf{E}(z,t) = E_0 \cos(kz - \omega t)\hat{\mathbf{x}} = \frac{E_0}{2}\left[e^{i(kz-\omega t)} + \text{h.c.}\right]\hat{\mathbf{x}} \\ \mathbf{B}(z,t) = B_0 \cos(kz - \omega t)\hat{\mathbf{y}} = \frac{E_0}{2c}\left[e^{i(kz-\omega t)} + \text{h.c.}\right]\hat{\mathbf{y}} \,. \end{cases} \tag{A.6}$$

Different choices for the potentials $\mathbf{A}$ and V can be made, one of the simplest being the so-called *radiation gauge*

$$\begin{cases} \mathbf{A}(z,t) = \frac{E_0}{2}\left[\frac{1}{i\omega}e^{i(kz-\omega t)} + \text{h.c.}\right]\hat{\mathbf{x}} \\ V(z,t) = 0 \,. \end{cases} \tag{A.7}$$

The spatial dependence of the vector potential can be treated with a series expansion of the term e^{ikz}. As a matter of fact, the size of the atomic orbitals is on the order of the Bohr radius $a_0 \simeq 0.53$ Å, which is much smaller than the wavelength $\lambda = 2\pi/k$ of the radiation needed to resonantly excite the atoms (in the visible or near-visible range). Since $a_0 \ll \lambda$, the spatial variation of the electromagnetic field on the electron unperturbed wavefunctions can be neglected, $kz \ll 1$, and the following zero-order *electric dipole approximation* can be made:

$$e^{ikz} \approx 1 \,. \tag{A.8}$$

In order to derive the interaction Hamiltonian a different gauge for the electromagnetic potentials can be more conveniently chosen. As a matter of fact, the measurable fields $\mathbf{B}$ and $\mathbf{E}$ associated with the potentials $\mathbf{A}$ and V are invariant under gauge transformations

$$\begin{cases} \mathbf{A} \to \mathbf{A}' = \mathbf{A} + \nabla\chi \\ V \to V' = V - \frac{\partial\chi}{\partial t} \,, \end{cases} \tag{A.9}$$

where χ is an arbitrary scalar function. Under the electric dipole approximation $\mathbf{A}$ does not depend on position, therefore it is possible to choose a gauge function $\chi = -\mathbf{r}\cdot\mathbf{A}$ which allows us to cancel the contribution of the vector potential:

$$\begin{cases} \mathbf{A}'(\mathbf{r},t) = 0 \\ V'(\mathbf{r},t) = \mathbf{r}\cdot\frac{\partial\mathbf{A}}{\partial t} = -\mathbf{r}\cdot\mathbf{E}(t) \,. \end{cases} \tag{A.10}$$

With this gauge choice the Hamiltonian in eqn (A.3) can be written as

$$\hat{H} = \frac{\mathbf{p}^2}{2m} + U(\mathbf{r}) + e\mathbf{r}\cdot\mathbf{E}(t) \,, \tag{A.11}$$

in which the interaction term corresponds to the potential energy $-\mathbf{d}\cdot\mathbf{E}(t)$ of an electric dipole $\mathbf{d} = -e\mathbf{r}$ in an oscillating electric field $\mathbf{E}(t)$.

In the following sections we will focus on the atom–field interaction within the electric dipole approximation. We anticipate that, when this approximation is released and more terms are taken into account in the expansion of eqn (A.8), new interaction terms will appear in eqn (A.11) corresponding to higher-order contributions. We will discuss these higher-order interaction terms in Sec. A.4.2.

A.1.2 Rotating wave approximation

It is easy to show that, on the basis spanned by the unperturbed electronic states $|g\rangle$ and $|e\rangle$, the dipole interaction Hamiltonian $\hat{H}_{int} = e\mathbf{r} \cdot \mathbf{E}(t)$ has off-diagonal matrix elements, i.e. it couples the ground and excited states according to:

$$\begin{aligned} \langle e|\hat{H}_{int}|g\rangle &= e \int d\mathbf{r}\, \psi_e^* \, (\mathbf{r} \cdot \mathbf{E}) \, \psi_g \\ &= e\frac{\mathbf{E_0}}{2} \left(e^{-i\omega t} + \text{h.c.}\right) \cdot \int d\mathbf{r}\, \psi_e^* \, \mathbf{r} \, \psi_g \\ &= e\frac{\mathbf{E_0}}{2} \left(e^{-i\omega t} + \text{h.c.}\right) \cdot \boldsymbol{\mu}_{eg} \\ &= \hbar\frac{\Omega}{2} \left(e^{-i\omega t} + \text{h.c.}\right) , \end{aligned} \tag{A.12}$$

where we have introduced the *transition dipole moment*

$$\boldsymbol{\mu}_{eg} \equiv \langle e \,|\mathbf{r}|\, g\rangle = \int d\mathbf{r}\, \psi_e^* \, \mathbf{r} \, \psi_g \, , \tag{A.13}$$

which only depends on the unperturbed wavefunctions, and the *Rabi frequency*

$$\Omega \equiv \frac{e\boldsymbol{\mu}_{eg} \cdot \mathbf{E_0}}{\hbar} \, , \tag{A.14}$$

which parametrizes the strength of the coupling with the electromagnetic field. On the contrary, the diagonal matrix elements of $\hat{H}_{int}$ are zero owing to the well-defined parity of the electronic orbitals.[1] Therefore the full Hamiltonian can be written as:

$$\begin{aligned} \hat{H} = \; & \hbar\omega_g|g\rangle\langle g| + \hbar\omega_e|e\rangle\langle e| + \\ & \hbar\frac{\Omega}{2} \left(e^{-i\omega t} + \text{h.c.}\right) |e\rangle\langle g| + \hbar\frac{\Omega^*}{2} \left(e^{-i\omega t} + \text{h.c.}\right) |g\rangle\langle e| \, , \end{aligned} \tag{A.15}$$

or, using the matrix representation of the Hamiltonian operator, as

$$\hat{H} = \begin{bmatrix} 0 & \hbar\frac{\Omega^*}{2} \left(e^{-i\omega t} + \text{h.c.}\right) \\ \hbar\frac{\Omega}{2} \left(e^{-i\omega t} + \text{h.c.}\right) & \hbar\omega_0 \end{bmatrix} . \tag{A.16}$$

In this latter representation the energy has been redefined by subtracting a constant energy term from the diagonal in such a way that the energy of the ground state is zero, while the energy of the excited state is $\hbar\omega_0 = \hbar(\omega_e - \omega_g)$.

We now consider the time-dependent Schrödinger equation for the Hamiltonian in eqn (A.15). By expanding a generic state $|\psi\rangle$ according to eqn (A.1), we obtain the

[1] The electronic orbitals $\psi(\mathbf{r})$ are eigenfunctions of the parity operator $\hat{P}$ with eigenvalues $(-1)^l$, l being the orbital angular momentum quantum number. As a consequence, the integrals $\boldsymbol{\mu}_{gg} \equiv \int d\mathbf{r}\, |\psi_g|^2 \, \mathbf{r}$ and $\boldsymbol{\mu}_{ee} \equiv \int d\mathbf{r}\, |\psi_e|^2 \, \mathbf{r}$ are identically zero, since their arguments are spatially antisymmetric functions.

following equations for the coefficients $a_g(t)$ and $a_e(t)$, describing the amplitudes of the ground and excited state, respectively:

$$\begin{cases} i\dot{a}_g = \frac{\Omega^*}{2}\left(e^{-i\omega t} + \text{h.c.}\right) a_e \\ i\dot{a}_e = \frac{\Omega}{2}\left(e^{-i\omega t} + \text{h.c.}\right) a_g + \omega_0 a_e \ , \end{cases} \tag{A.17}$$

where the dotted coefficients $\dot{a}_g$ and $\dot{a}_e$ indicate their time derivatives. This system of differential equations with time-dependent coefficients can be reduced to a system of equations with constant coefficients by applying the so-called *rotating wave approximation (RWA)*, which consists in writing the same equations in a rotating basis and dropping the fast-oscillating terms. The new basis can be chosen in different ways. It is particularly convenient to consider a new basis in which the ground state $|\tilde{g}\rangle$ is the same as $|g\rangle$, while the excited state $|\tilde{e}\rangle$ differs from $|e\rangle$ by a phase factor rotating at the same frequency ω as the driving field. With this choice, the state $|\psi\rangle$ can be expanded on the new basis according to:

$$\begin{cases} \tilde{a}_g = a_g \\ \tilde{a}_e = a_e e^{i\omega t} \ . \end{cases} \tag{A.18}$$

Substituting eqn (A.18) in eqn (A.17) we obtain the equations for the new coefficients:

$$\begin{cases} i\dot{\tilde{a}}_g = \frac{\Omega^*}{2}\left(1 + e^{-i2\omega t}\right) \tilde{a}_e \simeq \frac{\Omega^*}{2}\tilde{a}_e \\ i\dot{\tilde{a}}_e = \frac{\Omega}{2}\left(1 + e^{i2\omega t}\right) \tilde{a}_g - \delta\tilde{a}_e \simeq \frac{\Omega}{2}\tilde{a}_g - \delta\tilde{a}_e \ , \end{cases} \tag{A.19}$$

where we have defined the *detuning* $\delta = \omega - \omega_0$ as the difference between the radiation frequency ω and the atomic resonance frequency ω_0. The rotating wave approximation made in eqn (A.19) consists in dropping the fast oscillating terms with frequency 2ω, which are varying on a much faster timescale than the evolution of the amplitudes $\tilde{a}_g$ and $\tilde{a}_e$, which takes place on a timescale $\sim 1/|\Omega|$, as can be verified in the next section.[2] These equations describe the slow time-evolution of the amplitudes $\tilde{a}_g$ and $\tilde{a}_e$ induced by a rotating-wave Hamiltonian with matrix elements given by:

$$\hat{\tilde{H}} = \begin{bmatrix} 0 & \hbar\frac{\Omega^*}{2} \\ \hbar\frac{\Omega}{2} & -\hbar\delta \end{bmatrix} . \tag{A.20}$$

A.1.3 Coherent dynamics

Equation (A.19) is a system of two linear differential equations with time-independent coefficients, which we are going to solve for the initial condition

$$\begin{cases} \tilde{a}_g(0) = 1 \\ \tilde{a}_e(0) = 0 \end{cases} \tag{A.21}$$

describing an atom in the ground state at time $t = 0$. The analytic solution of the problem is

[2] In a typical experiment with alkali atoms $\Omega \approx 2\pi \times 10^6$ Hz, while $\omega \approx 2\pi \times 10^{14}$ Hz.

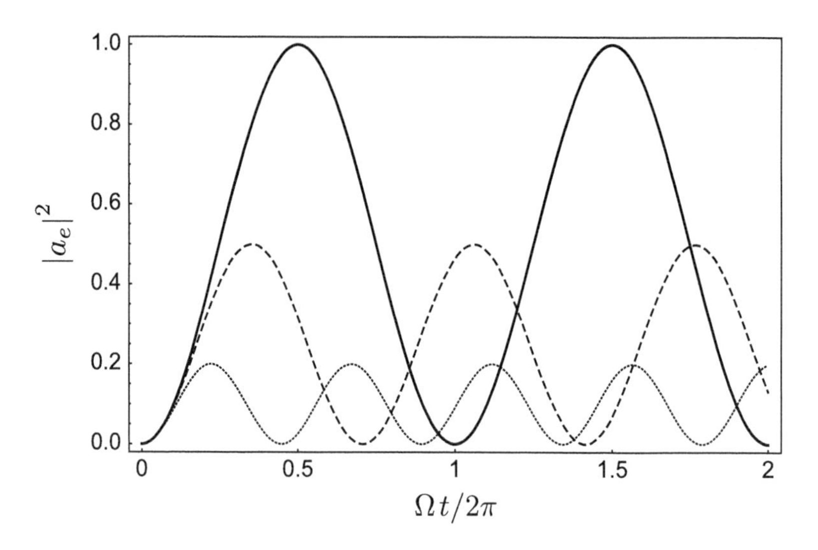

Fig. A.1 Rabi oscillations of the excited-state population $|a_e(t)|^2 = |\tilde{a}_e(t)|^2$ in eqn (A.24) for a two-level system interacting with a monochromatic near-resonant radiation field. The curves refer to different detunings: $\delta = 0$ (solid), $\delta = \Omega$ (dashed), $\delta = 2\Omega$ (dotted).

$$\begin{cases} \tilde{a}_g(t) = e^{\frac{i}{2}\delta t}\left(\cos\frac{\Omega' t}{2} - i\frac{\delta}{\Omega'}\sin\frac{\Omega' t}{2}\right) \\ \tilde{a}_e(e) = e^{\frac{i}{2}\delta t}\left(-i\frac{\Omega}{\Omega'}\sin\frac{\Omega' t}{2}\right) , \end{cases} \tag{A.22}$$

where we have defined the *generalized Rabi frequency* as

$$\Omega' \equiv \sqrt{\delta^2 + |\Omega|^2} . \tag{A.23}$$

From eqn (A.22) the populations of the upper and lower levels can be written as

$$\begin{cases} |\tilde{a}_g(t)|^2 = \cos^2\frac{\Omega' t}{2} + \frac{\delta^2}{\Omega'^2}\sin^2\frac{\Omega' t}{2} \\ |\tilde{a}_e(t)|^2 = \frac{|\Omega|^2}{\Omega'^2}\sin^2\frac{\Omega' t}{2} . \end{cases} \tag{A.24}$$

Figure A.1 shows the above solution for the excited state population $|a_e(t)|^2 = |\tilde{a}_e(t)|^2$ for different values of the detuning δ. The solid line shows the behaviour of the system at resonance ($\delta = 0$), where the excited state population oscillates sinusoidally between 0 and 1 at the resonant Rabi frequency Ω. This behaviour is called *Rabi flopping* or *Rabi oscillations*. Moving out of resonance ($\delta = \Omega$ dashed line, $\delta = 2\Omega$ dotted line) the oscillation frequency Ω' becomes larger and the amplitude

$$\frac{|\Omega|^2}{\Omega'^2} = \frac{1}{1 + \frac{\delta^2}{|\Omega|^2}} \tag{A.25}$$

drops according to a Lorentzian function of δ with half-width at half maximum $|\Omega|$.

We note that Rabi oscillations describe the unitary evolution of any isolated two-level quantum system driven only by a near-resonant coherent excitation. In particular, we have neglected any other interaction of the quantum system with the environment, which may cause decay of the population from the excited state (e.g. due to spontaneous emission) or decoherence of the quantum superposition in eqn (A.1) (e.g. due to collisions between the atoms). We will consider these effects in the next section.

A.2 Spontaneous emission

Microwave excitation of ground-state hyperfine transitions in alkali atoms can be very accurately described by the Rabi dynamics derived in the previous section,[3] owing to the negligible spontaneous emission from the higher-energy state $|e\rangle$ and to the long coherence times of the quantum superposition between $|g\rangle$ and $|e\rangle$, which is usually much larger than the timescale of the Rabi dynamics. For a dipole-allowed optical transition between different electronic states the situation is different, since the Rabi dynamics can be slower than the decay caused by spontaneous emission from the excited state, which has not been accounted for in the previous section. Spontaneous emission comes from the coupling of the atom with the vacuum state of the quantized electromagnetic field. A rigorous analysis of the problem (see e.g. Corney (2006)) shows that it can be described with a radiative decay rate

$$\gamma = \frac{e^2}{3\pi\epsilon_0\hbar c^3}\omega_0^3\mu_{eg}^2 \tag{A.26}$$

for the population of the upper energy level.

In order to include this effect in the time evolution of the two-level system, instead of using the Schrödinger equation for the amplitudes a_g and a_e, it is more convenient to use a different approach, which is based on the introduction of the *density operator*

$$\hat{\rho} = |\psi\rangle\langle\psi| \, . \tag{A.27}$$

For the pure quantum state $|\psi\rangle$ in eqn (A.1) the density operator can be represented by the *density matrix*

$$\hat{\rho} = \begin{bmatrix} \rho_{gg} & \rho_{ge} \\ \rho_{eg} & \rho_{ee} \end{bmatrix} = \begin{bmatrix} |a_g|^2 & a_e^* a_g \\ a_g^* a_e & |a_e|^2 \end{bmatrix} \, . \tag{A.28}$$

The diagonal terms ρ_{gg} and ρ_{ee} are called *populations* and measure the probability of detecting the atom in state $|g\rangle$ or state $|e\rangle$ respectively: since we are considering a closed two-level system, $\mathrm{Tr}(\rho) = \rho_{gg} + \rho_{ee} = 1$. The off-diagonal terms $\rho_{ge} = \rho_{eg}^*$ are called *coherences* and determine the strength of the electric dipole moment induced by the oscillating electric field (further discussed in Sec. A.3.2).

[3] Actually, this transition is not permitted by electric dipole selection rules, as will be explained in Sec. A.4 (it is a magnetic dipole transition). However, the Rabi mechanism has a much more general validity than the electric dipole approximation, since it holds for any coupling matrix element Ω connecting the two unperturbed states $|g\rangle$ and $|e\rangle$.

From the Schrödinger evolution of the pure quantum state $|\psi\rangle$ driven by the Hamiltonian $\hat{H}$ we derive the following equation of motion for the density operator:

$$\begin{aligned}\frac{d\hat{\rho}}{dt} &= |\dot{\psi}\rangle\langle\psi| + |\psi\rangle\langle\dot{\psi}| \\ &= \frac{1}{i\hbar}\hat{H}\,|\psi\rangle\langle\psi| - \frac{1}{i\hbar}\,|\psi\rangle\langle\psi|\,\hat{H} \\ &= \frac{1}{i\hbar}\left[\hat{H},\hat{\rho}\right] .\end{aligned} \tag{A.29}$$

Following the rotating-wave approximation introduced in Sec. A.1.2, we evaluate the evolution of an effective density matrix $\hat{\tilde{\rho}}$ for the slowly varying amplitudes $\tilde{a}_g$ and $\tilde{a}_e$ defined in eqn (A.18)

$$\hat{\tilde{\rho}} = \begin{bmatrix} \tilde{\rho}_{gg} & \tilde{\rho}_{ge} \\ \tilde{\rho}_{eg} & \tilde{\rho}_{ee} \end{bmatrix} = \begin{bmatrix} |\tilde{a}_g|^2 & \tilde{a}_e^*\tilde{a}_g \\ \tilde{a}_g^*\tilde{a}_e & |\tilde{a}_e|^2 \end{bmatrix} = \begin{bmatrix} |a_g|^2 & a_e^*a_g e^{-i\omega t} \\ a_g^*a_e e^{i\omega t} & |a_e|^2 \end{bmatrix} \tag{A.30}$$

under the time-independent rotating-wave Hamiltonian $\hat{\tilde{H}}$ of eqn (A.20). Applying the equation of motion in eqn (A.29) to $\hat{\tilde{H}}$ and $\hat{\tilde{\rho}}$ we obtain:

$$\begin{cases} \Delta\dot{\tilde{\rho}} = -i\left(\Omega\tilde{\rho}_{ge} - \Omega^*\tilde{\rho}_{eg}\right) \\ \dot{\tilde{\rho}}_{ge} = -i\delta\tilde{\rho}_{ge} - \frac{i}{2}\Omega^*\Delta\tilde{\rho} \\ \dot{\tilde{\rho}}_{eg} = i\delta\tilde{\rho}_{eg} + \frac{i}{2}\Omega\Delta\tilde{\rho} , \end{cases} \tag{A.31}$$

where we have defined the *population difference* $\Delta\tilde{\rho} = \tilde{\rho}_{ee} - \tilde{\rho}_{gg}$.

These equations are known as *Maxwell–Bloch equations* (or *optical Bloch equations*) and, in the above form, describe the same unitary dynamics as the one that we have derived in Sec. A.1.3. The advantage of these equations over the Schrödinger evolution for the amplitudes emerges when one introduces relaxation. It is possible to introduce "by hand" the spontaneous emission as a decay in the excited state population in the form (Metcalf and van Der Straten, 1999)

$$\begin{cases} \Delta\dot{\tilde{\rho}} = -i\left(\Omega\tilde{\rho}_{ge} - \Omega^*\tilde{\rho}_{eg}\right) - \gamma\left(\Delta\tilde{\rho} + 1\right) \\ \dot{\tilde{\rho}}_{ge} = -i\delta\tilde{\rho}_{ge} - \frac{i}{2}\Omega^*\Delta\tilde{\rho} - \frac{\gamma}{2}\tilde{\rho}_{ge} \\ \dot{\tilde{\rho}}_{eg} = i\delta\tilde{\rho}_{eg} + \frac{i}{2}\Omega\Delta\tilde{\rho} - \frac{\gamma}{2}\tilde{\rho}_{eg} . \end{cases} \tag{A.32}$$

Here we consider the time evolution of the excited state population $\rho_{ee}(t) = \tilde{\rho}_{ee}(t)$ starting from the initial condition

$$\begin{cases} \tilde{\rho}_{gg}(0) = 1 \\ \tilde{\rho}_{ee}(0) = \tilde{\rho}_{eg}(0) = \tilde{\rho}_{ge}(0) = 0 \end{cases} \tag{A.33}$$

corresponding to the $t = 0$ population of the ground state. The result of a numerical integration of eqns (A.32) is shown in Fig. A.2 for $\delta = 0$, a fixed value of γ, and different values of Ω. Two regimes can be distinguished: for $\Omega > \gamma$ (solid and dashed lines) the excited state population initially oscillates at the Rabi frequency Ω, then its evolution

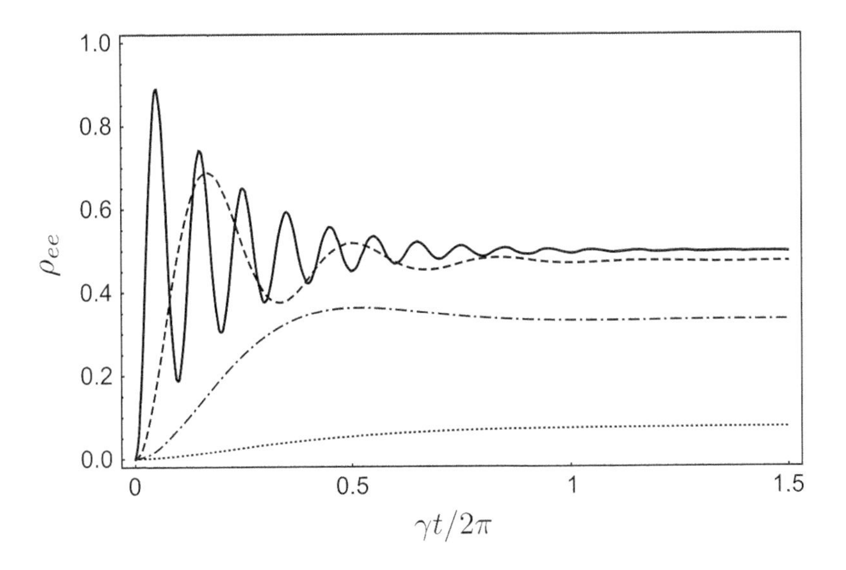

Fig. A.2 Excited-state population $\rho_{ee} = \tilde{\rho}_{ee}$ obtained from the integration of eqns (A.32). The curves are calculated on resonance $\delta = 0$, fixed γ, and different Rabi frequencies: $\Omega = 10\gamma$ (solid), $\Omega = 3\gamma$ (dashed), $\Omega = \gamma$ (dash-dotted), $\Omega = 0.3\gamma$ (dotted).

is damped to a stationary value $\tilde{\rho}_{ee} \approx 0.5$ after a time $t \approx \gamma^{-1}$ because of radiative decay. For $\Omega < \gamma$ the oscillation is overdamped and the stationary value is smaller.

The stationary value of the excited state population can be easily obtained by setting $\Delta\dot{\tilde{\rho}} = \dot{\tilde{\rho}}_{ge} = \dot{\tilde{\rho}}_{eg} = 0$ in eqns (A.32) and solving the resulting inhomogeneous linear system of algebraic equations

$$\begin{cases} -\gamma\Delta\tilde{\rho} - i\Omega\tilde{\rho}_{ge} + i\Omega^*\tilde{\rho}_{eg} = \gamma \\ -\frac{i}{2}\Omega^*\Delta\tilde{\rho} + \left(-i\delta - \frac{\gamma}{2}\right)\tilde{\rho}_{ge} = 0 \\ +\frac{i}{2}\Omega\Delta\tilde{\rho} + \left(i\delta - \frac{\gamma}{2}\right)\tilde{\rho}_{eg} = 0 \end{cases} \tag{A.34}$$

which yields the steady-state excited population

$$\tilde{\rho}_{ee} = \frac{1 + \Delta\tilde{\rho}}{2} = \frac{1}{2}\frac{\frac{|\Omega|^2}{2}}{\delta^2 + \frac{\gamma^2}{4} + \frac{|\Omega|^2}{2}} . \tag{A.35}$$

This equation can be rearranged in a different simpler form as

$$\tilde{\rho}_{ee} = \frac{1}{2}\frac{s}{1+s}\frac{1}{1 + \left(\frac{2\delta}{\gamma_s}\right)^2} , \tag{A.36}$$

which describes a Lorentzian function with peak value $s/2(1+s)$ and full width at half maximum

$$\gamma_s = \gamma\sqrt{1+s} , \tag{A.37}$$

where we have defined the *saturation parameter* at resonance as

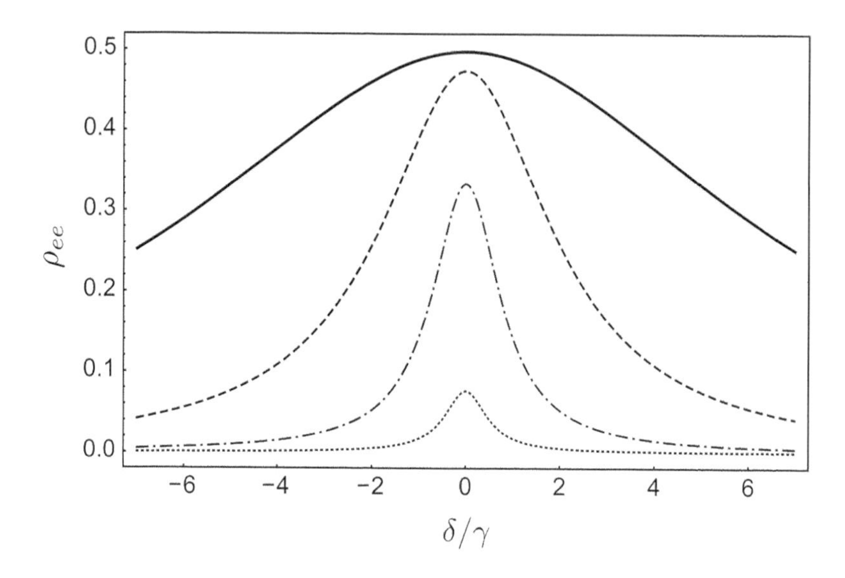

Fig. A.3 Asymptotic excited-state population $\rho_{ee} = \tilde{\rho}_{ee}$ in eqn (A.35) for $t \gg \gamma^{-1}$. The curves are calculated at fixed γ and different Rabi frequencies: $\Omega = 10\gamma$ (solid, $\gamma_s \simeq 14.2\gamma$), $\Omega = 3\gamma$ (dashed, $\gamma_s \simeq 4.4\gamma$), $\Omega = \gamma$ (dash-dotted, $\gamma_s \simeq 1.7\gamma$), $\Omega = 0.3\gamma$ (dotted, $\gamma_s \simeq 1.1\gamma$).

$$s = \frac{2|\Omega|^2}{\gamma^2}\,. \tag{A.38}$$

From eqn (A.37) we note that the linewidth depends on the field intensity, which is proportional to $|\Omega|^2$. This phenomenon is known as *power broadening* (or *saturation broadening*). At small intensities $s \ll 1$ the linewidth is dominated by the natural linewidth $\gamma_s \approx \gamma$, while for $s \gg 1$ it is dominated by the Rabi frequency $\gamma_s \approx \sqrt{2}\Omega$. We also observe that the maximal stationary value of $\tilde{\rho}_{ee}$ is $1/2$, that can be obtained at resonance only in the limit of very large field intensities $s \gg 1$: no population inversion can be obtained at steady state in a two-level system. A visualization of these results is presented in Fig. A.3, showing the asymptotic population of the excited state as a function of detuning for different values of Ω.

A.3 Spectroscopic observables

A.3.1 Absorption and fluorescence

Absorption and emission of light are the two main mechanisms which allow us to gather information on atomic structures. Here we consider an optical dipole-allowed transition in a two-level atom. Figure A.4 represents a typical spectroscopy experiment in which light is partially absorbed by an atomic gas, causing a reduction of the transmitted intensity, and partially scattered in different directions owing to the spontaneous emission of fluorescence photons.

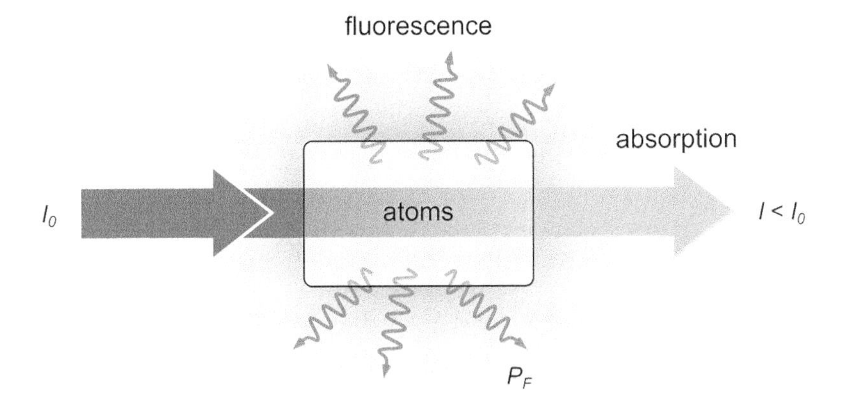

Fig. A.4 As a laser beam is sent through an atomic sample, its intensity is attenuated because of absorption and the missing energy is radiated in the form of spontaneously emitted fluorescence photons.

Fluorescence. The rate of energy emitted by a sample of N atoms as a consequence of the spontaneous decay from the excited state can be written as

$$P_F - \hbar\omega_0 \gamma N \tilde{\rho}_{ee} \ , \tag{A.39}$$

where $N\tilde{\rho}_{ee}$ is the number of excited atoms, $\hbar\omega_0$ is the energy of a fluorescence photon, γ is the rate of spontaneous emission. Using the expression for the steady-state excited population in eqn (A.36), one finds:

$$P_F = N\hbar\omega_0 \frac{\gamma}{2} \frac{s}{1+s} \frac{1}{1+\left(\frac{2\delta}{\gamma_s}\right)^2} \ . \tag{A.40}$$

This expression depends on the laser intensity, which is contained in the parameter s. Its dependence on s is shown in Fig. A.5a: for small intensities $s \ll 1$ the fluorescence increases linearly with s, while at large intensity $s \gg 1$ it saturates to a value which does not depend on the laser intensity. At saturation, in the case of resonant photons ($\delta = 0$), the fluorescence radiated power is $N\hbar\omega_0\gamma/2$, which corresponds to the average emission of a fluorescence photon per atom in a time $2/\gamma$.

The dependence of s on the laser intensity can be made explicit by defining a *saturation intensity* I_s by the following equation:

$$s = \frac{2|\Omega|^2}{\gamma^2} = \frac{I}{I_s} \ . \tag{A.41}$$

By using the definition of Rabi frequency and eqn (A.26) for the radiative decay rate, after some manipulations the saturation intensity can be written as

$$I_s = \frac{\hbar\gamma\omega_0^3}{12\pi c^2} \ , \tag{A.42}$$

which depends on the resonance frequency ω_0 and on the natural linewidth γ.

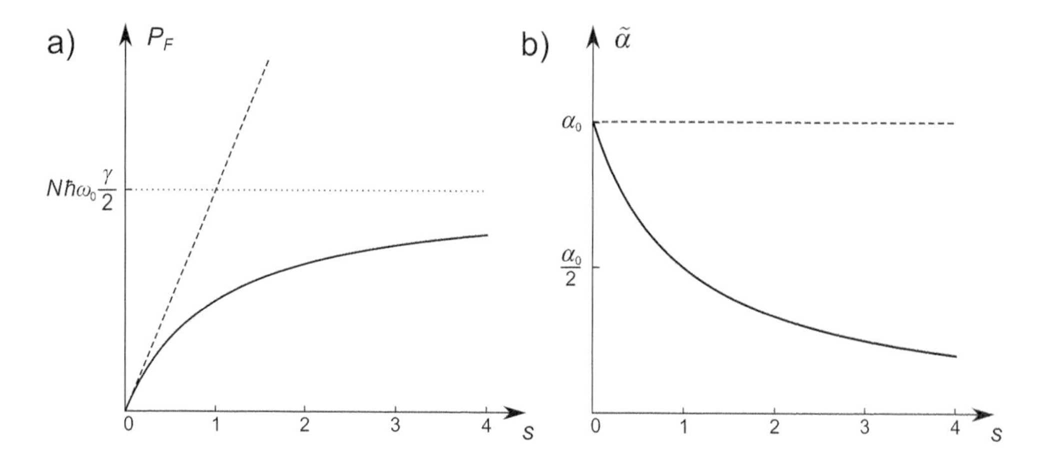

Fig. A.5 a) Fluorescence radiated power as a function of the saturation parameter s: for $s \ll 1$ the fluorescence power increases linearly with s, while for $s \gg 1$ it saturates to a constant value $N\hbar\omega_0\gamma/2$. b) Absorption coefficient as a function of s, showing the same saturation behaviour: at large s the absorption coefficient approaches zero and the system becomes transparent.

Absorption. At equilibrium, the average energy stored in the atoms is constant, therefore the rate of absorbed energy must be equal to the rate of energy re-emitted by fluorescence, because of energy conservation.[4] The decrease of laser intensity after a distance dz travelled in the gas is

$$\begin{aligned} dI &= -\frac{1}{A}P_F \\ &= -\frac{1}{A}N\hbar\omega_0\frac{\gamma}{2}\frac{s}{1+s}\frac{1}{1+\left(\frac{2\delta}{\gamma_s}\right)^2} \\ &= -dz\, n\hbar\omega_0\frac{\gamma}{2}\frac{s}{1+s}\frac{1}{1+\left(\frac{2\delta}{\gamma_s}\right)^2} \\ &= -\tilde{\alpha}I dz\ , \end{aligned} \tag{A.43}$$

where A is the laser beam cross section, n is the number of atoms per unit of volume, and we have defined the *absorption coefficient* $\tilde{\alpha}$ as[5]

[4]If relaxation processes are absent, the system does not reach a steady state and, *on average*, it does not absorb energy. As a matter of fact, the unitary evolution described by Rabi oscillations can be interpreted as a symmetric repetition of absorption and stimulated emission cycles, in which the average absorbed energy is equal to zero. On average, light can be absorbed only when spontaneous emission is considered: the energy of the atomic system is not conserved and at steady state the energy gain by absorption (minus the energy loss by stimulated emission) must equal the energy loss by spontaneous emission.

[5]We use the $\tilde{\alpha}$ notation in order to distinguish the absorption coefficient from the atomic polarizability (discussed in Sec. A.3.2), since they are usually denoted with the same symbol α.

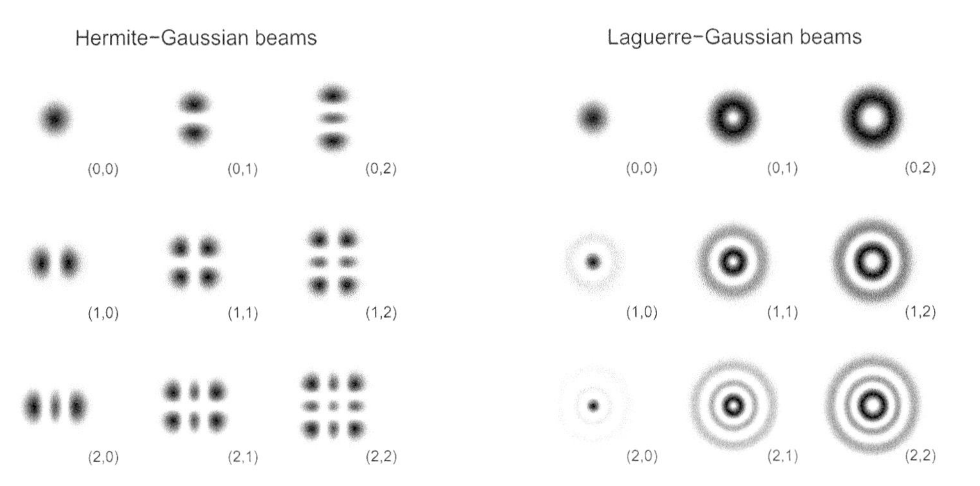

Fig. B.1 Solutions of the paraxial wave equation. Intensity distribution of the lowest-order Hermite–Gaussian and Laguerre–Gaussian transverse modes in a plane orthogonal to the propagation direction.

basis of solutions is provided by the *Hermite–Gaussian* beams TEM_{mn}, in which higher-order transverse modes are characterized by an increasing number $m+n$ of nodal lines along which the electric field is zero. For cylindrical symmetry the solutions can be given in terms of *Laguerre–Gaussian* beams, which possess a nontrivial azimuthal phase profile and are characterized by a nonzero orbital angular momentum. Examples of low-order Hermite–Gaussian and Laguerre–Gaussian modes are shown in Fig. B.1.

Here we consider only the fundamental transverse mode, the Gaussian TEM_{00} mode, which is both the lowest-order Hermite–Gaussian mode and the lowest-order Laguerre–Gaussian mode. A Gaussian TEM_{00} beam does not present any nodal points or lines and is a very good approximation for the spatial mode of many lasers. The dependence of the electric field amplitude on the spatial coordinates can be written as:

$$E(r,z) = E_0 \frac{w_0}{w(z)} \exp\left[\frac{-r^2}{w^2(z)}\right] \exp\left[-ikz - ik\frac{-r^2}{2R(z)+i\zeta(z)}\right], \tag{B.5}$$

where r denotes the distance from the beam axis. The following definitions for the beam radius $w(z)$, the wavefront curvature radius $R(z)$, and the *Gouy phase* $\zeta(z)$ hold:

$$w(z) = w_0\sqrt{1+\left(\frac{z}{z_R}\right)^2} \tag{B.6}$$

$$R(z) = z\left[1+\left(\frac{z_R}{z}\right)^2\right] \tag{B.7}$$

$$\zeta(z) = \arctan\left(\frac{z}{z_R}\right) . \tag{B.8}$$

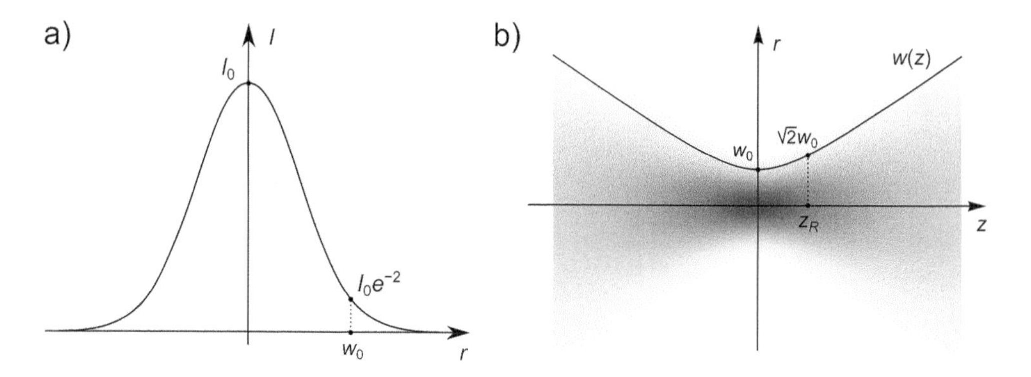

Fig. B.2 a) Intensity of a Gaussian beam at the beam waist $z = 0$ as a function of the distance r from the axis. b) The intensity of the Gaussian beam is plotted in greyscale as a function of r and z. The curved line shows the dependence of the beam radius $w(z)$ on z.

All the above quantities are expressed in terms of the *beam waist radius* w_0, which univocally determines all the properties of the Gaussian beam, and of the *Rayleigh length*

$$z_R = \frac{\pi w_0^2}{\lambda} . \tag{B.9}$$

From eqn (B.5) we can evaluate the beam intensity $I = \epsilon_0 c|E|^2/2$, which is described by a Gaussian function of the radial coordinate r

$$I(r,z) = I_0 \left(\frac{w_0}{w(z)}\right)^2 \exp\left[\frac{-2r^2}{w^2(z)}\right] , \tag{B.10}$$

the beam radius $w(z)$ representing the distance from the axis at which the intensity is reduced by a factor e^2. Figure B.2a shows a plot of the beam intensity as a function of r in the *beam waist* at $z = 0$, where the beam has the smallest radius w_0. Out of the beam waist the beam radius $w(z)$ increases with z according to eqn (B.6), reaching a value $\sqrt{2}w_0$ at a distance corresponding to the Rayleigh length z_R. This dependence is shown in Fig. B.2b together with a greyscale plot of the Gaussian beam intensity as a function of z and r.

Far from the beam waist, for $z \gg z_R$, the beam size $w(z) \simeq w_0 z/z_R$ increases linearly with z and the beam is characterized by a divergence angle

$$\theta = \frac{w(z)}{z} \simeq \frac{w_0}{z_R} = \frac{\lambda}{\pi w_0} \tag{B.11}$$

which increases as the beam waist size w_0 is reduced. This is a consequence of diffraction: the more the beam is focused at the beam waist, the more rapidly it spreads as it propagates out of the focus.[3]

[3]The angular spread of a wave diffracted by an aperture can be estimated from the uncertainty principle $\Delta x \Delta k_\perp \sim 1$, where Δx is the size of the aperture and $\Delta k_\perp$ is the transverse wavevector

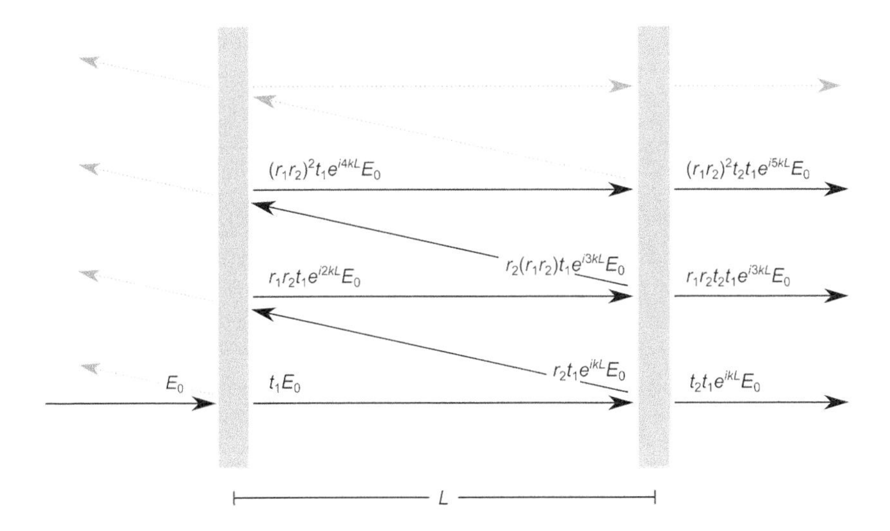

Fig. B.3 Scheme of a Fabry–Perot resonator formed by two plane mirrors separated by a distance L. The multiple reflections are designed at an angle for graphical convenience.

B.2 Optical resonators

Optical resonators (or *optical cavities*) are important systems in atomic and optical physics: besides constituting an essential part of the design of a laser, they provide frequency references for spectroscopy and tools for the stabilization and line-narrowing of the laser emission spectrum as well. They also represent a physical system in which the atom–photon interaction can be studied in regimes of strong coupling, as studied in cavity-QED experiments (Haroche and Raimond, 2006). The simplest example of optical cavity is provided by two partially reflective mirrors one in front of the other. This configuration takes the name of *Fabry–Perot resonator* from the French physicists Charles Fabry and Alfred Perot who proposed it at the end of the nineteenth century.

We consider two mirrors, separated by a distance L, characterized by a field reflectivity r_i and field transmittivity t_i ($i = 1, 2$). The reflectivity and transmittivity coefficients for the intensities are $R_i = |r_i|^2$ and $T_i = |t_i|^2$ respectively, with the condition $R_i + T_i = 1$ imposed by the conservation of energy. We assume that the space between the mirrors is empty, i.e. no absorption takes place and the index of refraction of the intra-cavity medium is 1. An incident beam (a plane wave) with electric field E_0 entering the cavity experiences an infinite sequence of partial reflections and transmissions from the two mirrors, as schematically represented in Fig. B.3. Although the reflections are drawn at an angle in order to graphically distinguish the different beams, we assume the situation of normal incidence, in which all the beams propagate along the same direction and overlap with each other. The figure indicates the attenuation of each partially reflected and transmitted beam, including the phase e^{ikL} acquired by the field because of its propagation over a distance L. The

component after the aperture. The divergence angle for an aperture size $\Delta x = w_0$ can be estimated as $\theta = \Delta k_\perp/k \sim 1/kw_0 = \lambda/2\pi w_0$, of the same order as the result in eqn (B.11).

total transmitted electric field can be calculated by summing over the (infinite) partial beams leaking from the cavity in the forward direction:

$$E_T = t_1 t_2 e^{ikL} E_0 \sum_{n=0}^{\infty} \left(r_1 r_2 e^{i2kL}\right)^n = t_1 t_2 e^{ikL} E_0 \frac{1}{1 - r_1 r_2 e^{i2kL}} , \tag{B.12}$$

where we have used the sum formula for the geometric series (which is convergent since $|r_1 r_2 e^{i2kL}| < 1$ and, more physically, because of energy conservation). In order to simplify the presentation of the results, we consider a particular case in which the field reflectivity and transmittivity coefficients are real-valued and the same for the two mirrors: $r_1 = r_2 = \sqrt{R}$ and $t_1 = t_2 = \sqrt{T}$. By taking the squared modulus of eqn (B.12) we can write the transmitted intensity as

$$\begin{aligned} I_T &= I_0 \frac{|t_1 t_2|^2}{|1 - r_1 r_2 e^{i2kL}|^2} = I_0 \frac{(1-R)^2}{|1 - Re^{i2kL}|^2} \\ &= I_0 \frac{(1-R)^2}{1 + R^2 - 2R\cos(2kL)} = I_0 \frac{(1-R)^2}{(1-R)^2 + 4R\sin^2(kL)} \\ &= I_0 \frac{1}{1 + \frac{4\mathcal{F}^2}{\pi^2}\sin^2(kL)} = \frac{1}{1 + \frac{4\mathcal{F}^2}{\pi^2}\sin^2(\pi\frac{\nu}{\mathrm{FSR}})} , \end{aligned} \tag{B.13}$$

where we have defined the *finesse* of the resonator as

$$\mathcal{F} = \frac{\pi\sqrt{R}}{1-R} \tag{B.14}$$

and its *free spectral range* as

$$\mathrm{FSR} = \frac{c}{2L} . \tag{B.15}$$

Looking at eqn (B.13) we observe that the transmitted intensity is maximal when the sine term at the denominator is null, which happens for radiation frequencies ν_n that satisfy the resonance condition

$$\nu_n = n\,\mathrm{FSR} , \tag{B.16}$$

n being an integer number. This condition corresponds to a constructive interference of the infinite number of transmitted partial waves, which are all summed coherently with the same phase.[4] The transmitted spectrum of the Fabry–Perot resonator is made up of a comb of peaks centred at equally spaced frequencies, separated by a frequency interval FSR which only depends on the *geometry* of the resonator. This makes a Fabry–Perot resonator a simple system for the calibration of laser scans in spectroscopy, e.g. for the determination of hyperfine structures or isotope shifts. The same physics also determines the uniform spacing of the emission frequencies in an optical frequency comb (discussed in Sec. 1.4.2).

The reflectivity of the mirrors, which enters the definition of the finesse $\mathcal{F}$, determines the width of the resonances. The transmitted intensity I_T is plotted in Fig. B.4a as a

[4] As a matter of fact, the phase shift of the field after a round trip of length $2L$ in the cavity is $\delta\phi = 2kL = 4\pi\nu L/c = 2\pi\nu/\mathrm{FSR}$, which is an integer multiple of 2π at the resonance frequencies ν_n given by eqn (B.16).

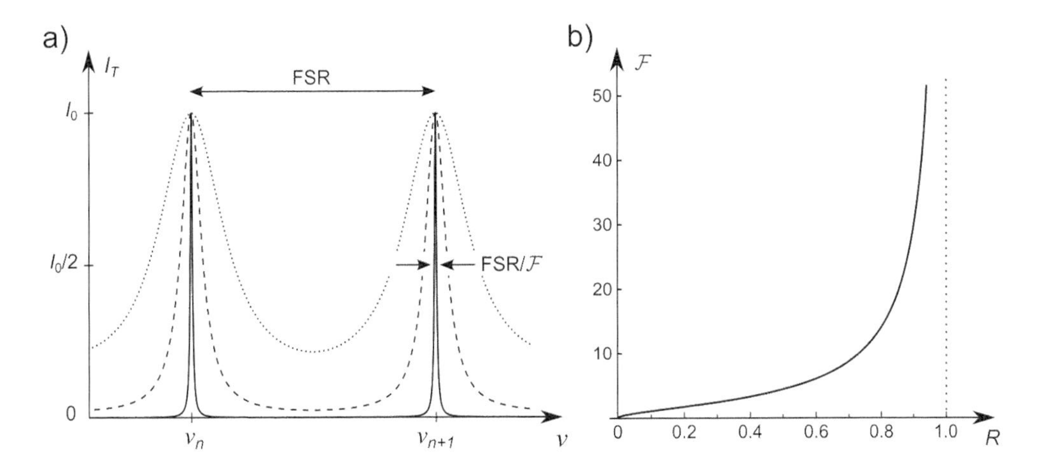

Fig. B.4 Fabry–Perot resonator. a) Transmitted intensity as a function of frequency for three different values of the resonator finesse $\mathcal{F} = 3$ (dotted), 10 (dashed), 100 (solid). b) Dependence of $\mathcal{F}$ on the mirror reflectivity R.

function of the laser frequency for three different values of $\mathcal{F}$. As the finesse is increased, the resonances become sharper: in the high-finesse limit $\mathcal{F} \gg 1$, their full-width at half maximum is given by

$$\delta\nu = \frac{\text{FSR}}{\mathcal{F}} \,. \tag{B.17}$$

The finesse of the resonator only depends on the mirror reflectivity R, according to eqn (B.14), plotted in Fig. B.4b.

Finally, we note that the intensity of light *inside* the optical cavity can be much larger than the intensity outside. Performing a similar analysis to the one carried out to derive eqn (B.12), it is possible to demonstrate that for $\mathcal{F} \gg 1$ the average intra-cavity intensity[5] at resonance is enhanced by a factor on the same order as the finesse:[6]

$$I_{IC} \simeq \frac{2\mathcal{F}}{\pi} I_0 \,. \tag{B.18}$$

In the presence of a dielectric medium inside the cavity, this enhancement of the intensity allows the investigation of strong nonlinear effects in light–matter interaction, which we are going to discuss in the next section.

[5]The average refers to the fact that inside the optical cavity light propagates both forwards and backwards, which produces a standing wave. The result is given for the spatially averaged intensity.

[6]This intensity enhancement does not violate the conservation of energy! Energy is accumulated inside the cavity during an initial transient time $\tau = \mathcal{F}L/\pi c$ after the laser is switched on (this time is given by the round-trip time $2L/c$ times the average number of round trips per photon $\mathcal{F}/2\pi$). At steady state, the energy stored inside the optical cavity does not change and the coupled light is entirely transmitted, which is in perfect agreement with energy conservation. When the laser is switched off, the accumulated energy is then released, again in a time τ.

B.3 Nonlinear optics

In this section we give a very brief introduction to nonlinear optics. We aim at giving only the essential information on this topic, leaving the interested reader to more specialized textbooks (Boyd, 2008).

Nonlinear optics requires the interaction of light with matter. As a matter of fact, the Maxwell equations form a system of partial differential equations which are linear in the electromagnetic fields:

$$\begin{cases} \nabla \cdot \mathbf{D} = \rho \\ \nabla \cdot \mathbf{B} = 0 \\ \nabla \times \mathbf{E} = -\frac{\partial \mathbf{B}}{\partial t} \\ \nabla \times \mathbf{H} = \mathbf{J} + \frac{\partial \mathbf{D}}{\partial t} \, . \end{cases} \tag{B.19}$$

Nonlinearities may come into play when the constitutive relations between the displacement field $\mathbf{D}$ and $\mathbf{E}$, and the magnetic field $\mathbf{H}$ and $\mathbf{B}$ are considered:

$$\mathbf{D} = \epsilon_0 \mathbf{E} + \mathbf{P} \tag{B.20}$$

$$\mathbf{H} = \frac{\mathbf{B}}{\mu_0} + \mathbf{M} \, . \tag{B.21}$$

These relations describe how the fields interact with matter by means of the macroscopic electric polarization $\mathbf{P}$ and the magnetization $\mathbf{M}$. We consider the usual case of a dielectric non-magnetic material, for which $\mathbf{M} = 0$. For the sake of illustration, we assume that the medium polarization $\mathbf{P}$ has the same direction as the electric field $\mathbf{E}$ and we write it as a series expansion in the electric field amplitude:[7]

$$\mathbf{P} = \epsilon_0 \chi^{(1)} \mathbf{E} + \epsilon_0 \chi^{(2)} E \mathbf{E} + \epsilon_0 \chi^{(3)} E^2 \mathbf{E} + \cdots \tag{B.22}$$

For weak fields only the first term is relevant: the polarization is proportional to the field and the linear electric susceptibility $\chi^{(1)}$ defines the index of refraction of the medium. The following terms in the expansion, described by nonlinear susceptibilities $\chi^{(i \geq 2)}$, become important only for strong fields and are responsible for nonlinear effects that we are going to describe in the following.

We first derive the wave equation for the electromagnetic field propagating in a non-magnetic dielectric medium characterized by the nonlinear polarization in eqn (B.22) in absence of charge or current field sources ($\rho = 0$ and $\mathbf{J} = 0$). From eqns (B.19), (B.20), and (B.22):

[7]Here we are deliberately oversimplifying the treatment. More generally, the medium polarization may not be parallel to the applied field and the susceptibilities are tensorial quantities. This is indeed the case for many nonlinear crystals, in which birefringence is important for achieving the phase-matching condition, but neglecting it does not spoil the main conclusions of the elementary introduction given in this section.

$$
\begin{aligned}
\nabla \times (\nabla \times \mathbf{E}) &= -\frac{\partial}{\partial t}(\nabla \times \mathbf{B}) = -\mu_0 \frac{\partial}{\partial t}(\nabla \times \mathbf{H}) \\
&= -\mu_0 \frac{\partial^2 \mathbf{D}}{\partial t^2} = -\mu_0 \epsilon_0 \frac{\partial^2}{\partial t^2}\left(\mathbf{E} + \frac{\mathbf{P}}{\epsilon_0}\right) \\
&= -\mu_0 \epsilon_0 \frac{\partial^2}{\partial t^2}\left(\mathbf{E} + \frac{\mathbf{P}_L}{\epsilon_0} + \frac{\mathbf{P}_{NL}}{\epsilon_0}\right) \\
&= -\mu_0 \epsilon_0 (1 + \chi^{(1)}) \frac{\partial^2 \mathbf{E}}{\partial t^2} - \mu_0 \frac{\partial^2 \mathbf{P}_{NL}}{\partial t^2} \\
&= -\frac{1}{v^2} \frac{\partial^2 \mathbf{E}}{\partial t^2} - \mu_0 \frac{\partial^2 \mathbf{P}_{NL}}{\partial t^2}\,, \qquad \text{(B.23)}
\end{aligned}
$$

where we have separated the polarization into a linear part $\mathbf{P}_L = \epsilon_0 \chi^{(1)} \mathbf{E}$ and into a nonlinear part $\mathbf{P}_{NL} = \epsilon_0 \chi^{(2)} E \mathbf{E} + \epsilon_0 \chi^{(3)} E^2 \mathbf{E} + \cdots$ containing the remaining terms of the expansion in eqn (B.22). As in the case of linear media, the linear part of the polarization leads to a redefinition of the dielectric constant $\epsilon = \epsilon_0(1 + \chi^{(1)})$, which is associated with an index of refraction $\eta = \sqrt{1 + \chi^{(1)}}$ and a propagation velocity $v = c/\eta = 1/\sqrt{\mu_0 \epsilon_0 (1 + \chi^{(1)})}$ inside the medium. The left-hand side of the above equation can be rewritten using the vectorial identity

$$
\nabla \times (\nabla \times \mathbf{E}) = \nabla (\nabla \cdot \mathbf{E}) - \nabla^2 \mathbf{E} \simeq -\nabla^2 \mathbf{E}\,, \qquad \text{(B.24)}
$$

where we have made the assumption $\nabla \cdot \mathbf{E} \simeq 0$.[8] Combining Eqs. (B.23) and (B.24) we obtain the nonlinear wave equation

$$
\left(\nabla^2 - \frac{1}{v^2} \frac{\partial^2}{\partial t^2}\right) \mathbf{E} = \mu_0 \frac{\partial^2 \mathbf{P}_{NL}}{\partial t^2}\,, \qquad \text{(B.25)}
$$

which differs from the ordinary wave equation in a linear medium for the presence of the nonlinear polarization $\mathbf{P}_{NL}$ which acts as a source term. Since this term depends nonlinearly on the field, it allows the generation of radiation at frequencies which are different from the frequency of the driving field. Below we consider some relevant cases.

Second-harmonic and sum/difference frequency generation. In the presence of a second-order nonlinearity eqn (B.25) becomes

$$
\left(\nabla^2 - \frac{1}{v^2} \frac{\partial^2}{\partial t^2}\right) E = \mu_0 \epsilon_0 \chi^{(2)} \frac{\partial^2 E^2}{\partial t^2}\,, \qquad \text{(B.26)}
$$

where, for the sake of illustration, we consider only the amplitude of the electric field, neglecting its polarization.

[8]The first Maxwell equation states that $\nabla \cdot \mathbf{D} = 0$ in the absence of free charges. In a homogeneous linear medium with dielectric constant ϵ one has $\mathbf{D} = \epsilon \mathbf{E}$, therefore also $\nabla \cdot \mathbf{E} = 0$. However, in a nonlinear medium, the divergence of $\mathbf{E}$ can be different from zero (also in isotropic media), but it is possible to show that in most of the cases its contribution to eqn (B.24) can be neglected (Boyd, 2008).

As a first example, we consider an electromagnetic wave with time dependence $E(t) = E_0 \cos(\omega t)$ entering the nonlinear medium. The right-hand side of eqn (B.26) can be recast as

$$\mu_0 \epsilon_0 \chi^{(2)} \frac{\partial^2 E^2}{\partial t^2} = \mu_0 \epsilon_0 \chi^{(2)} E_0^2 \frac{\partial^2}{\partial t^2} \cos^2(\omega t) \\ = -2\omega^2 \mu_0 \epsilon_0 \chi^{(2)} E_0^2 \cos(2\omega t) \,, \tag{B.27}$$

which represents a source of electromagnetic field oscillating at twice the frequency ω of the incident wave. This process, called *second-harmonic generation* (SHG) allows the conversion of energy from the fundamental beam at frequency ω to frequency-doubled radiation at frequency 2ω.

As a second example, we consider two electromagnetic waves $E_1(t) = E_0 \cos(\omega_1 t)$ and $E_2(t) = E_0 \cos(\omega_2 t)$ with different frequencies. The right-hand side of eqn (B.26) can now be written as

$$\mu_0 \epsilon_0 \chi^{(2)} \frac{\partial^2 E^2}{\partial t^2} = \mu_0 \epsilon_0 \chi^{(2)} E_0^2 \frac{\partial^2}{\partial t^2} \left[\cos(\omega_1 t) + \cos(\omega_2 t)\right]^2 \\ = -\mu_0 \epsilon_0 \chi^{(2)} E_0^2 \left[2\omega_1^2 \cos(2\omega_1 t) + 2\omega_2^2 \cos(2\omega_2 t)\right] \\ -\mu_0 \epsilon_0 \chi^{(2)} E_0^2 (\omega_1 + \omega_2)^2 \cos\left[(\omega_1 + \omega_2)t\right] \\ -\mu_0 \epsilon_0 \chi^{(2)} E_0^2 (\omega_1 - \omega_2)^2 \cos\left[(\omega_1 - \omega_2)t\right] \,. \tag{B.28}$$

In addition to the second-harmonic source terms oscillating at frequencies $2\omega_1$ and $2\omega_2$, there are two additional terms at frequencies $\omega_1 + \omega_2$ and $\omega_1 - \omega_2$ arising from the nonlinear mixing of the two frequencies. These two processes are called, respectively, *sum-frequency generation* (SFG) and *difference-frequency generation* (DFG).

Phase matching. The nonlinear effects described above are visualized in a more "quantum-optical" way in Fig. B.5 by examining the elementary processes involving photons. In SHG two photons with frequency ω are annihilated and a photon with frequency 2ω is created by the interaction with the nonlinear medium. In SFG two photons with frequencies ω_1 and ω_2 are annihilated and a photon with frequency $\omega_1 + \omega_2$ is generated. In DFG one photon with frequency ω_1 is annihilated and two photons, with frequencies $\omega_1 - \omega_2$ and ω_2 respectively, are generated.[9] In all these elementary processes energy conservation is satisfied.

For these processes to be efficient momentum conservation has to be satisfied as well. In the case of SHG momentum conservation implies that

$$k_{2\omega} = 2k_\omega \,. \tag{B.29}$$

While this condition would be automatically satisfied in vacuum, in a dielectric medium it is more difficult to achieve, because of dispersion, which makes the index of refraction

[9] As evident from Fig. B.5, the DFG process has a slightly different nature from SHG and SFG, since one of the input photons is amplified instead of being destroyed. The DFG process can be also interpreted as an *optical parametric amplifier* (OPA) in which a higher-frequency pump beam at ω_1 is used to amplify the *signal* wave at ω_2, generating at the same time an *idler* wave at $\omega_1 - \omega_2$. In an *optical parametric generator* the process is not stimulated and the conversion of photons at ω_1 into pairs of photons at ω_s and ω_i (with $\omega_s + \omega_i = \omega_1$) occurs spontaneously without any input photon at ω_2.

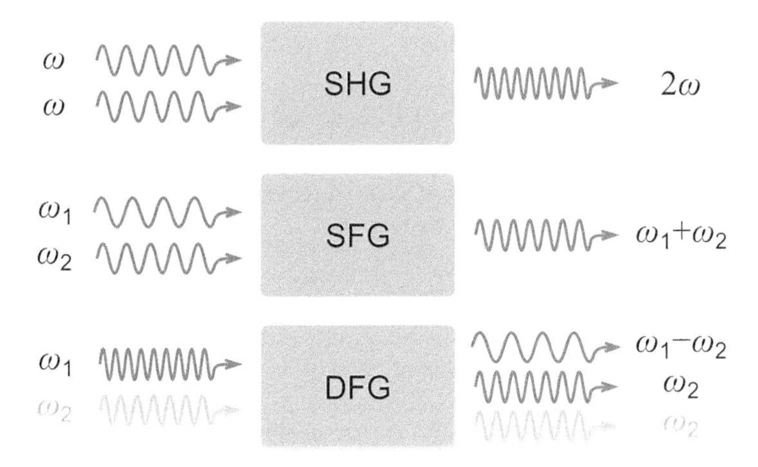

Fig. B.5 Representation of the elementary processes involved in second-harmonic generation (SHG), sum-frequency generation (SFG), and difference-frequency generation (DFG). In the latter case two photons are shaded in light grey, since the process can take place also spontaneously (see footnote 9).

$\eta(\omega)$ dependent on the radiation frequency ω. As a matter of fact, the above equation can be recast as a condition on the index of refraction:

$$\eta(2\omega) = \eta(\omega) \,. \tag{B.30}$$

This *phase-matching* condition requires the fundamental wave at frequency ω to travel at the same speed as the generated wave at frequency 2ω, in order to maintain a well-defined phase relation between them and achieve constructive interference of the radiation at frequency 2ω generated at different positions along the crystal.

We do not discuss the details of the methods to achieve phase matching (see Boyd (2008) for details). We just mention two possibilities. The first one is *birefringent phase matching*, in which a birefringent nonlinear crystal is used and the frequency-doubled radiation is polarized orthogonally to the radiation at fundamental frequency. In this way, the different indices of refraction of the crystal for the two polarizations are used to achieve the condition in eqn (B.30). A second possibility is *quasi phase matching*, which can be achieved in a periodically poled crystal, i.e. a nonlinear crystal in which the sign of the $\chi^{(2)}$ coefficient is spatially modulated: although eqn (B.30) is not satisfied (which means that the frequency-doubled radiation rapidly dephase with respect to the fundamental radiation), the modulation of $\chi^{(2)}$ periodically inverts the phase of the radiation at 2ω in such a way as to mantain, on average, a condition of constructive interference.

When (quasi-)phase matching is achieved, nonlinear optical processes become quite efficient. With lasers powers of a few hundreds of mW and nonlinear crystals placed in optical cavities to enhance the intracavity power (see Sec. B.2), it is possible to achieve SHG with a conversion efficiency of $\approx 50\%$ or even larger.

Third-order nonlinearity. We briefly mention the effect of a third-order nonlinearity, described by a nonlinear wave equation

$$\left(\nabla^2 - \frac{1}{v^2}\frac{\partial^2}{\partial t^2}\right) E = \mu_0 \epsilon_0 \chi^{(3)} \frac{\partial^2 E^3}{\partial t^2} \,. \tag{B.31}$$

If we consider an electromagnetic wave with time dependence $E(t) = E_0 \cos(\omega t)$, the right-hand side of eqn (B.31) can be written as

$$\begin{aligned} \mu_0 \epsilon_0 \chi^{(3)} \frac{\partial^2 E^3}{\partial t^2} &= \mu_0 \epsilon_0 \chi^{(3)} E_0^3 \frac{\partial^2}{\partial t^2} \cos^3(\omega t) \\ &= -\frac{1}{4}\omega^2 \mu_0 \epsilon_0 \chi^{(3)} E_0^3 \left[3\cos(\omega t) + 9\cos(3\omega t)\right] \,. \end{aligned} \tag{B.32}$$

While the second term oscillating at 3ω is responsible for third-harmonic generation, the first term describes a component of the nonlinear polarization oscillating at the same frequency as the driving field. Its effects can be treated as those of the linear component $\mathbf{P}_L$ of the atomic polarization and lead to a modification of the index of refraction η by an additional term which is proportional to E_0^2, and therefore to the field intensity I:

$$\eta' = \eta + \eta_K I \,. \tag{B.33}$$

This dependence of the refraction index on the laser intensity, known as *Kerr effect*, leads to a self-focusing behaviour of the laser beam and its control is important in the design of high-power laser cavities and in the operation of many pulsed mode-locked lasers (used e.g. for the implementation of frequency combs, see Sec. 1.4.2).

Appendix C
Bose–Einstein condensation

In this appendix we review the basic theory of Bose–Einstein condensation in a trapped ultracold gas. More advanced information can be found in Dalfovo *et al.* (1999) and Pitaevskii and Stringari (2004), on which this appendix is largely based.

C.1 Noninteracting Bose gas

In Bose–Einstein condensation (BEC) experiments, ultracold gases of neutral atoms are confined in harmonic trapping potentials provided by either magnetic (see Sec. 3.1.1) or optical traps (see Sec. 5.2.2). Here we consider the theory of Bose–Einstein condensation for a gas of N noninteracting identical bosons in a three-dimensional (3D) harmonic confining potential

$$V_{ext}(\mathbf{r}) = \frac{1}{2}m\left(\omega_x^2 x^2 + \omega_y^2 y^2 + \omega_z^2 z^2\right) , \tag{C.1}$$

where m is the atomic mass and ω_i are the (angular) trap frequencies. The eigenvalues of the single-particle Hamiltonian problem are

$$\epsilon_{n_x,n_y,n_z} = \left(n_x + \frac{1}{2}\right)\hbar\omega_x + \left(n_y + \frac{1}{2}\right)\hbar\omega_y + \left(n_z + \frac{1}{2}\right)\hbar\omega_z , \tag{C.2}$$

where n_i, are the quantum numbers identifying the 3D harmonic oscillator state. For a system in thermal equilibrium, the occupancy of the levels is described by the Bose–Einstein statistics, according to which the mean occupation number is given in the grand-canonical ensemble by (Huang, 1987)

$$f(n_x, n_y, n_z) = \frac{1}{e^{\beta(\epsilon_{n_x,n_y,n_z}-\mu)} - 1} , \tag{C.3}$$

where $\beta = (k_B T)^{-1}$ is the inverse reduced temperature and μ is the *chemical potential*, that accounts for the conservation of the total number of particles

$$N = \sum_{n_x,n_y,n_z} \frac{1}{e^{\beta(\epsilon_{n_x,n_y,n_z}-\mu)} - 1} . \tag{C.4}$$

In the classical limit (large T or small N) the mean occupation of the levels is much less than unity, $f(n_x, n_y, n_z) \ll 1\ \forall n_i$, the chemical potential is much smaller than the ground-state energy, $\mu \ll \epsilon_{0,0,0}$, and the Bose statistics reduces to the classical Maxwell–Boltzmann statistics. As the number of particles N is increased, μ increases as well in order to satisfy the normalization condition in eqn (C.4). However, there is an upper limit on the growth of μ, which is imposed by the condition that the occupation number in eqn (C.3) must be positive for all the quantum states: $\mu < \epsilon_{n_x,n_y,n_z}$ $\forall n_i$. Therefore, at most the chemical potential can approach the energy of the lowest-energy state: $\mu \simeq \epsilon_{0,0,0}$. When this happens, the occupation of the excited states becomes saturated and, increasing N, the new particles "condense" in the ground state, the occupation of which becomes macroscopic according to eqn (C.3). Substituting $\mu = \epsilon_{0,0,0}$ in eqn (C.4) we obtain

$$N = N_0 + \sum_{n_x,n_y,n_z} \frac{1}{e^{\beta\hbar(\omega_x n_x + \omega_y n_y + \omega_z n_z)} - 1}, \tag{C.5}$$

where we have isolated the population of the ground state N_0, that produces a divergence in the sum. If the level spacing $\hbar\omega_i$ is much smaller than $k_B T$, we can replace the sum with an integral (semiclassical approximation):[1]

$$N = N_0 + \int_0^\infty \frac{dn_x dn_y dn_z}{e^{\beta\hbar(\omega_x n_x + \omega_y n_y + \omega_z n_z)} - 1}. \tag{C.6}$$

By carrying out the integration, one finds that the number of atoms in the condensed fraction N_0 as a function of T is

$$N_0 = N\left[1 - \left(\frac{T}{T_C}\right)^3\right], \tag{C.7}$$

where the BEC *critical temperature* T_C is defined as

$$T_C = \frac{\hbar\omega_{ho}}{k_B}\left(\frac{N}{\zeta(3)}\right)^{1/3} = 0.94\frac{\hbar\omega_{ho}}{k_B}N^{1/3}, \tag{C.8}$$

in which $\zeta(n)$ is the Riemann function and $\omega_{ho} = (\omega_x\omega_y\omega_z)^{1/3}$ is the geometric average of the trapping frequencies. This derivation is strictly valid in the thermodynamic limit, that for a harmonically trapped gas corresponds to the limit $N \to \infty$, with $N\omega_{ho}^3$ constant. In Fig. C.1 we plot the condensate fraction as a function of temperature as obtained from eqn (C.7).

At $T = 0$ all the atoms occupy the harmonic oscillator ground state and the condensate density is given by

$$n^{ho}(\mathbf{r}) = N\left(\frac{m\omega_{ho}}{\pi\hbar}\right)^{3/2} \exp\left[-\frac{m}{\hbar}\left(\omega_x x^2 + \omega_y y^2 + \omega_z z^2\right)\right], \tag{C.9}$$

corresponding to the squared modulus of the Gaussian wavefunction of the harmonic oscillator ground state normalized with $\int d\mathbf{r}\, n^{ho}(\mathbf{r}) = N$. We note that, for a harmonically trapped gas, Bose–Einstein condensation occurs with a sudden narrowing of the density distribution both in momentum and coordinate space. This is different from the textbook description of a bosonic gas confined in a box, in which condensation occurs only in momentum space, while in coordinate space the condensed gas remains delocalized and cannot be spatially distinguished from the non-condensed component.

[1]This assumption corresponds to the requirement of a large number of levels to be thermally occupied. From the result of the calculation, given in eqn (C.8), at the critical temperature $\hbar\omega_i/k_B T_C \simeq N^{-1/3} = 0.01$ for $N = 10^6$, so the assumption is reasonably satisfied.

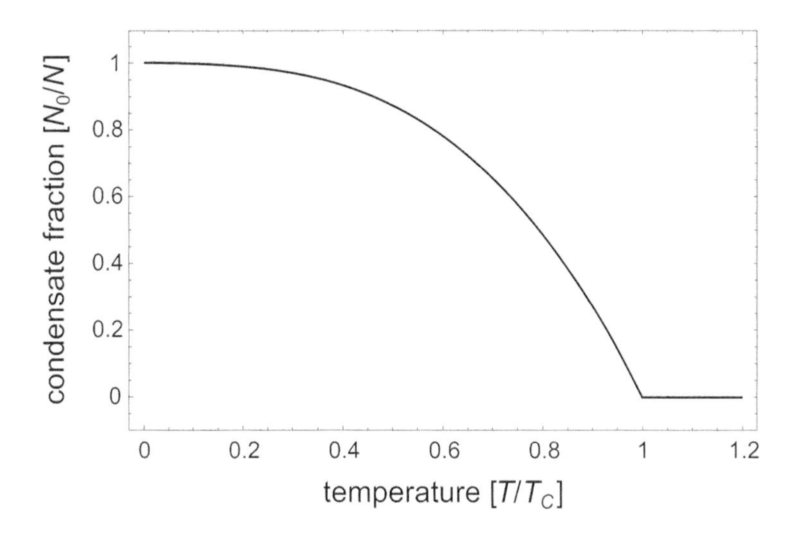

Fig. C.1 Condensate fraction as a function of temperature for a gas of noninteracting bosons confined in a 3D harmonic potential.

C.2 Effect of interactions

In the previous section we have considered the problem of Bose–Einstein condensation for a noninteracting gas, the thermodynamic behaviour of which is completely governed by quantum statistics. However, real systems are always characterized by interactions among their constituent particles. In strongly interacting systems the presence of interactions can blur the quantum effects and lead to a significant quantum depletion of the condensate phase even at zero temperature. This is the case of superfluid ^{4}He, in which only a small fraction of the liquid (typically around 10%) is Bose-condensed. In the case of weakly interacting systems, as for BEC of dilute gases, quantum depletion can be generally neglected. However, the presence of interactions strongly modifies the properties of the system, leading to superfluidity and to a nonlinear behaviour of the de Broglie matter-wave that enriches the multitude of observable phenomena.

In second quantization, the many-body Hamiltonian operator describing a system of N bosons in an external potential V_{ext} is given by

$$\hat{H} = \int d\mathbf{r}\ \hat{\Psi}^\dagger(\mathbf{r}) \left[-\frac{\hbar^2\nabla^2}{2m} + V_{ext}(\mathbf{r}) \right] \hat{\Psi}(\mathbf{r}) + \frac{1}{2} \iint d\mathbf{r}d\mathbf{r}'\ \hat{\Psi}^\dagger(\mathbf{r})\hat{\Psi}^\dagger(\mathbf{r}')V_{int}(\mathbf{r}-\mathbf{r}')\hat{\Psi}(\mathbf{r}')\hat{\Psi}(\mathbf{r}) , \quad \text{(C.10)}$$

where $\hat{\Psi}(\mathbf{r})$ ($\hat{\Psi}^\dagger(\mathbf{r})$) is the boson field annihilation (creation) operator and V_{int} describes low-energy binary s-wave collisions, that in an ultracold dilute gas are the only relevant interaction processes (see Sec. 3.1.4). The field operator can be written as

$$\hat{\Psi}(\mathbf{r}) = \sum_j \Psi_j(\mathbf{r})\hat{b}_j , \quad \text{(C.11)}$$

where $\Psi_j(\mathbf{r})$ is the wavefunction of the single-particle state j and $\hat{b}_j$ is the annihilation operator of a boson in state j. From a very general point of view, the field operator $\hat{\Psi}(\mathbf{r})$ can be written in Heisenberg representation as

$$\hat{\Psi}(\mathbf{r},t) = \Psi(\mathbf{r},t) + \delta\hat{\Psi}(\mathbf{r},t)\ , \tag{C.12}$$

where $\Psi(\mathbf{r},t) = \langle\hat{\Psi}(\mathbf{r},t)\rangle$ is the expectation value of the field operator and the fluctuations $\delta\hat{\Psi}(\mathbf{r},t)$ describe the quantum and thermal excitations.

Bose–Einstein condensation occurs when the single-particle ground state becomes macroscopically occupied. Using a standard approach in quantum field theory, when the number of particles in one state is macroscopic ($N_0 \gg 1$), the creation and annihilation operators can be substituted with *c-numbers*, thus recovering the limit of a classical field. With this assumption the expectation value $\Psi(\mathbf{r},t)$ naturally results from the macroscopic occupancy of the ground state, while $\delta\hat{\Psi}(\mathbf{r},t)$ describes the excitations. Following the Bogoliubov approximation (Dalfovo *et al.*, 1999) we neglect the contribution of $\delta\hat{\Psi}(\mathbf{r},t)$, thus the Heisenberg equation of motion for $\hat{\Psi}(\mathbf{r},t)$ becomes an equation for the classical field $\Psi(\mathbf{r},t)$:

$$i\hbar\frac{\partial}{\partial t}\Psi(\mathbf{r},t) = \left[-\frac{\hbar^2\nabla^2}{2m} + V_{ext}(\mathbf{r}) + \int d\mathbf{r}'\ \Psi^\dagger(\mathbf{r}',t)V_{int}(\mathbf{r}-\mathbf{r}')\Psi(\mathbf{r}',t)\right]\Psi(\mathbf{r},t)\ , \tag{C.13}$$

where the complex function $\Psi(\mathbf{r},t)$ stands for the condensate wavefunction. If the mean interparticle distance $n^{-1/3}$ is much larger than the range of the two-body potential V_{int} (diluteness condition), the latter can be substituted with the effective pseudopotential $g\delta(\mathbf{r}-\mathbf{r}')$, that is independent of the details of the two-body interaction. The scattering between two atoms can thus be entirely described by the scalar parameter

$$g = \frac{4\pi\hbar^2 a}{m}\ , \tag{C.14}$$

where a is the so-called *scattering length*, that is positive for repulsive interactions and negative for attractive interactions (see Sec. 3.1.4). With this assumption for the interaction potential, which is typically well satisfied, eqn (C.13) becomes

$$i\hbar\frac{\partial}{\partial t}\Psi(\mathbf{r},t) = \left[-\frac{\hbar^2\nabla^2}{2m} + V_{ext}(\mathbf{r}) + g|\Psi(\mathbf{r},t)|^2\right]\Psi(\mathbf{r},t)\ . \tag{C.15}$$

This equation is known as the *Gross–Pitaevskii equation*, after the two scientists that independently derived it in the early 1960s (Gross, 1961; Gross, 1963; Pitaevskii, 1961). In this equation, well verified in a great number of experiments, the effect of interactions is described by a *nonlinear* term for the condensate order parameter Ψ. This term, describing the condensate self-interaction, results from a *mean-field* approximation, in which the condensate wavefunction Ψ locally probes the average collisional potential $g|\Psi|^2$ produced by itself.

The presence of the interaction term in eqn (C.15) is important since it has an essential role in determining the superfluid properties of atomic Bose–Einstein condensates discussed in Secs. 3.2.2 and 3.2.3: among these, their collective excitations, the propagation of sound, the existence of a critical superflow velocity. The interaction term in eqn (C.15) is also responsible for a rich multitude of nonlinear phenomena. In the context of coherent matter-wave optics it plays the same role as the third-order susceptibility $\chi^{(3)}$ in nonlinear optics (see Appendix B.3), producing effects such as four-wave mixing and solitonic propagation (Rolston and Phillips, 2002). However, it may also constitute a limitation, in particular for experiments of atom interferometry requiring long-lived single-particle coherence (see Secs. 3.2.6 and 6.4.2).

C.2.1 BEC wavefunction

The stationary solutions of eqn (C.15) can be calculated using the ansatz $\Psi(\mathbf{r},t) = e^{-i\mu t/\hbar}\psi(\mathbf{r})$, where μ is the condensate chemical potential. Substituting this expression in eqn (C.15) we obtain the time-independent Gross–Pitaevskii equation

$$\left[-\frac{\hbar^2\nabla^2}{2m} + V_{ext}(\mathbf{r}) + g|\psi(\mathbf{r})|^2\right]\psi(\mathbf{r}) = \mu\psi(\mathbf{r})\,. \tag{C.16}$$

Repulsive interactions. We first consider positive values of the scattering length, $a > 0$, corresponding to repulsive interactions between the particles. The ground-state condensate wavefunction can be analytically determined by solving eqn (C.16) in the so-called *Thomas–Fermi approximation*, holding for $Na/a_{ho} \gg 1$ (where $a_{ho} = \sqrt{\hbar/m\omega_{ho}}$ is the average harmonic oscillator length) (Dalfovo *et al.*, 1999). In this regime of a large number of atoms the kinetic energy $\langle -\hbar^2\nabla^2/2m\rangle$ is much smaller than the interaction energy $\langle g|\psi|^2\rangle$ and therefore it can be neglected. With this assumption, well satisfied in most of the experiments, the differential equation (C.16) becomes an algebraic equation from which we obtain the condensate density

$$n(\mathbf{r}) = |\psi(\mathbf{r})|^2 = \max\left[\frac{\mu - V_{ext}(\mathbf{r})}{g}\,,\,0\right]. \tag{C.17}$$

In the case of an axially symmetric harmonic trap with $\omega_x = \omega_y = \omega_\perp$ (this is the geometry adopted in most of the experiments), the explicit shape of the condensate density distribution is an inverted parabola

$$n(\mathbf{r}) = \max\left[n_0\left(1 - \frac{x^2+y^2}{R_\perp^2} - \frac{z^2}{R_z^2}\right)\,,\,0\right], \tag{C.18}$$

with height and widths connected to the chemical potential by

$$n_0 = \frac{\mu}{g} = \frac{\hbar\omega_{ho}}{2g}\left(\frac{15Na}{a_{ho}}\right)^{2/5} \tag{C.19}$$

$$R_z = \sqrt{\frac{2\mu}{m\omega_z^2}} \tag{C.20}$$

$$R_\perp = \sqrt{\frac{2\mu}{m\omega_\perp^2}}\,. \tag{C.21}$$

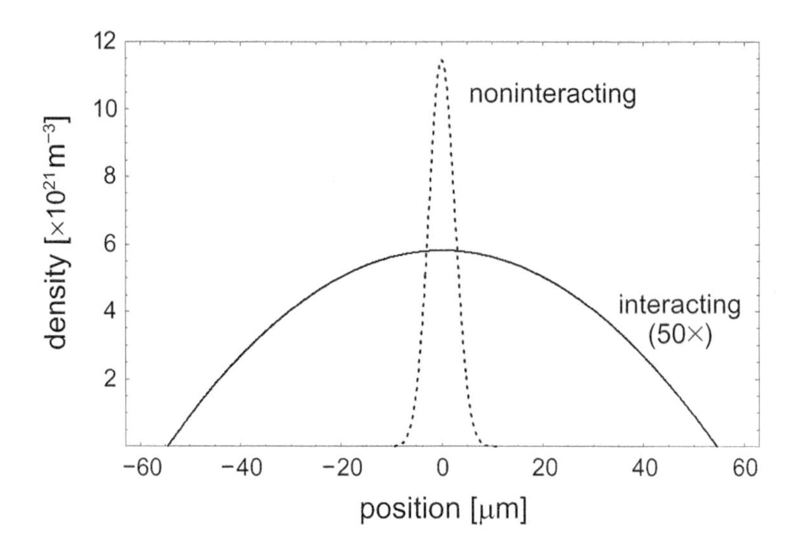

Fig. C.2 BEC density distribution for $N = 3 \times 10^5$, $m = 1.44 \times 10^{-25}$ kg (^{87}Rb mass), $\omega_z = 2\pi \times 9$ Hz, and $\omega_\perp = 2\pi \times 90$ Hz in the noninteracting case (dotted line) and for a repulsive interaction characterized by the scattering length $a = 5.7$ nm (solid line, value for ^{87}Rb). The density is evaluated along the trap axis (note the 50× magnification of the interacting density with respect to the noninteracting case).

The parabolic shape of the interacting BEC ground state is actually quite significant, since the change in shape from a Gaussian thermal distribution to a parabolic Thomas–Fermi distribution across the BEC transition is one the most evident signatures of Bose–Einstein condensation (see Sec. 3.2.1).

We note that the repulsive interaction term in eqn (C.16) has the effect of broadening the density distribution of the single-particle ground state. In Fig. C.2 we plot the condensate density for $N = 3 \times 10^5$, $m = 1.44 \times 10^{-25}$ kg (^{87}Rb mass), $\omega_z = 2\pi \times 9$ Hz, and $\omega_\perp = 2\pi \times 90$ Hz in the noninteracting case (dotted line) and for a repulsive interaction (solid line) with strength $a = 5.7$ nm (value for ^{87}Rb).[2] The ratio between the peak of the Thomas–Fermi density in eqn (C.19) and the peak of the noninteracting density in eqn (C.9) is

$$\frac{n_0}{n_0^{ho}} \approx \left(\frac{Na}{a_{ho}}\right)^{-3/5}, \tag{C.22}$$

corresponding to 1% for the parameters of Fig. C.2, where $Na/a_{ho} \simeq 10^3$.

Attractive interactions. BEC experiments are usually performed with atoms with repulsive interactions, because in this case condensates are stable for an arbitrarily large number of atoms.[3] If interactions are attractive (negative scattering length, $a < 0$)

[2]These numbers, holding for one of the experiments running at LENS (Florence), are representative of typical values for a ^{87}Rb BEC.

[3]This is true provided that the density is not too large to determine excessive three-body losses.

a different scenario emerges: the gas tends to occupy a smaller volume, increasing its density with respect to the noninteracting case, in such a way as to have a larger negative interaction energy.

For a very small number of atoms the zero-point kinetic energy term can still stabilize the system and a trapped condensate can be produced in a metastable state (Bradley *et al.*, 1997*a*). Above a critical value N_C, however, the attractive interactions between atoms cause a collapse of the condensate, that is eventually destroyed by collisional processes out of the mean-field description, including inelastic collisional processes (Gerton *et al.*, 2000; Roberts *et al.*, 2001; Donley *et al.*, 2001).

Appendix D
Constants and units

D.1 Fundamental constants

Table D.1 Values of some fundamental constants. Taken from Mohr *et al.* (2012).

Quantity	Symbol	Value
Speed of light in vacuum	c	$299\,792\,458\ \mathrm{m\,s^{-1}}$
Magnetic constant	μ_0	$4\pi \times 10^{-7}\ \mathrm{N\,A^{-2}}$
Electric constant	$\epsilon_0 = 1/\mu_0 c^2$	$8.854\,187\,817\ldots \times 10^{-12}\ \mathrm{F\,m^{-1}}$
Planck constant	h	$6.626\,069\,57(29) \times 10^{-34}\ \mathrm{J\,s}$
Reduced Planck constant	$\hbar = h/2\pi$	$1.054\,571\,726(47) \times 10^{-34}\ \mathrm{J\,s}$
Fine structure constant	$\alpha = e^2/4\pi\epsilon_0\hbar c$	$7.297\,352\,5698(24) \times 10^{-3}$
	α^{-1}	$137.035\,999\,074(44)$
Rydberg constant	$R_\infty = \alpha^2 m_e c/2h$	$10\,973\,731.568\,539(55)\ \mathrm{m^{-1}}$
	$R_\infty c$	$3.289\,841\,960\,364(17) \times 10^{15}\ \mathrm{Hz}$
	$R_\infty hc$	$13.605\,692\,53(30)\ \mathrm{eV}$
Boltzmann constant	k_B	$1.380\,6488(13) \times 10^{-23}\ \mathrm{J\,K^{-1}}$
Avogadro constant	N_A	$6.022\,141\,29(27) \times 10^{23}\ \mathrm{mol^{-1}}$
Newtonian constant	G	$6.673\,84(80) \times 10^{-11}\ \mathrm{m^3\,kg^{-1}\,s^{-2}}$

Table D.2 Values of other relevant quantities. Taken from Mohr *et al.* (2012).

Quantity	Symbol	Value
Elementary charge	e	$1.602\,176\,565(35) \times 10^{-19}\ \mathrm{C}$
Electron mass	m_e	$9.109\,382\,91(40) \times 10^{-31}\ \mathrm{kg}$
Proton mass	m_p	$1.672\,621\,777(74) \times 10^{-27}\ \mathrm{kg}$
Bohr radius	$a_0 = 4\pi\epsilon_0\hbar^2/m_e e^2$	$0.529\,177\,210\,92(17) \times 10^{-10}\ \mathrm{m}$
Bohr magneton	$\mu_B = e\hbar/2m_e$	$927.400\,968(20) \times 10^{-26}\ \mathrm{J\,T^{-1}}$
	$\mu_B/h = e/4\pi m_e$	$13.996\,245\,55(31)\ \mathrm{GHz\,T^{-1}}$
Nuclear magneton	$\mu_N = e\hbar/2m_p$	$5.050\,783\,53(11) \times 10^{-27}\ \mathrm{J\,T^{-1}}$
	$\mu_N/h = e/4\pi m_p$	$7.622\,593\,57(17)\ \mathrm{MHz\,T^{-1}}$
Electron g-factor	g_e	$2.002\,319\,304\,361\,53(53)$

D.2 Units and conversions

Table D.3 Conversions factors between energy, frequency, and temperature units. Taken from Mohr *et al.* (2012).

(1 eV)	$= 1.602\,176\,565(35) \times 10^{-19}$ J
(1 eV)/h	$= 241\,798.9348(53)$ GHz
(1 eV)/hc	$= 8065.544\,29(18)$ cm^{-1}
(1 eV)/k_B	$= 1.160\,4519(11) \times 10^{4}$ K
(1 J)	$= 6.241\,509\,34(14) \times 10^{18}$ eV
(1 J)/h	$= 1.509\,190\,311(67) \times 10^{24}$ GHz
(1 J)/hc	$= 5.034\,117\,01(22) \times 10^{22}$ cm^{-1}
(1 J)/k_B	$= 7.242\,9716(66) \times 10^{22}$ K
(1 GHz)h	$= 4.135\,667\,516(91) \times 10^{-6}$ eV
(1 GHz)h	$= 6.626\,069\,57(29) \times 10^{-25}$ J
(1 GHz)/c	$= 0.033\,356\,409\,519\ldots$ cm^{-1}
(1 GHz)h/k_B	$= 0.047\,992\,434(44)$ K
(1 cm^{-1})hc	$= 1.239\,841\,930(27) \times 10^{-4}$ eV
(1 cm^{-1})hc	$= 1.986\,445\,684(88) \times 10^{-23}$ J
(1 cm^{-1})c	$= 29.979\,245\,8$ GHz
(1 cm^{-1})hc/k_B	$= 1.438\,7770(13)$ K
(1 K)k_B	$= 8.617\,3324(78) \times 10^{-5}$ eV
(1 K)k_B	$= 1.380\,6488(13) \times 10^{-23}$ J
(1 K)k_B/h	$= 20.836\,618(19)$ GHz
(1 K)k_B/hc	$= 0.695\,034\,76(63)$ cm^{-1}

References

Abo-Shaeer, J. R., Raman, C., Vogels, J. M., and Ketterle, W. (2001). Observation of vortex lattices in Bose–Einstein condensates. *Science*, **292**, 476.

Aidelsburger, M., Atala, M., Nascimbène, S., Trotzky, S., Chen, Y.-A., and Bloch, I. (2011). Experimental realization of strong effective magnetic fields in an optical lattice. *Physical Review Letters*, **107**, 255301.

Aikawa, K., Akamatsu, D., Hayashi, M., Oasa, K., Kobayashi, J., Naidon, P., Kishimoto, T., Ueda, M., and Inouye, S. (2010). Coherent transfer of photoassociated molecules into the rovibrational ground state. *Physical Review Letters*, **105**, 203001.

Aikawa, K., Frisch, A., Mark, M., Baier, S., Rietzler, A., Grimm, R., and Ferlaino, F. (2012). Bose–Einstein condensation of erbium. *Physical Review Letters*, **108**, 210401.

Akatsuka, T., Takamoto, M., and Katori, H. (2008). Optical lattice clocks with non-interacting bosons and fermions. *Nature Physics*, **4**, 954.

Al Khawaja, U., Stoof, H. T. C., Hulet, R. G., Strecker, K. E., and Partridge, G. B. (2002). Bright soliton trains of trapped Bose–Einstein condensates. *Physical Review Letters*, **89**(20), 200404.

Alberti, A., Ferrari, G., Ivanov, V. V., Chiofalo, M. L., and Tino, G. M. (2010). Atomic wave packets in amplitude-modulated vertical optical lattices. *New Journal of Physics*, **12**, 065037.

Allen, L. and Eberly, H. J. (1987). *Optical Resonance and Two Level Atoms.* Dover Publications.

Altman, E., Demler, E., and Lukin, M. D. (2004). Probing many-body states of ultracold atoms via noise correlations. *Physical Review A*, **70**, 013603.

Altman, E., Hofstetter, W., Demler, E., and Lukin, M. D. (2003). Phase diagram of two-component bosons on an optical lattice. *New Journal of Physics*, **5**, 113.

Amole, C., Ashkezari, M. D., Baquero-Ruiz, M., Bertsche, W., Bowe, P. D., Butler, E., Capra, A., Cesar, C. L., Charlton, M., Deller, A., Donnan, P. H., Eriksson, S., Fajans, J., Friesen, T., Fujiwara, M. C., Gill, D. R., Gutierrez, A., Hangst, J. S., Hardy, W. N., Hayden, M. E., Humphries, A. J., Isaac, C. A., Jonsell, S., Kurchaninov, L., Little, A., Madsen, N., McKenna, J. T. K., Menary, S., Napoli, S. C., Nolan, P., Olchanski, K., Olin, A., Pusa, P., Rasmussen, C. Ø., Robicheaux, F., Sarid, E., Shields, C. R., Silveira, D. M., Stracka, S., So, C., Thompson, R. I., van der Werf, D. P., and Wurtele, J. S. (2012). Resonant quantum transitions in trapped antihydrogen atoms. *Nature*, **483**(7390), 439.

Amoretti, M., Amsler, C., Bonomi, G., Bouchta, A., Bowe, P., Carraro, C., Cesar, C. L., Charlton, M., Collier, M. J. T., Doser, M., Filippini, V., Fine, K. S., Fontana, A., Fujiwara, M. C., Funakoshi, R., Genova, P., Hangst, J. S., Hayano, R. S., Holzscheiter, M. H., Jø rgensen, L. V., Lagomarsino, V., Landua, R., Lindelöf, D., Lodi Rizzini, E., Macrì, M., Madsen, N., Manuzio, G., Marchesotti, M., Montagna, P., Pruys, H.,

Regenfus, C., Riedler, P., Rochet, J., Rotondi, A., Rouleau, G., Testera, G., Variola, A., Watson, T. L., and van der Werf, D. P. (2002). Production and detection of cold antihydrogen atoms. *Nature*, **419**(6906), 456.

Andersen, M. F., Ryu, C., Cladé, P., Natarajan, V., Vaziri, A., Helmerson, K., and Phillips, W. D. (2006). Quantized rotation of atoms from photons with orbital angular momentum. *Physical Review Letters*, **97**, 170406.

Anderson, B. P. and Kasevich, M. A. (1998). Macroscopic quantum interference from atomic tunnel arrays. *Science*, **282**, 1686.

Anderson, M. H., Ensher, J. R., Matthews, M. R., Wieman, C. E., and Cornell, E. A. (1995). Observation of Bose–Einstein condensation in a dilute atomic vapor. *Science*, **269**, 198.

Anderson, P. W. (1958). Absence of diffusion in certain random lattices. *Physical Review*, **109**(5), 1492.

Anderson, P. W. (1978). Local moments and localized states. *Reviews of Modern Physics*, **50**(2), 191.

Andreev, S.V., Balykin, V. I., Letokhov, V. S., and Minogin, V. G. (1982). Radiative slowing and reduction of the energy spread of a beam of sodium atoms to 1.5 K in an oppositely directed laser beam. *JETP Letters*, **34**, 442.

Andresen, G. B., Ashkezari, M. D., Baquero-Ruiz, M., Bertsche, W., Bowe, P. D., Butler, E., Cesar, C. L., Chapman, S., Charlton, M., Deller, A., Eriksson, S., Fajans, J., Friesen, T., Fujiwara, M. C., Gill, D. R., Gutierrez, A., Hangst, J. S., Hardy, W. N., Hayden, M. E., Humphries, A. J., Hydomako, R., Jenkins, M. J., Jonsell, S., Jø rgensen, L. V., Kurchaninov, L., Madsen, N., Menary, S., Nolan, P., Olchanski, K., Olin, A., Povilus, A., Pusa, P., Robicheaux, F., Sarid, E., Seif el Nasr, S., Silveira, D. M., So, C., Storey, J. W., Thompson, R. I., van der Werf, D. P., Wurtele, J. S., and Yamazaki, Y. (2010). Trapped antihydrogen. *Nature*, **468**(7324), 673.

Andresen, G. B., Ashkezari, M. D., Baquero-Ruiz, M., Bertsche, W., Bowe, P. D., Butler, E., Cesar, C. L., Charlton, M., Deller, A., Eriksson, S., Fajans, J., Friesen, T., Fujiwara, M. C., Gill, D. R., Gutierrez, A., Hangst, J. S., Hardy, W. N., Hayano, R. S., Hayden, M. E., Humphries, A. J., Hydomako, R., Jonsell, S., Kemp, S. L., Kurchaninov, L., Madsen, N., Menary, S., Nolan, P., Olchanski, K., Olin, A., Pusa, P., Rasmussen, C. Ø., Robicheaux, F., Sarid, E., Silveira, D. M., So, C., Storey, J. W., Thompson, R. I., van der Werf, D. P., Wurtele, J. S., and Yamazaki, Y. (2011). Confinement of antihydrogen for 1000 seconds. *Nature Physics*, **7**(7), 558.

Andrews, M. R., Kurn, D. M., Miesner, H.-J., Durfee, D. S., Townsend, C. G., Inouye, S., and Ketterle, W. (1997*a*). Propagation of sound in a Bose–Einstein condensate. *Physical Review Letters*, **79**(4), 553.

Andrews, M. R., Townsend, C. G., Miesner, H.-J., Durfee, D. S., Kurn, D. M., and Ketterle, W. (1997*b*). Observation of interference between two Bose condensates. *Science*, **275**, 637.

Antezza, M., Pitaevskii, L. P., and Stringari, S. (2005). New asymptotic behavior of the surface–atom force out of thermal equilibrium. *Physical Review Letters*, **95**, 113202.

Antezza, M., Pitaevskii, L. P., Stringari, S., and Svetovoy, V. (2006). Casimir–Lifshitz force out of thermal equilibrium and asymptotic nonadditivity. *Physical Review*

Letters, **97**, 223203.

Antognini, A., Biraben, F., Cardoso, J. M. R., Covita, D. S., Dax, A., Fernandes, L. M. P., Gouvea, A. L., Graf, T., Hänsch, T. W., Hildebrandt, M., Indelicato, P., Julien, L., Kirch, K., Kottmann, F., Liu, Y.-W., Monteiro, C. M. B., Mulhauser, F., Nebel, T., Nez, F., dos Santos, J. M. F., Schuhmann, K., Taqqu, D., Veloso, J. F. C. A., Voss, A., and Pohl, R. (2011). Illuminating the proton radius conundrum: The μHe^+ Lamb shift. *Canadian Journal of Physics*, **89**, 47.

Antognini, A., Nez, F., Schuhmann, K., Amaro, F. D., Biraben, F., Cardoso, J. M. R., Covita, D. S., Dax, A., Dhawan, S., Diepold, M., Fernandes, L. M. P., Giesen, A., Gouvea, A. L., Graf, T., Hänsch, T. W., Indelicato, P., Julien, L., Kao, C.-Y., Knowles, P., Kottmann, F., Le Bigot, E.-O., Liu, Y.-W., Lopes, J. A. M., Ludhova, L., Monteiro, C. M. B., Mulhauser, F., Nebel, T., Rabinowitz, P., dos Santos, J. M. F., Schaller, L. A., Schwob, C., Taqqu, D., Veloso, J. F. C. A., Vogelsang, J., and Pohl, R. (2013). Proton structure from the measurement of 2s-2p transition frequencies of muonic hydrogen. *Science*, **339**, 417.

Aoyama, T., Hayakawa, M., Kinoshita, T., and Nio, M. (2012). Tenth-order QED contribution to the electron g-2 and an improved value of the fine structure constant. *Physical Review Letters*, **109**, 111807.

Arecchi, F. T., Gatti, E., and Sona, A. (1966). Time distribution of photons from coherent and Gaussian sources. *Physics Letters*, **20**, 27.

Arimondo, E., Inguscio, M., and Violino, P. (1977). Experimental determinations of the hyperfine structure in the alkali atoms. *Reviews of Modern Physics*, **49**(1), 31.

Arimondo, E., Lew, H., and Oka, T. (1979). Deflection of a Na beam by resonant standing-wave radiation. *Physical Review Letters*, **43**(11), 753.

Arnoult, O., Nez, F., Julien, L., and Biraben, F. (2010). Optical frequency measurement of the 1s-3s two-photon transition in hydrogen. *The European Physical Journal D*, **60**, 243.

Ashby, N. (2002). Relativity and the Global Positioning System. *Physics Today*, **55**(5), 41.

Ashcroft, N. W. and Mermin, D. N. (1976). *Solid State Physics*. Saunders College Publishing.

Aspect, A., Arimondo, E., Kaiser, R., Vansteenkiste, N., and Cohen-Tannoudji, C. (1988). Laser cooling below the one-photon recoil energy by velocity-selective coherent population trapping. *Physical Review Letters*, **61**(7), 826.

Aspect, A. and Inguscio, M. (2009). Anderson localization of ultracold atoms. *Physics Today*, **62**(8), 30.

Aubry, S. and André, G. (1980). Analyticity breaking and Anderson localization in incommensurate lattices. *Annals of the Israel Physical Society*, **3**, 133.

Bakr, W. S., Gillen, J. I., Peng, A., Fölling, S., and Greiner, M. (2009). A quantum gas microscope for detecting single atoms in a Hubbard-regime optical lattice. *Nature*, **462**, 74.

Bakr, W. S., Peng, A., Tai, M. E., Ma, R., Simon, J., Gillen, J. I., Fölling, S., Pollet, L., and Greiner, M. (2010). Probing the superfluid-to-Mott insulator transition at the single-atom level. *Science*, **329**, 547.

Balykin, V. I., Letokhov, V. S., and Mushin, V. I. (1980). Observation of the cooling of free sodium atoms in a resonance laser field with a scanning frequency. *JETP Letters*, **29**, 560.

Balykin, V. I., Letokhov, V. S., Ovchinnikov, Yu. B., and Sidorov, A. I. (1987). Reflection of an atomic beam from a gradient of an optical field. *JETP Letters*, **45**, 353.

Baranov, M. A. (2008). Theoretical progress in many-body physics with ultracold dipolar gases. *Physics Reports*, **464**, 71.

Baranov, M. A., Dalmonte, M., Pupillo, G., and Zoller, P. (2012). Condensed matter theory of dipolar quantum gases. *Chemical Reviews*, **112**, 5012.

Barber, Z. W., Hoyt, C. W., Oates, C. W., Hollberg, L., Taichenachev, A. V., and Yudin, V. I. (2006). Direct excitation of the forbidden clock transition in neutral ^{174}Yb atoms confined to an optical lattice. *Physical Review Letters*, **96**, 083002.

Bardeen, J., Cooper, L. N., and Schrieffer, J. R. (1957). Microscopic theory of superconductivity. *Physical Review*, **106**, 162.

Bartenstein, M., Altmeyer, A., Riedl, S., Jochim, S., Chin, C., Hecker Denschlag, J., and Grimm, R. (2004). Collective excitations of a degenerate gas at the BEC–BCS crossover. *Physical Review Letters*, **92**, 203201.

Bassani, G. F. (ed.) (2006). *Ettore Majorana: Scientific Papers*. Springer.

Battesti, R., Cladé, P., Guellati-Khélifa, S., Schwob, C., Grémaud, B., Nez, F., Julien, L., and Biraben, F. (2004). Bloch oscillations of ultracold atoms: A tool for a metrological determination of h/m_{Rb}. *Physical Review Letters*, **92**(25), 253001.

Beaufils, Q., Chicireanu, R., Zanon, T., Laburthe-Tolra, B., Maréchal, E., Vernac, L., Keller, J.-C., and Gorceix, O. (2008). All-optical production of chromium Bose–Einstein condensates. *Physical Review A*, **77**, 061601.

Becker, C., Soltan-Panahi, P., Kronjäger, J., Dörscher, S., Bongs, K., and Sengstock, K. (2010). Ultracold quantum gases in triangular optical lattices. *New Journal of Physics*, **12**, 065025.

Bellini, M. and Hänsch, T. W. (2000). Phase-locked white-light continuum pulses: Toward a universal optical frequency-comb synthesizer. *Optics Letters*, **25**, 1049.

Ben Dahan, M., Peik, E., Reichel, J., Castin, Y., and Salomon, C. (1996). Bloch oscillations of atoms in an optical potential. *Physical Review Letters*, **76**(24), 4508.

Bergquist, J. C., Hulet, R. G., Itano, W. M., and Wineland, D. J. (1986). Observation of quantum jumps in a single atom. *Physical Review Letters*, **57**(14), 1699.

Bergquist, J. C., Itano, W. M., and Wineland, D.J. (1987). Recoilless optical absorption and Doppler sidebands of a single trapped ion. *Physical Review A*, **36**(1), 428.

Berman, P. R. (ed.) (1997). *Atom Interferometry*. Academic Press.

Bertoldi, A., Lamporesi, G., Cacciapuoti, L., de Angelis, M., Fattori, M., Petelski, T., Peters, A., Prevedelli, M., Stuhler, J., and Tino, G. M. (2006). Atom interferometry gravity-gradiometer for the determination of the Newtonian gravitational constant G. *The European Physical Journal D*, **40**(2), 271–279.

Best, Th., Will, S., Schneider, U., Hackermüller, L., van Oosten, D., Bloch, I., and Lühmann, D.-S. (2009). Role of interactions in ^{87}Rb–^{40}K Bose-Fermi mixtures in a 3D optical lattice. *Physical Review Letters*, **102**, 030408.

Bethe, H. A. and Salpeter, E. E. (1957). *Quantum Mechanics of One- and Two-Electron Atoms*. Academic Press.

Bethlem, H. L., Berden, G., Crompvoets, F. M. H., Jongma, R. T., van Roij, A. J. A., and Meijer, G. (2000). Electrostatic trapping of ammonia molecules. *Nature*, **406**, 491.

Bethlem, H. L., Berden, G., and Meijer, G. (1999). Decelerating neutral dipolar molecules. *Physical Review Letters*, **83**(8), 1558.

Billy, J., Josse, V., Zuo, Z., Bernard, A., Hambrecht, B., Lugan, P., Clément, D., Sanchez-Palencia, L., Bouyer, P., and Aspect, A. (2008). Direct observation of Anderson localization of matter waves in a controlled disorder. *Nature*, **453**, 891.

Biraben, F. (2009). Spectroscopy of atomic hydrogen. *The European Physical Journal Special Topics*, **172**(1), 109.

Biraben, F., Cagnac, B., and Grynberg, G. (1974). Experimental evidence of two-photon transition without Doppler broadening. *Physical Review Letters*, **32**(12), 643.

Biraben, F., Hänsch, T. W., Fischer, M., Niering, M., Holzwarth, R., Reichert, J., Udem, Th., Weitz, M., de Beauvoir, B., Schwob, C., Jozefoswki, L., Hilico, L., Nez, F., Julien, L., Acef, O., Zondy, J.-J., and Clairon, A. (2001). Precision spectroscopy of atomic hydrogen. In *The Hydrogen Atom. Precision Physics of Simple Atomic Systems* (ed. S. G. Karshenboim, F. S. Pavone, G. F. Bassani, M. Inguscio, and T. W. Hänsch), pp. 17–41. Springer-Verlag.

Bjorkholm, J. E., Freeman, R. R., Ashkin, A., and Pearson, D. B. (1978). Observation of focusing of neutral atoms by the dipole forces of resonance-radiation pressure. *Physical Review Letters*, **41**(20), 1361.

Blatt, R. and Roos, C. F. (2012). Quantum simulations with trapped ions. *Nature Physics*, **8**, 277.

Blatt, S., Ludlow, A. D., Campbell, G. K., Thomsen, J. W., Zelevinsky, T., Boyd, M. M., Ye, J., Baillard, X., Fouché, M., Le Targat, R., Brusch, A., Lemonde, P., Takamoto, M., Hong, F.-L., Katori, H., and Flambaum, V. V. (2008). New limits on coupling of fundamental constants to gravity using ^{87}Sr optical lattice clocks. *Physical Review Letters*, **100**, 140801.

Bloch, F. (1928). Über die Quantenmechanik der Elektronen in Kristallgittern. *Zeitschrift für Physik*, **52**, 555.

Bloch, I., Dalibard, J., and Zwerger, W. (2008). Many-body physics with ultracold gases. *Reviews of Modern Physics*, **80**, 885.

Bloch, I., Greiner, M., Mandel, O., Hänsch, T. W., and Esslinger, T. (2001). Sympathetic cooling of ^{85}Rb and ^{87}Rb. *Physical Review A*, **64**, 021402.

Blunden, P. G. and Sick, I. (2005). Proton radii and two-photon exchange. *Physical Review C*, **72**(5), 057601.

Bohn, J. L., Cavagnero, M., and Ticknor, C. (2009). Quasi-universal dipolar scattering in cold and ultracold gases. *New Journal of Physics*, **11**, 055039.

Boiron, D., Mennerat-Robilliard, C., Fournier, J.-M., Guidoni, L., Salomon, C., and Grynberg, G. (1999). Trapping and cooling cesium atoms in a speckle field. *The European Physical Journal D*, **7**, 373.

Bookjans, E. M., Hamley, C. D., and Chapman, M. S. (2011). Strong quantum spin correlations observed in atomic spin mixing. *Physical Review Letters*, **107**, 210406.

Borbely, J. S., George, M. C., Lombardi, L. D., Weel, M., Fitzakerley, D. W., and

Hessels, E. A. (2009). Separated oscillatory-field microwave measurement of the 2 $^3P_1 - 2\ ^3P_2$ fine-structure interval of atomic helium. *Physical Review A*, **79**, 060503.
Bordé, C. (1970). Spectroscopie d'absorption saturée de diverses molécules au moyen des lasers à gaz carbonique et à protoxyde d'azote. *Comptes rendus de l'Académie des Sciences B*, **271**, 371.
Bordé, Ch. J. (1989). Atomic interferometry with internal state labelling. *Physics Letters A*, **140**, 10.
Born, M. and Wolf, E. (1997). *Principles of Optics.* Cambridge University Press.
Bose, S. N. (1924). Plancks Gesetz und Lichtquantenhypothese. *Zeitschrift für Physik*, **26**, 178.
Bouchendira, R., Cladé, P., Guellati-Khélifa, S., Nez, F., and Biraben, F. (2011). New determination of the fine structure constant and test of the quantum electrodynamics. *Physical Review Letters*, **106**, 080801.
Bourdel, T., Khaykovich, L., Cubizolles, J., Zhang, J., Chevy, F., Teichmann, M., Tarruell, L., Kokkelmans, S. J. J. M. F., and Salomon, C. (2004). Experimental study of the BEC–BCS crossover region in lithium 6. *Physical Review Letters*, **93**, 050401.
Boyd, R. (2008). *Nonlinear Optics.* Academic Press.
Bradley, C. C., Sackett, C. A., and Hulet, R. G. (1997*a*). Bose–Einstein condensation of lithium: Observation of limited condensate number. *Physical Review Letters*, **78**(6), 985.
Bradley, C. C., Sackett, C. A., Tollett, J. J., and Hulet, R. G. (1995). Evidence of Bose–Einstein condensation in an atomic gas with attractive interactions. *Physical Review Letters*, **75**(9), 1687.
Bradley, C. C., Sackett, C. A., Tollett, J. J., and Hulet, R. G. (1997*b*). Evidence of Bose–Einstein condensation in an atomic gas with attractive interactions (errata). *Physical Review Letters*, **79**(6), 1170.
Bradley, M. P., Porto, J. V., Rainville, S., Thompson, J. K., and Pritchard, D. E. (1999). Penning trap measurements of the masses of ^{133}Cs, ^{87,85}Rb, and ^{23}Na with uncertainties <0.2 ppb. *Physical Review Letters*, **83**, 4510.
Breit, G. (1929). The effect of retardation on the interaction of two electrons. *Physical Review*, **34**(4), 553.
Britton, J. W., Sawyer, B. C., Keith, A. C., Wang, C.-C. J., Freericks, J. K., Uys, H., Biercuk, M. J., and Bollinger, J. J. (2012). Engineered two-dimensional Ising interactions in a trapped-ion quantum simulator with hundreds of spins. *Nature*, **484**, 489.
Brown, L. S. and Gabrielse, G. (1986). Geonium theory: Physics of a single electron or ion in a Penning trap. *Reviews of Modern Physics*, **58**(1), 233.
Brunello, A., Dalfovo, F., Pitaevskii, L., Stringari, S., and Zambelli, F. (2001). Momentum transferred to a trapped Bose–Einstein condensate by stimulated light scattering. *Physical Review A*, **64**, 063614.
Bücker, R., Grond, J., Manz, S., Berrada, T., Betz, T., Koller, C., Hohenester, U., Schumm, T., Perrin, A., and Schmiedmayer, J. (2011). Twin-atom beams. *Nature Physics*, **7**, 608.
Buluta, I. and Nori, F. (2009). Quantum simulators. *Science*, **326**, 108.

Burger, S., Bongs, K., Dettmer, S., Ertmer, W., Sengstock, K., Sanpera, A., Shlyapnikov, G. V., and Lewenstein, M. (1999). Dark solitons in Bose–Einstein condensates. *Physical Review Letters*, **83**(25), 5198.

Burt, E. A., Ghrist, R. W., Myatt, C. J., Holland, M. J., Cornell, E. A., and Wieman, C. E. (1997). Coherence, correlations, and collisions: What one learns about Bose–Einstein condensates from their decay. *Physical Review Letters*, **79**(3), 337.

Cacciapuoti, L. and Salomon, Ch. (2009). Space clocks and fundamental tests: The ACES experiment. *The European Physical Journal Special Topics*, **172**, 57.

Cadoret, M., de Mirandes, E., Cladé, P., Guellati-Khélifa, S., Schwob, C., Nez, F., Julien, L., and Biraben, F. (2008). Combination of Bloch oscillations with a Ramsey–Bordé interferometer: New determination of the fine structure constant. *Physical Review Letters*, **101**, 230801.

Campbell, C. J., Radnaev, A. G., Kuzmich, A., Dzuba, V. A., Flambaum, V. V., and Derevianko, A. (2012). Single-ion nuclear clock for metrology at the 19th decimal place. *Physical Review Letters*, **108**, 120802.

Campbell, G. K., Boyd, M. M., Thomsen, J. W., Martin, M. J., Blatt, S., Swallows, M. D., Nicholson, T. L., Fortier, T., Oates, C. W., Diddams, S. A., Lemke, N. D., Naidon, P., Julienne, P., Ye, J., and Ludlow, A. D. (2009). Probing interactions between ultracold fermions. *Science*, **324**, 360.

Campbell, G. K., Ludlow, A. D., Blatt, S., Thomsen, J. W., Martin, M. J., de Miranda, M. H. G., Zelevinsky, T., Boyd, M. M., Ye, J., Diddams, S. A., Heavner, T. P., Parker, T. E., and Jefferts, S. R. (2008). The absolute frequency of the ^{87}Sr optical clock transition. *Metrologia*, **45**, 539.

Cancio, P., Minardi, F., and Inguscio, M. (2001). Fine structure constant α and precision laser spectroscopy of helium. In *Recent Advances in Metrology and Fundamental Constants* (ed. T. J. Quinn, S. Leschiutta, and P. Tavella), pp. 217–238. IOS Press.

Cancio Pastor, P., Consolino, L., Giusfredi, G., De Natale, P., Inguscio, M., Yerokhin, V., and Pachucki, K. (2012). Frequency metrology of helium around 1083 nm and determination of the nuclear charge radius. *Physical Review Letters*, **108**, 143001.

Cancio Pastor, P., Giusfredi, G., De Natale, P., Hagel, G., de Mauro, C., and Inguscio, M. (2004). Absolute frequency measurements of the $2\,^3S_1 - 2\,^3P_{0,1,2}$ atomic helium transitions around 1083 nm. *Physical Review Letters*, **92**, 023001.

Cancio Pastor, P., Giusfredi, G., De Natale, P., Hagel, G., de Mauro, C., and Inguscio, M. (2006). Absolute frequency measurements of the $2\,^3S_1 - 2\,^3P_{0,1,2}$ atomic helium transitions around 1083 nm (errata). *Physical Review Letters*, **97**, 139903.

Capogrosso-Sansone, B., Trefzger, C., Lewenstein, M., Zoller, P., and Pupillo, G. (2010). Quantum phases of cold polar molecules in 2D optical lattices. *Physical Review Letters*, **104**, 125301.

Carboni, G., Gorini, G., Iacopini, E., Palffy, L., Palmonari, F., Torelli, G., and Zavattini, E. (1978). Measurement of the $2S_{1/2} - 2P_{1/2}$ splitting in the $(\mu\text{-}^4\text{He})^+$ muonic ion. *Physics Letters B*, **73**(2), 229.

Carnal, O. and Mlynek, J. (1991). Young's double-slit experiment with atoms: A simple atom interferometer. *Physical Review Letters*, **66**(21), 2689.

Carr, L. D., DeMille, D., Krems, R. V., and Ye, J. (2009). Cold and ultracold molecules: Science, technology and applications. *New Journal of Physics*, **11**, 055049.

Carusotto, I., Pitaevskii, L., Stringari, S., Modugno, G., and Inguscio, M. (2005). Sensitive measurement of forces at the micron scale using Bloch oscillations of ultracold atoms. *Physical Review Letters*, **95**, 093202.

Casalbuoni, R. and Nardulli, G. (2004). Inhomogeneous superconductivity in condensed matter and QCD. *Reviews of Modern Physics*, **76**, 263.

Casimir, H. B. G. and Polder, D. (1948). The influence of retardation on the London–van der Waals forces. *Physical Review*, **73**(4), 360.

Castin, Y. and Dalibard, J. (1997). Relative phase of two Bose–Einstein condensates. *Physical Review A*, **55**(6), 4330.

Castin, Y. and Dum, R. (1996). Bose–Einstein condensates in time dependent traps. *Physical Review Letters*, **77**(27), 5315.

Cataliotti, F. S., Burger, S., Fort, C., Maddaloni, P., Minardi, F., Trombettoni, A., Smerzi, A., and Inguscio, M. (2001). Josephson junction arrays with Bose–Einstein condensates. *Science*, **293**, 843.

Cataliotti, F. S., Cornell, E. A., Fort, C., Inguscio, M., Marin, F., Prevedelli, M., Ricci, L., and Tino, G. M. (1998). Magneto-optical trapping of Fermionic potassium atoms. *Physical Review A*, **57**(2), 1136.

Cataliotti, F. S., Fallani, L., Ferlaino, F., Fort, C., Maddaloni, P., and Inguscio, M. (2003). Superfluid current disruption in a chain of weakly coupled Bose–Einstein condensates. *New Journal of Physics*, **5**, 71.

Catani, J., De Sarlo, L., Barontini, G., Minardi, F., and Inguscio, M. (2008). Degenerate Bose–Bose mixture in a three-dimensional optical lattice. *Physical Review A*, **77**, 011603.

Cavalieri, S., Eramo, R., Materazzi, M., Corsi, C., and Bellini, M. (2002). Ramsey-type spectroscopy with high-order harmonics. *Physical Review Letters*, **89**(13), 133002.

Chiba, T. (2011). The constancy of the constants of nature: Updates. *Progress of Theoretical Physics*, **126**(6), 993.

Chin, C., Bartenstein, M., Altmeyer, A., Riedl, S., Jochim, S., Hecker Denschlag, J., and Grimm, R. (2004). Observation of the pairing gap in a strongly interacting Fermi gas. *Science*, **305**, 1128.

Chin, C., Grimm, R., Julienne, P., and Tiesinga, E. (2010). Feshbach resonances in ultracold gases. *Reviews of Modern Physics*, **82**, 1225.

Chiow, S., Kovachy, T., Chien, H.-C., and Kasevich, M. A. (2011). 102 $\hbar k$ large area atom interferometers. *Physical Review Letters*, **107**, 130403.

Chotia, A., Neyenhuis, B., Moses, S. A., Yan, B., Covey, J. P., Foss-Feig, M., Rey, A. M., Jin, D. S., and Ye, J. (2012). Long-lived dipolar molecules and Feshbach molecules in a 3D optical lattice. *Physical Review Letters*, **108**, 080405.

Chou, C. W., Hume, D. B., Koelemeij, J. C. J., Wineland, D. J., and Rosenband, T. (2010*a*). Frequency comparison of two high-accuracy Al^+ optical clocks. *Physical Review Letters*, **104**, 070802.

Chou, C. W., Hume, D. B., Rosenband, T., and Wineland, D. J. (2010*b*). Optical clocks and relativity. *Science*, **329**, 1630.

Chu, S. (1998). The manipulation of neutral particles. *Reviews of Modern Physics*, **70**(3), 685.

Chu, S., Bjorkholm, J. E., Ashkin, A., and Cable, A. (1986). Experimental observation

of optically trapped atoms. *Physical Review Letters*, **57**(3), 314.
Chu, S., Hollberg, L., Bjorkholm, J. E., Cable, A., and Ashkin, A. (1985). Three-dimensional viscous confinement and cooling of atoms by resonance radiation pressure. *Physical Review Letters*, **55**(1), 48.
Cingöz, A., Yost, D. C., Allison, T. K., Ruehl, A., Fermann, M. E., Hartl, I., and Ye, J. (2012). Direct frequency comb spectroscopy in the extreme ultraviolet. *Nature*, **482**, 68.
Cirac, J. I., Lewenstein, M., Mølmer, K., and Zoller, P. (1998). Quantum superposition states of Bose–Einstein condensates. *Physical Review A*, **57**(2), 1208.
Cirac, J. I. and Zoller, P. (1995). Quantum computations with cold trapped ions. *Physical Review Letters*, **74**, 4091.
Cladé, P., de Mirandes, E., Cadoret, M., Guellati-Khélifa, S., Schwob, C., Nez, F., Julien, L., and Biraben, F. (2006). Determination of the fine structure constant based on Bloch oscillations of ultracold atoms in a vertical optical lattice. *Physical Review Letters*, **96**, 033001.
Cladé, P., Guellati-Khélifa, S., Nez, F., and Biraben, F. (2009). Large momentum beam splitter using Bloch oscillations. *Physical Review Letters*, **102**, 240402.
Clairon, A., Laurent, P., Santarelli, G., Ghezali, S., Lea, S. N., and Bahoura, M. (1995). A cesium fountain frequency standard: Preliminary results. *IEEE Transactions on Instrumentation and Measurement*, **44**(2), 128.
Clément, D., Fabbri, N., Fallani, L., Fort, C., and Inguscio, M. (2009*a*). Exploring correlated 1D Bose gases from the superfluid to the Mott-insulator state by inelastic light scattering. *Physical Review Letters*, **102**, 155301.
Clément, D., Fabbri, N., Fallani, L., Fort, C., and Inguscio, M. (2009*b*). Multi-band spectroscopy of inhomogeneous Mott-insulator states of ultracold bosons. *New Journal of Physics*, **11**, 103030.
Cline, R. W., Smith, D. A., Greytak, T. J., and Kleppner, D. (1980). Magnetic confinement of spin-polarized atomic hydrogen. *Physical Review Letters*, **45**(26), 2117.
Cohen-Tannoudji, C. (1992). Atomic motion in laser light. In *Fundamental Systems in Quantum Optics* (ed. J. Dalibard, J.-M. Raymond, and J. Zinn-Justin), pp. 1–164. North-Holland.
Cohen-Tannoudji, C. and Guéry-Odelin, D. (2011). *Advances in Atomic Physics: An Overview*. World Scientific.
Cohen-Tannoudji, C. N. (1998). Manipulating atoms with photons. *Reviews of Modern Physics*, **70**(3), 707.
Consolino, L., Taschin, A., Bartolini, P., Bartalini, S., Cancio, P., Tredicucci, A., Beere, H. E., Ritchie, D. A., Torre, R., Vitiello, M. S., and De Natale, P. (2012). Phase-locking to a free-space terahertz comb for metrological-grade terahertz lasers. *Nature Communications*, **3**, 1040.
Corkum, P. B. (1993). Plasma perspective on strong-field multiphoton ionization. *Physical Review Letters*, **71**(13), 1994.
Cornell, E. A. and Wieman, C. E. (2002). Nobel Lecture: Bose–Einstein condensation in a dilute gas, the first 70 years and some recent experiments. *Reviews of Modern Physics*, **74**, 875.

Corney, A. (2006). *Atomic and Laser Spectroscopy.* Oxford University Press.

Cronin, A. D., Schmiedmayer, J., and Pritchard, D. E. (2009). Optics and interferometry with atoms and molecules. *Reviews of Modern Physics*, **81**, 1051.

Cundiff, S. T. and Ye, J. (2003). Colloquium: Femtosecond optical frequency combs. *Reviews of Modern Physics*, **75**, 325.

Dalfovo, F., Giorgini, S., Pitaevskii, L. P., and Stringari, S. (1999). Theory of Bose–Einstein condensation in trapped gases. *Reviews of Modern Physics*, **71**(3), 463.

Dalibard, J. and Cohen-Tannoudji, C. (1989). Laser cooling below the Doppler limit by polarization gradients: Simple theoretical models. *Journal of the Optical Society of America B*, **6**(11), 2023.

Dalibard, J., Gerbier, F., Juzeliūnas, G., and Öhberg, P. (2011). Colloquium: Artificial gauge potentials for neutral atoms. *Reviews of Modern Physics*, **83**, 1523.

Damski, B., Zakrzewski, J., Santos, L., Zoller, P., and Lewenstein, M. (2003). Atomic Bose and Anderson glasses in optical lattices. *Physical Review Letters*, **91**(8), 080403.

Danzl, J. G., Haller, E., Gustavsson, M., Mark, M. J., Hart, R., Bouloufa, N., Dulieu, O., Ritsch, H., and Nägerl, H.-C. (2008). Quantum gas of deeply bound ground state molecules. *Science*, **321**, 1062.

Danzl, J. G., Mark, M. J., Haller, E., Gustavsson, M., Hart, R., Aldegunde, J., Hutson, J. M., and Nägerl, H.-C. (2010). An ultracold high-density sample of rovibronic ground-state molecules in an optical lattice. *Nature Physics*, **6**, 265.

Davis, K. B., Mewes, M.-O., Andrews, M. R., van Druten, N. J., Durfee, D. S., Kurn, D. M., and Ketterle, W. (1995). Bose–Einstein condensation in a gas of sodium atoms. *Physical Review Letters*, **75**(22), 3969.

de Beauvoir, B., Nez, F., Julien, L., Cagnac, B., Biraben, F., Touahri, D., Hilico, L., Acef, O., Clairon, A., and Zondy, J. J. (1997). Absolute frequency measurement of the 2S – 8S/D transitions in hydrogen and deuterium: New determination of the Rydberg constant. *Physical Review Letters*, **78**, 440.

de Miranda, M. H. G., Chotia, A., Neyenhuis, B., Wang, D., Quéméner, G., Ospelkaus, S., Bohn, J. L., Ye, J., and Jin, D. S. (2011). Controlling the quantum stereodynamics of ultracold bimolecular reactions. *Nature Physics*, **7**, 502.

Dehmelt, H. (1990). Experiments with an isolated subatomic particle at rest. *Reviews of Modern Physics*, **62**(3), 525.

Dehmelt, H. G. (1975). Proposed 10^{14} $\delta\nu < \nu$ laser fluorescence spectroscopy on Tl^{+} monoion oscillator II (spontaneous quantum jumps). *Bulletin of the American Physical Society*, **20**, 60.

Deiglmayr, J., Grochola, A., Repp, M., Mörtlbauer, K., Glück, C., Lange, J., Dulieu, O., Wester, R., and Weidemüller, M. (2008). Formation of ultracold polar molecules in the rovibrational ground state. *Physical Review Letters*, **101**, 133004.

Deissler, B., Zaccanti, M., Roati, G., D'Errico, C., Fattori, M., Modugno, M., Modugno, G., and Inguscio, M. (2010). Delocalization of a disordered bosonic system by repulsive interactions. *Nature Physics*, **6**, 354.

DeMarco, B. and Jin, D. S. (1999). Onset of Fermi degeneracy in a trapped atomic gas. *Science*, **285**, 1703.

DeMarco, B., Rohner, H., and Jin, D. S. (1999). An enriched ^{40}K source for fermionic atom studies. *Review of Scientific Instruments*, **70**(4), 1967.

Demtröder, W. (2003). *Laser Spectroscopy.* Springer-Verlag.

Deng, L., Hagley, E. W., Wen, J., Trippenbach, M., Band, Y., Julienne, P. S., Simsarian, J. E., Helmerson, K., Rolston, S. L., and Phillips, W. D. (1999). Four-wave mixing with matter waves. *Nature*, **398**, 218.

Derevianko, A., Dzuba, V. A., and Flambaum, V. V. (2012). Highly charged ions as a basis of optical atomic clockwork of exceptional accuracy. *Physical Review Letters*, **109**, 180801.

Derevianko, A. and Katori, H. (2011). Colloquium: Physics of optical lattice clocks. *Reviews of Modern Physics*, **83**, 331.

DeSalvo, B. J., Yan, M., Mickelson, P. G., Martinez de Escobar, Y. N., and Killian, T. C. (2010). Degenerate Fermi gas of ^{87}Sr. *Physical Review Letters*, **105**, 030402.

Dettmer, S., Hellweg, D., Ryytty, P., Arlt, J. J., Ertmer, W., Sengstock, K., Petrov, D. S., Shlyapnikov, G. V., Kreutzmann, H., Santos, L., and Lewenstein, M. (2001). Observation of phase fluctuations in elongated Bose–Einstein condensates. *Physical Review Letters*, **87**(16), 160406.

Deutsch, C., Ramirez-Martinez, F., Lacroûte, C., Reinhard, F., Schneider, T., Fuchs, J. N., Piéchon, F., Laloë, F., Reichel, J., and Rosenbusch, P. (2010). Spin self-rephasing and very long coherence times in a trapped atomic ensemble. *Physical Review Letters*, **105**, 020401.

Dicke, R. H. (1953). The effect of collisions upon the Doppler width of spectral lines. *Physical Review*, **89**(2), 472.

Diddams, S. A., Udem, Th., Bergquist, J. C., Curtis, E. A., Drullinger, R. E., Hollberg, L., Itano, W. M., Lee, W. D., Oates, C. W., Vogel, K. R., and Wineland, D. J. (2001). An optical clock based on a single trapped ^{199}Hg^{+} ion. *Science*, **293**, 825.

Diedrich, F., Bergquist, J. C., Itano, W. M., and Wineland, D. J. (1989). Laser cooling to the zero-point energy of motion. *Physical Review Letters*, **62**(4), 403.

Dimopoulos, S. and Geraci, A. A. (2003). Probing submicron forces by interferometry of Bose–Einstein condensed atoms. *Physical Review D*, **68**, 124021.

Donley, E. A., Claussen, N. R., Cornish, S. L., Roberts, J. L., Cornell, E. A., and Wieman, C. E. (2001). Dynamics of collapsing and exploding Bose–Einstein condensates. *Nature*, **412**, 295.

Drake, G. W. F. (1989). High precision calculations for the Rydberg states of helium. In *Long Range Casimir Forces: Theory and Recent Experiments on Atomic Systems* (ed. F. S. Levin and D. A. Micha). Plenum.

Drake, G. W. F. (2004). Helium. Relativity and QED. *Nuclear Physics A*, **737**, 25.

Drake, G. W. F. (2006). High precision calculations for helium. In *Atomic, Molecular & Optical Physics Handbook* (ed. G. W. F. Drake), pp. 199–219. Springer.

Drever, R. W. P., Hall, J. L., Kowalski, F. V., Hough, J., Ford, G. M., Munley, A. J., and Ward, H. (1983). Laser phase and frequency stabilization using an optical resonator. *Applied Physics B*, **31**, 97.

Drullinger, R. E., Wineland, D. J., and Bergquist, J. C. (1980). High-resolution optical spectra of laser cooled ions. *Applied Physics*, **22**, 365.

Duan, L.-M., Demler, E., and Lukin, M. D. (2003). Controlling spin exchange interactions of ultracold atoms in optical lattices. *Physical Review Letters*, **91**(9),

090402.

Duan, L.-M., Sørensen, A., Cirac, J. I., and Zoller, P. (2000). Squeezing and entanglement of atomic beams. *Physical Review Letters*, **85**(19), 3991.

Durfee, D. S., Shaham, Y. K., and Kasevich, M. A. (2006). Long-term stability of an area-reversible atom-interferometer Sagnac gyroscope. *Physical Review Letters*, **97**, 240801.

Eikema, K. S. E., Ubachs, W., Vassen, W., and Hogervorst, W. (1993). First laser excitation of the ^{4}He 1 ^{1}S – 2 ^{1}P resonance line at 58 nm. *Physical Review Letters*, **71**(11), 1690.

Eikema, K. S. E., Ubachs, W., Vassen, W., and Hogervorst, W. (1996). Precision measurements in helium at 58 nm: Ground state Lamb shift and the 1 ^{1}S – 2 ^{1}P transition isotope shift. *Physical Review Letters*, **76**(8), 1216.

Eikema, K. S. E., Ubachs, W., Vassen, W., and Hogervorst, W. (1997). Lamb shift measurement in the 1 ^{1}S ground state of helium. *Physical Review A*, **55**(3), 1866.

Einstein, A. (1924). Quantentheorie des idealen einatomigen Gases. *Sitzungsberichte der Preussischen Akademie der Wissenschaften*, **261**.

Einstein, A. (1925). Quantentheorie des idealen einatomigen Gases. *Sitzungsberichte der Preussischen Akademie der Wissenschaften*, **3**.

Eramo, R., Cavalieri, S., Corsi, C., Liontos, I., and Bellini, M. (2011). Method for high-resolution frequency measurements in the extreme ultraviolet regime: Random-sampling Ramsey spectroscopy. *Physical Review Letters*, **106**, 213003.

Ertmer, W., Blatt, R., Hall, J. L., and Zhu, M. (1985). Laser manipulation of atomic beam velocities: Demonstration of stopped atoms and velocity reversal. *Physical Review Letters*, **54**(10), 996.

Eschner, J., Morigi, G., Schmidt-Kaler, F., and Blatt, R. (2003). Laser cooling of trapped ions. *Journal of the Optical Society of America B*, **20**(5), 1003.

Essen, L. and Parry, J. V. L. (1955). An atomic standard of frequency and time interval: A caesium resonator. *Nature*, **176**, 280.

Esslinger, Tilman (2010). Fermi–Hubbard physics with atoms in an optical lattice. *Annual Review of Condensed Matter Physics*, **1**, 129.

Esslinger, T., Bloch, I., and Hänsch, T. W. (1998). Bose–Einstein condensation in a quadrupole–Ioffe-configuration trap. *Physical Review A*, **58**(4), R2664.

Estève, J., Gross, C., Weller, A., Giovanazzi, S., and Oberthaler, M. K. (2008). Squeezing and entanglement in a Bose–Einstein condensate. *Nature*, **455**, 1216.

Evenson, K. M., Inguscio, M., and Jennings, D. A. (1985). Point contact diode at laser frequencies. *Journal of Applied Physics*, **57**(3), 956.

Evenson, K. M., Wells, J. S., Petersen, F. R., Danielson, B. L., and Day, G. W. (1973). Accurate frequencies of molecular transitions used in laser stabilization: The 3.39-μm transition in CH_4 and the 9.33- and 10.18-μm transitions in CO_2. *Applied Physics Letters*, **22**, 192.

Evenson, K. M., Wells, J. S., Petersen, F. R., Danielson, B. L., Day, G. W., Barger, R. L., and Hall, J. L. (1972). Speed of light from direct frequency and wavelength measurements of the methane-stabilized laser. *Physical Review Letters*, **29**(19), 1346.

Fabbri, N., Clément, D., Fallani, L., Fort, C., Modugno, M., van Der Stam, K. M. R., and Inguscio, M. (2009). Excitations of Bose–Einstein condensates in a one-dimensional periodic potential. *Physical Review A*, **79**, 043623.

Fallani, L., Cataliotti, F. S., Catani, J., Fort, C., Modugno, M., Zawada, M., and Inguscio, M. (2003). Optically induced lensing effect on a Bose–Einstein condensate expanding in a moving lattice. *Physical Review Letters*, **91**(24), 240405.

Fallani, L., De Sarlo, L., Lye, J. E., Modugno, M., Saers, R., Fort, C., and Inguscio, M. (2004). Observation of dynamical instability for a Bose–Einstein condensate in a moving 1D optical lattice. *Physical Review Letters*, **93**(14), 140406.

Fallani, L., Fort, C., and Inguscio, M. (2008). Bose–Einstein condensates in disordered potentials. *Advances in Atomic, Molecular, and Optical Physics*, **56**, 119.

Fallani, L. and Inguscio, M. (2011). Ultracold atoms in bichromatic lattices. In *Nano Optics and Atomics: Transport of Light and Matter Waves* (ed. R. Kaiser, D. S. Wiersma, and L. Fallani), pp. 179–231. IOS Press.

Fallani, L., Lye, J. E., Guarrera, V., Fort, C., and Inguscio, M. (2007). Ultracold atoms in a disordered crystal of light: Towards a Bose glass. *Physical Review Letters*, **98**, 130404.

Fano, U. (1961*a*). Effects of configuration interaction on intensities and phase shifts. *Physical Review*, **124**(6), 1866.

Fano, U. (1961*b*). Quantum theory of interference effects in the mixing of light from phase-independent sources. *American Journal of Physics*, **29**(8), 539.

Fattori, M., D'Errico, C., Roati, G., Zaccanti, M., Jona-Lasinio, M., Modugno, M., Inguscio, M., and Modugno, G. (2008*a*). Atom interferometry with a weakly interacting Bose–Einstein condensate. *Physical Review Letters*, **100**, 080405.

Fattori, M., Roati, G., Deissler, B., D'Errico, C., Zaccanti, M., Jona-Lasinio, M., Santos, L., Inguscio, M., and Modugno, G. (2008*b*). Magnetic dipolar interaction in a Bose–Einstein condensate atomic interferometer. *Physical Review Letters*, **101**, 190405.

Fedichev, P. O., Kagan, Yu., Shlyapnikov, G. V., and Walraven, J. T. M. (1996). Influence of nearly resonant light on the scattering length in low-temperature atomic gases. *Physical Review Letters*, **77**, 2913.

Feldmann, J., Leo, K., Shah, J., Miller, D. A. B., Cunningham, J. E., Meier, T., von Plessen, G., Schulze, A., Thomas, P., and Schmitt-Rink, S. (1992). Optical investigation of Bloch oscillations in a semiconductor superlattice. *Physical Review B*, **46**(11), 7252.

Ferrari, G., Cancio, P., Drullinger, R., Giusfredi, G., Poli, N., Prevedelli, M., Toninelli, C., and Tino, G. M. (2003). Precision frequency measurement of visible intercombination lines of strontium. *Physical Review Letters*, **91**, 243002.

Ferrari, G., Poli, N., Sorrentino, F., and Tino, G. (2006). Long-lived Bloch oscillations with bosonic Sr atoms and application to gravity measurement at the micrometer scale. *Physical Review Letters*, **97**, 060402.

Feshbach, H. (1958). Unified theory of nuclear reactions. *Annals of Physics*, **5**, 357.

Feynman, R. P. (1982). Simulating physics with computers. *International Journal of Theoretical Physics*, **21**(6-7), 467.

Feynman, R. P. (1988). *QED: The Strange Theory of Light and Matter.* Princeton University Press.

Fischer, M., Kolachevsky, N., Zimmermann, M., Holzwarth, R., Udem, Th., Hänsch,

T. W., Abgrall, M., Grünert, J., Maksimovic, I., Bize, S., Marion, H., Pereira Dos Santos, F., Lemonde, P., Santarelli, G., Laurent, P., Clairon, A., Salomon, C., Haas, M., Jentschura, U. D., and Keitel, C. H. (2004). New limits on the drift of fundamental constants from laboratory measurements. *Physical Review Letters*, **92**(23), 230802.

Fisher, M. P. A. (1990). Quantum phase transitions in disordered two-dimensional superconductors. *Physical Review Letters*, **65**, 923.

Fisher, M. P. A., Weichman, P. B., Grinstein, G., and Fisher, D. S. (1989). Boson localization and the superfluid–insulator transition. *Physical Review B*, **40**(1), 546.

Fixler, J. B., Foster, G. T., McGuirk, J. M., and Kasevich, M. A. (2007). Atom interferometer measurement of the Newtonian constant of gravity. *Science*, **315**, 74.

Fölling, S., Gerbier, F., Widera, A., Mandel, O., Gericke, T., and Bloch, I. (2005). Spatial quantum noise interferometry in expanding ultracold atom clouds. *Nature*, **434**, 481.

Foot, C. J. (2005). *Atomic Physics*. Oxford University Press.

Foreman, S. M., Marian, A., Ye, J., Petrukhin, E. A., Gubin, M. A., Mücke, O. D., Wong, F. N. C., Ippen, E. P., and Kärtner, F. X. (2005). Demonstration of a HeNe/CH_4-based optical molecular clock. *Optics Letters*, **30**, 570.

Fort, C., Bambini, A., Cacciapuoti, L., Cataliotti, F. S., Prevedelli, M., Tino, G. M., and Inguscio, M. (1998). Cooling mechanisms in potassium magneto-optical traps. *The European Physical Journal D*, **3**, 113.

Fort, C., Prevedelli, M., Minardi, F., Cataliotti, F. S., Ricci, L., Tino, G. M., and Inguscio, M. (2000). Collective excitations of a ^{87}Rb Bose condensate in the Thomas-Fermi regime. *Europhysics Letters*, **49**, 8.

Fortágh, J. and Zimmermann, C. (2005). Toward atom chips. *Science*, **307**, 860.

Friebe, J., Pape, A., Riedmann, M., Moldenhauer, K., Mehlstäubler, T., Rehbein, N., Lisdat, C., Rasel, E., Ertmer, W., Schnatz, H., Lipphardt, B., and Grosche, G. (2008). Absolute frequency measurement of the magnesium intercombination transition $^1S_0 - {}^3P_1$. *Physical Review A*, **78**, 033830.

Fried, D. G., Killian, T. C., Willmann, L., Landhuis, D., Moss, S. C., Kleppner, D., and Greytak, T. J. (1998). Bose–Einstein condensation of atomic hydrogen. *Physical Review Letters*, **81**(18), 3811.

Friedenauer, A., Schmitz, H., Glueckert, J. T., Porras, D., and Schaetz, T. (2008). Simulating a quantum magnet with trapped ions. *Nature Physics*, **4**, 757.

Fukuhara, T., Takasu, Y., Kumakura, M., and Takahashi, Y. (2007). Degenerate Fermi gases of ytterbium. *Physical Review Letters*, **98**, 030401.

Gabrielse, G., Bowden, N. S., Oxley, P., Speck, A., Storry, C. H., Tan, J. N., Wessels, M., Grzonka, D., Oelert, W., Schepers, G., Sefzick, T., Walz, J., Pittner, H., Hänsch, T. W., and Hessels, E. A. (2002). Background-free observation of cold antihydrogen with field-ionization analysis of its states. *Physical Review Letters*, **89**, 213401.

Gabrielse, G., Hanneke, D., Kinoshita, T., Nio, M., and Odom, B. (2006). New determination of the fine structure constant from the electron g value and QED. *Physical Review Letters*, **97**, 030802.

Gavish, U. and Castin, Y. (2005). Matter-wave localization in disordered cold atom lattices. *Physical Review Letters*, **95**, 020401.

Gemelke, N., Zhang, X., Hung, C.-L., and Chin, C. (2009). In situ observation of

incompressible Mott-insulating domains in ultracold atomic gases. *Nature*, **460**, 995.
George, M. C., Lombardi, L. D., and Hessels, E. A. (2001). Precision microwave measurement of the 2 ^{3}P$_1$ – 2 ^{3}P$_0$ interval in atomic helium: A determination of the fine-structure constant. *Physical Review Letters*, **87**(17), 173002.
Gerbier, F., Fölling, S., Widera, A., Mandel, O., and Bloch, I. (2006). Probing number squeezing of ultracold atoms across the superfluid–Mott insulator transition. *Physical Review Letters*, **96**, 090401.
Gerritsma, R., Kirchmair, G., Zähringer, F., Solano, E., Blatt, R., and Roos, C. F. (2010). Quantum simulation of the Dirac equation. *Nature*, **463**, 68.
Gerritsma, R., Lanyon, B. P., Kirchmair, G., Zähringer, F., Hempel, C., Casanova, J., García-Ripoll, J. J., Solano, E., Blatt, R., and Roos, C. F. (2011). Quantum simulation of the Klein paradox with trapped ions. *Physical Review Letters*, **106**, 060503.
Gerry, C. C. and Knight, P. L. (2005). *Introductory Quantum Optics*. Cambridge University Press.
Gerton, J. M., Strekalov, D., Prodan, I., and Hulet, R. G. (2000). Direct observation of growth and collapse of a Bose–Einstein condensate with attractive interactions. *Nature*, **408**, 692.
Gill, P. (2011). When should we change the definition of the second? *Philosophical Transactions of the Royal Society A: Mathematical, Physical and Engineering Sciences*, **369**, 4109.
Giltner, D. M., McGowan, R. W., and Lee, S. A. (1995). Atom interferometer based on Bragg scattering from standing light waves. *Physical Review Letters*, **75**(14), 2638.
Giorgini, S., Pitaevskii, L. P., and Stringari, S. (2008). Theory of ultracold atomic Fermi gases. *Reviews of Modern Physics*, **80**, 1215.
Giovannetti, V., Lloyd, S., and Maccone, L. (2004). Quantum-enhanced measurements: Beating the standard quantum limit. *Science*, **306**, 1330.
Giusfredi, G., Cancio Pastor, P., De Natale, P., Mazzotti, D., de Mauro, C., Fallani, L., Hagel, G., Krachmalnicoff, V., and Inguscio, M. (2005). Present status of the fine structure frequencies of the 2 ^{3}P helium level. *Canadian Journal of Physics*, **83**, 301.
Glauber, R. J. (1963). The quantum theory of optical coherence. *Physical Review*, **130**, 2529.
Glauber, R. J. (2006). Nobel Lecture: One hundred years of light quanta. *Reviews of Modern Physics*, **78**, 1267.
Gohle, C., Udem, Th., Herrmann, M., Rauschenberger, J., Holzwarth, R., Schuessler, H. A., Krausz, F., and Hänsch, T. W. (2005). A frequency comb in the extreme ultraviolet. *Nature*, **436**(7048), 234.
Goldenberg, H. M., Kleppner, D., and Ramsey, N. F. (1960). Atomic hydrogen maser. *Physical Review Letters*, **5**, 361.
Goodman, J. W. (2008). *Speckle Phenomena in Optics: Theory and Applications*. Roberts and Company Publishers.
Góral, K., Santos, L., and Lewenstein, M. (2002). Quantum phases of dipolar bosons in optical lattices. *Physical Review Letters*, **88**, 170406.

Gordon, J. P., Zeiger, H. J., and Townes, C. H. (1954). Molecular microwave oscillator and new hyperfine structure in the microwave spectrum of NH_3. *Physical Review*, **95**, 282.

Gordon, J. P., Zeiger, H. J., and Townes, C. H. (1955). The maser: New type of microwave amplifier, frequency standard, and spectrometer. *Physical Review*, **99**, 1264.

Gould, P. L., Ruff, G. A., and Pritchard, D. E. (1986). Diffraction of atoms by light: The near-resonant Kaptiza–Dirac effect. *Physical Review Letters*, **56**(8), 827.

Greiner, M. (2003). Ultracold quantum gases in three-dimensional optical lattice potentials. Phd thesis, Ludwig-Maximilians-Universität München.

Greiner, M., Bloch, I., Mandel, O., Hänsch, T. W., and Esslinger, T. (2001). Exploring phase coherence in a 2D lattice of Bose–Einstein condensates. *Physical Review Letters*, **87**(16), 160405.

Greiner, M., Mandel, O., Esslinger, T., Hänsch, T. W., and Bloch, I. (2002*a*). Quantum phase transition from a superfluid to a Mott insulator in a gas of ultracold atoms. *Nature*, **415**, 39.

Greiner, M., Mandel, O., Hänsch, T. W., and Bloch, I. (2002*b*). Collapse and revival of the matter wave field of a Bose–Einstein condensate. *Nature*, **419**, 51.

Greiner, M., Regal, C. A., and Jin, D. S. (2003). Emergence of a molecular Bose–Einstein condensate from a Fermi gas. *Nature*, **426**, 537.

Griesmaier, A., Werner, J., Hensler, S., Stuhler, J., and Pfau, T. (2005). Bose–Einstein condensation of chromium. *Physical Review Letters*, **94**, 160401.

Grimm, R., Weidemüller, M., and Ovchinnikov, Y. B. (2000). Optical dipole traps for neutral atoms. *Advances in Atomic, Molecular and Optical Physics*, **42**, 95.

Grinchuk, V. A., Kuzin, E. F., Nagaeva, M. L., Ryabenko, G. A., Kazantsev, A. P., Surdutovich, G. I., and Yakovlev, V. P. (1981). Scattering of an atomic beam by a short light pulse. *Physics Letters A*, **86**(3), 136.

Gross, C., Strobel, H., Nicklas, E., Zibold, T., Bar-Gill, N., Kurizki, G., and Oberthaler, M. K. (2011). Atomic homodyne detection of continuous-variable entangled twin-atom states. *Nature*, **480**, 219.

Gross, C., Zibold, T., Nicklas, E., Estève, J., and Oberthaler, M. K. (2010). Nonlinear atom interferometer surpasses classical precision limit. *Nature*, **464**, 1165.

Gross, E. P. (1961). Structure of a quantized vortex in boson systems. *Il Nuovo Cimento*, **20**(3), 454.

Gross, E. P. (1963). Hydrodynamics of a superfluid condensate. *Journal of Mathematical Physics*, **4**(2), 195.

Grynberg, G., Aspect, A., and Fabre, C. (2010). *Introduction to quantum optics: from the semi-classical approach to quantized light*. Cambridge University Press.

Guéna, J., Li, R., Gibble, K., Bize, S., and Clairon, A. (2011). Evaluation of Doppler shifts to improve the accuracy of primary atomic fountain clocks. *Physical Review Letters*, **106**, 130801.

Günter, K., Stöferle, T., Moritz, H., Köhl, M., and Esslinger, T. (2006). Bose–Fermi mixtures in a three-dimensional optical lattice. *Physical Review Letters*, **96**, 180402.

Gupta, S., Hadzibabic, Z., Zwierlein, M. W., Stan, C. A., Dieckmann, K., Schunck, C. H., Van Kempen, E. G. M., Verhaar, B. J., and Ketterle, W (2003). Radiofrequency spectroscopy of ultracold fermions. *Science*, **300**, 1723.

Gustavson, T. L., Bouyer, P., and Kasevich, M. A. (1997). Precision rotation measurements with an atom interferometer gyroscope. *Physical Review Letters*, **78**(11), 2046.

Gustavsson, M., Haller, E., Mark, M. J., Danzl, J. G., Rojas-Kopeinig, G., and Nägerl, H.-C. (2008). Control of interaction-induced dephasing of Bloch oscillations. *Physical Review Letters*, **100**, 080404.

Hadzibabic, Z., Krüger, P., Cheneau, M., Battelier, B., and Dalibard, J. (2006). Berezinskii–Kosterlitz–Thouless crossover in a trapped atomic gas. *Nature*, **441**, 1118.

Hadzibabic, Z., Stan, C. A., Dieckmann, K., Gupta, S., Zwierlein, M. W., Görlitz, A., and Ketterle, W. (2002). Two-species mixture of quantum degenerate Bose and Fermi gases. *Physical Review Letters*, **88**(16), 160401.

Hadzibabic, Z., Stock, S., Battelier, B., Bretin, V., and Dalibard, J. (2004). Interference of an array of independent Bose–Einstein condensates. *Physical Review Letters*, **93**, 180403.

Häffner, H., Hänsel, W., Roos, C. F., Benhelm, J., Chek-al Kar, D., Chwalla, M., Körber, T., Rapol, U. D., Riebe, M., Schmidt, P. O., Becher, C., Gühne, O., Dür, W., and Blatt, R. (2005). Scalable multiparticle entanglement of trapped ions. *Nature*, **438**, 643.

Häffner, H., Roos, C. F., and Blatt, R. (2008). Quantum computing with trapped ions. *Physics Reports*, **469**, 155.

Hall, J. (2006). Nobel Lecture: Defining and measuring optical frequencies. *Reviews of Modern Physics*, **78**(4), 1279.

Hanbury Brown, R. and Twiss, R. Q. (1956*a*). A test of a new type of stellar interferometer on Sirius. *Nature*, **178**, 1046.

Hanbury Brown, R. and Twiss, R. Q. (1956*b*). Correlation between photons in two coherent beams of light. *Nature*, **177**, 27.

Hanneke, D., Fogwell, S., and Gabrielse, G. (2008). New measurement of the electron magnetic moment and the fine structure constant. *Physical Review Letters*, **100**, 120801.

Hänsch, T. W. (1989). High resolution spectroscopy of hydrogen. In *The Hydrogen Atom* (ed. G. F. Bassani, M. Inguscio, and T. W. Hänsch), pp. 93–102. Springer-Verlag.

Hänsch, T. W. (2006). Nobel Lecture: Passion for precision. *Reviews of Modern Physics*, **78**(4), 1297.

Hänsch, T. W., Harvey, K. C., Meisel, G., and Schawlow, A. L. (1974). Two-photon spectroscopy of Na 3s-4d without Doppler broadening using a cw dye laser. *Optics Communications*, **11**, 50.

Hänsch, T. W., Lee, S. A., Wallenstein, R., and Wieman, C. (1975). Doppler-free two-photon spectroscopy of hydrogen 1S-2S. *Physical Review Letters*, **34**(6), 307.

Hänsch, T. W., Levenson, M. D., and Schawlow, A. L. (1971). Complete hyperfine structure of a molecular iodine line. *Physical Review Letters*, **26**, 946.

Hänsch, T. W. and Schawlow, A. L. (1975). Cooling of gases by laser radiation. *Optics Communications*, **13**(I), 68.

Hänsch, T. W., Shanin, I. S., and Schawlow, A. L. (1972). Optical resolution of the Lamb shift in atomic hydrogen by laser saturation spectroscopy. *Nature*, **235**, 63.

Hänsel, W, Hommelhoff, P, Hänsch, T. W., and Reichel, J (2001). Bose–Einstein condensation on a microelectronic chip. *Nature*, **413**, 498.

Harber, D. M., Obrecht, J. M., McGuirk, J. M., and Cornell, E. A. (2005). Measurement of the Casimir–Polder force through center-of-mass oscillations of a Bose–Einstein condensate. *Physical Review A*, **72**, 033610.

Haroche, S. (1976). Quantum beat spectroscopy. In *High Resolution Spectroscopy* (ed. K. Shimoda). Springer-Verlag.

Haroche, S. and Raimond, J.-M. (2006). *Exploring the Quantum: Atoms, Cavities, and Photons*. Oxford University Press.

Haussmann, R., Rantner, W., Cerrito, S., and Zwerger, W. (2007). Thermodynamics of the BCS–BEC crossover. *Physical Review A*, **75**, 023610.

Hayano, R. S., Hori, M., Horváth, D., and Widmann, E. (2007). Antiprotonic helium and CPT invariance. *Reports on Progress in Physics*, **70**, 1995.

Hecht, C. E. (1959). The possible superfluid behaviour of hydrogen atom gases and liquids. *Physica*, **25**, 1159.

Heisenberg, W. (1926). Über die Spektra von Atomsystemen mit zwei Elektronen. *Zeitschrift für Physik*, **39**, 499.

Hellweg, D., Cacciapuoti, L., Kottke, M., Schulte, T., Sengstock, K., Ertmer, W., and Arlt, J. J. (2003). Measurement of the spatial correlation function of phase fluctuating Bose–Einstein condensates. *Physical Review Letters*, **91**(1), 010406.

Hemmerich, A. and Hänsch, T. W. (1993). Two-Dimensional Atomic Crystal Bound by Light. *Physical Review Letters*, **70**(4), 410.

Hess, H. F. (1986). Evaporative cooling of magnetically trapped and compressed spin-polarized hydrogen. *Phys. Rev. B*, **34**, 3476.

Hess, H. F., Kochanski, G. P., Doyle, J. M., Masuhara, N., Kleppner, D., and Greytak, T. J. (1987). Magnetic Trapping of Spin-Polarized Atomic Hydrogen. *Physical Review Letters*, **59**(6), 672.

Higbie, J. M., Sadler, L. E., Inouye, S., Chikkatur, A. P., Leslie, S. R., Moore, K. L., Savalli, V., and Stamper-Kurn, D. M. (2005). Direct nondestructive imaging of magnetization in a spin-1 Bose–Einstein gas. *Physical Review Letters*, **95**, 050401.

Hinkley, N., Sherman, J. A., Phillips, N. B., Schioppo, M., Lemke, N. D., Beloy, K., Pizzocaro, M., Oates, C. W., and Ludlow, A. D. (2013). An atomic clock with 10^{-18} instability. Preprint available at http://arxiv.org/abs/1305.5869.

Hocker, L. O., Sokoloff, D. R., Daneu, V., Szoke, A., and Javan, A. (1968). Frequency mixing in the infrared and far-infrared using a metal-to-metal point contact diode. *Applied Physics Letters*, **12**, 401.

Hodapp, T. W., Gerz, C., Furtlehner, C., Westbrook, C. I., Phillips, W. D., and Dalibard, J. (1995). Three-dimensional spatial diffusion in optical molasses. *Applied Physics B*, **60**, 135.

Hodgman, S. S., Dall, R. G., Manning, A. G., Baldwin, K. G. H., and Truscott, A. G. (2011). Direct measurement of long-range third-order coherence in Bose–Einstein condensates. *Science*, **331**, 1046.

Hofferberth, S., Lesanovsky, I., Fischer, B., Schumm, T., and Schmiedmayer, J. (2007).

Non-equilibrium coherence dynamics in one-dimensional Bose gases. *Nature*, **449**, 324.

Hofferberth, S., Lesanovsky, I., Fischer, B., Verdu, J., and Schmiedmayer, J. (2006). Radiofrequency-dressed-state potentials for neutral atoms. *Nature Physics*, **2**, 710.

Hofferberth, S., Lesanovsky, I., Schumm, T., Imambekov, A., Gritsev, V., Demler, E., and Schmiedmayer, J. (2008). Probing quantum and thermal noise in an interacting many-body system. *Nature Physics*, **4**, 489.

Hollberg, L., Oates, C. W., Curtis, E. A., Ivanov, E. N., Diddams, S. A., Udem, T., Robinson, H. G., Bergquist, J. C., Rafac, R. J., Itano, W. M., Drullinger, R. E., and Wineland, D. J. (2001). Optical frequency standards and measurements. *IEEE Journal of Quantum Electronics*, **37**(12), 1502.

Hollberg, L., Oates, C. W., Wilpers, G., Hoyt, C. W., Barber, Z. W., Diddams, S. A., Oskay, W. H., and Bergquist, J. C. (2005). Optical frequency/wavelength references. *Journal of Physics B: Atomic, Molecular and Optical Physics*, **38**, S469.

Hori, M., Sótér, A., Barna, D., Dax, A., Hayano, R., Friedreich, S., Juhász, B., Pask, T., Widmann, E., Horváth, D., Venturelli, L., and Zurlo, N. (2011). Two-photon laser spectroscopy of antiprotonic helium and the antiproton-to-electron mass ratio. *Nature*, **475**, 484.

Hu, H., Strybulevych, A., Page, J. H., Skipetrov, S. E., and van Tiggelen, B. A. (2008). Localization of ultrasound in a three-dimensional elastic network. *Nature Physics*, **4**, 945.

Huang, K. (1987). *Statistical Mechanics*. John Wiley & Sons.

Huber, A., Gross, B., Weitz, M., and Hänsch, T. W. (1999). High-resolution spectroscopy of the 1S-2S transition in atomic hydrogen. *Physical Review A*, **59**(3), 1844.

Hudson, J. J., Kara, D. M., Smallman, I. J., Sauer, B. E., Tarbutt, M. R., and Hinds, E. A. (2011). Improved measurement of the shape of the electron. *Nature*, **473**, 493.

Hugbart, M., Retter, J. A., Gerbier, F., Varón, A. F., Richard, S., Thywissen, J. H., Clément, D., Bouyer, P., and Aspect, A. (2005). Coherence length of an elongated condensate. *The European Physical Journal D*, **35**, 155.

Hughes, V. W. (1969). The fine-structure constant. *Comments on Atomic and Molecular Physics*, **1**, 5.

Inguscio, M. (1994). High-resolution and high-sensitivity spectroscopy using semiconductor diode lasers. In *Frontiers in Laser Spectroscopy* (ed. T. W. Hänsch and M. Inguscio), pp. 41–59. North-Holland, Elsevier.

Inguscio, M. (2006). Majorana "spin-flip" and ultra-low temperature atomic physics. *Proceedings of Science* (EMC2006), 008.

Inguscio, M., Ketterle, W., and Salomon, C. (ed.) (2008). *Ultracold Fermi Gases*. IOS Press.

Inguscio, M., Stringari, S., and Wieman, C. E. (ed.) (1999). *Bose–Einstein Condensation in Atomic Gases*. IOS Press.

Inouye, S., Andrews, M. R., Stenger, J., Miesner, H.-J., Stamper-Kurn, D. M., and Ketterle, W. (1998). Observation of Feshbach resonances in a Bose–Einstein condensate. *Nature*, **392**, 151.

Ising, E. (1925). Beitrag zur theorie des ferromagnetismus. *Zeitschrift für Physik*, **31**,

253.

Itano, W. M., Bergquist, J. C., Bollinger, J. J., Gilligan, J. M., Heinzen, D. J., Moore, F. L., Raizen, M. G., and Wineland, D. J. (1993). Quantum projection noise: Population fluctuations in two-level systems. *Physical Review A*, **47**(5), 3554.

Itano, W. M., Lewis, L. L., and Wineland, D. J. (1982). Shift of $2S_{1/2}$ hyperfine splittings due to blackbody radiation. *Physical Review A*, **25**, 1233.

Ivanov, V., Alberti, A., Schioppo, M., Ferrari, G., Artoni, M., Chiofalo, M. L., and Tino, G. M. (2008). Coherent delocalization of atomic wave packets in driven lattice potentials. *Physical Review Letters*, **100**, 043602.

Jaksch, D., Bruder, C., Cirac, J. I., Gardiner, C. W., and Zoller, P. (1998). Cold bosonic atoms in optical lattices. *Physical Review Letters*, **81**(15), 3108.

Jaksch, D. and Zoller, P. (2003). Creation of effective magnetic fields in optical lattices: The Hofstadter butterfly for cold neutral atoms. *New Journal of Physics*, **5**, 56.

Jaskula, J.-C., Bonneau, M., Partridge, G. B., Krachmalnicoff, V., Deuar, P., Kheruntsyan, K. V., Aspect, A., Boiron, D., and Westbrook, C. I. (2010). Sub-Poissonian number differences in four-wave mixing of matter waves. *Physical Review Letters*, **105**(19), 190402.

Jeffery, A., Elmquist, R. E., Shields, J. Q., Lee, L. H., Cage, M. E., Shields, S. H., and Dziuba, R. F. (1998). Determination of the von Klitzing constant and the fine-structure constant through a comparison of the quantized Hall resistance and the ohm derived from the NIST calculable capacitor. *Metrologia*, **35**(2), 83.

Jeltes, T., McNamara, J. M., Hogervorst, W., Vassen, W., Krachmalnicoff, V., Schellekens, M., Perrin, A., Chang, H., Boiron, D., Aspect, A., and Westbrook, C. I. (2007). Comparison of the Hanbury Brown–Twiss effect for bosons and fermions. *Nature*, **445**, 402.

Jendrzejewski, F., Bernard, A., Müller, K., Cheinet, P., Josse, V., Piraud, M., Pezzé, L., Sanchez-Palencia, L., Aspect, A., and Bouyer, P. (2012). Three-dimensional localization of ultracold atoms in an optical disordered potential. *Nature Physics*, **8**, 398.

Jessen, P. S., Gerz, C., Lett, P. D., Phillips, W. D., Rolston, S. L., Spreeuw, R. J. C., and Westbrook, C. I. (1992). Observation of quantized motion of Rb atoms in an optical field. *Physical Review Letters*, **69**(1), 49.

Jiang, Y. Y., Ludlow, A. D., Lemke, N. D., Fox, R. W., Sherman, J. A., Ma, L.-S., and Oates, C. W. (2011). Making optical atomic clocks more stable with 10^{16} level laser stabilization. *Nature Photonics*, **5**, 158.

Jin, D. S., Ensher, J. R., Matthews, M. R., Wieman, C. E., and Cornell, E. A. (1996). Collective excitations of a Bose–Einstein condensate in a dilute gas. *Physical Review Letters*, **77**(3), 420.

Jo, G.-B., Shin, Y., Will, S., Pasquini, T., Saba, M., Ketterle, W., Pritchard, D. E., Vengalattore, M., and Prentiss, M. (2007). Long phase coherence time and number squeezing of two Bose–Einstein condensates on an atom chip. *Physical Review Letters*, **98**, 030407.

Jochim, S., Bartenstein, M., Altmeyer, A., Hendl, G., Riedl, S., Chin, C., Hecker Denschlag, J., and Grimm, R. (2003). Bose–Einstein condensation of molecules. *Science*, **302**, 2101.

Jones, D. J., Diddams, S. A., Ranka, J. K., Stentz, A., Windeler, R. S., Hall, J. L., and Cundiff, S. T. (2000). Carrier-envelope phase control of femtosecond mode-locked lasers and direct optical frequency synthesis. *Science*, **288**, 635.

Jördens, R., Strohmaier, N., Günter, K., Moritz, H., and Esslinger, T. (2008). A Mott insulator of fermionic atoms in an optical lattice. *Nature*, **455**, 204.

Kagan, Yu., Svistunov, B. V., and Shlyapnikov, G. V. (1985). Bose-condensation effect on inelastic processes in a gas. *JETP Letters*, **42**, 209.

Kandula, D. Z., Gohle, C., Pinkert, T., Ubachs, W., and Eikema, K. S. E. (2010). Extreme ultraviolet frequency comb metrology. *Physical Review Letters*, **105**, 063001.

Kandula, D. Z., Gohle, C., Pinkert, T. J., Ubachs, W., and Eikema, K. S. E. (2011). XUV frequency-comb metrology on the ground state of helium. *Physical Review A*, **84**, 062512.

Kara, D. M., Smallman, I. J., Hudson, J. J., Sauer, B. E., Tarbutt, M. R., and Hinds, E. A. (2012). Measurement of the electron's electric dipole moment using YbF molecules: Methods and data analysis. *New Journal of Physics*, **14**, 103051.

Karshenboim, S. G. (2005). Precision physics of simple atoms: QED tests, nuclear structure and fundamental constants. *Physics Reports*, **422**(1-2), 1.

Kasevich, M. and Chu, S. (1991). Atomic interferometry using stimulated Raman transitions. *Physical Review Letters*, **67**(2), 181.

Kasevich, M. and Chu, S. (1992). Laser cooling below a photon recoil with three-level atoms. *Physical Review Letters*, **69**(12), 1741.

Kasevich, M. A., Riis, E., Chu, S., and DeVoe, R. G. (1989). Rf spectroscopy in an atomic fountain. *Physical Review Letters*, **63**(6), 612.

Katori, H. (2011). Optical lattice clocks and quantum metrology. *Nature Photonics*, **5**, 203.

Katori, H., Takamoto, M., Palchikov, V. G., and Ovsiannikov, V. D. (2003). Ultrastable optical clock with neutral atoms in an engineered light shift trap. *Physical Review Letters*, **91**(17), 173005.

Keith, D. W., Ekstrom, C. R., Turchette, Q. A., and Pritchard, D. E. (1991). An interferometer for atoms. *Physical Review Letters*, **66**(21), 2693.

Kessler, T., Hagemann, C., Grebing, C., Legero, T., Sterr, U., Riehle, F., Martin, M. J., Chen, L., and Ye, J. (2012). A sub-40-mHz-linewidth laser based on a silicon single-crystal optical cavity. *Nature Photonics*, **6**, 687.

Ketterle, W. (2002). Nobel Lecture: When atoms behave as waves: Bose–Einstein condensation and the atom laser. *Reviews of Modern Physics*, **74**, 1131.

Ketterle, W., Durfee, D. S., and Stamper-Kurn, D. M. (1999). Making, probing and understanding Bose–Einstein condensates. In *Bose–Einstein Condensation in Atomic Gases* (ed. M. Inguscio, S. Stringari, and C. E. Wieman), pp. 67–176. IOS Press.

Ketterle, W. and van Druten, N. J. (1996). Evaporative cooling of trapped atoms. *Advances in Atomic, Molecular, and Optical Physics*, **37**, 181.

Ketterle, W. and Zwierlein, M. W. (2008). Making, probing and understanding ultracold Fermi gases. In *Ultracold Fermi Gases* (ed. M. Inguscio, W. Ketterle, and C. Salomon), pp. 95–287. IOS Press.

Khaykovich, L., Schreck, F., Ferrari, G., Bourdel, T., Cubizolles, J., Carr, L. D., Castin, Y., and Salomon, C. (2002). Formation of a matter-wave bright soliton.

Science, **296**, 1290.
Kim, K., Chang, M.-S., Islam, R., Korenblit, S., Duan, L.-M., and Monroe, C. (2009). Entanglement and tunable spin–spin couplings between trapped ions using multiple transverse modes. *Physical Review Letters*, **103**, 120502.
Kim, K., Chang, M.-S., Korenblit, S., Islam, R., Edwards, E. E., Freericks, J. K., Lin, G.-D., Duan, L.-M., and Monroe, C. (2010). Quantum simulation of frustrated Ising spins with trapped ions. *Nature*, **465**, 590.
Kinast, J., Hemmer, S. L., Gehm, M. E., Turlapov, A., and Thomas, J. E. (2004). Evidence for superfluidity in a resonantly interacting Fermi gas. *Physical Review Letters*, **92**, 150402.
Kinast, J., Turlapov, A., Thomas, J. E., Chen, Q., Stajic, J., and Levin, K. (2005). Heat capacity of a strongly interacting Fermi gas. *Science*, **307**, 1296.
King, B. E., Wood, C. S., Myatt, C. J., Turchette, Q. A., Leibfried, D., Itano, W. M., Monroe, C., and Wineland, D. J. (1998). Cooling the collective motion of trapped ions to initialize a quantum register. *Physical Review Letters*, **81**(7), 1525.
King, J. A., Webb, J. K., Murphy, M. T., and Carswell, R. F. (2008). Stringent null constraint on cosmological evolution of the proton-to-electron mass ratio. *Physical Review Letters*, **101**, 251304.
Kitagawa, M. and Ueda, M. (1993). Squeezed spin states. *Physical Review A*, **47**(6), 5138.
Kitching, J., Knappe, S., Liew, L., Moreland, J., Schwindt, P. D. D., Shah, V., Gerginov, V., and Hollberg, L. (2005). Microfabricated atomic frequency references. *Metrologia*, **42**, S100.
Kleppner, D., Greytak, T. J., Killian, T. C., Fried, D. G., Willmann, L., Landhuis, D., and Moss, S. C. (1999). Bose–Einstein condensation of atomic hydrogen. In *Bose–Einstein Condensation in Atomic Gases* (ed. M. Inguscio, S. Stringari, and C. E. Wieman), pp. 177–199. IOS Press.
Köhl, M., Moritz, H., Stöferle, T., Günter, K., and Esslinger, T. (2005). Fermionic atoms in a three dimensional optical lattice: Observing Fermi surfaces, dynamics, and interactions. *Physical Review Letters*, **94**, 080403.
Kolachevsky, N., Matveev, A., Alnis, J., Parthey, C. G., Karshenboim, S. G., and Hänsch, T. W. (2009*a*). Measurement of the 2S hyperfine interval in atomic hydrogen. *Physical Review Letters*, **102**, 213002.
Kolachevsky, N., Matveev, A., Alnis, J., Parthey, C. G., Steinmetz, T., Wilken, T., Holzwarth, R., Udem, T., and Hänsch, T. W. (2009*b*). Testing the stability of the fine structure constant in the laboratory. *Space Science Reviews*, **148**, 267.
Kolbe, D., Scheid, M., and Walz, J. (2012). Triple resonant four-wave mixing boosts the yield of continuous coherent vacuum ultraviolet generation. *Physical Review Letters*, **109**, 063901.
Kondov, S. S., McGehee, W. R., Zirbel, J. J., and DeMarco, B. (2011). Three-dimensional Anderson localization of ultracold matter. *Science*, **334**, 66.
Kozuma, M., Deng, L., Hagley, E. W., Wen, J., Lutwak, R., Helmerson, K, Rolston, S. L., and Phillips, W. D. (1999). Coherent splitting of Bose–Einstein condensed atoms with optically induced Bragg diffraction. *Physical Review Letters*, **82**(5), 871.
Kraft, S., Vogt, F., Appel, O., Riehle, F., and Sterr, U. (2009). Bose–Einstein

condensation of alkaline earth atoms: ^{40}Ca. *Physical Review Letters*, **103**, 130401.

Krems, R. V., Stwalley, W. C., and Friedrich, B. (ed.) (2009). *Cold Molecules: Theory, Experiment, Applications.* CRC Press.

Kuga, T., Torii, Y., Shiokawa, N., Hirano, T., Shimizu, Y., and Sasada, H. (1997). Novel optical trap of atoms with a doughnut beam. *Physical Review Letters*, **78**(25), 4713.

Kuklov, A. B. and Svistunov, B. V. (2003). Counterflow superfluidity of two-species ultracold atoms in a commensurate optical lattice. *Physical Review Letters*, **90**(10), 100401.

Kusch, P. and Foley, H. M. (1948). The magnetic moment of the electron. *Physical Review*, **74**(3), 250.

Laburthe Tolra, B., O'Hara, K. M., Huckans, J. H., Phillips, W. D., Rolston, S. L., and Porto, J. V. (2004). Observation of reduced three-body recombination in a correlated 1D degenerate Bose gas. *Physical Review Letters*, **92**(19), 190401.

Lahaye, T., Koch, T., Fröhlich, B., Fattori, M., Metz, J., Griesmaier, A., Giovanazzi, S., and Pfau, T. (2007). Strong dipolar effects in a quantum ferrofluid. *Nature*, **448**, 672.

Lahaye, T., Menotti, C., Santos, L., Lewenstein, M., and Pfau, T. (2009). The physics of dipolar bosonic quantum gases. *Reports on Progress in Physics*, **72**, 126401.

Lahaye, T., Metz, J., Fröhlich, B., Koch, T., Meister, M., Griesmaier, A., Pfau, T., Saito, H., Kawaguchi, Y., and Ueda, M. (2008). d-wave collapse and explosion of a dipolar Bose–Einstein condensate. *Physical Review Letters*, **101**, 080401.

Lamb, W. E., Schleich, W. P., Scully, M. O., and Townes, C. H. (1999). Laser physics: Quantum controversy in action. *Reviews of Modern Physics*, **71**, 263.

Lamb Jr, W. E. (1957). Microwave technique for determining the fine structure of the helium atom. *Physical Review*, **105**, 559.

Lamb Jr, W. E. and Retherford, R. C. (1947). Fine structure of the hydrogen atom by a microwave method. *Physical Review*, **72**, 241.

Lamporesi, G., Bertoldi, A., Cacciapuoti, L., Prevedelli, M., and Tino, G. M. (2008). Determination of the Newtonian gravitational constant using atom interferometry. *Physical Review Letters*, **100**, 050801.

Landau, L. D. (1941). The theory of superfluidity of Helium II. *Journal of Physics - USSR*, **5**, 71.

Landau, L. D. and Lifshitz, E. M. (1977). *Quantum Mechanics: Non-Relativistic Theory.* Pergamon Press.

Landini, M., Roy, S., Carcagní, L., Trypogeorgos, D., Fattori, M., Inguscio, M., and Modugno, G. (2011). Sub-Doppler laser cooling of potassium atoms. *Physical Review A*, **84**, 043432.

Landragin, A., Courtois, J.-Y., Labeyrie, G., Vansteenkiste, N., Westbrook, C. I., and Aspect, A. (1996). Measurement of the van der Waals force in an atomic mirror. *Physical Review Letters*, **77**(8), 1464.

Lang, F., Winkler, K., Strauss, C., Grimm, R., and Hecker Denschlag, J. (2008). Ultracold triplet molecules in the rovibrational ground state. *Physical Review Letters*, **101**, 133005.

Lanyon, B. P., Hempel, C., Nigg, D., Müller, M., Gerritsma, R., Zähringer, F.,

Schindler, P., Barreiro, J. T., Rambach, M., Kirchmair, G., Hennrich, M., Zoller, P., Blatt, R., and Roos, C. F. (2011). Universal digital quantum simulation with trapped ions. *Science*, **334**, 57.

Larson, D. J., Bergquist, J. C., Bollinger, J. J., Itano, W. M., and Wineland, D. J. (1986). Sympathetic cooling of trapped ions: A laser-cooled two-species nonneutral ion plasma. *Physical Review Letters*, **57**(1), 70.

Laughlin, R. B. (1999). Nobel Lecture: Fractional quantization. *Reviews of Modern Physics*, **71**(4), 863.

Laurent, Ph., Abgrall, M., Jentsch, Ch., Lemonde, P., Santarelli, G., Clairon, A., Maksimovic, I., Bize, S., Salomon, Ch., Blonde, D., Vega, J. F., Grosjean, O., Picard, F., Saccoccio, M., Chaubet, M., Ladiette, N., Guillet, L., Zenone, I., Delaroche, Ch., and Sirmain, Ch. (2006). Design of the cold atom PHARAO space clock and initial test results. *Applied Physics B*, **84**, 683.

Le Targat, R., Baillard, X., Fouché, M., Brusch, A., Tcherbakoff, O., Rovera, G. D., and Lemonde, P. (2006). Accurate optical lattice clock with ^{87}Sr atoms. *Physical Review Letters*, **97**, 130801.

Leggett, A. J. (1980). Diatomic molecules and Cooper pairs. In *Modern Trends in the Theory of Condensed Matter* (ed. A. Pekalski and J. A. Przystawa), pp. 13–27. Springer-Verlag.

Leggett, A. J. (2006). What DO we know about high T_C? *Nature Physics*, **2**, 134.

Leibfried, D., Blatt, R., Monroe, C., and Wineland, D. (2003*a*). Quantum dynamics of single trapped ions. *Reviews of Modern Physics*, **75**, 281.

Leibfried, D., DeMarco, B., Meyer, V., Lucas, D., Barrett, M., Britton, J., Itano, W. M., Jelenković, B., Langer, C., Rosenband, T., and Wineland, D. J. (2003*b*). Experimental demonstration of a robust, high-fidelity geometric two ion-qubit phase gate. *Nature*, **422**, 412.

Leibfried, D., Knill, E., Seidelin, S., Britton, J., Blakestad, R. B., Chiaverini, J., Hume, D. B., Itano, W. M., Jost, J. D., Langer, C., Ozeri, R., Reichle, R., and Wineland, D. J. (2005). Creation of a six-atom "Schrödinger cat" state. *Nature*, **438**, 639.

Leindecker, N., Marandi, A., Byer, R. L., Vodopyanov, K. L., Jiang, J., Hartl, I., Fermann, M., and Schunemann, P. G. (2012). Octave-spanning ultrafast OPO with 2.6–6.1µm instantaneous bandwidth pumped by femtosecond Tm-fiber laser. *Optics Express*, **20**(7), 7046.

Lenef, A., Hammond, T. D., Smith, E. T., Chapman, M. S., Rubenstein, R. A., and Pritchard, D. E. (1997). Rotation sensing with an atom interferometer. *Physical Review Letters*, **78**(5), 760.

Leo, K., Bolivar, P. H., Brüggemann, F., and Schwedler, R. (1992). Observation of Bloch oscillations in a semiconductor superlattice. *Solid State Communications*, **84**(10), 943.

Leroux, I. D., Schleier-Smith, M. H., and Vuletić, V. (2010). Orientation-dependent entanglement lifetime in a squeezed atomic clock. *Physical Review Letters*, **104**, 250801.

Letokhov, V. S. and Lebedev, P. N. (1968). Narrowing of the Doppler width in a standing light wave. *JETP Letters*, **7**, 272.

Letokhov, V. S., Minogin, V. G., and Pavlik, B. D. (1976). Cooling and trapping of

atoms and molecules by a resonant laser field. *Optics Communications*, **19**(1), 72.
Lett, P. D., Watts, R. N., Westbrook, C. I., Phillips, W. D., Gould, P. L., and Metcalf, H. J. (1988). Observation of atoms laser cooled below the Doppler limit. *Physical Review Letters*, **61**(2), 169.
Levenson, M. D. and Bloembergen, N. (1974). Observation of two-photon absorption without Doppler broadening on the 3S-5S transition in sodium vapor. *Physical Review Letters*, **32**(12), 645.
Levi, F., Calosso, C., Calonico, D., Lorini, L., Bertacco, E. K., Godone, A., Costanzo, G. A., Mongino, B., Jefferts, S. R., Heavner, T. P., and Donley, E. A. (2010). Cryogenic fountain development at NIST and INRIM: Preliminary characterization. *IEEE Transactions on Ultrasonics, Ferroelectrics, and Frequency Control*, **57**(3), 600.
Lewenstein, M., Sanpera, A., and Ahufinger, V. (2012). *Ultracold Atoms in Optical Lattices*. Oxford University Press.
Lifshitz, E. M. (1956). The theory of molecular attractive forces between solids. *Soviet Physics JETP*, **2**(1), 73.
Lin, Y.-J., Compton, R. L., Jiménez-García, K., Porto, J. V., and Spielman, I. B. (2009). Synthetic magnetic fields for ultracold neutral atoms. *Nature*, **462**, 628.
Lin, Y.-J., Jiménez-García, K., and Spielman, I. B. (2011). Spin–orbit-coupled Bose–Einstein condensates. *Nature*, **471**, 83.
Liontos, I., Cavalieri, S., Corsi, C., Eramo, R., Kaziannis, S., Pirri, A., Sali, E., and Bellini, M. (2010). Ramsey spectroscopy of bound atomic states with extreme-ultraviolet laser harmonics. *Optics Letters*, **35**, 832.
Lloyd, S. (1996). Universal quantum simulators. *Science*, **273**(5278), 1073.
Lodewyck, J., Westergaard, P. G., and Lemonde, P. (2009). Nondestructive measurement of the transition probability in a Sr optical lattice clock. *Physical Review A*, **79**, 061401.
Louchet-Chauvet, A., Appel, J., Renema, J. J., Oblak, D., Kjaergaard, N., and Polzik, E. S. (2010). Entanglement-assisted atomic clock beyond the projection noise limit. *New Journal of Physics*, **12**, 065032.
Lu, M., Burdick, N. Q., and Lev, B. L. (2012). Quantum degenerate dipolar Fermi gas. *Physical Review Letters*, **108**, 215301.
Lu, M., Burdick, N. Q., Youn, S. H., and Lev, B. L. (2011). Strongly dipolar Bose–Einstein condensate of dysprosium. *Physical Review Letters*, **107**, 190401.
Lucioni, E., Deissler, B., Tanzi, L., Roati, G., Zaccanti, M., Modugno, M., Larcher, M., Dalfovo, F., Inguscio, M., and Modugno, G. (2011). Observation of subdiffusion in a disordered interacting system. *Physical Review Letters*, **106**, 230403.
Lücke, B., Scherer, M., Kruse, J., Pezzé, L., Deuretzbacher, F., Hyllus, P., Topic, O., Peise, J., Ertmer, W., Arlt, J., Santos, L., Smerzi, A., and Klempt, C. (2011). Twin matter waves for interferometry beyond the classical limit. *Science*, **334**, 773.
Ludlow, A. D., Boyd, M. M., Zelevinsky, T., Foreman, S. M., Blatt, S., Notcutt, M., Ido, T., and Ye, J. (2006). Systematic study of the ^{87}Sr clock transition in an optical lattice. *Physical Review Letters*, **96**, 033003.
Ludlow, A. D., Zelevinsky, T., Campbell, G. K., Blatt, S., Boyd, M. M., de Miranda, M. H. G., Martin, M. J., Thomsen, J. W., Foreman, S. M., Ye, J., Fortier, T. M.,

Stalnaker, J. E., Diddams, S. A., Le Coq, Y., Barber, Z. W., Poli, N., Lemke, N. D., Beck, K. M., and Oates, C. W. (2008). Sr lattice clock at 1×10^{-16} fractional uncertainty by remote optical evaluation with a Ca clock. *Science*, **319**, 1805.

Lye, J. E., Fallani, L., Modugno, M., Wiersma, D. S., Fort, C., and Inguscio, M. (2005). Bose–Einstein condensate in a random potential. *Physical Review Letters*, **95**, 070401.

Ma, J., Wang, X., Sun, C. P., and Nori, F. (2011). Quantum spin squeezing. *Physics Reports*, **509**, 89.

Maddaloni, P., Cancio, P., and De Natale, P. (2009). Optical comb generators for laser frequency measurement. *Measurement Science and Technology*, **20**(5), 052001.

Maddaloni, P., Malara, P., Gagliardi, G., and De Natale, P. (2006). Mid-infrared fibre-based optical comb. *New Journal of Physics*, **8**(11), 262.

Madison, K. W., Chevy, F., Wohlleben, W., and Dalibard, J. (2000). Vortex formation in a stirred Bose–Einstein condensate. *Physical Review Letters*, **84**(5), 806.

Maiman, T. H. (1960). Stimulated optical radiation in ruby. *Nature*, **187**, 493.

Majorana, E. (1932). Atomi orientati in campo magnetico variabile. *Nuovo Cimento*, **9**, 43.

Malara, P., Maddaloni, P., Gagliardi, G., and De Natale, P. (2008). Absolute frequency measurement of molecular transitions by a direct link to a comb generated around 3μm. *Optics Express*, **16**, 8242.

Mandel, L. and Wolf, E. (1995). *Optical Coherence and Quantum Optics*. Cambridge University Press.

Margolis, H. S., Barwood, G. P., Huang, G., Klein, H. A., Lea, S. N., Szymaniec, K., and Gill, P. (2004). Hertz-level measurement of the optical clock frequency in a single $^{88}Sr^{+}$ ion. *Science*, **306**, 1355.

Marin, F., Minardi, F., Pavone, F. S., and Inguscio, M. (1994). Precision measurement of the isotope shift of the $2\ ^3S_1 - 3\ ^3P_0$ transition in helium. *Physical Review A*, **49**(3), 1523.

Martin, M. J. and Ye, J. (2011). High-precision laser stabilization via optical cavities. In *Optical Coatings and Thermal Noise in Precision Measurement* (ed. G. M. Harry, T. Bodiya, and R. DeSalvo), p. 237. Cambridge University Press.

Martin, P. J., Oldaker, B. G., Miklich, A. H., and Pritchard, D. E. (1988). Bragg scattering of atoms from a standing light wave. *Physical Review Letters*, **60**(6), 515.

Martinez de Escobar, Y. N., Mickelson, P. G., Yan, M., DeSalvo, B. J., Nagel, S. B., and Killian, T. C. (2009). Bose–Einstein condensation of ^{84}Sr. *Physical Review Letters*, **103**, 200402.

Masuhara, N., Doyle, J. M., Sandberg, J. C., Kleppner, D., Greytak, T. J., Hess, H. F., and Kochanski, G. P. (1988). Evaporative cooling of spin-polarized atomic hydrogen. *Physical Review Letters*, **61**(8), 935.

Matthews, M. R., Anderson, B. P., Haljan, P. C., Hall, D. S., Wieman, C. E., and Cornell, E. A. (1999). Vortices in a Bose–Einstein condensate. *Physical Review Letters*, **83**(13), 2498.

Maussang, K., Marti, G. E., Schneider, T., Treutlein, P., Li, Y., Sinatra, A., Long, R., Estève, J., and Reichel, J. (2010). Enhanced and reduced atom number fluctuations in a BEC splitter. *Physical Review Letters*, **105**, 080403.

Mazzotti, D., Cancio, P., Giusfredi, G., De Natale, P., and Prevedelli, M. (2005). Frequency-comb-based absolute frequency measurements in the mid-infrared with a difference-frequency spectrometer. *Optics Letters*, **30**, 997.

McKay, D. C. and DeMarco, B. (2011). Cooling in strongly correlated optical lattices: Prospects and challenges. *Reports on Progress in Physics*, **74**, 054401.

McNamara, J. M., Jeltes, T., Tychkov, A. S., Hogervorst, W., and Vassen, W. (2006). Degenerate Bose–Fermi mixture of metastable atoms. *Physical Review Letters*, **97**, 080404.

Meschede, D., Jhe, W., and Hinds, E. A. (1990). Radiative properties of atoms near a conducting plane: An old problem in a new light. *Physical Review A*, **41**(3), 1587.

Metcalf, H. J. and van Der Straten, P. (1999). *Laser Cooling and Trapping*. Springer-Verlag.

Mewes, M.-O., Andrews, M. R., Van Druten, N. J., Kurn, D. M., Durfee, D. S., Townsend, C. G., and Ketterle, W. (1996). Collective excitations of a Bose–Einstein condensate in a magnetic trap. *Physical Review Letters*, **77**(6), 988.

Miesner, H.-J., Stamper-Kurn, D. M., Andrews, M. R., Durfee, D. S., Inouye, S., and Ketterle, W. (1998). Bosonic stimulation in the formation of a Bose–Einstein condensate. *Science*, **279**, 1005.

Migdall, A. L., Prodan, J. V., Phillips, W. D., Bergeman, T. H., and Metcalf, H. J. (1985). First observation of magnetically trapped neutral atoms. *Physical Review Letters*, **54**(24), 2596.

Milburn, G. J., Corney, J., Wright, E. M., and Walls, D. F. (1997). Quantum dynamics of an atomic Bose–Einstein condensate in a double-well potential. *Physical Review A*, **55**(6), 4318.

Minardi, F. and Inguscio, M. (2001). Helium fine structure, from Breit interaction to precise laser spectroscopy. In *The Gregory Breit Centennial Symposium* (ed. V. W. Hughes, F. Iachello, and D. Kusnezov). World Scientific.

Modugno, G., Ferlaino, F., Heidemann, R., Roati, G., and Inguscio, M. (2003). Production of a Fermi gas of atoms in an optical lattice. *Physical Review A*, **68**, 011601.

Modugno, G., Ferrari, G., Roati, G., Brecha, R. J., Simoni, A., and Inguscio, M. (2001). Bose–Einstein condensation of potassium atoms by sympathetic cooling. *Science*, **294**, 1320.

Modugno, G., Modugno, M., Riboli, F., Roati, G., and Inguscio, M. (2002). Two atomic species superfluid. *Physical Review Letters*, **89**(19), 190404.

Mohr, P. J., Taylor, B. N., and Newell, D. B. (2008). CODATA recommended values of the fundamental physical constants: 2006. *Reviews of Modern Physics*, **80**, 633.

Mohr, P. J., Taylor, B. N., and Newell, D. B. (2012). CODATA recommended values of the fundamental physical constants: 2010. *Reviews of Modern Physics*, **84**, 1527.

Monroe, C., Meekhof, D. M., King, B. E., Itano, W. M., and Wineland, D. J. (1995). Demonstration of a fundamental quantum logic gate. *Physical Review Letters*, **75**(25), 4714.

Monroe, C., Swann, W., Robinson, H., and Wieman, C. (1990). Very cold trapped atoms in a vapor cell. *Physical Review Letters*, **65**(13), 1571.

Monz, T., Schindler, P., Barreiro, J. T., Chwalla, M., Nigg, D., Coish, W. A., Harlander,

M., Hänsel, W., Hennrich, M., and Blatt, R. (2011). 14-qubit entanglement: Creation and coherence. *Physical Review Letters*, **106**, 130506.

Morsch, O., Müller, J. H., Cristiani, M., Ciampini, D., and Arimondo, E. (2001). Bloch oscillations and mean-field effects of Bose–Einstein condensates in 1D optical lattices. *Physical Review Letters*, **87**(14), 140402.

Moskowitz, E., Gould, P. L., Atlas, S. R., and Pritchard, D. E. (1983). Diffraction of an atomic beam by standing-wave radiation. *Physical Review Letters*, **51**(5), 370.

Mott, N. F. (1990). *Metal Insulator Transitions* (2nd edn). Taylor & Francis.

Mount, B. J., Redshaw, M., and Myers, E. G. (2010). Atomic masses of ^{6}Li, ^{23}Na, ^{39,41}K, ^{85,87}Rb, and ^{133}Cs. *Physical Review A*, **82**, 042513.

Mueller, P., Sulai, I. A., Villari, A. C. C., Alcántara-Núñez, J. A., Alves-Condé, R., Bailey, K., Drake, G. W. F., Dubois, M., Eléon, C., Gaubert, G., Holt, R. J., Janssens, R. V. F., Lecesne, N., Lu, Z.-T., O'Connor, T. P., Saint-Laurent, M.-G., Thomas, J.-C., and Wang, L.-B. (2007). Nuclear charge radius of ^{8}He. *Physical Review Letters*, **99**, 252501.

Müller, H., Chiow, S., Herrmann, S., and Chu, S. (2009). Atom interferometers with scalable enclosed area. *Physical Review Letters*, **102**, 240403.

Müller, H., Chiow, S.-W., Long, Q., Vo, C., and Chu, S. (2006). A new photon recoil experiment: Towards a determination of the fine structure constant. *Applied Physics B*, **84**, 633.

Murphy, M. T., Flambaum, V. V., Muller, S., and Henkel, C. (2008). Strong limit on a variable proton-to-electron mass ratio from molecules in the distant universe. *Science*, **320**, 1611.

Murphy, M. T., Webb, J. K., and Flambaum, V. V. (2003). Further evidence for a variable fine-structure constant from Keck/HIRES QSO absorption spectra. *Monthly Notices of the Royal Astronomical Society*, **345**(2), 609.

Myatt, C. J., Burt, E. A., Ghrist, R. W., Cornell, E. A., and Wieman, C. E. (1997). Production of two overlapping Bose–Einstein condensates by sympathetic cooling. *Physical Review Letters*, **78**(4), 586.

Nagourney, W., Sandberg, J., and Dehmelt, H. (1986). Shelved optical electron amplifier: Observation of quantum jumps. *Physical Review Letters*, **56**(26), 2797.

Nair, R. R., Blake, P., Grigorenko, A. N., Novoselov, K. S., Booth, T. J., Stauber, T., Peres, N. M. R., and Geim, A. K. (2008). Fine structure constant defines visual transparency of graphene. *Science*, **320**, 1308.

Neuhauser, W., Hohenstatt, M., Toschek, P., and Dehmelt, H. (1978). Optical-sideband cooling of visible atom cloud confined in parabolic well. *Physical Review Letters*, **41**(4), 233.

Neuhauser, W., Hohenstatt, M., Toschek, P. E., and Dehmelt, H. (1980). Localized visible Ba^{+} mono-ion oscillator. *Physical Review A*, **22**(3), 1137.

Nez, F, Biraben, F, Felder, R, and Millerioux, Y (1993). Optical frequency determination of the hyperfine components of the $5S_{1/2} - 5D_{3/2}$ two-photon transitions in rubidium. *Optics Communications*, **102**, 432–438.

Ni, K.-K., Ospelkaus, S., de Miranda, M. H. G., Pe'er, A., Neyenhuis, B., Zirbel, J. J., Kotochigova, S., Julienne, P. S., Jin, D. S., and Ye, J. (2008). A high phase-space-density gas of polar molecules. *Science*, **322**, 231.

Nicholson, T. L., Martin, M. J., Williams, J. R., Bloom, B. J., Bishof, M., Swallows, M. D., Campbell, S. L., and Ye, J. (2012). Comparison of two independent Sr optical clocks with 1×10^{-17} stability at 10^3 s. *Physical Review Letters*, **109**, 230801.

Novoselov, K. S. (2011). Nobel Lecture: Graphene: Materials in the flatland. *Reviews of Modern Physics*, **83**, 837.

Noziéres, P. and Pines, D. (1990). *The Theory of Quantum Liquids.* Addison-Wesley.

Obrecht, J. M., Wild, R. J., Antezza, M., Pitaevskii, L. P., Stringari, S., and Cornell, E. A. (2007). Measurement of the temperature dependence of the Casimir–Polder force. *Physical Review Letters*, **98**, 063201.

O'Dell, D. H. J., Giovanazzi, S., and Kurizki, G. (2003). Rotons in gaseous Bose–Einstein condensates irradiated by a laser. *Physical Review Letters*, **90**(11), 110402.

O'Hara, K. M., Hemmer, S. L., Gehm, M. E., Granade, S. R., and Thomas, J. E. (2002). Observation of a strongly interacting degenerate Fermi gas of atoms. *Science*, **298**, 2179.

Orzel, C., Tuchman, A. K., Fenselau, M. L., Yasuda, M., and Kasevich, M. A. (2001). Squeezed states in a Bose–Einstein condensate. *Science*, **291**, 2386.

Ospelkaus, S., Ni, K.-K., Wang, D., de Miranda, M. H. G., Neyenhuis, B., Quéméner, G., Julienne, P. S., Bohn, J. L., Jin, D. S., and Ye, J. (2010). Quantum-state controlled chemical reactions of ultracold potassium–rubidium molecules. *Science*, **327**, 853.

Ospelkaus, S., Ospelkaus, C., Wille, O., Succo, M., Ernst, P., Sengstock, K., and Bongs, K. (2006). Localization of bosonic atoms by fermionic impurities in a three-dimensional optical lattice. *Physical Review Letters*, **96**, 180403.

Ott, H., de Mirandes, E., Ferlaino, F., Roati, G., Modugno, G., and Inguscio, M. (2004). Collisionally induced transport in periodic potentials. *Physical Review Letters*, **92**, 160601.

Ott, H., Fortagh, J., Schlotterbeck, G., Grossmann, A., and Zimmermann, C. (2001). Bose–Einstein condensation in a surface microtrap. *Physical Review Letters*, **87**, 230401.

Ovchinnikov, Yu. B., Manek, I., and Grimm, R. (1997). Surface trap for Cs atoms based on evanescent-wave cooling. *Physical Review Letters*, **79**(12), 2225.

Pachucki, K. (2006). Improved theory of helium fine structure. *Physical Review Letters*, **97**, 013002.

Pachucki, K. and Yerokhin, V. A. (2010). Fine structure of helium-like ions and determination of the fine structure constant. *Physical Review Letters*, **104**, 070403.

Papp, S. B., Pino, J. M., Wild, R. J., Ronen, S., Wieman, C. E., Jin, D. S., and Cornell, E. A. (2008). Bragg spectroscopy of a strongly interacting ^{85}Rb Bose–Einstein condensate. *Physical Review Letters*, **101**, 135301.

Parker, T. E. (2010). Long-term comparison of caesium fountain primary frequency standards. *Metrologia*, **47**, 1.

Parthey, C. G., Matveev, A., Alnis, J., Bernhardt, B., Beyer, A., Holzwarth, R., Maistrou, A., Pohl, R., Predehl, K., Udem, T., Wilken, T., Kolachevsky, N., Abgrall, M., Rovera, D., Salomon, C., Laurent, P., and Hänsch, T. W. (2011). Improved measurement of the hydrogen 1s-2s transition frequency. *Physical Review Letters*, **107**, 203001.

Partridge, G. B., Li, W., Kamar, R. I., Liao, Y.-A., and Hulet, R. G. (2006). Pairing

and phase separation in a polarized Fermi gas. *Science*, **311**, 503.
Pasienski, M., McKay, D., White, M., and DeMarco, B. (2010). A disordered insulator in an optical lattice. *Nature Physics*, **6**, 677.
Paul, W. (1990). Electromagnetic traps for charged and neutral particles. *Reviews of Modern Physics*, **62**(3), 531.
Pavone, F. S., Marin, F., De Natale P., Inguscio, M., and Biraben, F. (1994). First pure frequency measurement of an optical transition in helium: Lamb shift on the 2 $^3S_1$ metastable level. *Physical Review Letters*, **73**(1), 42.
Pedri, P., Pitaevskii, L., Stringari, S., Fort, C., Burger, S., Cataliotti, F. S., Maddaloni, P., Minardi, F., and Inguscio, M. (2001). Expansion of a coherent array of Bose–Einstein condensates. *Physical Review Letters*, **87**(22), 220401.
Pereira Dos Santos, F., Léonard, J., Wang, J., Barrelet, C. J., Perales, F., Rasel, E., Unnikrishnan, C. S., Leduc, M., and Cohen-Tannoudji, C. (2001). Bose–Einstein condensation of metastable helium. *Physical Review Letters*, **86**(16), 3459.
Peters, A., Chung, K. Y., and Chu, S. (1999). Measurement of gravitational acceleration by dropping atoms. *Nature*, **400**, 849.
Peters, A., Chung, K. Y., and Chu, S. (2001). High-precision gravity measurements using atom interferometry. *Metrologia*, **38**, 25.
Petersen, M., Chicireanu, R., Dawkins, S. T., Magalhaes, D. V., Mandache, C., Le Coq, Y., Clairon, A., and Bize, S. (2008). Doppler-free spectroscopy of the $^1S_0$ – $^3P_0$ optical clock transition in laser-cooled fermionic isotopes of neutral mercury. *Physical Review Letters*, **101**, 183004.
Petrich, W., Anderson, M. H., Ensher, J. R., and Cornell, E. A. (1995). Stable, tightly confining magnetic trap for evaporative cooling of neutral atoms. *Physical Review Letters*, **74**, 3352.
Petrov, D. S., Salomon, C., and Shlyapnikov, G. V. (2004). Weakly bound dimers of fermionic atoms. *Physical Review Letters*, **93**, 090404.
Pezzé, L., Collins, L. A., Smerzi, A., Berman, G. P., and Bishop, A. R. (2005). Sub-shot-noise phase sensitivity with a Bose–Einstein condensate Mach–Zehnder interferometer. *Physical Review A*, **72**, 043612.
Pezzé, L. and Smerzi, A. (2009). Entanglement, nonlinear dynamics, and the Heisenberg limit. *Physical Review Letters*, **102**, 100401.
Phillips, W. D. (1998). Laser cooling and trapping of neutral atoms. *Reviews of Modern Physics*, **70**(3), 721.
Phillips, W. D. and Metcalf, H. (1982). Laser deceleration of an atomic beam. *Physical Review Letters*, **48**(9), 596.
Pichanick, F. M. J., Swift, R. D., Johnson, C. E., and Hughes, V. W. (1968). Experiments on the 2 ^{3}P state of helium. I. A measurement of the 2 $^3P_1$ – 2 $^3P_2$ fine structure. *Physical Review*, **169**, 55.
Pitaevskii, L. P. (1961). Vortex lines in an imperfect Bose gas. *Soviet Physics JETP*, **13**(2), 451.
Pitaevskii, L. P. and Stringari, S. (2004). *Bose–Einstein Condensation.* Oxford University Press.
Pohl, R., Antognini, A., Nez, F., Amaro, F. D., Biraben, F., Cardoso, J. M. R., Covita, D. S., Dax, A., Dhawan, S., Fernandes, L. M. P., Giesen, A., Graf, T., Hänsch, T. W.,

Indelicato, P., Julien, L., Kao, C.-Y., Knowles, P., Le Bigot, E.-O., Liu, Y.-W., Lopes, J. A. M., Ludhova, L., Monteiro, C. M. B., Mulhauser, F., Nebel, T., Rabinowitz, P., dos Santos, J. M. F., Schaller, L. A., Schuhmann, K., Schwob, C., Taqqu, D., Veloso, J. F. C. A., and Kottmann, F. (2010). The size of the proton. *Nature*, **466**, 213.

Poli, N., Wang, F.-Y., Tarallo, M. G., Alberti, A., Prevedelli, M., and Tino, G. M. (2011). Precision measurement of gravity with cold atoms in an optical lattice and comparison with a classical gravimeter. *Physical Review Letters*, **106**, 038501.

Porsev, S. G. and Derevianko, A. (2004). Hyperfine quenching of the metastable $^3P_{0,2}$ states in divalent atoms. *Physical Review A*, **69**, 042506.

Pritchard, D. E. (1983). Cooling neutral atoms in a magnetic trap for precision spectroscopy. *Physical Review Letters*, **51**(15), 1336.

Prodan, J., Migdall, A., Phillips, W. D., So, I., Metcalf, H., and Dalibard, J. (1985). Stopping atoms with laser light. *Physical Review Letters*, **54**(10), 992.

Pu, H. and Meystre, P. (2000). Creating macroscopic atomic Einstein–Podolsky–Rosen states from Bose–Einstein condensates. *Physical Review Letters*, **85**(19), 3987.

Raab, E. L., Prentiss, M., Cable, A., Chu, S., and Pritchard, D. E. (1987). Trapping of neutral sodium atoms with radiation pressure. *Physical Review Letters*, **59**(23), 2631.

Raghavan, S., Smerzi, A., Fantoni, S., and Shenoy, S. R. (1999). Coherent oscillations between two weakly coupled Bose–Einstein condensates: Josephson effects, π oscillations, and macroscopic quantum self-trapping. *Physical Review A*, **59**(1), 620.

Raizen, M. G., Salomon, C., and Niu, Q. (1997). New light on quantum transport. *Physics Today*, **50**(7), 30.

Raman, C., Köhl, M., Onofrio, R., Durfee, D. S., Kuklewicz, C. E., Hadzibabic, Z., and Ketterle, W. (1999). Evidence for a critical velocity in a Bose–Einstein condensed gas. *Physical Review Letters*, **83**(13), 2502.

Ramsey, N. F. (1950). A molecular beam resonance method with separated oscillating fields. *Physical Review*, **78**(6), 695.

Ramsey, N. F. (1956). *Molecular Beams.* Oxford University Press.

Ramsey, N. F. (1965). The atomic hydrogen maser. *Metrologia*, **1**, 7.

Ramsey, N. F. (1990). Experiments with separated oscillatory fields and hydrogen masers. *Reviews of Modern Physics*, **62**(3), 541.

Ramsey, N. F. (1993). Experiments with trapped hydrogen atoms and neutrons. *Hyperfine Interactions*, **81**, 97.

Ranka, J. K., Windeler, R. S., and Stentz, A. J. (2000). Visible continuum generation in air-silica microstructure optical fibers with anomalous dispersion at 800 nm. *Optics Letters*, **25**, 25.

Rasel, E. M., Oberthaler, M. K., Batelaan, H., Schmiedmayer, J., and Zeilinger, A. (1995). Atom wave interferometry with diffraction gratings of light. *Physical Review Letters*, **75**(14), 2633.

Regal, C. A., Greiner, M., and Jin, D. S. (2004). Observation of resonance condensation of fermionic atom pairs. *Physical Review Letters*, **92**, 040403.

Regan, B. C., Commins, E. D., Schmidt, C. J., and DeMille, D. (2002). New limit on the electron electric dipole moment. *Physical Review Letters*, **88**, 071805.

Richard, S., Gerbier, F., Thywissen, J. H., Hugbart, M., Bouyer, P., and Aspect,

A. (2003). Momentum spectroscopy of 1D phase fluctuations in Bose–Einstein condensates. *Physical Review Letters*, **91**, 010405.

Riedel, M. F., Böhi, P., Li, Y., Hänsch, T. W., Sinatra, A., and Treutlein, P. (2010). Atom-chip-based generation of entanglement for quantum metrology. *Nature*, **464**, 1170.

Riehle, F., Kisters, Th., Witte, A., Helmcke, J., and Bordé, Ch. J. (1991). Optical Ramsey spectroscopy in a rotating frame: Sagnac effect in a matter-wave interferometer. *Physical Review Letters*, **67**(2), 177.

Roati, G., de Mirandes, E., Ferlaino, F., Ott, H., Modugno, G., and Inguscio, M. (2004). Atom interferometry with trapped Fermi gases. *Physical Review Letters*, **92**(23), 230402.

Roati, G., D'Errico, C., Fallani, L., Fattori, M., Fort, C., Zaccanti, M., Modugno, G., Modugno, M., and Inguscio, M. (2008). Anderson localization of a non-interacting Bose–Einstein condensate. *Nature*, **453**, 895.

Roati, G., Riboli, F., Modugno, G., and Inguscio, M. (2002). Fermi–Bose quantum degenerate ^{40}K–^{87}Rb mixture with attractive interaction. *Physical Review Letters*, **89**(15), 150403.

Robert, A., Sirjean, O., Browaeys, A., Poupard, J., Nowak, S., Boiron, D., Westbrook, C. I., and Aspect, A. (2001). A Bose–Einstein condensate of metastable atoms. *Science*, **292**, 461.

Roberts, J. L., Claussen, N. R., Cornish, S. L., Donley, E. A., Cornell, E. A., and Wieman, C. E. (2001). Controlled collapse of a Bose–Einstein condensate. *Physical Review Letters*, **86**(19), 4211.

Rohde, H., Gulde, S. T., Roos, C. F., Barton, P. A., Leibfried, D., Eschner, J., Schmidt-Kaler, F., and Blatt, R. (2001). Sympathetic ground-state cooling and coherent manipulation with two-ion crystals. *Journal of Optics B: Quantum and Semiclassical Optics*, **3**, S34.

Rolston, S. L. and Phillips, W. D. (2002). Nonlinear and quantum atom optics. *Nature*, **416**, 219.

Rom, T., Best, Th., van Oosten, D., Schneider, U., Fölling, S., Paredes, B., and Bloch, I. (2006). Free fermion antibunching in a degenerate atomic Fermi gas released from an optical lattice. *Nature*, **444**, 733.

Rosenband, T., Hume, D. B., Schmidt, P. O., Chou, C. W., Brusch, A., Lorini, L., Oskay, W. H., Drullinger, R. E., Fortier, T. M., Stalnaker, J. E., Diddams, S. A., Swann, W. C., Newbury, N.R., Itano, W. M., Wineland, D. J., and Bergquist, J. C. (2008). Frequency ratio of Al^+ and Hg^+ single-ion optical clocks; Metrology at the 17th decimal place. *Science*, **319**, 1808.

Rosenband, T., Schmidt, P. O., Hume, D. B., Itano, W. M., Fortier, T. M., Stalnaker, J. E., Kim, K., Diddams, S. A., Koelemeij, J. C. J., Bergquist, J. C., and Wineland, D. J. (2007). Observation of the $^1S_0$-$^3P_0$ clock transition in $^{27}Al^+$. *Physical Review Letters*, **98**, 220801.

Roth, R. and Burnett, K. (2003). Phase diagram of bosonic atoms in two-color superlattices. *Physical Review A*, **68**, 023604.

Sachdev, S. (2000). *Quantum Phase Transitions*. Cambridge University Press.

Sachdev, S., Sengupta, K., and Girvin, S. M. (2002). Mott insulators in strong electric fields. *Physical Review B*, **66**, 075128.

Sage, J. M., Sainis, S., Bergeman, T., and DeMille, D. (2005). Optical production of ultracold polar molecules. *Physical Review Letters*, **94**, 203001.

Sagnac, G. (1913). L'éther lumineux démontré par l'effet du vent relatif d'éther. *Comptes Rendus*, **157**, 708.

Sakurai, J. J. (1993). *Modern Quantum Mechanics.* Addison-Wesley Publishing Company.

Sandoghdar, V., Sukenik, C. I., Hinds, E. A., and Haroche, S. (1992). Direct measurement of the van der Waals interaction between an atom and its images in a micron-sized cavity. *Physical Review Letters*, **68**(23), 3432.

Santarelli, G., Laurent, Ph., Lemonde, P., Clairon, A., Mann, A. G., Chang, S., Luiten, A. N., and Salomon, C. (1999). Quantum projection noise in an atomic fountain: A high stability cesium frequency standard. *Physical Review Letters*, **82**(23), 4619.

Santos, L., Shlyapnikov, G. V., and Lewenstein, M. (2003). Roton-maxon spectrum and stability of trapped dipolar Bose–Einstein condensates. *Physical Review Letters*, **90**(25), 250403.

Sapienza, R., Costantino, P., Wiersma, D., Ghulinyan, M., Oton, C. J., and Pavesi, L. (2003). Optical analogue of electronic Bloch oscillations. *Physical Review Letters*, **91**(26), 263902.

Sauter, Th., Neuhauser, W., Blatt, R., and Toschek, P. E. (1986). Observation of quantum jumps. *Physical Review Letters*, **57**(14), 1696.

Schawlow, A. L. (1982). Spectroscopy in a new light. *Reviews of Modern Physics*, **54**, 697.

Schawlow, A. L. and Townes, C. H. (1958). Infrared and optical masers. *Physical Review*, **112**(6), 1940.

Scheel, S. and Hinds, E. A. (2011). Atoms at micrometer distances from a macroscopic body. In *Atom Chips* (ed. J. Reichel and V. Vuletic), pp. 121–146. Wiley-VCH.

Schellekens, M., Hoppeler, R., Perrin, A., Viana Gomes, J., Boiron, D., Aspect, A., and Westbrook, C. I. (2005). Hanbury Brown Twiss effect for ultracold quantum gases. *Science*, **310**, 648.

Schliesser, A., Picqué, N., and Hänsch, T. W. (2012). Mid-infrared frequency combs. *Nature Photonics*, **6**, 440.

Schmidt, P. O., Rosenband, T., Langer, C., Itano, W. M., Bergquist, J. C., and Wineland, D. J. (2005). Spectroscopy using quantum logic. *Science*, **309**, 749.

Schmidt-Kaler, F., Häffner, H., Riebe, M., Gulde, S., Lancaster, G. P. T., Deuschle, T., Becher, C., Roos, C. F., Eschner, J., and Blatt, R. (2003). Realization of the Cirac–Zoller controlled-NOT quantum gate. *Nature*, **422**, 408.

Schneider, U., Hackermüller, L., Will, S., Best, Th., Bloch, I., Costi, T. A., Helmes, R. W., Rasch, D., and Rosch, A. (2008). Metallic and insulating phases of repulsively interacting fermions in a 3D optical lattice. *Science*, **322**, 1520.

Schreck, F., Ferrari, G., Corwin, K. L., Cubizolles, J., Khaykovich, L., Mewes, M.-O., and Salomon, C. (2001*a*). Sympathetic cooling of bosonic and fermionic lithium gases towards quantum degeneracy. *Physical Review A*, **64**, 011402.

Schreck, F., Khaykovich, L., Corwin, K. L., Ferrari, G., Bourdel, T., Cubizolles, J., and Salomon, C. (2001*b*). Quasipure Bose–Einstein condensate immersed in a Fermi

sea. *Physical Review Letters*, **87**, 080403.
Schrödinger, E. (1926). Quantisierung als Eigenwertproblem. *Annalen der Physik*, **79**, 361.
Schumm, T., Hofferberth, S., Andersson, L. M., Wildermuth, S., Groth, S., Bar-Joseph, I., Schmiedmayer, J., and Krüger, P. (2005). Matter-wave interferometry in a double well on an atom chip. *Nature Physics*, **1**, 57.
Schwartz, C. (1964). Fine structure of helium. *Physical Review*, **134**(5A), A1181.
Schwartz, T., Bartal, G., Fishman, S., and Segev, M. (2007). Transport and Anderson localization in disordered two-dimensional photonic lattices. *Nature*, **446**, 52.
Schwob, C., Jozefowski, L., de Beauvoir, B., Hilico, L., Nez, F., Julien, L., Biraben, F., Acef, O., and Clairon, A. (1999). Optical frequency measurement of the $2S - 12D$ transitions in hydrogen and deuterium: Rydberg constant and Lamb shift determinations. *Physical Review Letters*, **82**(25), 4960.
Scully, M. O. and Zubairy, M. S. (1997). *Quantum Optics*. Cambridge University Press.
Sebby-Strabley, J., Brown, B. L., Anderlini, M., Lee, P. J., Phillips, W. D., Porto, J. V., and Johnson, P. R. (2007). Preparing and probing atomic number states with an atom interferometer. *Physical Review Letters*, **98**, 200405.
Series, G.W. (1957). *Spectrum of Atomic Hydrogen*. Oxford University Press.
Series, G.W. (1988). *The Spectrum of Atomic Hydrogen: Advances*. World Scientific.
Setija, I. D., Werij, H. G. C., Luiten, O. J., Reynolds, M. W., Hijmans, T. W., and Walraven, J. T. M. (1993). Optical cooling of atomic hydrogen in a magnetic trap. *Physical Review Letters*, **70**(15), 2257.
Sherson, J. F., Weitenberg, C., Endres, M., Cheneau, M., Bloch, I., and Kuhr, S. (2010). Single-atom-resolved fluorescence imaging of an atomic Mott insulator. *Nature*, **467**, 68.
Shimizu, F. (2001). Specular reflection of very slow metastable neon atoms from a solid surface. *Physical Review Letters*, **86**(6), 987.
Shin, Y., Saba, M., Pasquini, T. A., Ketterle, W., Pritchard, D. E., and Leanhardt, A. E. (2004). Atom interferometry with Bose–Einstein condensates in a double-well potential. *Physical Review Letters*, **92**(5), 050405.
Shin, Y., Zwierlein, M. W., Schunck, C. H., Schirotzek, A., and Ketterle, W. (2006). Observation of phase separation in a strongly interacting imbalanced Fermi gas. *Physical Review Letters*, **97**, 030401.
Shlyapnikov, G. V., Walraven, J. T. M., Rahmanov, U. M., and Reynolds, M. W. (1994). Decay kinetics and Bose condensation in a gas of spin-polarized triplet helium. *Physical Review Letters*, **73**(24), 3247.
Shuman, E. S., Barry, J. F., and DeMille, D. (2010). Laser cooling of a diatomic molecule. *Nature*, **467**, 820.
Sick, I. (2003). On the rms-radius of the proton. *Physics Letters B*, **576**, 62.
Silvera, I. F. and Walraven, J. T. M. (1980). Stabilization of atomic hydrogen at low temperature. *Physical Review Letters*, **44**(3), 164.
Simon, J., Bakr, W. S., Ma, R., Tai, M. E., Preiss, P. M., and Greiner, M. (2011). Quantum simulation of antiferromagnetic spin chains in an optical lattice. *Nature*, **472**, 307.

Smerzi, A., Trombettoni, A., Kevrekidis, P. G., and Bishop, A. R. (2002). Dynamical superfluid–insulator transition in a chain of weakly coupled Bose–Einstein condensates. *Physical Review Letters*, **89**(17), 170402.

Smiciklas, M. and Shiner, D. (2010). Determination of the fine structure constant using helium fine structure. *Physical Review Letters*, **105**, 123001.

Smith, P. W. and Hänsch, T. (1971). Cross-relaxation effects in the saturation of the 6328-Å neon-laser line. *Physical Review Letters*, **26**(13), 740.

Snadden, M. J., McGuirk, J. M., Bouyer, P., Haritos, K. G., and Kasevich, M. A. (1998). Measurement of the Earth's gravity gradient with an atom interferometer-based gravity gradiometer. *Physical Review Letters*, **81**(5), 971.

Söding, J., Guéry-Odelin, D., Desbiolles, P., Ferrari, G., and Dalibard, J. (1998). Giant spin relaxation of an ultracold cesium gas. *Physical Review Letters*, **80**(9), 1869.

Soltan-Panahi, P., Struck, J., Hauke, P., Bick, A., Plenkers, W., Meineke, G., Becker, C., Windpassinger, P., Lewenstein, M., and Sengstock, K. (2011). Multi-component quantum gases in spin-dependent hexagonal lattices. *Nature Physics*, **7**, 434.

Sørensen, A., Duan, L.-M., Cirac, J. I., and Zoller, P. (2001). Many-particle entanglement with Bose–Einstein condensates. *Nature*, **409**, 63.

Srianand, R., Chand, H., Petitjean, P., and Aracil, B. (2004). Limits on the time variation of the electromagnetic fine-structure constant in the low energy limit from absorption lines in the spectra of distant quasars. *Physical Review Letters*, **92**(12), 121302.

Steinhardt, P. J. (1987). *The Physics of Quasicrystals*. World Scientific.

Steinhauer, J., Ozeri, R., Katz, N., and Davidson, N. (2002). Excitation spectrum of a Bose–Einstein condensate. *Physical Review Letters*, **88**, 120407.

Stellmer, S., Tey, M. K., Huang, B., Grimm, R., and Schreck, F. (2009). Bose–Einstein condensation of strontium. *Physical Review Letters*, **103**, 200401.

Stenger, J., Inouye, S., Chikkatur, A. P., Stamper-Kurn, D. M., Pritchard, D. E., and Ketterle, W. (1999). Bragg spectroscopy of a Bose–Einstein condensate. *Physical Review Letters*, **82**(23), 4569.

Stock, S., Hadzibabic, Z., Battelier, B., Cheneau, M., and Dalibard, J. (2005). Observation of phase defects in quasi-two-dimensional Bose–Einstein condensates. *Physical Review Letters*, **95**, 190403.

Stöferle, T., Moritz, H., Schori, C., Köhl, M., and Esslinger, T. (2004). Transition from a strongly interacting 1D superfluid to a Mott insulator. *Physical Review Letters*, **92**(13), 130403.

Stringari, S. (1996). Collective excitations of a trapped Bose-condensed gas. *Physical Review Letters*, **77**(12), 2360.

Stwalley, W. C. and Nosanow, L. H. (1976). Possible "new" quantum systems. *Physical Review Letters*, **36**(15), 910.

Sukenik, C. I., Boshier, M. G., Cho, D., Sandoghdar, V., and Hinds, E. A. (1993). Measurement of the Casimir–Polder force. *Physical Review Letters*, **70**(5), 560.

Swallows, M. D., Bishof, M., Lin, Y., Blatt, S., Martin, M. J., Rey, A. M., and Ye, J. (2011). Suppression of collisional shifts in a strongly interacting lattice clock. *Science*, **331**, 1043.

Tai, K., Hasegawa, A., and Tomita, A. (1986). Observation of modulational instability in optical fibers. *Physical Review Letters*, **56**(2), 135.

Taichenachev, A. V., Yudin, V. I., Oates, C. W., Hoyt, C. W., Barber, Z. W., and Hollberg, L. (2006). Magnetic field-induced spectroscopy of forbidden optical transitions with application to lattice-based optical atomic clocks. *Physical Review Letters*, **96**, 083001.

Taie, S., Takasu, Y., Sugawa, S., Yamazaki, R., Tsujimoto, T., Murakami, R., and Takahashi, Y. (2010). Realization of a SU(2)×SU(6) system of fermions in a cold atomic gas. *Physical Review Letters*, **105**, 190401.

Takamoto, M., Hong, F.-L., Higashi, R., and Katori, H. (2005). An optical lattice clock. *Nature*, **435**, 321.

Takasu, Y., Maki, K., Komori, K., Takano, T., Honda, K., Kumakura, M., Yabuzaki, T., and Takahashi, Y. (2003). Spin-singlet Bose–Einstein condensation of two-electron atoms. *Physical Review Letters*, **91**, 040404.

Tarbutt, M. R., Hudson, J. J., Sauer, B. E., and Hinds, E. A. (2009). Preparation and manipulation of molecules for fundamental physics tests. In *Cold Molecules* (ed. R. V. Krems, W. C. Stwalley, and B. Friedrich), Chapter 15, p. 555. CRC Press.

Tarruell, L., Greif, D., Uehlinger, T., Jotzu, G., and Esslinger, T. (2012). Creating, moving and merging Dirac points with a Fermi gas in a tunable honeycomb lattice. *Nature*, **483**, 302.

Tey, M., Stellmer, S., Grimm, R., and Schreck, F. (2010). Double-degenerate Bose–Fermi mixture of strontium. *Physical Review A*, **82**, 011608.

Thalhammer, G., Barontini, G., De Sarlo, L., Catani, J., Minardi, F., and Inguscio, M. (2008). Double species Bose–Einstein condensate with tunable interspecies interactions. *Physical Review Letters*, **100**, 210402.

Theis, M., Thalhammer, G., Winkler, K., Hellwig, M., Ruff, G., Grimm, R., and Hecker Denschlag, J. (2004). Tuning the scattering length with an optically induced Feshbach resonance. *Physical Review Letters*, **93**, 123001.

Thomann, I., Bartels, A., Corwin, K. L., Newbury, N.R., Hollberg, L., Diddams, S. A., Nicholson, J. W., and Yan, M. F. (2003). 420-MHz Cr:forsterite femtosecond ring laser and continuum generation in the 1–2-μm range. *Optics Letters*, **28**, 1368.

Tiesinga, E., Verhaar, B. J., and Stoof, H. T. C. (1993). Threshold and resonance phenomena in ultracold ground-state collisions. *Physical Review A*, **47**(5), 4114.

Tino, G. M., Barsanti, M., de Angelis, M., Gianfrani, L., and Inguscio, M. (1992). Spectroscopy of the 689 nm intercombination line of strontium using an extended-cavity InGaP/InGaAIP diode laser. *Applied Physics B*, **55**, 397.

Townes, C. H. (1972). Production of coherent radiation by atoms and molecules. In *Nobel Lectures, Physics 1963–1970*, p. 58. Elsevier Publishing Company.

Townes, C. H. and Schawlow, A. L. (1975). *Microwave Spectroscopy.* Dover Publications.

Trotzky, S., Cheinet, P., Fölling, S., Feld, M., Schnorrberger, U., Rey, A. M., Polkovnikov, A., Demler, E. A., Lukin, M. D., and Bloch, I. (2008). Time-resolved observation and control of superexchange interactions with ultracold atoms in optical lattices. *Science*, **319**, 295.

Truscott, A. G., Strecker, K. E., McAlexander, W. I., Partridge, G. B., and Hulet,

R. G. (2001). Observation of Fermi pressure in a gas of trapped atoms. *Science*, **291**, 2570.

Udem, Th., Huber, A., Gross, B., Reichert, J., Prevedelli, M., Weitz, M., and Hänsch, T. W. (1997). Phase-coherent measurement of the hydrogen 1s-2s transition frequency with an optical frequency interval divider chain. *Physical Review Letters*, **79**(14), 2646.

Udem, Th., Reichert, J., Holzwarth, R., and Hänsch, T. W. (1999). Absolute optical frequency measurement of the cesium D1 line with a mode-locked laser. *Physical Review Letters*, **82**, 3568.

Ungar, P. J., Weiss, D. S., Riis, E., and Chu, S. (1989). Optical molasses and multilevel atoms: Theory. *Journal of the Optical Society of America B*, **6**(11), 2058.

Uzan, J.-P. (2011). Varying constants, gravitation and cosmology. *Living Reviews in Relativity*, **14**, 2.

van de Meerakker, S. Y. T., Bethlem, H. L., and Meijer, G. (2008). Taming molecular beams. *Nature Physics*, **4**, 595–602.

van Dyck, R. S., Schwinberg, P. B., and Dehmelt, H. G. (1987). New high-precision comparison of electron and positron g factors. *Physical Review Letters*, **59**(1), 26.

van Rooij, R., Borbely, J. S., Simonet, J., Hoogerland, M. D., Eikema, K. S. E., Rozendaal, R. A., and Vassen, W. (2011). Frequency metrology in quantum degenerate helium: Direct measurement of the $2\ {}^3S_1 - 2\ {}^1S_0$ transition. *Science*, **333**, 196.

van Weerdenburg, F., Murphy, M. T., Malec, A. L., Kaper, L., and Ubachs, W. (2011). First constraint on cosmological variation of the proton-to-electron mass ratio from two independent telescopes. *Physical Review Letters*, **106**, 180802.

van Zoest, T., Gaaloul, N., Singh, Y., Ahlers, H., Herr, W., Seidel, S. T., Ertmer, W., Rasel, E., Eckart, M., Kajari, E., Arnold, S., Nandi, G., Schleich, W. P., Walser, R., Vogel, A., Sengstock, K., Bongs, K., Lewoczko-Adamczyk, W., Schiemangk, M., Schuldt, T., Peters, A., Könemann, T., Müntinga, H., Lämmerzahl, C., Dittus, H., Steinmetz, T., Hänsch, T. W., and Reichel, J (2010). Bose–Einstein condensation in microgravity. *Science*, **328**, 1540.

Vasilenko, L. S., Chebotayev, V. P., and Shishaev, A. V. (1970). Line shape of two-photon absorption in a standing-wave field in a gas. *JETP Letters*, **12**, 113.

Vassen, W., Cohen-Tannoudji, C., Leduc, M., Boiron, D., Westbrook, C. I., Truscott, A., Baldwin, K., Birkl, G., Cancio, P., and Trippenbach, M. (2012). Cold and trapped metastable noble gases. *Reviews of Modern Physics*, **84**, 175.

Veeravalli, G., Kuhnle, E., Dyke, P., and Vale, C. (2008). Bragg spectroscopy of a strongly interacting Fermi gas. *Physical Review Letters*, **101**, 250403.

Vengalattore, M., Higbie, J. M., Leslie, S. R., Guzman, J., Sadler, L. E., and Stamper-Kurn, D. M. (2007). High-resolution magnetometry with a spinor Bose–Einstein condensate. *Physical Review Letters*, **98**, 200801.

Verkerk, P., Lounis, B., Salomon, C., Cohen-Tannoudji, C., Courtois, J.-Y., and Grynberg, G. (1992). Dynamics and spatial order of cold cesium atoms in a periodic optical potential. *Physical Review Letters*, **68**(26), 3861.

Viteau, M., Chotia, A., Allegrini, M., Bouloufa, N., Dulieu, O., Comparat, D., and Pillet, P. (2008). Optical pumping and vibrational cooling of molecules. *Science*, **321**, 232.

von Klitzing, K., Dorda, G., and Pepper, M. (1980). New method for high-accuracy determination of the fine-structure constant based on quantized Hall resistance. *Physical Review Letters*, **45**(6), 494.

Wang, D.-W., Lukin, M. D., and Demler, E. (2004*a*). Disordered Bose–Einstein condensates in quasi-one-dimensional magnetic microtraps. *Physical Review Letters*, **92**(7), 076802.

Wang, L.-B., Mueller, P., Bailey, K., Drake, G. W. F., Greene, J. P., Henderson, D., Holt, R. J., Janssens, R. V. F., Jiang, C. L., Lu, Z.-T., O'Connor, T. P., Pardo, R. C., Rehm, K. E., Schiffer, J. P., and Tang, X. D. (2004*b*). Laser spectroscopic determination of the ^{6}He nuclear charge radius. *Physical Review Letters*, **93**, 142501.

Waschke, C., Roskos, H. G., Schwedler, R., Leo, K., Kurz, H., and Köhler, K. (1993). Coherent submillimeter-wave emission from Bloch oscillations in a semiconductor superlattice. *Physical Review Letters*, **70**(21), 3319.

Webb, J. K., King, J. A., Murphy, M. T., Flambaum, V. V., Carswell, R. F., and Bainbridge, M. B. (2011). Indications of a spatial variation of the fine structure constant. *Physical Review Letters*, **107**, 191101.

Weber, C., Barontini, G., Catani, J., Thalhammer, G., Inguscio, M., and Minardi, F. (2008). Association of ultracold double-species bosonic molecules. *Physical Review A*, **78**, 061601.

Weinstein, J. D., DeCarvalho, R., Guillet, T., Friedrich, B., and Doyle, J. M. (1998). Magnetic trapping of calcium monohydride molecules at millikelvin temperatures. *Nature*, **395**, 148.

Weiss, D. S., Young, B. C., and Chu, S. (1993). Precision measurement of the photon recoil of an atom using atomic interferometry. *Physical Review Letters*, **70**(18), 2706.

Weitenberg, C., Endres, M., Sherson, J. F., Cheneau, M., Schauss, P., Fukuhara, T., Bloch, I., and Kuhr, S. (2011). Single-spin addressing in an atomic Mott insulator. *Nature*, **471**, 319.

Wicht, A., Hensley, J. M., Sarajlic, E., and Chu, S. (2002). A preliminary measurement of the fine structure constant based on atom interferometry. *Physica Scripta*, **T102**, 82.

Wieder, I. and Lamb Jr, W. E. (1957). Fine structure of the 2 ^{3}P and 3 ^{3}P states in helium. *Physical Review*, **107**(1), 125.

Wiersma, D. S., Bartolini, P., Lagendijk, A., and Righini, R. (1997). Localization of light in a disordered medium. *Nature*, **390**, 671.

Wildermuth, S., Hofferberth, S., Lesanovsky, I., Haller, E., Andersson, L. M., Groth, S., Bar-Joseph, I., Krüger, P., and Schmiedmayer, J. (2005). Microscopic magnetic-field imaging. *Nature*, **435**, 440.

Wilpers, G., Oates, C. W., Diddams, S. A., Bartels, A., Fortier, T. M., Oskay, W. H., Bergquist, J. C., Jefferts, S. R., Heavner, T. P., Parker, T. E., and Hollberg, L. (2007). Absolute frequency measurement of the neutral ^{40}Ca optical frequency standard at 657 nm based on microkelvin atoms. *Metrologia*, **44**, 146.

Wineland, D., Ekstrom, P., and Dehmelt, H. (1973). Monoelectron oscillator. *Physical Review Letters*, **31**(21), 1279.

Wineland, D. J., Bollinger, J. J., Itano, W. M., and Heinzen, D. J. (1994). Squeezed atomic states and projection noise in spectroscopy. *Physical Review A*, **50**, 67.

Wineland, D. J. and Dehmelt, H. G. (1975). Proposed 10^{14} $\delta\nu < \nu$ laser fluorescence spectroscopy on Tl^+ monoion oscillator III (sideband cooling). *Bulletin of the American Physical Society*, **20**, 637.

Wineland, D. J., Drullinger, R. E., and Walls, F. L. (1978). Radiation-pressure cooling of bound resonant absorbers. *Physical Review Letters*, **40**(25), 1639.

Wineland, D. J. and Itano, W. M. (1981). Spectroscopy of a single Mg^+ ion. *Physics Letters A*, **82**(2), 75.

Wing, W. H. (1994). On neutral particle trapping in quasistatic electromagnetic fields. *Progress in Quantum Electronics*, **8**, 181.

Witte, S., Zinkstok, R. Th., Ubachs, W., Hogervorst, W., and Eikema, K. S. E. (2005). Deep-ultraviolet quantum interference metrology with ultrashort laser pulses. *Science*, **307**, 400.

Wu, B. and Niu, Q. (2001). Landau and dynamical instabilities of the superflow of Bose–Einstein condensates in optical lattices. *Physical Review A*, **64**, 061603.

Wynands, R. and Weyers, S. (2005). Atomic fountain clocks. *Metrologia*, **42**, S64.

Yamazaki, T., Morita, N., Hayano, R. S., Widmann, E., and Eades, J. (2002). Antiprotonic helium. *Physics Reports*, **366**, 183.

Yasuda, M. and Shimizu, F. (1996). Observation of two-atom correlation of an ultracold neon atomic beam. *Physical Review Letters*, **77**, 3090.

Ye, J. and Cundiff, S. T. (ed.) (2005). *Femtosecond Optical Frequency Comb: Principle, Operation and Applications.* Springer.

Ye, J., Kimble, H. J., and Katori, H. (2008). Quantum state engineering and precision metrology using state-insensitive light traps. *Science*, **320**, 1734.

Young, B. C., Cruz, F. C., Itano, W. M., and Bergquist, J. C. (1999). Visible lasers with subhertz linewidths. *Physical Review Letters*, **82**(19), 3799.

Zelevinsky, T., Farkas, D., and Gabrielse, G. (2005). Precision measurement of the three $2\ ^3P_J$ helium fine structure intervals. *Physical Review Letters*, **95**, 203001.

Zeppenfeld, M., Englert, B. G. U., Glöckner, R., Prehn, A., Mielenz, M., Sommer, C., van Buuren, L. D., Motsch, M., and Rempe, G. (2012). Sisyphus cooling of electrically trapped polyatomic molecules. *Nature*, **491**, 570.

Zobay, O. and Garraway, B. M. (2001). Two-dimensional atom trapping in field-induced adiabatic potentials. *Physical Review Letters*, **86**(7), 1195.

Zwerger, W. (2003). Mott–Hubbard transition of cold atoms in optical lattices. *Journal of Optics B: Quantum and Semiclassical Optics*, **5**, S9.

Zwierlein, M. W., Abo-Shaeer, J. R., Schirotzek, A., Schunck, C. H., and Ketterle, W. (2005). Vortices and superfluidity in a strongly interacting Fermi gas. *Nature*, **435**, 1047.

Zwierlein, M. W., Schirotzek, A., Schunck, C. H., and Ketterle, W. (2006). Fermionic superfluidity with imbalanced spin populations. *Science*, **311**, 492.

Zwierlein, M. W., Stan, C. A., Schunck, C. H., Raupach, S. M. F., Gupta, S., Hadzibabic, Z., and Ketterle, W. (2003). Observation of Bose–Einstein condensation of molecules. *Physical Review Letters*, **91**, 250401.

Zwierlein, M. W., Stan, C. A., Schunck, C. H., Raupach, S. M. F., Kerman, A. J., and Ketterle, W. (2004). Condensation of pairs of fermionic atoms near a Feshbach resonance. *Physical Review Letters*, **92**, 120403.

Index